MATERIALS SCIENCE AND TECHNOLOGIES

RESIN COMPOSITES: PROPERTIES, PRODUCTION AND APPLICATIONS

MATERIALS SCIENCE AND TECHNOLOGIES

MATERIALS SCIENCE AND TECHNOLOGIES

RESIN COMPOSITES: PROPERTIES, PRODUCTION AND APPLICATIONS

DEBORAH B. SONG
EDITOR

Nova Science Publishers, Inc.
New York

For permission to use material from this book please contact us:
Telephone 631-231-7269; Fax 631-231-8175
Web Site: http://www.novapublishers.com

LIBRARY OF CONGRESS CATALOGING-IN-PUBLICATION DATA

Resin composites : properties, production, and applications / editors,
Deborah B. Song.
p. cm.
Includes index.
ISBN 978-1-61209-129-7 (hardcover)
1. Gums and resins. I. Song, Deborah B.
TP978.R34 2011
620.1'924--dc22

2010047089

Published by Nova Science Publishers, Inc. † New York

CONTENTS

PREFACE

Composite resins are types of synthetic resins which are used in dentistry as restorative material or adhesives. Synthetic resins evolved as restorative materials since they were insoluble, aesthetic, and insensitive to dehydration and were inexpensive. This book presents topical research in the study of resin composites, including water immersion and impact damage effects on resin composites; fabrication and evaluation of bioactive dental composites; nanostructured organosilicate composites; the electromagnetic properties of a composite made of metal particles dispersed in resin and posterior composite resin restoration.

Chapter 1 - In recent years the use of composite materials, based on polymeric resins, has increased tremendously. However, most modern engineering composite structures have not been in service long enough for their properties and long-term environmental stability to be fully understood. As such, the effect of water immersion on the impact-damaged composites with regard to their residual compression strength is a subject of importance that has received significant attention in recent years. Most of the studies reported in open literature so far either investigate specific problems or review certain aspects on the subject. The literature lacks a document that reviews holistically all the complex mechanisms involved and reports the state of the art as far as the accumulated knowledge. The current paper is an effort to provide a review into the effects of water immersion on the residual post-impact compressive strength of laminated composites and the differences induced in strength retention by the timing of the impact with respect to the exposure. The survey in particular touches two major fields that have widely been researched, that of environmental degradation of composites and that of impact damage induced strength reduction. The spread of the issues involved allows only a selection of the most closely related work to the subject under review from the separate fields to be reported here. In more detail, general aspects of degradation of composites due to environmental exposure are presented herein while issues regarding natural and accelerated testing are discussed. In addition, damage in the form of impact on laminated composites and its significance is presented as well as the test methods for the assessment of residual compressive strength after impact. The issue of damage detection and characterisation via non destructive techniques is also considered. Furthermore, highlights of prediction models for residual strength after environmental exposure or damage are provided.

Chapter 2 - With the exception of a small portion of the inner ear, all hard tissues of the human body are formed of calcium phosphate(s), CaPs. CaPs also occur in pathological calcifications (dental and urinary calculi, atherosclerotic lesions). The atomic arrangements of

CaPs are built up around an orthophosphate (PO_4) network that gives stability to their structures. It is, therefore, of no surprise that structural and chemical properties of CaPs have been extensively studied A total of eleven crystalline and one amorphous form of CaP with different Ca/P molar ratios, solubilities and crystallographic properties have been identified, seven of which are biologically-relevant. The general rule in CaP family is: the lower the Ca/PO_4 molar ratio, the more acidic and water-soluble the CaP. The *in vivo* presence of small peptides, proteins and inorganic additives considerably influences CaP precipitation, making it difficult to predict the possible phases that may form. The least soluble, hydroxyapatite (HAP), preferentially formed under neutral and basic conditions, is usually non-stoichiometric. In normal *in vivo* calcifications, formation of apatite is reportedly preceded by an amorphous calcium phosphate (ACP) precursor that spontaneously converts into thermodynamically favored HAP.

Chapter 3 - The present paper is devoted to organosilicate nanocomposites, representing interest as a new class of materials, with a high chemical resistance and mechanical durability in combination with optimal thermophysical properties, which provides effective increase of service characteristics of composite materials. Modern representations on obtaining and study of structure, chemical and physico-mechanical properties of organosilicates composites are considered.

The authors discuss technological problems of modifying soluble alkaline sodium silicate by organic reagents, physico-chemical processes of obtaining stable in aggregate sols based on organosilicate precursors, methods of realizing directed and controllable in time sol-gel transition of stabilized combined systems with the purpose of management of structure and properties of new multifunctional organosilikate nanocomposites. The technological features of formation of high-dispersed hybrid products from organosilicate solutions and homogeneous highly elastic gels are considered.

The possibility to control nanocomposite properties under thermal effect allowing to form highly elastic gels or highly dispersed xerogels is shown. This also concerns microwave heating that results in polymerization of silicate matrix. On the basis of the received experimental data the mechanisms of formation of organosilicate nanostructured materials and correlation of dependences are considered: structure and properties of the initial reagents → reaction technology of combining components → physical and chemical conditions of sol-gel transition → nanophase structure of a product → property of a material. The features of the nanophase structure of organosilicate composites depending on structure and technology of the formation product are considered. The properties of the initial components of combined systems and received hybrid nanocomposites are compared.

The promises of using the developed hybrid fillers for modifying thermoplastic and lubricant materials, and homogeneous highly elastic gels as waterproofing screens, grouting solutions and sealants are shown. The efficiency of using hybrid fillers in thermoplastics and greases depending on their structure and composition is estimated. The influence of dispersed organosilicate fillers on polyolefins, polyamide 6, phenilone and PTFE operation properties is considered.

Chapter 4 - The frequency dependences of the relative complex permeability μ_r^* and relative complex permittivity ε_r^* for the composite made of metal particles (aluminum particles) dispersed in polystyrene resin were measured in the frequency range from 1 MHz to 40 GHz. The volume mixture ratio and particle size of aluminum were varied and the dependences of the volume mixture ratio and particle size on the frequency dependences of

μ_r^* and ε_r^* were investigated. In addition, theoretical values of the real and imaginary part of μ_r^*, μ_r' and μ_r", for the composite made of metal particles dispersed in polystyrene resin were calculated using Maxwell's equations.

The measured value of the real part of ε_r^* was independent of frequency and both real and imaginary part of ε_r^* increased with increasing the volume mixture ratio of aluminum particle. The measured value of μ_r' was found to decrease with increasing frequency in the low frequency range and became constant in the high frequency range. Meanwhile, the measured value of μ_r" increased with frequency, had a maximum and decreased with increasing frequency. Moreover, at high frequencies where the skin depth is much smaller than the radius of aluminum, μ_r' was found to depend only on the volume mixture ratio of aluminum, whereas μ_r" was determined by both the volume mixture ratio and the aluminum particle size. These results almost agreed with the calculated values. Thus, the frequency dependences of μ_r' and μ_r" were found to be predicted by the theoretical calculation and to be controlled by the volume mixture ratio and particle size of aluminum.

For the application of this composite to an electromagnetic wave absorber, the return loss of this composite was calculated from the measured values of μ_r^* and ε_r^*. The return loss of the composite made of aluminum particles dispersed in polystyrene resin was less than -20 dB (absorption of 99% of an electromagnetic wave power) in the frequency range from 1 to 40 GHz when a suitable volume mixture ratio, particle size, and sample thickness were selected. Therefore, this composite can be used as an electromagnetic wave absorber in the gigahertz range. Furthermore, the absorption characteristics of such composite can be tailored based on the ability to control the values of μ_r' and μ_r" independently by adjusting the volume mixture ratio and aluminum particle size.

Chapter 5 - The development of composites based on conductive phases dispersed in polymeric matrices has led to important advances in analytical electrochemistry. Composite materials based on different forms of carbon as conductive phase have played a leading role in the analytical electrochemistry field, particularly in sensor devices. These materials combine the electrical properties of graphite and carbon nanotubes with the ease of processing of plastics (epoxy, methacrylate, Teflon, etc.). They show attractive electrochemical, physical, mechanical and economical features compared to the classic conductors (gold, platinum, graphite, etc.). Considering carbon composite in generally, their electrochemical properties present improvements over conventional solid carbon electrode, such as glassy carbon. The properties of these composites based are described, along with their application to the construction of electrochemical sensors.

The carbon-based composites exhibit interesting advantages, such as easy surface renewal, as well as low background current. Depending also on the conductive load, composites can behave as microelectrode arrays which are known to provide efficient mass transport of the electroactive species due to radial diffusion on the spaced carbon particles. Such improved mass transport favors the sensitive electroanalysis of a variety of reagents, including electrocatalysts, enzymes and chemical recognition agents. Moreover, the carbon surface chemistry also influences significantly the electron transfer processes at these electrodes.

During the past few decades, the electrochemical properties of different graphite powder composite materials based on different kinds of polymeric matrices were studied in detail. Nowadays, high interest is focused on composites based on carbon nanotubes (CNTs). They are attractive materials due to their remarkable mechanical and electrical properties. They have a highly accessible surface area, low resistance, high mechanical and chemical stability and their performance had been found to be superior to the other kinds of carbon material. The main drawback in CNT composite materials reside in the lack of homogeneity of the different commercial CNT lots due to different amount of impurities in the nanotubes, as well as dispersion in their diameter/length and state of aggregation (isolated, ropes, bundles). These variations are difficult to quantify and make mandatory a previous electrochemical characterization of the composite, before being used as a chemical sensor.

Another important point of consideration is the optimization of the conducting material (graphite or CNTs) loading in the composite materials for improving their electrochemical properties and analytical applications. Therefore, in this chapter the author will describe the strategy to find the optimum composite proportions for obtaining high electrode sensitivity, low limit of detection and fast response. Compositions of composites can be characterized by percolation theory, electrochemical impedance spectroscopy, cyclic voltammetry, scanning electron microscopy, atomic force microscopy and chronoamperometry.

Moreover, the optimized carbon-based composite electrodes can be integrated in a continuous flow analytical system, as a flow injection analysis (FIA) or a miniaturized device, to take advantage of all the benefits provided by these automatization techniques.

Chapter 6 - A review of recent results on a fabrication of epoxy resin composites with metal nanoparticles using viscous properties of polymer is reported. Preparation of metal nanoparticles realized during thermal vacuum evaporation of silver onto the surface of epoxy resin at a viscosity from 20 to 120 Pa·s) having room tempetature, which is well below the glass transition temperature of the polymer. Additionally, for synthesis of metal nanoparticles the ion irradiation of viscous polymer matrix is used. The viscous epoxy resin is implanted by silver ions with diferent doses. As a result, epoxy resin layers containing silver nanoparticles in their volume are fabricated. Various types of disperse structures formed by metallic nanoparticles in the polymer are detected. The morphology of the composite material is found to be controlled by the polymer viscosity and the metal deposition time. The use of the viscous state of epoxy resin increases the diffusion coefficient of silver impurity, which stimulates the nucleation and growth of nanoparticles and allows a high filling factor of metal in the polymer to be achieved. Mechanisms of metal nanoparticle growth in viscous epoxy resin are discussed.

Chapter 7 - In the present investigation a novel technique have been developed to fabricate nanocomposite materials containing SC-15 epoxy resin and K-10 montmorillonite clay. A high intensity ultrasonic liquid processor was used to obtain a homogeneous molecular mixture of epoxy resin and nano clay. The clays were infused into the part A of SC-15 (Diglycidylether of Bisphenol A) through sonic cavitations and then mixed with part B of SC-15 (cycloaliphatic amine hardener) using a high speed mechanical agitator. The trapped air and reaction volatiles were removed from the mixture using high vacuum. DMA, TGA and 3-point bending tests were performed on unfilled, 1wt. %, 2wt. %, 3% and 4wt. % clay filled SC-15 epoxy to identify the loading effect on thermal and mechanical properties of the composites. The flexural results indicate that 2.0 wt% loading of clay in epoxy resin showed the highest improvement in flexural strength as compared to the neat systems. DMA studies also revealed that 2.0 wt% doped system exhibit the highest storage modulus and Tg as compared to neat and other loading percentages. However, TGA results show that thermal stability of composite is insensitive to the clay content. After that, the nanophased matrix with 2 wt.% clay was then utilized in a Vacuum Assisted Resin Transfer Molding (VARTM) set up with satin weave carbon preforms to fabricate laminated composites. The resulting composites have been evaluated by TGA, DNA, and flexural test, and 5°C increasing in glass transition temperature, 6°C increasing in decomposition temperature and 13.5% improvement in flexural strength were observed in nanocomposite. Based on the experimental result, a linear damage model has been combined with the Weibull distribution function to establish a constitutive equation for neat and nanophased carbon/epoxy.

Chapter 8 - This chapter deals with the main causes of chemical ageing in organic matrix composites: hydrolysis, essentially in polyesters and polyamides, and oxidation in all kinds of polymer matrices. The first section is devoted to common aspects of chemical degradation of organic matrices. It is shown that chain scission and, at a lesser extent, crosslinking are especially important because they induce embrittlement at low conversions. Quantitative relationships between structural parameters and mechanical properties are briefly examined. The second section deals with diffusion–reaction coupling. In the cases of hydrolysis and oxidation, kinetics can be limited by respectively water and oxygen diffusion. Then, degradation is confined in a superficial layer, that can carry important consequences on use properties. The third section is devoted to hydrolysis. The kinetic equations are presented in both cases of non equilibrated and equilibrated hydrolysis. Structure–stability relationships are briefly examined. Osmotic cracking process, very important in the case of glass fiber/unsaturated polyester composites, is described. The last section is devoted to thermal oxidation. The simplest kinetic models are presented. The main gravimetric behaviours are explained. A mechanism is proposed for the "spontaneous cracking" in the superficial layer of oxidized samples.

Chapter 9 - The composites have been widely used over the last years, providing highly aesthetic restorations. The silorane composite is a new technology that replaces the conventionally used methacrylate resin matrix within conventional dental composite, thereby providing lower polymerization shrinkage, excellent marginal integrity, up to 9 minutes operatory light stability and low water sorption that substantially decreased exogenic staining. These qualities of the composite alleviate clinical problems such as marginal staining, microleakage, secondary caries, enamel micro-cracks and post-operative sensitivity. Only the specific self-etching adhesive must be applied to bond the silorane composite to the tooth enamel and dentin. The technique of silorane composite is different from that used in those methacrylate composites, which becomes important to prevent problems during its use. This paper aims to describe properties, application technique of new silorane-based composites and the difference between methacrylate- and silorane-based composites. After reading this paper, the reader will be able to apply such technique in their daily practice with maximum performance of these restorative materials.

Chapter 10 - Different approaches used for the simulation of woven reinforcement forming are investigated. Especially several methods based on geometrical and finite element approximations are presented. Some are based on continuous modelling, while others called discrete or mesoscopic approaches, model the components of the dry or prepregs woven fabric. Continuum semi discrete finite element made of woven unit cells under biaxial tension and in-plane shear is detailed. In continuous approaches, the difficulty lies in the necessity to take the strong specificity of the fibre directions into account. The fibres directions must be strictly followed during the large strains of the fabric. In the case of geometrical or continuum approaches (semi-discrete) the directions of the fibres are naturally followed because the fibres are modeled. Explicitly, however, modeling each component at the mesoscopic scale can lead to high numerical cost. During mechanical simulation of composite woven fabric forming, where large displacement and relative rotation of fibres are possible, severe mesh distortions occur after a few incremental loads. Hence an automatic mesh generation with remeshing capabilities is essential to carry out the finite element analysis. Some numerical simulations of forming process are proposed and compared with the experimental results in order to demonstrate the efficiency of the proposed approaches.

Chapter 11 - Plants and their exudates are used worldwide for the treatment of several diseases and novel drugs continue to be developed through phytochemical research. There are more than 20,000 species of high vegetables, used in traditional medicines that are sources of potential new drugs. Following the modern medicine and drug research advancing, chemically synthesized drugs have replaced plants as the source of most medicinal agents in industrialized countries. However, in developing countries, the majority of the world's population cannot afford pharmaceutical drugs and use their own plant based indigenous medicines. Several exudates from plants are well-known in folk medicine since ancient time, and they are today employed also for practical uses.

Dragon's blood is a deep red resin, which has been used as a famous traditional medicine since ancient times by many cultures. Dragon's blood is a non-specific name for red resinous exudations from quite different plant species endemic to various regions around the globe that belong to the genera *Dracaena* (Africa) and *Daemonorops* (South-East Asia), more rarely also to the genera *Pterocarpus* and *Croton* (both South America).

Dracaena draco L. is known as the dragon's blood tree, and it's endemic to the Canary Islands and Morocco. Phytochemical studies of resins obtained from incisions of the trunk of *D. draco*, have led to the isolation of flavans, along with homoisoflavans, homoisoflavones, chalcones and dihydrochalcones. Dragon's blood has been used for diverse medical applications in folk medicine and artistic uses. It has astringent effect and has been used as a hemostatic and antidiarrhetic drug.

Frankincense, also known as *Olibanum*, is an old-known oleogum resin obtained from the bark of trees belonging to the genera *Boswellia*. There are 43 different reported species in India, Arabian Peninsula and North Africa. The importance of these plants is related to the use of extracts and essential oils of resin in traditional medicine like Ayurvedic and Chinese. Extracts from *B. serrata* resin are currently used in India for the treatment of rheumatic diseases and ulcerative colitis. Furthermore, the extracts and essential oils of frankincense have been used as antiseptic agents in mouthwash, in the treatment of cough and asthma and as a fixative in perfumes, soaps, creams, lotions and detergents. In ancient Egypt the resin was used in mummification balms and unguents. Today frankincense is one of the most commonly used resins in aromatherapy.

The biological activity of frankincense resins is due to the pentacyclic triterpenic acids, α- and β- boswellic acids and their derivatives, which showed a well documented anti-inflammatory and immunomodulatory activities.

Manna is an exudate from the bark of *Fraxinus* trees (Oleaceae). Originally it was only collected from trees with damaged bark, but later in southern Italy and northern Sicily plantations were established for manna production, in which the bark is intentionally damaged for exudation and collection of manna.

In July-August a vertical series of oblique incisions are made in the bark on alternate sides of the trunk. A glutinous liquid exudes from this cut, hardens as it oxidises in the air into a yellowish crystalline mass with a bittersweet taste, and is then harvested. Manna is still produced in Sicily, mainly in the Castelbuono and Pollina areas, from *Fraxinus ornus* and *Fraxinus angustifolia* trees. The main component of manna is mannitol; it also contains glucose, fructose, maltotriose, mannotetrose, minerals and some unknown constituents.

Manna is a mild laxative and an excellent purgative, it is suitable in cases of digestive problems, in atonic or spastic constipation. It's useful as expectorant, fluidifier, emollient and sedative in coughs; as a decongestant in chronic bronchitis, laryngitis and tonsillitis; in hypertonic solutions it acts as a dehydrating agent in the treatment of wounds and ulcers. It can be used as a sweetener in cases of diabetes as it does not affect glycemia levels or cause glycosuria; in addition it is also a cholagogue as it promotes the flow of the contents of the gall bladder and bile ducts and so stimulates bile production.

Chapter 12 - Many traditional concepts in dentistry have changed since the 1950s. In 1955, Buonocore was the forerunner of a new phase in dentistry. With the advent of the technique that involved etching enamel, the implementation of bonding procedures began to be incorporated into daily practice; this enabled the execution of more conservative restorative techniques that required minimal removal of healthy tooth structure. Later, in 1963, Bowen developed composite resin restorations with esthetic characteristics including conservation of tooth structure. The improvement of the physical and mechanical properties, and esthetics of composite resins and made it possible to directly manufacture restorations on teeth by freehand; although the cosmetic results were very satisfactory, the process requires a great deal of manual skill on the part of the professional who performs it.

In: Resin Composites: Properties, Production and Applications ISBN: 978-1-61209-129-7
Editor: Deborah B. Song

Chapter 1

REVIEW ON THE WATER IMMERSION AND IMPACT DAMAGE EFFECTS ON THE RESIDUAL COMPRESSIVE STRENGTH OF COMPOSITES

***K. Berketis*[1]* *and D. Tzetzis*[2]**

[1] Spectrum Labs SA, Efplias 49, Piraeus, Greece.

[2]Queen Mary College, Department of Materials, University of London, E1 4NS, UK

ABSTRACT

In recent years the use of composite materials, based on polymeric resins, has increased tremendously. However, most modern engineering composite structures have not been in service long enough for their properties and long-term environmental stability to be fully understood. As such, the effect of water immersion on the impact-damaged composites with regard to their residual compression strength is a subject of importance that has received significant attention in recent years. Most of the studies reported in open literature so far either investigate specific problems or review certain aspects on the subject. The literature lacks a document that reviews holistically all the complex mechanisms involved and reports the state of the art as far as the accumulated knowledge. The current paper is an effort to provide a review into the effects of water immersion on the residual post-impact compressive strength of laminated composites and the differences induced in strength retention by the timing of the impact with respect to the exposure. The survey in particular touches two major fields that have widely been researched, that of environmental degradation of composites and that of impact damage induced strength reduction. The spread of the issues involved allows only a selection of the most closely related work to the subject under review from the separate fields to be reported here. In more detail, general aspects of degradation of composites due to environmental exposure are presented herein while issues regarding natural and accelerated testing are discussed. In addition, damage in the form of impact on laminated composites and its significance is presented as well as the test methods for the assessment of residual compressive strength after impact. The issue of damage detection and characterisation via non destructive techniques is also considered. Furthermore,

* Corresponding author

highlights of prediction models for residual strength after environmental exposure or damage are provided.

1. Introduction

Most modern engineering Polymer Matrix Composites (PMC) have not been in service long enough for their properties and long-term environmental stability to be fully understood. This problem has been traditionally dealt by over-design, thus negating to some extent the high specific stiffness and strength of optimally designed composites. Despite this, in aqueous environments, where corrosion is of prime concern, PMCs have found great success as replacement for metals. Another concern with typical laminate PMCs is the significant reduction in residual strength after out-of-plane impact. The greatest reduction in strength is observed when impact damaged plates are subjected to in-plane compression loading. The effect of water immersion on the impact-damaged composites with regard to their residual compression strength is a subject of importance for most marine and other structures in similar environments. Due to the complexity of such cases, no verified strength prediction models exist to date. Further basic understanding of the interactions involved, mainly via targeted experimental work, is needed.

2. Environmental Degradation

Glass fibre Reinforced Polymer (GRP) composites have found extensive use in structures exposed to weathering and typical process and chemical industry environments. The number of possible fibre/matrix combinations, the complexity of their microstructure and the variability at the processing stage makes it difficult to predict the effect of the environmental exposure. The ability to predict the long-term properties of GRPs is very important if their full potential is to be exploited. The evaluation of the long term stability of GRPs under hygrothermal and other possible service environments has attracted much attention. The amount of published work is large but to some extent is quite fragmented and discontinuous. This can be mainly attributed to the lack of specific composites durability test standards.

A thorough review on the subject of environmental durability of GRPs is presented by Schutte [1]. The focus of this work is the effects of the environment, water in particular, on the interface. A point strongly expressed is that there is a lack of understanding of the synergy between the different mechanisms of deterioration of the composites parts. Also, the difficulty in understanding from fundamental material properties, for each case, which will be the weakest component, is pointed out. Another review study of the effects of temperature and environment on polymer matrix composites, with a focus on mechanical properties is presented by Hancox [2]. A list of the applicable standards, relevant to accelerated environmental testing is presented here. The problem of the reliability of the long-term property prediction methods based on short-term accelerated tests is pointed out.

The kinetic aspects of water sorption in glassy polymers have been closely studied by many authors [3-6]. The complexity of the models proposed varies, but the number of the variables involved for each model, some of them theoretical and some obtained after long-term experiments, point out both the difficulty and the significance of this subject. To date, no prediction model based on first principles exists.

2.1. Fick's Diffusion Model

The analogy between heat conduction and diffusion of liquids in isotropic solids was recognized by Fick who first set a quantitative basis for the diffusion theory in homogeneous materials. Since Fick's model is based on a model that describes physical change, it is valid only when physical changes are assumed to take place. A further assumption is that the temperature inside the material approaches equilibrium much faster than the moisture concentration and so the energy and mass transfer equations are decoupled. Based on the mathematical expression derived by Fourier for heat conduction, the rate of transfer of diffusing substance through unit area of a section is proportional to the concentration gradient measured normal to that section [7]:

$$Q = -D\frac{\partial C}{\partial z} \tag{1}$$

where Q is the rate of transfer per unit area section, C is the concentration of diffusing substance, z is the space coordinate measured normal to the section and D is the diffusion coefficient.

If the diffusion coefficient is independent of concentration, then the relationship between M_t, the mass at time t, and M_∞, the mass at saturation is, according to Fick:

$$\frac{M_t}{M_\infty} = 1 - \frac{8}{\pi^2}\sum_{n=0}^{\infty}\frac{1}{(2n+1)^2}\exp\left(-\frac{D(2n+1)^2\pi^2 t}{h^2}\right) \tag{2}$$

where h is the thickness of the plate, and n is the number of layers.

The diffusion coefficient and the maximum or equilibrium mass uptake can then be used to calculate hygrothermal stresses and strains in composites using classical laminate theory (CLT).

A large volume of work [6, 9-11], related to diffusion of liquids in composites uses Fick's law, Figure 1, to calculate from the initial linear part the slope, the diffusion coefficient D. There are several assumptions that should be made in order to use Fick's law. It is really meant for homogeneous materials, so typically orthotropic laminates are a special case. The material must also be in equilibrium with its surroundings. The existence of voids, microcracks or diffusion due to capillary action at fibre interfaces, seriously limits the validity of the law. Serious limitation on the application of Fick's law is introduced by irreversible swelling caused mainly by immersion in liquids at high temperature, or solvents.

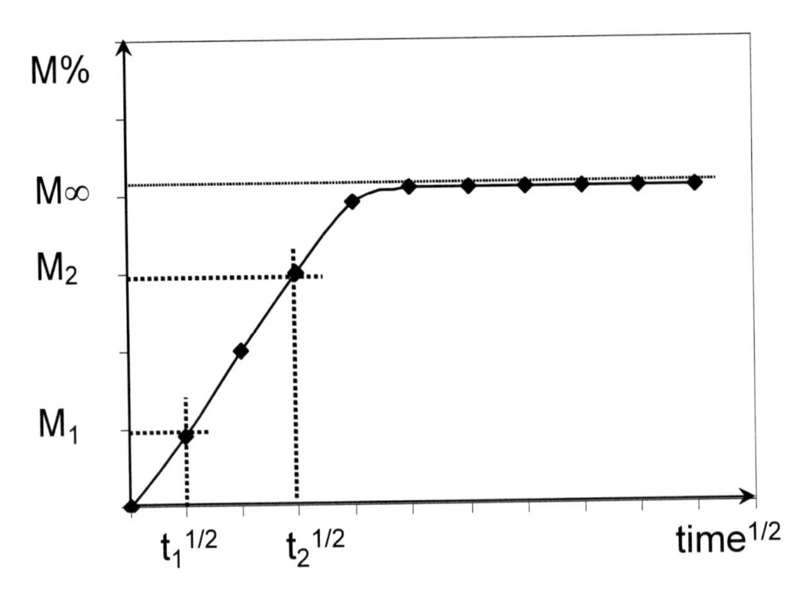

Figure 1. A schematic of a typical water absorption curve following Fick's law, where M_∞ is the equilibrium mass uptake [8].

2.2. Non-Fickian Diffusion Models

There are circumstances where the diffusion process can not adequately be described by Fick's law. The main assumption for use of Fick's model is that diffusion causes physical changes only to take place. In essence that means that diffusion should be a fully reversible process. In the case of crosslinked polymers and reinforced composites, swelling due to diffusion is an allowable physical change when upon drying is reversible. Under elevated temperature or aggressive environment exposure, chemical changes can take place, such as resin hydrolysis. In such a case the validity of Fick's model is reduced. Various diffusion models have been developed to accommodate the discrepancies from Fick's model. A comparative study of water absorption theories has been performed by Bonniau and Bunsell [4]. The differences in the behaviour of the materials when immersion or exposure at 100% RH takes place are pointed out in this study as different degradation mechanisms are found to be in action. This view is reinforced by others [12], as maximum mass uptake is found to be mainly dependent on relative humidity and not on temperature when specimens are exposed to humidity, whereas the effects due to immersion are clearly temperature dependent.

2.2.1 Multiple Stage Diffusion Models

Morii et al [13], having studied the weight change with time of randomly oriented glass reinforced polyester immersed in water, postulated a three phase behaviour as shown in Figure 2. In phase I, only the resin absorbed water, so a very fast, linear with time absorption occurred. From this linear part the diffusion coefficient is to be obtained. In stage II, the fibre/matrix interphase absorbed water. In stage III, water absorption in the interphase and matrix dissolution from the interphase occurred simultaneously.

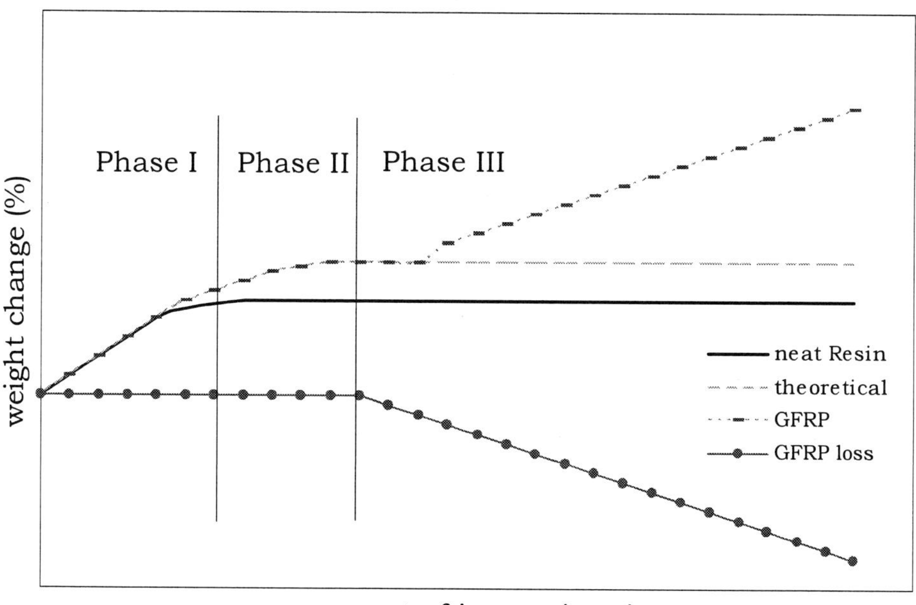

Figure 2. Schematic representation of the multi stage [13] weight change process for GFRP material immersed in hot water. The theoretical curve is Fick's law. The GFRP loss is a mechanism triggered with some incubation time in the beginning. Phase I is the initial quasi-linear part from which the diffusion coefficient D is obtained. Phase II marks the region where momentarily some equilibrium is reached. In phase III the matrix dissolution and the interface degradation produce weight loss and gain respectively. The ratio of GFRP loss curve and the GFRP curve sums the final water uptake curve.

Gurtin et al [14] proposed a model where moisture is allowed to exist in two phases, flowing through the matrix and trapped around fibres. The first phase, considered as a free phase diffuses according to Fick's law. The second phase does not diffuse. A dimensionless length is introduced to take into account for the length of the specimen. They notice that a two step saturation plateau can sometimes be observed but not always.

Taking into account the viscoelastic nature of polymers, Cai and Weitsman [15] suggested a diffusion model that utilizes the concept of a time-dependant boundary condition for saturation. The longer the exposure the more significant the relaxation effects become for the matrix and so the initial top boundary is raised. In a similar manner, Pritchard and Speake [5] resolved the anomalous Fickian diffusion as the sum of two terms, a Fickian term M_F with the expression presented previously and a resin relaxation term M_R. The relaxation term has been expressed as:

$$M_R = M_{R\max}\left(1 - e^{-kt^2}\right) \quad (3)$$

where

$M_{R\max}$ is the maximum water uptake due to resin relaxation,
k is the Arrhenius rate constant,
t is the time.

Their model has found extensive use for extrapolation of water absorption curves at low temperatures.

2.2.2 Langmuir Type

In a similar fashion to Gurtin et al [14], Carter and Kibler [16] noted that there is no a priori reason why moisture diffusion should follow a simple theory model. They proposed a model based on the Langmuir adsorption theory as shown in Figure 3. After some time, in a unit volume of resin there are n mobile water molecules per unit volume with a diffusion coefficient D_γ and become bound at a rate per unit volume γn. Equilibrium is established as at the same time N bound molecules become mobile at a rate per unit volume βN. The number densities of bound and unbound molecules at equilibrium n_∞ and N_∞ respectively are dependant upon relative humidity H, which can reach up to 100% for complete immersion.

$$\gamma n_\infty(H) = \beta N_\infty(H) \tag{4}$$

The long term solubility of water, expressed as percent by weight of the dry resin is

$$m_\infty(H) = \left(\frac{100M_w}{N_A\rho_r}\right)\cdot\left[n_\infty(H) + N_\infty(H)\right] \tag{5}$$

where M_w is the molecular weight of water, N_A is the Avogadro's number and ρ_r is the density of the dry resin.

$$\kappa = \frac{\pi^2 D_\gamma}{(h)^2} \tag{6}$$

where h is the thickness of the dry resin plate. When 2γ and 2β are both small compared to κ then the following approximation applies:

$$m_t \approx m_\infty\left\{\frac{\beta}{\gamma+\beta}e^{-\gamma t}\left[1 - \frac{8}{\pi^2}\sum_{l=1}^{\infty(odd)}\frac{e^{-\kappa l^2 t}}{l^2}\right] + \frac{\beta}{\gamma+\beta}\left(e^{-\beta t} - e^{-\gamma t}\right) + \left(1 - e^{-\beta t}\right)\right\} \tag{7}$$

where l is a positive integer. For a long exposure time t such that $\kappa t >> 1$ Eq. 7 can be reduced to the following expression

$$m_t \approx m_\infty\left[1 - \frac{\gamma}{\gamma+\beta}e^{-\beta t}\right] \tag{8}$$

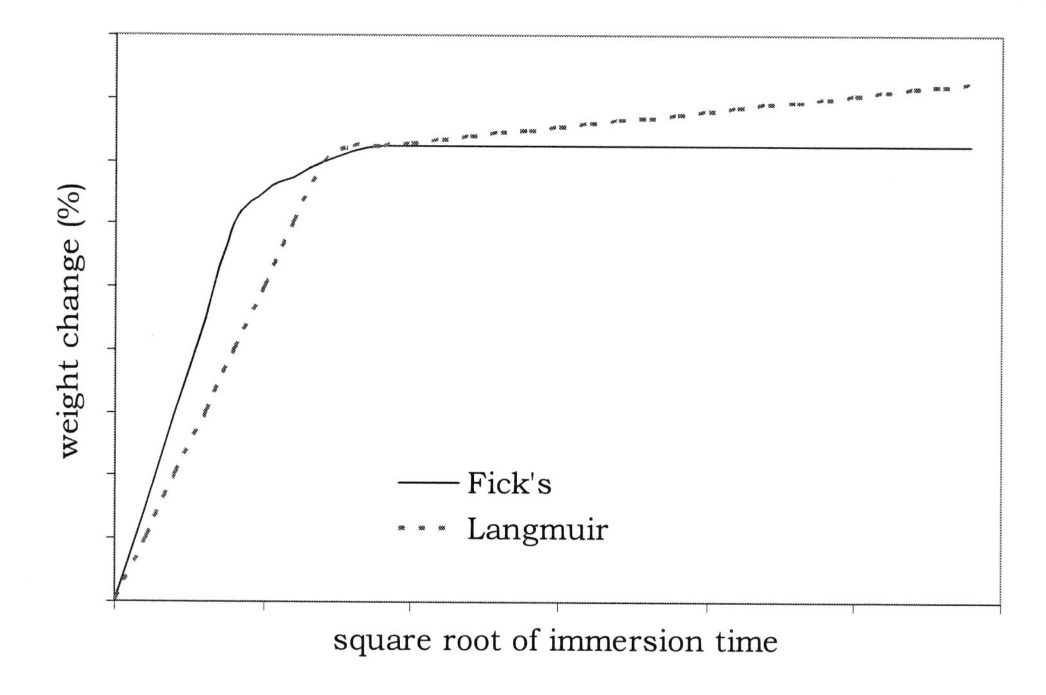

Figure 3. A schematic of Fick's law and the model based on the Langmuir adsorption.

Common problem to the proposed non-Fickian models presented is the need for extensive experimentation. In essence these models are to explain and interpret, post mortem the behaviour observed under some scientific framework rather than to actually predict it. Nevertheless they are used in practice.

2.3. Arrhenius-Type Temperature Dependence

Diffusion is a process driven mainly by the difference in concentrations of a substance. But diffusion is also a process dependant upon temperature. Under some circumstances, especially when the diffusion process is reversible and a monotonous relationship applies, the diffusion coefficient can follow an Arrhenius-type equation,

$$D = D_0 \exp\left(-\frac{E_a}{RT}\right) \tag{9}$$

where:

E_a is the activation energy,

R the universal gas constant,

T the temperature in K,

D_0 the diffusivity, a temperature independent material constant.

If no other parallel mechanisms are triggered by the change of temperature, the application of this equation can minimize the amount of testing needed for material characterization at different temperatures.

2.4. The Effects of Water Immersion on E-glass

E-glass fibres, E for electrical, are the most commonly used fibres for low cost - high volume production. For E-glass fibres the typical composition, by weight % is seen on Table 1 [17]. When exposed to aqueous solutions glass fibres that contain alkaline modifiers such as Na+ or K+ they are subject to chemical attack. For exposure in low pH solutions, pH<9 the reaction of glass degradation is:

$$Si - O - R + H^+ \rightarrow Si - OH + R^+ \tag{10}$$

where R can be Na or K. Depending on the concentrations of Na or K in the glass, there can be large increase of the pH in the solution due to the increase of the alkali content. If the pH of the solution is about 9 or above then the following reaction results:

$$Si - O - Si + NaOH \rightarrow Si - O - Na + Si - OH \tag{11}$$

this can lead to very fast reactions rates with catastrophic results for glass reinforced composites.

Table 1. Typical composition of E-glass fibres by weight% [17]

SiO2	A12O3	CaO	MgO	Na2O	K2O	B2O3	Fe2O3	TiO2
52-56	12-16	16-25	0-5	0-2	0-2	5-10	0-0.8	0-1.5

Schutte et al [18] performed single-fibre fragmentation tests as a part of a study of the effects of water exposure on composites. Based on the assumption that the alkali content of the glass would make it susceptible to environmental attack, they postulated that if the fibres were attacked in the distilled water used for the experiments, then leached ions should have been traceable in the water. The inductively coupled plasma mass spectrometry (ICP-MS) analysis was used, as it is a very sensitive instrument that can detect concentrations down to few ppm. Results have shown measurably higher contents of Ca, Mg and K in the water after exposure signifying the dissolution of the fibres. This was related to a strength reduction of the fibres.

Another detailed study of the effects of moisture and other environments on E-glass and other modified glass filaments is presented by Metcalfe and Schmitz [19]. They have pointed out that no strength loss was observed for glasses exposed to hot alkaline solutions but significant strength loss was observed for specimens exposed to acidic solutions. They also pointed out that there seems to be an incubation period during which no strength loss occurs.

2.5. The Effects of Moisture on the Matrix of Composites

There has been a lot of work regarding the effects of the environment on unsaturated polyester. The esterification process is a partially reversible process in the presence of absorbed water, under given conditions. This process is called hydrolysis and the by-products of this reduction, can be ethanoic (acetic) acid, styrene, glycols, free amines or a mixture of these.

UP are a very broad class of industrial thermoset resins. They are formulated using different prepolymers to obtain the required properties. Bélan et al [20, 21] studied the structure and assessed the hydrolytic stability of various unsaturated polyester resin prepolymers. The comparison showed that the relative order in terms of hydrolytic stability is propylene glycol (PG) < neopentyl glycol (NPG) and maleic acid (MA) < isophthalic acid (PA), the last performing best. A significant practical finding was that for industrial grades which are copolymers of these prepolymers, the hydrolysis rate was lower than the theoretically determined from the rule of mixtures.

The effects of water immersion on the polyester matrix of composites can be reversible or irreversible. Plasticisation and dimensional changes such as swelling or shrinking occur are reversible effects. Irreversible changes include microcracking and hydrolysis. Ashbee et al [22] among others studied the dimensional changes of thin polyester resin films immersed in water at 20° C, 60° C and 100° C. It was observed that initially resin swelling occurred but this was followed by shrinkage. This trend was more pronounced at the highest temperature of 100° C where after about 5 minutes a maximum swelling was measured. At 60° C it took about 30 minutes for the maximum swelling. At 20° C, resin swelling was observed after about several hundred hours, but no shrinkage was measured after up to 1500 hours. Dimensional changes have been also been reported to be temperature and moisture dependent [23], although this work was performed on epoxy resins.

A significant factor that determines the rate and the maximum level of water absorption in the polyester resin is the state of cure of the resin. It is more difficult for water to diffuse into a well cured resin. A relevant study by Benameur et al [24] has found that the maximum amount of water absorbed at equilibrium, decreases with increasing the initial state of cure. It was also found that the diffusivity and the rate of diffusion decrease with increasing state of cure. The practical implication of this point is that when comparing diffusion profiles obtained from the study of the same type of resin, produced at different conditions, results may not be directly comparable and discrepancies may be found. Post-curing the resin before water immersion was also found to be important.

Often encountered in literature are direct comparisons of different classes of resins and their respective composites, for practical reasons. The pair of resins most frequently investigated is polyester and vinylester [25]. Depending mainly on the temperature of immersion, water diffusion is found to follow Fickian or multistage behaviour. Vinylester resin composites are found to be more resistant to environmental degradation. This is mainly due to the resin structure of the vinylester which only has unsaturated double bonds on its ends, unlike polyester that has some unsaturated double bonds along the length of the chain. These double bonds are attacked by water causing plasticization and hydrolysis. Boinard et al [26] amongst others [6, 8, 27] studied the differences of glass reinforced polyester and vinylester resins with regard to the water uptake after immersion in distilled water at 30° C and 60° C. They performed gravimetric, dynamic mechanical thermal analysis (DMTA) and porosimetry measurements. Fickian sorption was not evident for any of the two resin systems. The rate of water absorption for polyester was found to be double that of vinylester. It has to be noted though that polyester plates were manufactured by wet lay up achieving a 50% reinforcement weight, whereas vinylester plates were manufactured by infusion achieving a 70% reinforcement weight. Abeysinghe et al [28] used six different polyester resins and one vinylester resin to assess the effects of the interaction of physical and chemical processes due to short and long term immersion in water. The superiority of the vinylester resin against the polyesters is reaffirmed. The significance of osmotic disc cracks on accelerating the hydrolysis process is noted. It is postulated that the presence of glycol in the polyester resins is the cause of these cracks.

The nucleation and the parameters controlling osmotic cracking in hydrothermally aged polyester are further discussed by Gautier et al [29]. Three parameters were identified as the main reasons for cracking due to osmotic pressure. The three parameters were a) the high fraction of monomeric and catalyst residues, b) the high fraction of ester functions in side chains and c) the high reactivity of ester functions. Depending on the glycol fraction of the resin and the temperature of the exposure, osmotic cracking has been observed to be instantaneous or with a leading incubation time. It was not possible to predict a change-over point though for passing from one behaviour type to the other though.

Studying the resin degradation process itself can be done in real-time or more typically after pre-set test points in time. A direct approach to real-time degradation monitoring was performed by Jayet et al [30] who monitored the hydrolytic degradation, in situ. The degradation process of a glass reinforced polyester resin was investigated by having a piezoelectric ceramic element embedded which measured the change of the electric impedance due to water sorption and physical dimension changes. The technique can be used for life-long durability monitoring after a calibration procedure. Although a solution like this provides real-time data, with no specimen sampling needed for data extraction, it is invasive and the existence of wires and connectors could act as an external source for degradation. Also a series of calibration runs have to be performed in order to obtain the constants for durability predictions.

An indirect approach to degradation monitoring was followed by Hakarainen et al [31] to study the degradation of polyester GFRP exposed to 40° and 60° C 80% RH for up to 6 years. Because the samples were stored in an air assisted oven, the analysis was not from volatiles from the environment but from samples taken from the test specimens. Gas Chromatography–Mass Spectrometry (GC-MS) analysis was the technique employed to study the samples. Principal Component Analysis (PCA) was used to find which of the degradation products were not related to the temperature and time. The aging time was then predicted successfully based on a correlation of the amounts of degradation products that were found to change with time. The test methodology is typical but there is really no actual prediction involved but a correlation of data with an empirically fitted curve.

Apart from optical changes, in the form of loss of translucency, caused to water immersed composites and neat resins due to swelling and possibly microcracking, changes in colour have been observed. Change in colour of polyester resin composites after hot water immersion in particular has been regularly observed. A detailed study of this phenomenon has been presented by Tomokumi and Shubata [32]. In their work they identified the various factors that could be the cause of yellowing. The factors considered where 1) the different glycols and di-acids, 2) promoters and additives, 3) counter metal in redox accelerators and 4) optimal resin systems. The significant finding of their study is that yellowing is not caused so much from heat alone but that the main cause of yellowing is hydrolysis.

2.6. Water Immersion Effects on the Interface

The effects of water immersion upon the contact condition at the interface between reinforcing fibres and the supporting matrix is the subject of this section. Surface coupling agents, or sizings, are typically used on glass fibre fabrics to be used for composites manufacture. These typically silane based coupling agents have a two-fold function. They are added on fibres at the manufacturing stage to protect the vulnerable fibre surface from scratches formed by friction. Their second function is that they form a chemical bond between the glass surface and the resin chains, Figure 4. For different types of resins different end groups are added to the sizing to facilitate the required properties. Water immersion and subsequently water absorption can cause swelling of the matrix which in turn creates stresses at the interface.

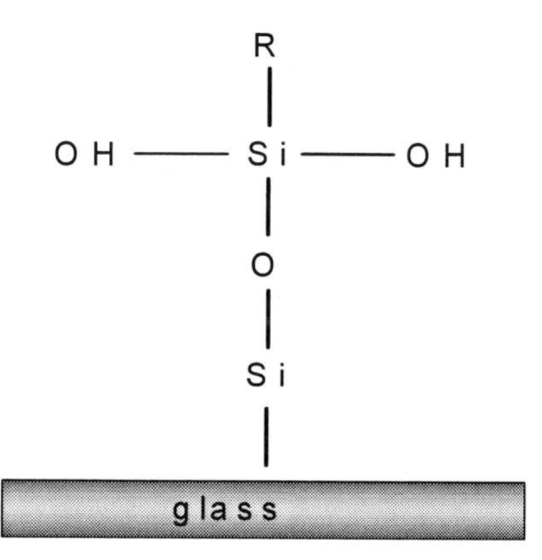

Figure 4. Chemical linkage between silane coupling agent and glass fibre. R group, depends on the resin to be used.

The Inter-Laminar Shear Strength (ILSS) test is a good measure of the interfacial strength. Varying the composition or removing altogether the sizing from glass fibres changes the ILSS measured. This idea was extended [33] to deduce the effect of water sorption and the paths followed depending on the interface condition. Glass/polyester specimens with varying interfacial strengths were tested for ILSS in order to deduce the effect of water absorption. Different initial interfacial strengths were obtained from specimens that the glass fibres had the sizing removed, had the normal silane-based sizing and fibres with a polydimethylosiloxane (PDMS) coating. When the interface was strong, as in the case with the silane sizing on, the diffusion was matrix dominated, that is to say for successive absorption experiments the maximum water uptake was increased. When the interface was weak, as in the other two cases, the water found an easier path in the interface and less damage was caused in the matrix. In successive absorption cycles the maximum water uptake level does not significantly change.

Weight change measurements are very common tools in the study of degradation of polymer composites [4, 6, 34]. The idea of studying the weight changes specifically at the interface was demonstrated by Morii et al [13, 35, 36] in order to evaluate interfacial damage. An analytical and experimental evaluation of the weight change in the interface of FRPs was performed. The weight changes in composites and neat resin were measured simultaneously and positively related to the volume of debonding at the interface.

The Fourier-transformed infrared spectroscopy is a technique that allows the detection of molecules according to the frequency of the infrared radiation being absorbed. Its use was demonstrated by Ikuta et al in a study on the hygrothermal degradation of the interface region [37]. GFRP material was immersed in water at 80° C and 95° C. At 80° C they found that no hydrolysis of the interface resin happened after up to 3000 hours of immersion, changes such as swelling and extraction of polymer by water permeation were evident. The finding signifies that an incubation period exists before hydrolysis becomes active at 80° C. At 95° C, after 10 hours, hydrolysis was evident by a change of the spectrum around 1670-1530 cm-1, the characteristic of carboxylate group produced through hydrolysis. The much shorter time that took for hydrolysis products to be found in this case can be linked with the possibility that the resin Tg was very near this temperature. Another interesting finding was that the rate of drop of bending modulus is higher at 95° C but after about 4000h of immersion the bending modulus measured was at a common minimum level for both temperatures.

The complexity of the degradation mechanisms and their interaction at the interface level are not limited to glass fibres [38]. The study by Gaur et al included Kevlar 49, AS4 carbon and E-glass fibres embedded in an epoxy resin and two thermoplastic resins. It was suggested that there was no unique trend and that the effects of hydrothermal ageing, after 24 hours at 88° C, were different for the various combination of fibres and resins. For the epoxy specimens tested it was found that the bond strength reduction due to the environmental exposure was fully reversible after drying at 110° C for 24 hours.

A way of reducing the complexity of the study of the condition at the matrix/fibre interface was the reduction of the problem to the single fibre level [39]. The use of the single fibre fragmentation technique to quantitatively assess the interfacial adhesion performance for composites that have been exposed to an environment that adversely effected the interfacial bond strength was proposed. The single fibre fragmentation test has also been used by Schutte et al [18] to study the differences in interfacial bond degradation under thermal and hygrothermal exposure. After exposure at 75° C dry air for 1632 hours, the interfacial strength did not show considerable difference from the control samples. On the other hand, after exposure for the same length of time in distilled water kept at same temperature, a significant drop was found in interfacial strength and the strength of the fibres themselves.

The problem of the reliability and the actual significance of the results obtained by methods assessing the global mechanical response of specimens as opposed to micromechanical modes was studied by Gautier et al [40]. The significance of the role of the interface and the failure modes associated with it when composites are immersed in water are studied. It was proposed that micromechanical tests and microscopic observations were more useful to ascertain the damaged condition. Differential swelling was found to be responsible for debonding and osmosis for the formation of cracks on the matrix. Both these mechanisms reduced the interlaminar strength properties of the materials tested after water ageing.

2.7. The Effect of Stress and Damage on the Absorption Rate

Damage or applied stresses in composites can have a significant role in enhancing the degradation state of composites that have been subjected to exposure or immersion in liquids. In this section work that deals with this subject is reviewed.

In a study by Veazie et al the effects of 5000 hours exposure to salt water of impacted S2 glass/5250-4 epoxy resin specimens on the tensile strength [41] were investigated. Under the environment studied, it was found that there was a linear increase of the amount of water absorbed with increasing impact energy, though the actual rate of absorption with time was not different.

The net result of water absorption on a damaged composite is not always an increase of the maximum absorption level. The effect of matrix cracks introduced by tensile forces, on the moisture uptake of glass/epoxy laminates at 70° C 85% relative humidity were investigated by Lundgren et al [42]. It was found that no significant difference was measured among the moisture uptake specimens of different crack densities. It was concluded that crack closure took place early in the experiment and this was attributed to swelling. Along similar lines, Autran et al studied in parallel standard and low styrene emission unsaturated polyester resins for the effects of mechanical stresses on the water ageing behaviour [43]. The conclusion was that the continuous application of mechanical stress did not have any effects either on the diffusion coefficient or on the maximum absorption level.

Different types of damage, depending on the environmental boundary condition imposed, can cause considerably different effects. These effects can sometimes be modelled with established diffusion models but sometimes not. A case for the effects of water absorption in damaged composite pipes is presented by Perreux et al [44]. Based on the Langmuir's diffusion model [16], a new method for the mechanical damage kinetics to be coupled with water absorption is proposed. Another model was proposed for moisture diffusion in the presence of cracks caused by static biaxial or fatigue loads [45]. The model indicates that effective diffusivity and maximum saturation level can be expressed as quadratic functions of crack density, assuming that there is no damage evolution.

Studies point out that there are differences between the behaviour of composites and their respective neat resin. Janas and McCullough performed a series of experiments with unfilled neat polyester resin and glass-filled specimens exposed at 60° C immersed in water under tension [46]. They found that whereas in the case of neat resin, there was no change in terms of maximum water uptake because of the applied stress, the glass filled specimens have shown an increase in the maximum water uptake.

2.8. Comparing the Effect of Different Solvent Types on Ageing

A subject of great practical interest is that of the exposure of composites to different types of solvents that could be encountered in service, Table 2. Depending on the exposure environment substantial differences in behaviour in terms of absorption and residual properties can be found [47]. Most studies on this field compare the effects of fresh or deionized water exposure or immersion with some other liquid or gaseous atmosphere either for a common given time or for the time it takes to reach similar level of some mechanical property deterioration. In such a study by Gutierrez et al, the exposure of polyester composites in fresh water at 60° C shows similar results as aging in sea-water at 70° C [48].

Table 2. Studies concentrating on the effects of different solvent types on composites ageing

Time	Material	Environment	Test Type	Ref.
< 1300 h	isopolyster films	distilled water, salt solution, artificial concrete pore solution pH 13.5 at ambient, 60° and 90° C		[9]
< 6 months	glass/epoxy, glass/polyster	lubricating oil, fuel, seawater, wet heat	tensile, flexural, ILSS	[49]
	glass epoxy, glass/vinylester	lubricating oil, salt water, diesel fuel oil, antifreeze, indolene and humid air		[6]
< 5000h	glass/epoxy, caron/epoxy	38°C water, alkaline solution pH 8.5, pH 10.5		[11]
< 12 months		synthetic sea water, sea water, deionized water, 55% RH all at 23°C		[47]
<270 days	glass/polyster, glass/vinylester	5% NaC1, 10% NaOH, 1,1,1-trichloroethane and toluene at 25°C	tensile, flexural	[50, 51]

A comparison of the results of exposed thin isopolyester films and dogbone shaped specimens in distilled water, salt solution and artificial concrete pore solution with a pH of 13.5 for the latter, at ambient temperature, 60° C and 90° C, for up to 1300h was performed by Chin et al [9]. At ambient temperature a 0.5% weight increase was found for all environments with equilibrium having been reached at about the first 20 hours. At 60° C the weight increase for all environments does not reach above 0.6%. For distilled water the behavior is typical Fickian, where a plateau is reached. For the other two environments a weight loss starts at approximately 100h for the pore solution and about 215h for the salt solution. This weight loss was linked to a chemical attack on the matrix.

Exposure to an elevated temperature environment does not necessarily cause an immediate reduction in mechanical properties [49]. Siddaramah et al studied differences in tensile strength, flexural strength and ILSS of glass fibre reinforced epoxy and polyester composites in lubricating oil, fuel, seawater and wet heat. A marginal increase was found for the short-term heat ageing but for the rest of the environments a reduction in all the properties tested was found. A study in similar exposure environments, lubricating oil, salt-water, diesel fuel oil, antifreeze, indolene and humid air for a 6 months exposure period on glass reinforced polyester and vinylester composites was performed by Springer et al [6]. As expected, degradation was more serious for polyester compared to vinylester composites. According to this work, the worst degradation is caused by indolene and antifreeze at 93° C.

The effects of different chemical environments exposure on glass-reinforced polyesters and vinylesters were also studied by Sonawala et al [50, 51]. They measured tensile and flexural strength after immersion in 5% NaCl, 10% NaOH, 1,1,1-trichloroethane and toluene. Immersion temperature in the solutions was at 25° C. A sigmoidal non-fickian diffusion was found in the case of brine immersion. Tensile strength dropped to ~70 % the initial after 270 days in brine, to ~10 % after 50 days in NaOH, to ~ 94 % in trichloroethane in 270 days and to ~65 % after 270 days in toluene.

Results were presented for glass and carbon reinforced epoxy composites for up to 5000 hours, at 38° C in water and alkaline solutions of pH of 8.5 and 10.5 by Tsotsis et al [11]. It was found that changes in pH had little influence on the absorption behaviour of the composites.

Wu et al presented results for a period of up to 12 months at 23° C for specimens exposed to 55 % RH, synthetic sea water, sea water and deionized water [47]. Substantial differences in mechanical properties were reported based on the different exposure environments. Deionized water showed the fastest diffusion to midplane, resulting the more severe drop in interlaminar shear strength. A very important finding of the study was that although the test temperature was ambient, irreversible mechanical damage and chemical degradation were evident. Typically the effects of ambient temperature exposure are reversible, but this finding emphasizes that composites due to manufacture variability for example, can behave in non typical way.

The concept of combining damage in the form of impact or drilled holes and degradation due to liquid absorption was investigated by Sala [52]. Liquids used for immersion were water, Skydrol, fuel and dichloromethane.

2.9. Methods to Reduce the Effects of the Environment

Various methods of reducing the effects of the environmental exposure on composites have been proposed over the time. Typically a gel-coat is used as a diffusion barrier on structures expected to be in continuous direct contact with water. A variation to this common practice was proposed [53] by the use of resin hybrids for laminate construction, where two different types of resin, one with higher moisture resistance would be used as barrier on the outer plies and the inner would be used to make up the rest of the laminate. This idea avoided the merely protective but otherwise parasitic gel-coat like layer.

Another way of reducing the effects of the environment on the composite is by modification of the resin. For unsaturated polyester resins, the hydrolytic stability could be enhanced by modification of the chain ends [54]. The addition of isocyanate, to control chain ends, decreased the weight loss. Addition of dicyclopentadiene did not decrease weight loss but it did limit the hydrolysis rate. Further protection of the unsaturated polyester resins can be realized by increasing their hydrophobic character [55]. The method concerned the effects of surface treatment of the resin with tetrafluoromethane (CF4) microwave plasma. The CF4 treatment decreased the surface energy and the water content measured through the resin and thus reduced the degradation.

Modification of the main composite structure does not always have positive results. Sonowala et al tried to incorporate polyethylene terephthalate or C-glass surface veils to protect polyester and vinylester laminates immersed in 5 % NaCl and 10 % NaOH solutions

[50]. Immersion time of up to 270 days was tested at 27° C. It was found that no substantial benefit was introduced by the veils in terms of mechanical properties retention. It was concluded that as the veils produced more interfaces they actually enhanced absorption through wicking.

3. Natural and Accelerated Testing

3.1. General Issues

For practical reasons the testing time available for the evaluation of the long term stability of GFRPs in their natural service environment is limited. The common practice is to accelerate the degradation process by changing one or more environmental variables sequentially or in parallel. Typically the most common accelerating factor is the increase of temperature. Other accelerating factors include the increase of pressure or the use of stronger solutions of the test medium or a combination of all these.

An extensive review on the subject of accelerated test methods as used to determine the long-term behavior of FRPs is presented by Bank et al [56]. The main focus of the study is on materials and practices used for infrastructure applications. One of the main concerns presented in this study is the absence, to a high degree, of studies relating the results obtained from accelerated tests with real time exposure.

The effect of a given accelerating factor at a given level does not necessarily allow for correlation with the same factor at a different level. The reason behind this is that otherwise dormant mechanisms can be triggered or chemical changes can take place at some time with the possibility that an incubation time can exist for these, rendering the process of following up the experiment very difficult. The difficulty of choosing appropriate conditions for accelerated degradation tests was stressed by Schutte [1]. Given the difficulties of controlling the test conditions over extended times and getting uniform moisture conditioning of samples, a way to overcome moisture testing was proposed [57]. It was suggested that moisture exposure could be substituted by an increase in the test temperature such that would produce a change of Tg equivalent to that caused by moisture exposure. Another procedure to overcome long testing times was proposed by Ciriscioli et al. The suggestion was to spike shortly above the expected saturation level and then allow time for equilibrium condition to be reached [34].

3.2. Statistical Analysis and Long-Term Testing

Accelerated testing is a necessary evil for it is difficult to relate quantitatively the damage caused in a short period to that which would normally be found after long-term natural ageing. Also the multitude of the possible combinations of resin, resin additives, fibre surface treatments and fibre systems cause significant burden as to whether the use of existing data to relate to a future system is valid.

An interesting method of planning long-term accelerated degradation testing is to use statistical analysis to relate the rate of environmental degradation to a combination of environmental and manufacturing parameters. In [58] McMillan et al performed a study on the rate of environmental degradation of resin transfer moulding (RTM) composites subject to combinations of environmental and manufacturing parameters. Taguchi statistical methods for the design of the experiments were demonstrated. The usefulness of the method was in reducing the amount of experimentation and in pointing out the trends and the interrelations of some of the parameters shown. However some disadvantages of the method were indicated at the same time. Problems of non-linear parameters with strong interactions and the fact that the method was sensitive to changes in the factor levels and not their absolute value were highlighted.

3.3. Long-Term Accelerated, Natural Ageing and Verification Studies

Long-term accelerated alongside natural ageing verification studies are a scarcity. The amount of time, planning and the cost involved make them generally prohibitive. Gutierrez et al [48] presented long term data from the company DCN. It was postulated that there is always a similarity between the natural and accelerated aging curve and that the same type curve is always obtained. For the systems studied it was reported that 1000 hours of accelerated conditioning at 60° C in distilled water, caused the same degradation as 15 years in water at ambient temperature.

Results from a parallel long term study in sea water and under accelerated conditions in the laboratory have been reported by Davies et al [59]. Specimens of glass reinforced isophthalic polyester resin were immersed in sea water for durations of up to 24 months and in parallel in distilled water at 16° C, 50° C for up to 12 months and in seawater kept at 50° C for up to 3 months. Tensile, flexure and ILSS tests were performed. An accelerating factor of around 4 to 5 was applied for laboratory exposure, over ambient conditions, for a span of up to 3 months laboratory exposure. The modulus was found to be a monotonous decreasing function for both environments. Strength though was not consistent showing a marked increase for natural seawater exposure after 12 months.

Kotsikos et al presented long-term accelerated data for polyester GRP specimens immersed in seawater at 40° C and 60° C for periods of up to 20000 hours [27]. The tests performed, for flexural strength, tensile strength and ILSS showed typical trends of an exponential decline reaching a plateau and correlated well with the semi-empirical formula proposed by Pritchard et al [5]. Strength results presented after exposure at 60° C were overall lower than for material kept at 40° C.

Ashbee and Wyatt measured swelling and found that 3 days water immersion at 100° C resulted in very similar level conditioning as for material conditioned for 16 weeks at 60° C [22]. Gellert et al performed ageing tests of glass fibre polyester, phenolic and vinylester specimens in seawater kept at 30° C for more than 2 years [60]. The temperature was considered the top limit of natural ageing for tropical waters in Australia. After 485 days in water, saturation was reached by the phenolic resin specimens. The drop measured in ILSS at that time was between 12 to 21%. Flexural strength losses in the range of 15 to 21% were reported.

The moisture uptake by sandwich composites after one or two sided immersion in deionized water was reported by Feichtinger et al [61]. Exposure time of 14 years was reached for 4 types of sandwich structures, two containing cross linked PVC and two balsa wood as core, whilst the skins were orthopthalic polyester coated with pigmented orthopthalic polyester gel coat. After 14 years of exposure in water no signs of moisture diffusion in to the cores of the laminates were found. A maximum of 1.2 % in weight moisture was measured for the skins. A correlation between accelerated testing and room temperature is made. One year at 65° C immersion in deionized water correlates well with the blistering performance of between 16 to 17 years of exposure in room temperature.

The long-term durability of GFRP composite pipes was investigated after service of 5 years by Chekalkin et al [62]. The pipes had been in operation at a pressure of 0.6 MPa at 70° C and at a pressure of 0.6 MPa at 20° C. The first pipe was operating with weak acid and the second with paper pulp. Based on the findings, an increase of the safety factor is proposed in order to estimate the residual life of the pipes.

Perreux et al worked on the durability of glass-epoxy pipes. Along with the pipe specimens they had some material in plate form. Specimens were fully immersed in distilled water at three temperatures at up to 60° C [63]. An important finding of the study was that diffusion from the inside wall of the pipes was almost negligible, thus rendering it not important consideration for practical applications. The pipes were also found to be absorbing less water, per weight, compared to the plate specimens.

In Figure 5. the difference in absorption rate and total water absorbed for immersion of pipes at three different temperatures are seen. The rate of water absorption at 20° C was very near the one at 40° C. For immersion at 60° C a sharper initial rate was found. At 60° C a peak value of about 2% weight increase was found before weight reduction started. This point was not reached for immersion at the two lower temperatures after a total of 10 years immersion.

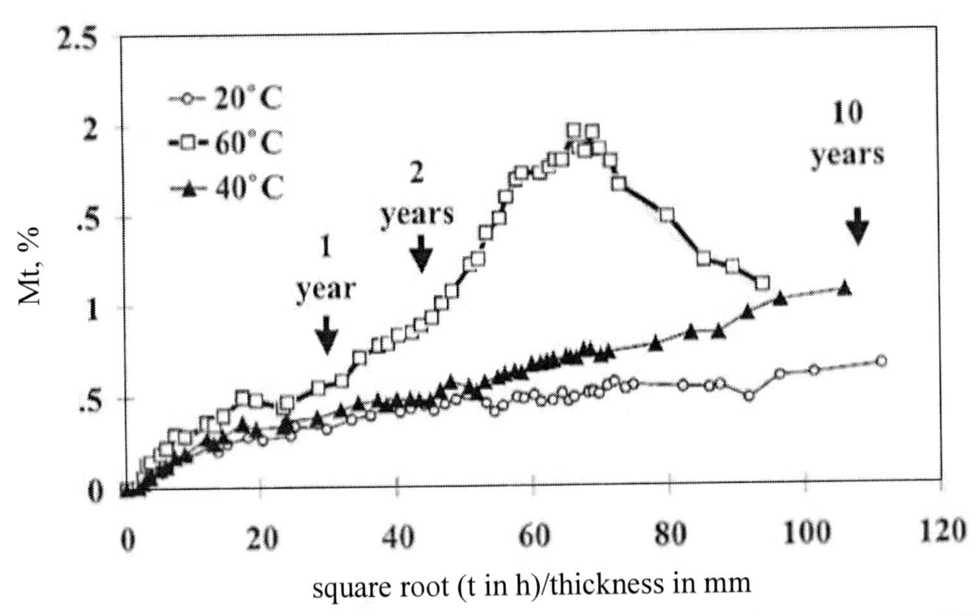

Figure 5. Absorption of composite panels after 10 years immersion at the three temperatures [63].

A comparative study of typical marine use laminates for the effects of long-term seawater immersion on the flexural strength and mode I interlaminar fracture toughness was presented by Kootsookos and Mouritz [64]. Significant chemical degradation of the polyester matrix was presented at 30° C from the first ~2 months of a ~2 year long immersion. Fully cured matrix composite show markedly better behaviour than incomplete cure (88%) composites which have a faster chemical degradation rate. The trend is repeated regardless of the fibre type used as reinforcement, glass or carbon.

4. Impact Damage

Composites are particularly vulnerable to out of plane impact damage which can be introduced in various ways to laminates during the manufacturing process or later on in the service environment. In general, detailed reviews can be found in [65-68] on the effects of impact on composites, however the present review considers only some aspects concerning the low velocity impact.

4.1. Impact Velocity

Impact can be separated in three areas of study, according to the initial striker velocity. There is low-velocity, ballistic and hypervelocity impact, although there are not universally accepted distinct changeover values [66]. A representative case for a low-velocity impact test is one performed using a drop-weight tower. Typical drop-weight test instruments have a maximum height of about 5 m and the maximum velocity that can be obtained from this height is about 9.8 m/s. Tests performed below about this velocity can be considered as low-velocity. When the striker's velocity is about 20 m/s and above can be considered ballistic impact.

The influence of velocity in low-velocity impact testing was studied by Rydin et al [69]. Woven and non-woven E-glass reinforced vinylester specimens were used. Results show that impact velocity determines the initial loading slope and time taken to reach maximum load. The inelastic damage accumulation is only marginally influenced by the initial velocity.

4.2. Striker Characteristics

Low velocity impact testing is experimentally performed using mostly instrumented drop weight towers involving sometimes very large mass impactors, of the order of hundred of kgs.

The effects of different shapes of striker were studied by Guillaumat [70]. The study simulated the accidental fall of a tool on a composite beam. It was found that for 5 different shapes, hemispherical, flat, pointed, linear and rounded, there is no difference on the global mechanical response of the cross-ply glass reinforced polyester plates. The difference encountered was in the size of the damage induced. The most severe damage was caused by the pointed end which caused partial penetration. The contact force corresponding to the onset of laminate damage though was found to be independent of the striker size [71]. The effect of striker shape on the perforation and penetration of thick FRPs was also studied by Wen [72]. A model was proposed to evaluate the depth of penetration, depending on the striker shape.

4.3. Damage Initiation- Damage Development

Work on the observation of the damage initiation and progression was presented by Guillaumat [73]. A series of experiments was performed using polyester resin with two different stacking sequences of glass reinforcement. Using the resin burn-off and de-ply technique the extent of delaminations and fibre breaks were studied. Multiple impact tests were also performed, on specimens which showed that local damage induced by impact did not have large influence on the global behaviour. This finding though was rather an artefact related to the boundary conditions and the specimen geometry used for the tests.

Wu and Shyu showed that delamination behaviour and strain history of static and impact tests were similar, if the vibration effects of the later were excluded [71]. They also found that the resilience increased after repeated loading.

4.4. Fibre Surface Treatment

A study was performed by Hirai et al, to assess the effects of fibre surface treatment on impact response [74]. After comparing five different types of surface treatment for E-glass fabric, substantial differences have been found. Two coupling agents were used γ-methacryloxypropyltrimethoxysilane (γ-MPS) and γ-glycidoxypropyltrimethoxtsilane (γ-GPS). The γ-MPS was used in three concentration levels and methanol washed. It was concluded that an increase in the concentration of □-MPS silane improved the damage resistance of the laminates.

Varelidis et al studied the effects of coating glass fibres with polyamide on the ILSS of polyester composites [75]. Laminates were immersed in water at 30°, 40° and 60° C. The use of polyamide, increased the maximum water uptake level when compared to uncoated specimens. Diffusivity also increased with increasing the amount of coating. ILSS was decreased as the amount of polyamide coating increased.

4.5. Thickness of Laminates

In out-of-plane impact, the thickness of specimens plays an important role. It was shown that the effect of thickness was much more significant than the in-plane dimensional effect [76-78]. In the work by Raju et al, bending rigidity was utilized as the parameter according to which the in-plane and the thickness effects were compared, the former being proportional to the third power of thickness but only on the first power of in-plane dimension [76].

Small variations of the launch height of the impact striker along with specimen thickness effects were considered by Belingardi et al [79]. By varying the drop height of the impact striker, thus changing the velocity of impact, it was deduced that the there was no sensitivity to strain rate effects for the region of velocities and GFRP specimen thicknesses studied. Also it was shown that the saturation energy values increased as the laminate thickness increased and that the damage degree increased with increasing specific impact energy. Here the specific impact energy was considered as energy per unit volume, which included the thickness effect. The term saturation energy was defined [80] as the maximum energy bearable by the material without perforation. The damage degree was defined as the ratio between the total energy transformed, stored and dissipated compared to the dissipated part of it.

As part of a study on the residual compressive strength of GFRP plates after impact, Zhou also considered specimen thickness effects [81]. It was concluded that there was not much difference in residual strength between thin and thick specimens, with thick specimens being just slightly better. Another very interesting finding was that even for completely delaminated specimens, about 30% strength retention was found to be the lower bound limit.

Sutherland and Guedes Soares studied the effects of specimen thickness, impactor kinetic energy and velocity of impact for glass woven-roving polyester laminates [78]. Larger damage size was found in thicker composites for the same amount of energy when compared to thinner specimens, a finding that is opposite to that presented by Zhou and Davies [82]. The group used a hemispherical striker whereas the second group used a flat striker.

4.6. Volume Fraction of Laminates

Few systematic comparisons of the effect of volume fraction on impact behaviour exist. Khan et al studied the effect of variation of reinforcement volume fraction on impact [83]. E-glass satin weave fabric was used in an LY 5052 epoxy resin with different volume fractions. It was reported that there was a minimum delamination damage area at a fibre volume fraction of 0.5. Below $V_f = 0.5$ matrix cracking was the dominant damage mechanism and above that the decreasing ILSS became dominant.

4.7. Fibre Architecture

There have been many studies focusing on the effects of the fibre architecture on the impact properties of plates. Bibo and Hogg [84] produced a thorough review of the role of reinforcement architecture in impact. Composites produced from non-crimped fabric performs showed an advantage in terms of impact damage containment and residual compressive strength when compared to prepreg type composites. Woven 2-D fabric showed much better compression after impact strength, than prepreg composites, mainly because of the much lesser damage extent due to impact.

The impact resistance and the damage characteristics of GFRP polyester reinforced with non-crimp, woven fabrics and discontinuous non-woven mat were investigated by Shyr and Pan [85]. Thin more flexible specimens sustained mainly delamination damage whereas thicker more rigid samples sustained fracture dominated damage. The non-crimp fabric reinforced specimens showed good impact resistance efficiency.

4.8. Stacking Sequence of Laminates

The general idea repeated by many authors, is that using ±45° layers on the outer-most layers and 0° in the center of the laminates makes the plates flexible which in turn means that fewer fibres are broken during impact and the extent of damage on the main load-bearing fibres at 0° is not so significant. The use of ±45° layers though as outer-most can potentially cause early stability problems in compression after impact (CAI) so a careful balance must be made.

Rydin et al studied the effect of stacking angle between woven glass fabric reinforcement mats in RTM vinylester matrix composites under impact and static penetration [86]. The four different stacking angles that were studied were 0°, 20°, 30° and 45°. A general agreement between static and dynamic behaviour was found. The trend showed that from least to more penetration resistant the stacking angles were 0°, 20°, 45° and finally 30°.

A damage resistance parameter was formulated by Fuos et al in such a way as to account for changes in stacking sequence [87,88].

$$\mathrm{DR} = \int^{2\pi} \int_{a=0} \mathrm{R}(\varphi)\partial \mathrm{R}\partial\varphi \qquad (12)$$

The parameter R(φ) is a measure of the damage radius measured from the point of impact at angle φ. The expression worked for moderately thick laminates containing interface angles below 60°. The formulation of the expression allowed for internal damage only such as delamination and matrix cracking. There was no provision for the effect of fibre breakage or back-face damage.

4.8.1 Woven Fabrics

Davallo et al compared woven and Chopped Strand Mat (CSM) E-glass reinforced laminates under low energy impact loads [89]. The superior behaviour of the woven laminates was evident. Woven reinforcement laminates contained damage mainly by matrix cracking and delaminations whereas, in the case of CSM, significant fibre fracture occurred locally causing significant drop in the tensile strength.

4.8.2 Non Crimp Fabrics

Karbhari looked at the differences in impact behaviour of RTM woven, RTM non crimped fabric and prepreg glass reinforced composites [90]. The concept of Inelastic Energy Curves (IEC) was used to obtain more information of the impact events so that changes in impact behaviour could be traced back to the reinforcement characteristics. IECs are formed by plotting the returned energy on the striker versus the impact energy. It was concluded that better impact performance was achieved by composites using non crimp fabric performs.

4.8.3 Stitching

The effects of stitching on glass woven and non-crimp fabrics in composites in compression after impact were studied by Yang et al [91]. It was found that stitching had beneficial results in residual compressive strength as delaminations caused by prior impact were not able to grow. Hosur et al produced work on the effects of stitching on woven glass fabric composites in impact loading. It was found that the shape of damage caused by impact changed due to stitching [92]. In un-stitched specimens the shape of damage caused by impact was conical in the thickness direction. After stitching, a cylindrical more concentrated damage area bound by the stitches was observed.

4.9. Impact Combined with Environmental Exposure

The combination of impact damage with environmental exposure can yield very interesting results. Since the main effect of the environment is on the interphase and the matrix, residual compressive strength is significantly degraded. Results of residual tension strength are affected less by environmental exposure. Veazie et al studied the post-impact behaviour of S2 glass/5250-4 bismaleimide resin and the effects of 5000 hours salt water ageing on tensile properties [41]. Salt water exposed specimens shown a small decrease in the tensile strength and stiffness even without impact damage. For specimens impacted at low energy impact levels and then aged there was virtually no loss in tensile strength observed. Only for higher impact energy levels that have been water exposed, specimens shown a decrease of less than 10% in tensile strength compared to dry impacted specimens.

The impact response of Kevlar and carbon reinforced epoxy specimens that had undergone accelerated immersion in various liquids was analyzed by Sala [52]. It was reported that impact damage on specimens did not lead to anomalous diffusion. In fact a positive correlation was observed between impact damage extent and increase of fluid uptake.

Strait et al compared two glass reinforced epoxy systems for the effects of water immersion on impact resistance [93]. One system was conventional epoxy resin with cross-ply and quasi-isotropic E-glass reinforcement and the other was woven E-glass fibre with a rubber toughened epoxy resin formulated for marine applications. It was concluded that the peak load and the total energy absorbed were significantly reduced for both systems after environmental exposure hence an overall significant reduction in impact resistance was observed.

Studies on the combined effects of impact and environment on the residual compressive strength, useful as they might be, they are very scarce. The scarcity is mainly due the understanding that complex interactions occur and the results obtained are not easily interpreted. Komai et al investigated the influence of water absorption on the impact bahaviour of CFRP quasi-isotropic epoxy matrix composites [94]. Specimens immersed in deionized water at 80° C for 2 months were subsequently impacted and their residual compression strength against a set of dry impacted specimens was assessed. Results showed that for same impact energy the delamination area of wet samples was larger. Also the impact induced delamination threshold level dropped significantly compared to dry specimens.

Ishai et al studied the impact and residual compressive strength of separate GFRP-CFRP skins and complete sandwich panels after water immersion at 25°, 50° and 75° C [95]. It was found that impact damage size was only marginally changed by hygrothermal exposure. Also the impact damage depth was deeper for specimens with free edges than for those specimens where the foam was pre-encapsulated. One other finding was that environmental exposure proved beneficial for residual strength after impact. For pre-encapsulated sandwich samples, when moisture gain was below 1% and the exposure temperature below 50° C, the residual strength was higher than that of the reference – unexposed samples.

The effect of water immersion on the impact behaviour of two woven glass-aramid fibre reinforced epoxy laminate composites was presented by Imielinska et al [96]. Immersion in water at 70° C was reported for a length of ~52 days. One laminate was made with a hybrid fabric whereas the second type was interlayer. Impact energies of up to 32 J were tested. Results have shown that the hybrid reinforced laminate was better in terms of peak load sustained when impacted dry but when wet no significant difference was shown between this and the interlayer laminates dry or wet.

5. Compression After Impact

The compression after impact test (CAI) is universally accepted as a means of assessing the damage tolerance of composites. It is vital that structures do not fail catastrophically and the CAI test gives an indication as to how damaged composites behave. The matrix dominated hot wet compression after impact test in particular, can be considered as the most severe evaluation method of the structural performance for polymer composites [97].

5.1. CAI Fixture Designs

Any compression after impact test rig design has as a goal to allow a region of pure compression to exist on a given sample geometry. Plate specimens need to be supported in such a way as to allow unconstrained compression but also enough support so that global plate buckling is not the cause of early failure. In the section below widely used CAI test rig designs are reviewed.

5.1.1. NASA CAI Fixture

In the NASA CAI fixture [98], which is shown in Figure 6, the specimen is supported from 4 sides. The specimen height starts at 254 to 318 mm. Width is 178 mm which is trimmed to 127 mm after impact. The thickness is 6.35 mm with lay-up $(45/0/\text{-}45/90)_{6s}$. The striker is hemispherical with diameter 12.7 mm. The mass is set to 4.5 kg and the drop height to 0.61 m so the energy of 27 J is obtained. The support is 127 mm square clamping the specimen. Loading is at 1.27 mm/min.

Figure 6. The NASA CAI test fixture with specimen.

The NASA CAI fixture has been partly surpassed by the Boeing fixture due to the very large specimen size, which translates to high specimen manufacturing cost and the difficulty in testing.

5.1.2. BOEING CAI Fixture

The Boeing standard [99] describes the test sample having dimensions of 4" by 6" (152 mm x 102 mm) with a thickness between 4 mm to 5 mm with lay-up $(-45/0/45/90)_{ns}$. The sample is to be impacted prior to compression with a hemispherical striker of 15.75 mm diameter. Mass of the striker is variable from 4.5 kg - 6.8 kg with the specimen being supported by a frame with a cut-out square of 127 mm x 76 mm. The specimen is clamped on the steel frame from the top at 4 points, Figure 7.

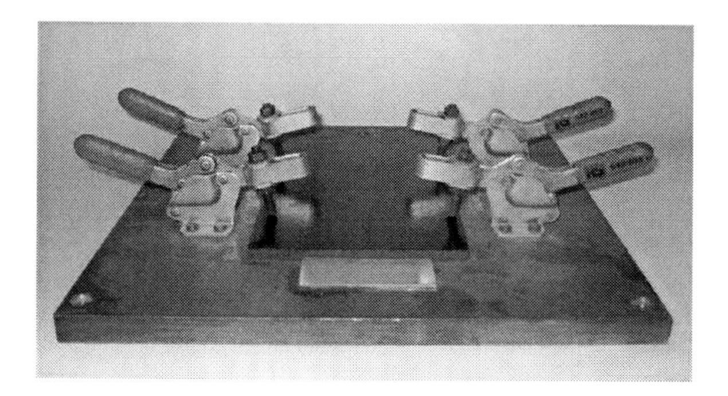

Figure 7. The clamping arrangement of specimens to be impacted for the Boeing CAI test.

The sample is clamped from top and bottom and simply supported on the vertical sides in the compression test fixture, Figure 8. The loading rate for compression is set at 0.5 mm/min. The top hat of the rig is smaller than the side supports lateral distance. The vertical sides of the specimens are held in place by knife-edge like supports. The unsupported gauge length is 5 mm which brings maximum possible compressive strain to around 3.3%.

Figure 8. The Boeing CAI test fixture without specimen. The top and bottom clamping plates are curved to provide a variant of the standard test fixture to test curved panels.

The design is quite robust as the side supports can be fixed to different positions along the base plate so that samples of different dimensions as well as curved panels can be tested, Figure 8. This feature has been explored and the design has been accepted by SACMA [100] for CAI testing. The striker diameter in this case is 15.88 mm, mass is set to 5 kg. The thickness is set to 3 mm with same lay-up as for Boeing. The loading rate is increased here at 1.27 mm / min.

5.1.3. JIS Miniaturized CAI Fixture

The Japanese Institute of Standards introduced along with the acceptance for the full size Boeing test rig a miniaturized rig [101]. This fixture is like a miniaturized Boeing CAI rig in terms of support. The specimen sample is set to 80 mm height by 50 mm width and a thickness of 2.5 mm.

5.1.4. AITM - EFA CAI Fixture

Along the lines of the Boeing CAI rig, Airbus Industries in Europe made their own metric "equivalent" design [102]. The sample size of 100 x 150 mm with thickness of 4 mm and lay-up $(45/0/\text{-}45/90)_{4s}$ correlates directly to the one used by Boeing. The hemispherical striker used for impact has a diameter of 16 mm, the mass ranges from 1 to 3 kg or 4 to 6 kg. The energy levels obtained are 9 J, 12 J, 16 J, 20 J and 25 J or 30 J and 40 J. Impact support is given by a steel frame of 125 mm x 75 mm clamped at 4 points. End loading is at a rate of 0.5 mm/min.

The specimen clamping arrangement differs from the Boeing CAI rig, Figure 9. The side supports run the full height of the specimen. For this reason the top clamping plate is smaller than the specimen width, to fit between the two side support knife edges. The area receiving the compressive load on top is smaller than the bottom area. This clamping arrangement eliminates misalignment problems but can cause uneven loading. The testing procedure has been accepted and adopted by the Eurofighter consortium [103].

Figure 9. The AITM CAI test fixture with impacted specimen inserted.

5.1.5. BAE Systems CAI Fixture

BAE systems CAI rig differs substantially from the rigs presented previously as there are no side supports, Figure 10. The test rig and the testing procedure and specifications are described by Habib [104]. The test fixture can accommodate substantially larger panels than those test fixtures mentioned earlier, with nominal plate dimensions reaching 250 mm in width by 300 mm in height respectively. As the test fixture has no side supports but front and backside supports, the general dimensions of samples can vary considerably so it is a very robust design. Both bottom and top of the plates are clamped. The top plate has no guides so it can be prone to cause early buckling. Due to the large dimensions of the plates that can be accommodated, the impact energy and subsequently the damaged region can be proportionately large.

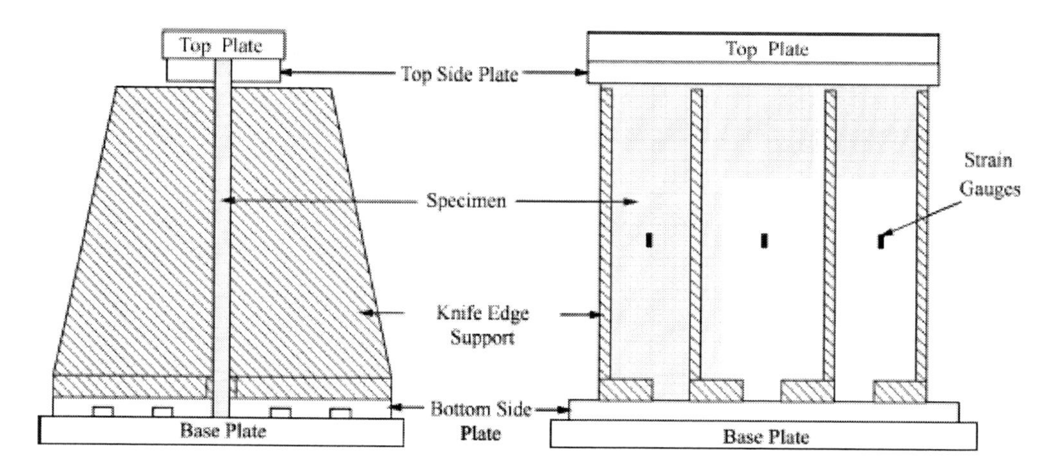

Figure 10. The BAE CAI test fixture, [104].

5.1.6. Miniaturized QMUL CAI Fixture

When the size of the specimen used is large, the cost related and the equipment needed for the test, make it expensive to perform. In an effort to overcome this problem some miniaturized CAI designs have evolved. Queen Mary and Westfield College and Hexcel collaborated [105] in order to evaluate a new design for a miniaturized CAI. The design reduced the sample size to 55 x 89 mm but kept the sides ratio of the AITM CAI rig. An extensive series of tests has proven the design to be representative of the results obtained with the large sample CAI rig used by AITM.

The top plate of the first version of the minaturized compression rig was not guided so misalignment problems arose. Also as the plate was larger than the side support distance, this allowed only limited stable gauge length. The bottom of the specimen was not clamped here as it was considered that crushing effects would not be important.

At the early stages of the current research project investigating the effects of pre and post environmental exposure impact damage on the residual compressive strength of composites, it was realised that a new CAI rig was needed. Based on existing sample dimensions, a new test fixture was designed, Figure 11. The main improvements on the old design were the incorporation of a set of a guided top plate, through the use of positioning sliders and ball bearings and the rearrangement of the side and top/bottom supports to accommodate longer gauge length.

Figure 11. The miniaturized QMUL CAI test fixture with specimen.

6. Non Destructive Testing

Non-destructive testing techniques (NDT) are used to detect any service induced damage and manufacturing process defects in a non invasive way. When composites are translucent, such as in the case of thin glass reinforced panels with un-pigmented resin, visual inspection is possible. Using a strong back-light the difference in translucence between a damaged and an undamaged area can be easily detected and quantified in terms of damage size/shape and intensity. Defect size of ~1 mm can normally be easily detected. Another commonly used technique used is tapping. A coin or any other stiff object can be used to tap the structure's surface. Sound travels through a different path, shorter or longer, in a damaged or un-damaged area. This is a very crude technique with limited applicability but is inexpensive and easy to use.

6.1. Ultrasonics

Based on the principle of measuring the difference of amplitude of the ultrasonic signal or the time of flight from different sound paths, as used in the tapping method, different ultrasonic NDT methods have evolved. Typically a high powered piezoelectric transducer is used to generate ultrasonic frequency waves. The range of the frequencies produced varies from about 50 kHz to about 20 MhZ. On the lower end of this range the power that can be obtained is high and less dense materials, such as polymeric foams can be tested. At the higher end of the range, the power available is reduced and so it is suited mainly to dense metallic materials. The resolution also increases with increasing frequency.

An important consideration about ultrasonic instruments is the contact medium, or in some cases its absence. Typical methods of operation with ultrasonic instruments will be presented next.

6.1.1. Water Coupled Immersion

There are two main configurations used in ultrasonic inspection in immersion mode, the Pulse-Echo and the Through-Transmission, Figure 12. The Pulse-Echo technique involves the use of a single piezoelectric probe that acts as both a transmitter of ultrasonic frequency waves and as receiver. The probe is moved in a raster pattern, parallel to the specimen and the bottom of the tank, using a step or servo motor driven gantry. At the bottom of the immersion tank lays a slab of glass acting as an acoustic mirror. A few studies [106,107] have been published using this technique for damage detection and characterization in composites.

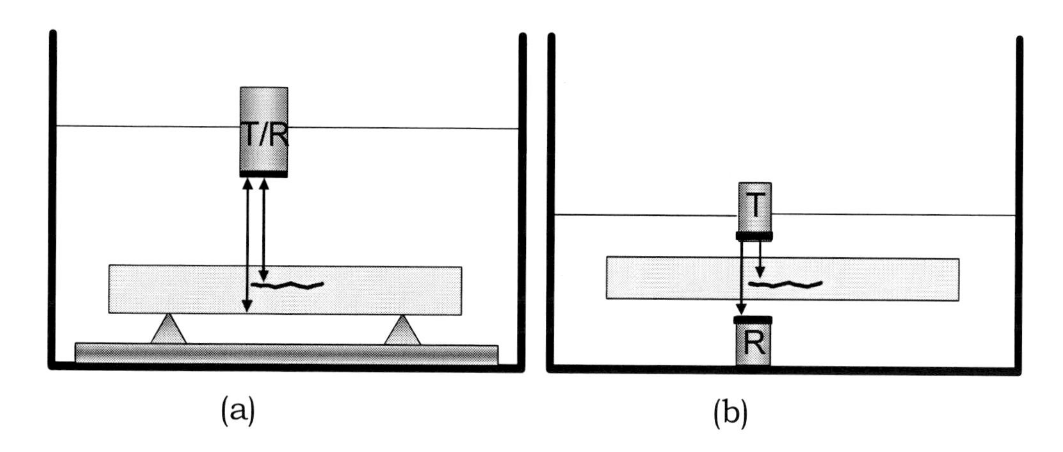

Figure 12. Arrangement of (a) pulse-echo and (b) through transmission ultrasonic scanning set-up in water immersion tanks. In (a) a probe that is both transmitter and receiver (T/R) is used. At the bottom of the tank rests a glass slab acting as acoustic mirror. The specimen rests on top of supports. In (b) two probes facing each other where the one is transmitter (T) and the other receiver (R) are used. There is no need for acoustic mirror in this case.

In the Pulse-Echo mode of operation it is possible to extract two types of information from the signal received regarding the damage state of samples. If the amplitude of the response wave is recorded alone, then a two-dimensional damage intensity map of the sample can be produced. The second type of measurement is that of the Time of Flight. In this case, the time taken for a specific path is recorded and compared to a reference "clean" path obtained within the thickness of the material. The duration of the reference signal is the time it takes for the signal to go to the back side of the specimen, be reflected and return to the probe. Given that the velocity of wave transmission through a particular type of material is known a depth calibration can be made and so a depth related dimension about damage is produced. Combining the amplitude signal levels with the Time of Flight data a pseudo-three-dimensional representation of damage can be obtained.

In the case of Through Transmission, signal transmission is one way only from the transmitter probe to the receiver. This limits the operation to signal amplitude only measurements. The signal level available though is stronger due to the reduced length path and the reduced overall losses at the interfaces so thicker sections can be scanned. Another advantage of this method is that curved panels and tubes can be scanned with suitable support gantries that keep both probes facing each other. From a practical point of view though extra space in needed for the probe holder or the specimen holder and it is assumed that the part to be scanned is accessible on both sides.

6.1.2. Direct Contact Systems

Under many circumstances, such as for in-situ work, immersion systems are impractical. Special probe types that have a hard-wearing cover can be used for direct contact with the test piece. Also, an acoustic coupling gel can be used. A typical contact system is the ANDSCAN, where a single probe mounted on bar is moved across the surface of the composite specimen and gel is used for acoustic coupling. The bar is linked to a system of two or three position encoders, for x-y-rotation that which digitize the relative position and a C-Scan or a D-scan can be produced. C-Scan is a 2 dimensional, x-y representation of the amplitude or time of flight of ultrasonic waves through a material. A D-Scan can be produced only with time of flight data. Time of flight data can be obtained using the pulse-echo or double through transmission. A D-scan is a 3 dimensional x-y-z representation of the structure where the z-axis (thickness) being made equivalent to the relative time of flight. Essentially it can be constructed by stacking C-Scans obtained from sequential time-gate positions. The use of an ANDSCAN contact ultrasonic system for damage evaluation on GFRP, KFRP and GFRP specimens is demonstrated amongst other authors by Scarponi and Briotti [108].

6.1.3. Non-Contact

Non-Contact ultrasonics have evolved over the last 20 years as a practical and useful alternative NDT to traditional contact ultrasonics. In situations where contact is not desired such as for sterilized components or where it is completely excluded such as for inspecting hydrophilic materials, for microelectronics inspection or high temperature component testing, non-contact systems are ideal.

At the interface between two materials some portion of sound is transmitted and some is reflected. The ratio depends on how close the acoustic impedances of the two materials are. Also naturally the more interfaces available the greater the loss of signal. To overcome this problem, high pressures need to be used so transmitter probes at high voltages are used. Receiver probes are pre-amplified and very sensitive. The range of probe frequencies used for non-contact ultrasonics is limited by the increasing attenuation of air at increasing frequencies. The 1 MHz is considered as the top limit and more typically 400 kHz probes are used [109].

6.1.4.Laser Ultrasonics

The laser ultrasonic method is another form of non-contact NDT. When a pulsed low energy laser hits the surface of a composite plate, an instantaneous temperature differential is created at that point. This local temperature difference becomes the cause for elastic stress waves to be created and to propagate. These waves can then be detected and their intensity map is used to obtain information in a similar manner to conventional ultrasonics. The use of laser generated and detected ultrasonics for measuring stiffness of composites is demonstrated by Audoin [110]. CFRP plates were inserted in an oven and the change in stiffness due to the temperature rise was measured by the laser ultrasonic through transmission system. The difficulty faced due to the non-isotropic nature of the material, to obtain stiffness data was indicated. The solution was given by extensive signal filtering and post-processing.

6.1.5. Acoustography (acousto-ultrasonics)

The use of acoustography has been employed successfully for the real-time study of impact damage growth during fatigue tests. An ultrasonic frequency sensitive liquid crystal panel is used to convert the signal received from an ultrasonic source into image. This image is recorded by CCD camera and the progress of the damage area growth can be monitored. The use of the acoustography technique has been demonstrated by Chen et al [111,112].

6.1.6. Acoustic Emission

When a composite material is stressed, characteristic sound signals are emitted by various mechanisms such as fibre breaking or matrix cracking. For simple materials, such as single fibre micro-composites, this method is easy to be used. Its use is more complicated for larger parts because the characteristic sound patterns of the various components become too complex. Kotsikos et al [113] used the acoustic emission AE technique to follow the damage accumulation due to the environmentally assisted fatigue testing. Glass fibre reinforced plates have been immersed to sea water and then their fatigue bahaviour under bending was compared to dry specimens.

6.1.7. Scanning Acoustic Microscopy

The use of Scanning Acoustic Microscopy allows for imaging of delaminated areas in thin composite plates, using a highly focused ultrasonic probe. Typical frequency for a probe used for SAM would be 50 MHz. Its use has been demonstrated by Komai et al for dry and wet CFRP plates [94].

6.1.8. Ultrasonics for Quantitative Evaluation of Composite Properties

The anisotropy of typical laminate composites has been a serious problem for the development of material properties quantitative uses of ultrasonic NDT techniques. Relatively little work exists in the field and is mainly in developmental stage.

Hong et al evaluated the performance of stress wave factors for quantitative evaluation of the properties of hygrothermally degraded specimens [114]. Ultrasonic stress wave propagation through specimens at various stages of the degradation process is preformed. The efficiency of the wave energy dissipation is used as a way to quantitatively judge the state of the degraded material.

The propagation velocities of longitudinal and transverse waves in laminates at different frequencies were measured by Tauchert et al [115]. The laminates used had woven fabric reinforcement and so were assumed to be orthotropic. Results for carbon reinforced plates elastic moduli were within 4% true of the results obtained from static tests.

Choi et al used a scanning acoustic microscope to measure the local ultrasonic velocity on the surface of glass reinforced and neat resin thermoplastic polyester specimens [116]. Some had been exposed outdoors for 11 months. It was found that the elastic surface-waves velocity for specimens being subject to weathering was much smaller than that for the unexposed. Based on the assumption that the Poisson's ratio and the density of samples did not drastically change, a change of the elastic modulus due to the weathering was reasoned to cause the change.

The case of very thin laminates where the thickness of the plate is less than the spatial length of the ultrasonic pulse is very special as delaminations in such plates are extremely difficult if not impossible to detect using conventional methods. This problem has been discussed and two methods have been proposed by Kazÿs et al [117]. The first technique used to produce C-scan from filtered A-scans involved the use of the difference in reflected signals from the reference and delaminated areas being exposed using the L1 norm deconvolution algorithm. The second technique was based on the spectral analysis of the differential signature of the signal obtained from the area under investigation.

Wooh et al also worked on a signal processing techniques for improved imaging of thin composite laminates [118]. Having noted the limitations of the conventional broadband ultrasonic signals, they proposed a homomorphic deconvolution technique to use signals from a typical system and to filter these to obtain as much information as possible for the through thickness direction. Also built-in in their algorithm was a surface-follower mode that reduced the problems faced when slightly curved panels were to be used.

A proposal was made by Nesvijski that composites should be seen as stochastic materials and that ultrasonic data analysis could provide the effective dynamic elastic modulus [119]. Based on the assumed stochastic nature of the materials, ultrasonic wave response signals could be used to estimate an upper and a lower boundary value for the composite moduli, so a calculated rather than a unique value for moduli would be obtained.

6.2. Other NDT Methods

Most ultrasonic NDT methods have been well established but some other non ultrasonic methods have evolved to augment or replace them. A practical problem of most typical ultrasonic NDT techniques is that a C-Scan requires raster scanning and consequently takes a lot of time to complete. Even if phased arrays of probes are used, where many probes are connected in series and so larger surfaces can be covered, time is a major issue. Full-field techniques, such as shearography or thermography have evolved to exploit the time problem posed by ultrasonics. In this section alternative to ultrasonic NDT methods are reviewed.

6.2.1. Dielectrics

Pethrick has over the years produced a large volume of work for the detection and quantification of water uptake on composites using the change in dielectric properties [120]. Measuring the differences in frequency and time response of dry samples to that of samples being exposed to the environment a relation with mechanical property degradation was obtained.

The possibilities of using dielectrics for condition monitoring both in manufacture level and later on, were investigated by Apicella et al [121] studied. The degree of cure and the changes of viscosity were associated with measured changes in dielectric. Also changes in the dielectric were associated with hygrothermal stability as in the case of Jayet et al [30].

6.2.2. Nuclear Magnetic Resonance Imaging

Due to their polar nature, water molecules in the presence of a strong magnetic field can be polarized. The Nuclear Magnetic Resonance Imaging (NMRI) [27] technique has been successfully employed to visualize the extent of water diffusion in composite plates after prolonged water exposure.

6.2.3. Shearography

Digital laser shearography has been used to detect the impact damage on boat hulls and disbonds in honeycomb aircraft structures. It is a full-field NDT method that can provide results almost instantly. Its use has been demonstrated by several authors [122-125] as a convenient means of damage evaluation especially on large structures.

6.2.4. Eddy-Current

Since the Eddy-current effect works on electrically conductive materials only, as far as polymer matrix composites are concerned this technique is available only for inspection of carbon reinforced composites, with Vf higher than 40% [126], or other conductive fibre composites. The Eddy-current technique has been investigated by Khadetskii and Martynovich [127] for its applicability in detecting near edge delaminations of CFRP composites. The analysis of the eddy-current field generated near edges is very complicated and there is not an analytical solution so calculations were performed using the finite difference method.

6.2.5. Thermal Techniques – Thermography

The thermographic technique normally employs a thermo-sensitive camera, operating near and at the infrared spectrum, to visualize the temperature distribution along the surface of an object. The method is based on the fact that flawed and flawless areas have different surface temperatures due to the variations in their thermophysical characteristics. It can be used as a one sided method and no physical contact with the test specimen is required. Pulse heaters are required to provide the thermal differentials needed. Due to the large difference in heat capacity of water compared to common structural materials, one of the main uses of thermography in the composites field is the detection of trapped water in sandwich structures [128]. Also a disbond between the core and the skins can be detected, as for a heat sensitive IR camera it appears as local discontinuity in the heat transfer path.

6.2.6. Vibrothermography

Vibrothermography is a real-time non-destructive technique applied to composites [129]. A mechanical vibration is applied onto a component that causes local thermal excitation at points with defects, due to the local friction. A heat sensitive camera is then used to locate the damage and with proper calibration the size and the relevant position of the defect, through the thickness is possible to be found. The application of this method is relatively limited by the fact that mechanical vibrations can be effective over limited range, so it is mostly a technique used at the laboratory.

6.2.7. X- Rays

X-Ray radiography is another NDT method that has been used to visualize the damage extent after impact on composites [130,131]. Impact and in some cases combined hygrothermal cycling damage are measured as projected damage, as X-Ray is a 2-D only intensity measuring method. X-Ray radiography is a well used NDT technique in metallic structures, such as aircraft frames. The use of radiography is somehow restricted due to the potential health and safety issues.

Capitalizing on the X-Ray radiography technique, Microtomography was developed. A sample is virtually sliced or sectioned by radiographs. Images are captured by X-Ray photo-sensitive sensors. The images produced by the consecutive scans are reconstructed in order to construct a 3-D image of the internals of the specimen. The method has been used with success for CFRP laminates [132]. The size of the sample normally accepted for analysis by such equipment is small, generally in the range of ~40 mm by ~40 mm. The method is currently an experimental technique that can be used for tasks such as analysis of the spatial distribution and alignment of fibres in the matrix. This can assist in modeling representative volume elements but not for large scale damage detection and quantification.

7. Prediction of Residual Compressive Strength After Impact

The prediction of strength for multidirectional composites under in-plane compressive loads is not an easy task by any means. The existence of some damage complicates this prediction even more. The amount of available publications dealing with the subject shows not only the huge interest on the subject but also the fact that a universally acceptable method does not exist yet. A detailed review on the subject of strength prediction in compression has been presented by Aramah [133], so here only prediction models that relate directly or indirectly to impact and environmental degradation effects will be considered.

7.1. Soutis-Curtis

The model proposed by Soutis and Curtis [131] for the prediction of residual compressive strength after impact is based on previous work by Soutis et al [134,135]. The initial work was based on the prediction of compressive strength of carbon fibre-epoxy laminates containing a hole. Based on extensive experimental study of the failure process near the edge of the hole, a cohesive damage zone model is proposed. In this model, the damage around the open hole is represented as a line crack loaded on its faces. The fracture model is based on two criteria. In the case where there is stable crack growth, microbuckling occurs over a distance l from the hole site when the average stress over this distance reaches the critical stress of the unnotched laminate:

$$\sigma_{un} = \frac{1}{l}\int_{R}^{R+l} \sigma_{xx}(0,y)dy \qquad (13)$$

where

R is the radius of the hole.

The second criterion is a fracture mechanics based expression for the case where unstable microbuckling occurs. Unstable microbuckling occurs when the stress intensity factor KI at the tip of the buckled region is equal to the material fracture toughness KC. For a microbuckle with length d, from the hole edge, the stress intensity factor at the tip is expressed as

$$K_I = \sigma^{\infty} \sqrt{\pi l} f\left(\frac{d}{R}\right) \tag{14}$$

where R is the radius of the hole.

For the case where impact damage is present on a plate rather than a hole, the hole radius is substituted with the equivalent impact damage radius. Since there is no clear boundary of the through thickness impact damage, care must be taken to include the most intense part of the damage and to exclude the back-face fibre pull-out area [136]. The use of X-Ray radiography for determining the correct impact damage size was proposed, rather than ultrasonic C-Scan.

7.2. Zhou

After extensive experimental work on the field of impact and CAI of composites, an interesting concept was proposed by Zhou for the estimation of residual compressive strength [137]. It was postulated that the impact force measured from an instrumented impact testing machine could be used to deduce the residual compressive strength. It was shown that the ratio of threshold force to maximum force measured was closely correlated with residual normalized strength, when these both were plotted against incident kinetic energy (IKE), Figure 13.

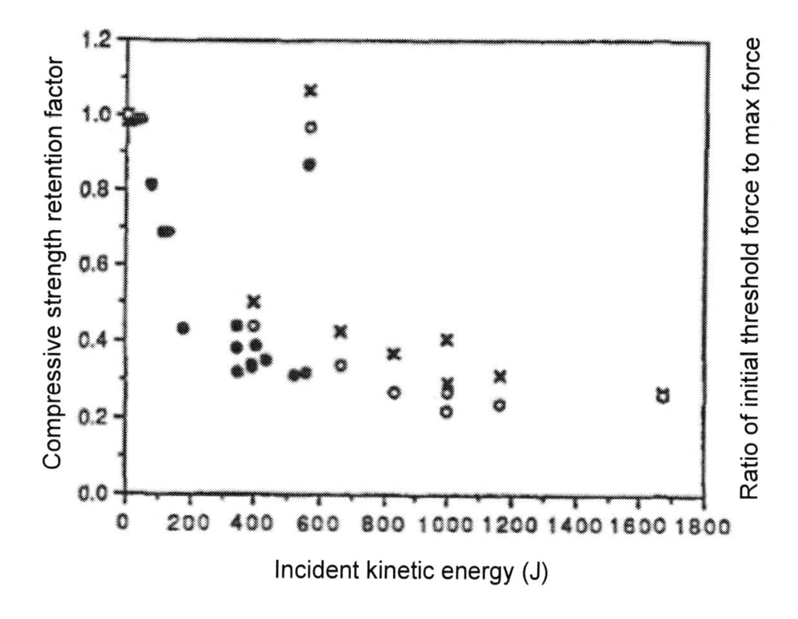

Figure 13. IKE and initial force to maximum force as a measure of residual compressive strength [137]. Markers x signify CAI data from large plates, empty circles mark the force ratio of large plates and filled circles signify the force ratio for small plates.

The idea behind this concept is that the CAI test and the difficulties associated with it can be bypassed. A generalization of this concept could be used for compressive strength prediction after environmental exposure and impact. The problem of course lies in the fact that extensive experimentation, for proofing purposes would be needed again to assess the baseline behavior after environmental exposure.

7.3. Whitney and Nuismer

Whitney and Nuismer proposed two stress fracture criteria, namely the Point Stress Criterion PSC and the Average Stress Criterion ASC [138]. The main idea behind these is that there is some characteristic distance d0 or α0 respectively ahead of the damaged area where the stress is equal to or higher than the strength of the undamaged material, Figure 14. It is further assumed that these characteristic distances d0 and α0 take the form of a material property which is independent of the laminate geometry and the stress distribution.

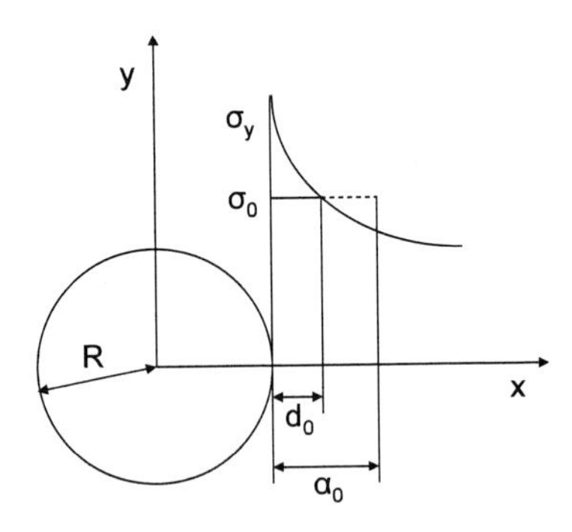

Figure 14. Schematic of the characteristic distances d_0 and α_0 for the PSC and ASC respectively.

7.3.1. The Point Stress Criterion

For the PSC, the characteristic dimension d_0 represents the distance over which the material must be critically stressed in order to find a sufficient flaw size to initiate failure.

$$\frac{\sigma_n}{\sigma_0} = \frac{2}{2 + \xi_1^{\,2} + \xi_1^{\,4}} \qquad (15)$$

where σ_n is the laminate damaged strength, σ_0 is the undamaged laminate strength and

$$\xi_1 = \frac{R}{R + d_0} \qquad (16)$$

7.3.2. The Average Stress Criterion

The ASC assumes that failure occurs when the average stress, over some distance α0, equals the undamaged strength. The ASC is based upon the assumption that the material is able to redistribute the local stress concentrations and is expressed as:

$$\frac{\sigma_n}{\sigma_0} = \frac{2(1-\xi_2)}{(2-\xi_2{}^2-\xi_2{}^4)} \quad (17)$$

where

$$\xi_2 = \frac{R}{R+\alpha_0} \quad (18)$$

The use of the Whitney and Nuismer stress fracture criteria has been demonstrated by Aramah [133-139] for NCF glass-epoxy specimens with holes, soft inserts and impacted specimens in compression. The specimens were immersed in distilled water at 93° C for 6 months. Half the specimens have been subject to impact or a central hole and then immersed in water. The other half were first immersed in water undamaged and then were been subjected to the same impact energy and hole size as the other set. The experimental results as well as the Whitney-Nuismer approximations to these results using the ASC and the PSC are shown in Figure 15.

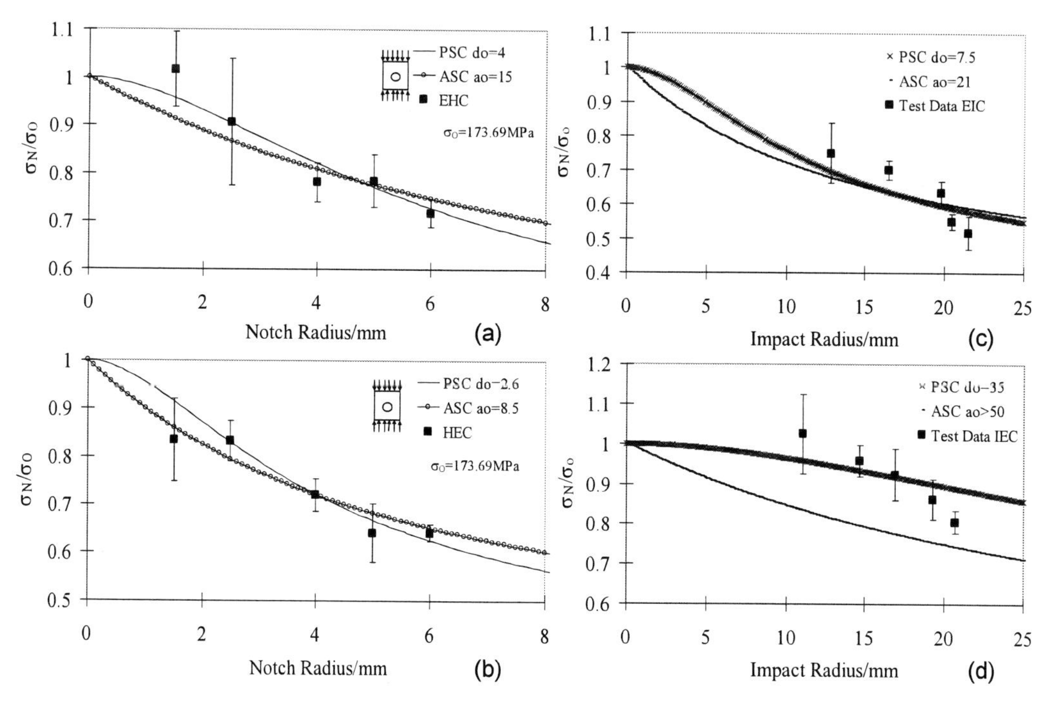

Figure 15. Normalized CAI strength for specimens with holes (a), (b) and impact (c), (d). Results in (a) and (c) refer to strength measured on specimens immersed in water and then damaged. Results in (b) and (d) refer to strength measured after water immersion, with existing damage [133,139].EHC is for water immersion then hole drilled and compressed, HEC is for hole drilled and then water immersion and finally compressed. EIC is for specimen immersed then impacted and finally compressed, IEC is for impacted immersed in water and finally compressed.

The ASC Whitney-Nuismer criterion is found to describe better the results in the case where specimens contain holes. However both criteria appear to provide insufficient description of the residual strength versus impact damage radius for both the cases of impact before or after water immersion. The interaction of the environment with impact damage, as was reported previously, is not easy to predict. Impact damage does not create specific boundaries as holes or inserts do. For impact damage a variation of intensity from the centre to the edges is found. This is something that the ASC and PSC can not sufficiently cope with.

7.4. Caprino

A model based on linear elastic fracture mechanics LEFM concepts is proposed for the prediction of residual compressive strength after impact [140]. The main idea of this model is that of equalizing the impact area with a hole or a reduced stiffness elastic inclusion. The model assumes that there is a characteristic defect dimension C_0, which needs to be found experimentally.

$$\frac{\sigma_r}{\sigma_0} = \left(\frac{R_0}{R}\right)^m \tag{19}$$

where:

σ_r represents the residual strength of the notched laminate,
σ_0 is the initial undamaged strength,
R is the dimension of the notch,
m is an experimentally obtained parameter.

When R≤R0 no strength reduction is associated with the composite. This can also be rewritten taking logarithms on each side so that a linear relationship can be plot. For the equality case the residual strength becomes equal to the initial strength. It is further assumed that if there is a relationship between R and E, with U being the impact energy, it can be expressed as a power law as

$$R = kU^n \tag{20}$$

where k and n are experimentally deduced constants with n>0. Substituting into Eq. 19 then:

$$\frac{\sigma_r}{\sigma_0} = \left(\frac{U_0}{U}\right)^a \tag{21}$$

with

$$a = mn \tag{22}$$

and $\mathbf{U}_0$ is given from

$$R_0 = kU_0^n \tag{23}$$

which represents the maximum energy level the composite can withstand without associated strength reduction.

This model although it is not environmental degradation specific, the parameter R_0 if corrected for the effects of environmental exposure could be used for correlation. The word prediction is not used intentionally as in reality extensive experimentation would be needed to obtain the exponential parameters, in which case the situation becomes a curve fitting exercise.

7.5. Environmental Degradation Specific Strength Prediction Models

As it has already been mentioned above, models that relate the action of the environment with the residual strength are rarities due to the fact that there is no universally accepted way to describe the behavior of composites under a specific environment. In essence most of the environmental degradation prediction models are efforts to give physical meaning, post-mortem, to the sequence of events that led to some final condition. A series of tests is performed and then based on the results of these tests some extrapolations in the near or far region, subject to some assumptions can be made.

A well used strength reduction model due to environmental exposure is that of Pritchard and Speake [5]. Having had tested glass reinforced polyester and plain resin immersed in water at temperatures of 30° C to 100°, the best fits for the data produced followed the form:

$$p = a\left(1 - e^{-b\exp(-cM_t)}\right) + d \tag{24}$$

where p is the residual property, while a,b,c and d are empirical constants. Mt the true water absorption, is defined as:

$$M_t = M_F + M_R \tag{25}$$

where MF and MR are the Fickian (MF) and a relaxation term (MR) respectively for water absorption, and are defined as:

$$M_F = M_{F\max}\left\{1 - \frac{8}{\pi^2}\sum_{j=0}^{\infty}\frac{1}{(2j+1)^2}\exp\left[-\left(\frac{Dt}{h^2}\right)\pi^2(2j+1)^2\right]\right\} \tag{26}$$

$$M_R = M_{R\max}\left(1 - e^{-k_r t^2}\right) \tag{27}$$

where h is the sample thickness, D is the diffusivity, kr is the relaxation rate constant, t is the time and M_{Fmax} and M_{Rmax} are the maximum for Fickian and relaxation term diffusion respectively. Tensile and ILSS strength have been predicted by extrapolation for up to 30 years using Eq. 23.

Phani and Bose performed a series of flexure tests on CSM glass reinforced polyester specimens immersed in distilled water [141]. They found that the following equation described well their results:

$$\sigma_{br} = \sigma_{br}(t) - \sigma_{br\infty} = (\sigma_{br0} - \sigma_{br\infty})\exp\left(-\frac{t}{t_{(T)}}\right) \tag{28}$$

where

σbr is the residual bending strength,
σbr(t) is the strength at time t,
σ br ∞ is the bending strength at infinite time,
t(T) is a characteristic time, dependent on temperature.

Combining Eq. 27 with the Arrhenius type equation the strength degradation with time was expressed as:

$$\sigma_{br}(t) = (\sigma_{br0} - \sigma_{br\infty})\exp\left[-\frac{t}{t_0}\exp\left(-\frac{E_a}{RT}\right)\right] + \sigma_{br\infty} \tag{29}$$

where

t_0 is a characteristic time constant,
R is the gas constant,
Ea is the activation energy,
T is the absolute temperature in K.

They assumed that temperature only changes the rate constant so that the time-temperature superposition principle is valid. Based on this assumption they were able to produce a master curve for estimating the residual strength for different test temperatures. Their results, based on an immersion period of ~20 days maximum fit well with the master curve for a range of temperatures between 25° C and 100° C. The only issue here is the assumption that the water exposure monotonically reduces strength. It has been shown that depending on the loading mode of course, this assumption is not to be always valid [5].

Upadhyay and Prucz presented a theoretical parametric model for the assessment of residual strength after moisture exposure [142]. They instigated that when laminates have been subjected to moisture absorption the mode of failure is such that debonding at the interface between the reinforcement and the matrix always precedes. They have proposed three models for longitudinal tension σ_L, transverse tension σ_T and pure shear τ_{LT}, being respectively:

$$\sigma_L = \frac{2\tau_{b0}(E_f V_f + E_r V_r)}{E_{b0}} \cdot \frac{(1 - s_b M)}{(1 - e_b M)} \tag{30}$$

$$\sigma_T = 2\tau_{b0}(1 - s_b M) \tag{31}$$

$$\tau_{LT} = \tau_{b0}(1 - s_b M) \tag{32}$$

where

τ_{b0} is shear strength of the bonding layer,
E_f and E_r are the moduli of the unexposed fibres and matrix,
V_f and V_r are the volume fractions of the unexposed fibres and matrix,
E_{b0} is the modulus of the unexposed bonding material,
s_b and e_b are constants related to the bonding material condition,
M is the moisture absorption.

The model shows well the reduced effect of moisture on tensile loading. In shear and transverse loading the considerable effect of moisture uptake on the residual strength is also shown. The main assumption of this model is that the properties of the vulnerable to environmental attack matrix and bonding material change linearly with moisture concentration.

Qi and Herszberg [130] proposed a method for the determination of the residual compressive strength after impact and hygrothermal cycling of CFRP laminate composites. The approach assumes the knowledge of the deteriorated stiffness and the strength due to the action of the environment as given, so that the stiffness at time t after exposure is

$$E_H = (1 - D_H)E_0 \tag{33}$$

and

$$D_H = 1 - \frac{\sigma_H}{\sigma_0} \tag{34}$$

where E_H and E_0 are the stiffnesses after and before exposure respectively and D_H is the ratio of decrease of the strength of the material, with σ_H and σ_0 being the strength after and before exposure. The strength reduction is obtained from experiment. The damage was found to be circular and therefore it was postulated that it could be modeled as a soft inclusion area whose modulus would be

$$M_r = \frac{E^*}{E_0} \tag{35}$$

and

$$M_r = \exp(-\lambda 2R) \tag{36}$$

where Mr is the modulus retention ratio, E^* and E_0 are the reduced modulus of the damaged area and the undamaged modulus respectively, R is the damage radius and λ represents a damage intensity factor mainly related to the reinforcement. The Whitney Nuismer Point Stress failure criterion [138] has been used for the prediction of the CAI strength, Figure 16.

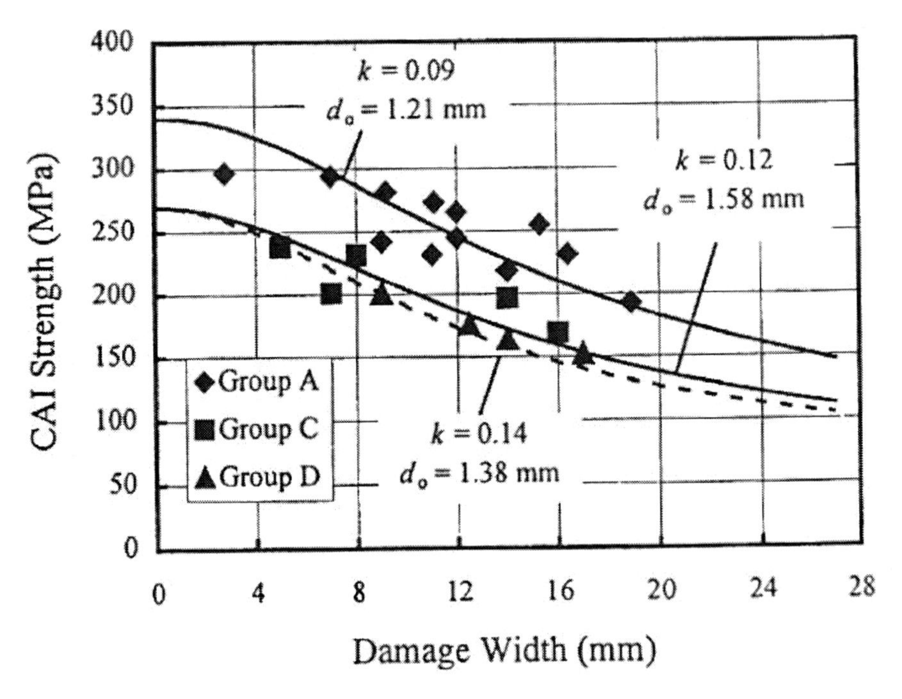

Figure 16. Comparison of experimental results and the semi-empirical prediction for cycled and non-cycled laminates. A, C, D refer to dry impacted specimen, wet impacted and hygrothermaly cycled and finally wet hygrothermaly cycled impacted specimens respectively [130].

A semi-empirical model approach was proposed by Papanicolaou for the prediction of compressive strength after impact of carbon fibre reinforced composites, that have been subject to hygrothermal aging [143,144]. The data for the model were obtained from Komai et al [94]. The model takes the final form as

$$\frac{\sigma_n}{\sigma_0} = 4m\frac{10\alpha+1}{U^{\alpha}} \tag{37}$$

where σ_n and σ_0 are the residual and initial strength respectively, U is the impact energy, α is a constant linked to energy absorption capacity and m is the ratio of summations, linked to the bending stiffness mismatch of a laminate [145] as

$$m = \frac{\sum_{\kappa=1}^{n} \left(\overline{M_{\kappa}}\right)_0 \left[Q_{xx,\kappa}\left(z_{\kappa}^{3} - z_{\kappa-1}^{3}\right)\right]}{\sum_{\kappa=1}^{n} \left[Q_{xx,\kappa}\left(z_{\kappa}^{3} - z_{\kappa-1}^{3}\right)\right]} \tag{38}$$

where $Q_{xx,\kappa}$ is the x-direction stiffness matrix term, z_κ is the distance of the κ-lamina from the middle plane of the lamina and $\left(\overline{M_{\kappa}}\right)_0$ is the mean value of the bending stiffness mismatching coefficient of the κ-lamina. A series of experiments must be performed in order to establish the material constants, both in dry condition and after environmental exposure.

Good agreement was found with the available data, Figure 17, although the model fails to describe the condition where after impact the residual strength is higher than the original undamaged plate strength.

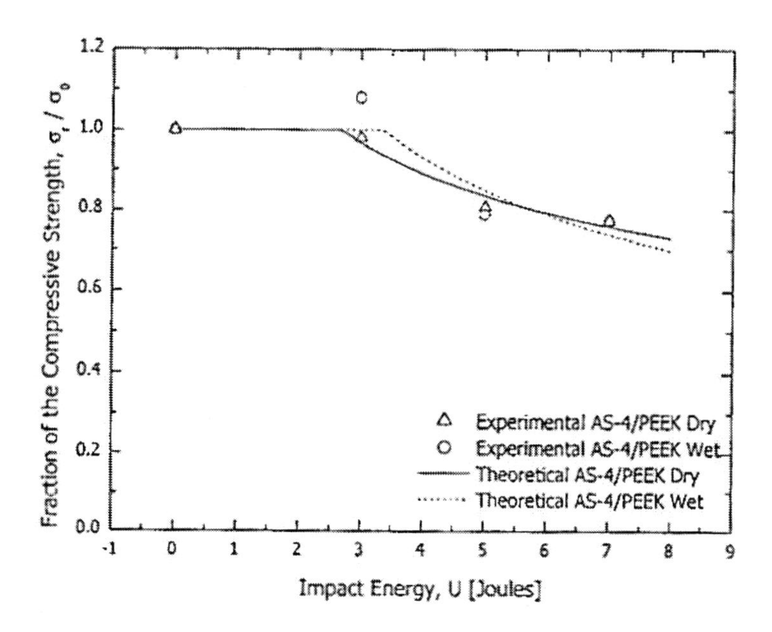

Figure 17. Normalized residual compressive strength, for dry and wet specimens, with respect to impact energy [143].

A generic concept for strength degradation over time of environmental exposure was proposed by Guedes et al [146]. The classical laminate plate theory was modified with terms to include time-related behavioral changes. Schapery's constitutive equations for viscoelastic materials were modified and then used to produce a program, LAMFLU, to solve problems related to long-term behavior of composites under in-plane and bending loads. One of the main findings of this work was that the combined time-temperature-moisture superposition principle (TTMSP) was not valid, at least for the material system and conditions under consideration. A comparison of other methodologies for the solution of long-term behavior models and prediction solutions were presented.

In summary, models that have been used for prediction of the residual compression after impact strength with or without the effect of environmental exposure have been presented in this last part. Models relating the effect of the environment on residual strength are scarce and are of semi-empirical nature. This is because there have not been relations with respect to time, of some measurable material parameters that could be used to predict the degradation status. Hence the normal practice is accelerated degradation testing for data collection and then curve fitting of the data to produce master curves.

The amount of experimental work available for the degradation of compressive strength with increasing impact damage size reveals the great interest and the complexity of the subject. The models that have been reviewed provide a broad basis for estimation of the residual compressive strength after impact. Zhou´s concept [137] is a very versatile tool for the quick estimation of residual compressive strength. The cohesive zone model proposed by Soutis [131] for the residual post impact compressive strength is based on a solid fracture mechanics approach which means that well defined material tests provide the input for the calculations but the extension from a hole to the impact area needs fine tuning to provide accurate results. The Whitney and Nuismer approach has proved a well used model whose single parameter fitting though proves to be a problem when the material becomes notch sensitive [133]. The model proposed by Caprino [140] has an inherent simplicity and for this has been extensively used. Its main limitation is due to the curve fitting nature which requires extensive experimentation prior to any actual prediction.

CONCLUSION

Based on the work reviewed here some conclusions can be drawn. Polyester composites have been shown to be susceptible to water diffusion on immersion or exposure to vapor. Both physical and chemical changes have been recorded. Many models have been proposed to describe the diffusion process but none can be used for prediction from first principles. Mechanical strength in general has been shown to be reduced after water immersion. Connection between the physical state of the material, after water absorption with some strength reduction model does not exist.

Out of plane impact has been shown to cause delaminations, matrix cracking and fibre breaking in varied proportions. Damaged plates subjected to in plane compression have considerably reduced strength. There are some models linking damage size to residual compressive strength, with varied levels of assumptions and existing data prerequisites.

There is very little work on combined environmental exposure, damage and their consequences on compressive strength. The interaction mechanism between impact damage and environmental exposure needs to be further studied. The multitude of material combinations available, the finished product quality and the possible different environmental conditions dictate that an extensive experimental database of material behaviours has to be kept and expanded.

REFERENCES

[1] Schutte, C.L., Environmental durability of glass-fiber composites. *Materials Science and Engineering*, 1994. R13: p. 265-324.

[2] Hancox, N.L., Overview of effects of temperature and environment on performance of polymer matrix composite properties. *Plastics, Rubber and Composites Processing and Applications*, 1998. 27(3): p. 97-106.

[3] Camino, G., et al., Kinetic aspects of water sorption in polyester-resin/glass-fibre composites. *Composites Science and Technology*, 1997. 57: p. 1469-1482.

[4] Bonniau, P. and A.R. Bunsell, A comparative study of water absorption theories applied to glass epoxy composites, in Environmental Effects on Composite Materials, G.S. Springer, Editor. 1984, Technomic: Lancaster,PA. p. 209-229.

[5] Pritchard, G. and S.D. Speake, The use of water absorption kinetic data to predict laminate property changes. *Composites*, 1987. 18(3): p. 227-232.

[6] Springer, G.S., B.A. Sanders, and R.W. Tung, Environmental effects on glass fiber reinforced polyester and vinylester composites. *J. of Composite Materials*, 1980. 14: p. 213-232.

[7] Crank, J., The mathematics of diffusion. 2nd ed. 1975: Oxford University Press.

[8] Springer, G.S., B.A. Sanders, and R.W. Tung, Environmental Effects on Glass Fiber Reinforced Polyester and Vinylester Composites, in Environmental Effects on Composite Materials, G.S. Springer, Editor. 1988, Technomic.

[9] Chin, J.W., T. Nguyen, and K. Aouadi, Sorption and diffusion of water, salt water and concrete pore solution in composite matrices. *J. of Applied Polymer Science*, 1999. 71: p. 483-492.

[10] Harper, J.F. and M. Naeem, The moisture absorption of glass fibre reinforced vinylester and polyester composites. *Materials & Design*, 1989. 10(6): p. 297-300.

[11] Tsotsis, T.K. and S.M. Lee, Long-Term durability of carbon and glass epoxy composite materials in wet environments. *J. of Reinforced Plastics and Composites*, 1997. 16(17): p. 1609-1621.

[12] Loos, A.C. and G.S. Springer, eds. Moisture Absorption of Graphite-Epoxy Composition Immersed in Liquids and Humid Air. Environmental Effects on Composite Materials, ed. G.S. Springer. Vol. 1. 1981.

[13] Morii, T., et al., Weight change mechanism of randomly oriented GFRP panel immersed in hot water. *Composites Structures*, 1993. 25: p. 95-100.

[14] Gurtin, M.E. and C. Yatomi, On a model for two phase diffusion in composite materials. *J. of Composite Materials*, 1979. 13: p. 126-130.

[15] Cai, L.W. and Y. Weitsman, Non-Fickian moisture diffusion in polymeric composites. *J. of Composite Materials*, 1994. 28(2): p. 130-154.

[16] Carter, H.G. and K.G. Kibler, Langmuir- type model for anomalous moisture diffusion in composite resins. *J. of Composite Materials*, 1978. 12: p. 118-131.

[17] Dwight, D.W., Glass Fiber Reinforcements, in Comprehensive Composite Materials. 2000, Elsevier Science Ltd. p. 231-261.

[18] Schutte, C.L., et al., The use of a single-fibre fragmentation test to study environmental durability of interfaces/interphases between DGEBA/mPDA epoxy and glass fibre: the effect of moisture. *Composites*, 1994. 25(7): p. 617-624.

[19] Metcalfe, A.G. and G.K. Schmitz, Mechanism of stress corrosion in E-glass fibres. *Glass Technology*, 1972. 13(1): p. 5-16.
[20] Belan, F., et al., Hydrolytic stability of unsaturated polyester prepolymers. *Composites Science and Technology*, 1996. 56: p. 733-737.
[21] Belan, F., et al., Relationship between the structure and hydrolysis rate of unsaturated polyester prepolymers. *Polymer Degradation and Stability*, 1997. 56: p. 301-309.
[22] Ashbee, K.H.G. and R.C. Wyatt, Water damage in glass fibre/resin composites. *Proceedings of the Royal Society*, 1969. 312(A): p. 553-564.
[23] Cairns, D.S. and D.F. Adams, Moisture and thermal expansion properties of unidirectional composite materials and the epoxy matrix, in Environmental Effects on Composite Materials, G.S. Springer, Editor. 1981, Technomic Publishing Co.: CT, USA. p. 300-316.
[24] Benameur, T., R. Granger, and J.M. Vergnaud, Correlations between the state of cure of unsaturated polyester coatings and their behaviour to water at 60 C. Polymer Testing, 1995. 14: p. 35-44.
[25] Weitsman, Y., Effects of Fluids on Polymeric Composites - *A Review, in Comprehensive Composite Materials*. 2000, Elsevier. p. 369 – 401.
[26] Boinard, E., et al., Influence of resin chemistry on water uptake and environmental ageing in glass fibre reinforced composites-polyester and vinylester laminates. *J. of Materials Science*, 2000. 35: p. 1931-1937.
[27] Kotsikos, G., et al. Modelling the properties of ageing marine laminates. in ACMC/SAMPE Conference. on marine composites. 2003. University of Plymouth, UK: ACMC Plymouth.
[28] Abeysinghe, H.P., et al., Degradation of crosslinked resins in water and electrolyte solutions. *Polymer*, 1982. 23: p. 1785-1790.
[29] Gautier, L., et al., Osmotic cracking nucleation in hydrothermal-aged polyester matrix. *Polymer*, 2000. 41: p. 2481-2490.
[30] Jayet, Y., et al., Monitoring the hydrolytic degradation of composites by a piezoelectric method. *Ultrasonics*, 1996. 34: p. 397-400.
[31] Hakkarainen, M., G. Gallet, and S. Karlsson, Prediction by multivariate data analysis of long-term properties of glassfiber reinforced polyester composites. Polymer Degradation and Stability, 1999. 64: p. 91-99.
[32] Tomokumi, H. and O. Shibata. Non-yellowing Resin Sytem during Hot Water Immersion. in COMPOSITES 2004, Convention and Trade Show Composites Fabricators Association. 2004. USA.
[33] Pavlidou, S. and C.D. Papaspyrides, The effect of hygrothermal history on water sorption and interlaminar shear strength of glass/plyester composites with different interfacial strength. *Composites Part A*, 2003. 34A: p. 1117-1124.
[34] Ciriscioli, P.R., et al., Accelerated Environmental Testing of Composites. *J. of Composite Materials*, 1987. 21: p. 225-242.
[35] Morii, T., et al., Weight-change analysis of the interphase in hygrothermally aged FRP: consideration of debonding. *Composites Science and Technology*, 1997. 57: p. 985-990
[36] Morii, T., et al., Weight changes of a randomly orietated GRP panel in hot water. *Composites Science and Technology*, 1993. 49: p. 209-216.

[37] Ikuta, N., et al., Evaluation of degradation behaviour of glass fibre-reinforced plastics in hot water by Fourier-transformed infrared spectroscopy. *Journal of Materials Science Letters*, 1993. 12: p. 1577-1578.

[38] Gaur, U., C.T. Chou, and B. Miller, Effect of hydrothermal ageing on bond strength. *Composites*, 1994. 25(7): p. 609-612.

[39] Wagner, H.D. and A. Lustiger, Effect of water on the mechanical adhesion of the glass/epoxy interphase. *Composites*, 1994. 25(7): p. 613-616.

[40] Gautier, L., B. Mortaigne, and V. Bellenger, Interface damage study of hydrothermally aged glass-fibre-reinforced polyester composites. *Composites Science and Technology*, 1999. 59: p. 2329-2337.

[41] Veazie, D.R., S.W. Park, and M. Zhou, Post-impact behaviour of polymeric composites and the effects of salt water aging on tensile strength. *Advanced Composite Letters*, 1999. 8(5): p. 257-264.

[42] Lundgren, J.-E. and P. Gudmundson, Moisture absorption in glass-fibre/epoxy laminates with transverse matrix cracks. *Composites Science and Technology*, 1999. 59: p. 1983-1191.

[43] Autran, M., et al., Influence of Mechanical Stresses on Hydrolytic Ageing of Standard and Low Styrene Unsaturated Polyester Composites. *J. of Applied Polymer Science*, 2002. 84: p. 2185-2195.

[44] Perreux, D. and C. Suri, A study of the coupling between the phenomena of water absorption and damage in glass/epoxy composite pipes. *Composites Science and Technology*, 1997. 57: p. 1403-1413.

[45] Roy, S., et al., Modeling of moisture diffusion in the presence of bi-axial damage in polymer matrix composite laminates. *International Journal of Solids and Structures*, 2001. 38: p. 7627-7641.

[46] Janas, V.F. and R.L. McCullogh, Moisture absorption in unfilled and glass-filled, cross-linked polyester. *Composites Science and Technology*, 1987. 29: p. 293-315.

[47] Wu, L., et al., Short-Term effects of sea water on E-glass/Vinylester composites. *J. of Applied Polymer Science*, 2002. 84: p. 2760-2767.

[48] Gutierrez, J., F. Le Lay, and P. Hoarau. A study of the aging of glass fibre-resin composites in a marine environment. in Nautical construction with composite materials, IFREMER. 1992. Paris, France.

[49] Siddaramaiah, et al., Effect of Aggressive Environments on Composite Properties. *J. of Applied Polymer Science*, 1999. 73: p. 795-799.

[50] Sonawala, S.P. and R.J. Spotnak, Degradation kinetics of glass-reinforced polyesters in chemical environments Part I Aqueous solutions. *Journal of Materials Science*, 1996. 31: p. 4745-4756.

[51] Sonawala, S.P. and R.J. Spotnak, Degradation kinetics of glass-reinforced polyesters in chemical environments Part II Organic solvents. *Journal of Materials Science*, 1996. 31: p. 4757-4765.

[52] Sala, G., Composite degradation due to fluid absorption. *Composites: Part B*, 2000. 31: p. 357-373.

[53] Collings, T.A., The use of resin hybrids to control moisture absorption in fibre-reinforced plastics. *Composites*, 1991. 22(5): p. 369-372.

[54] Belan, F., V. Bellenger, and B. Mortaigne, Hydrolytic stability of unsaturated polyester networks with controlled chain ends. *Polymer Degradation and Stability*, 1997. 56: p. 93-102.

[55] Marais, S., M. Metayer, and F. Poncin-Epaillard, Effect of a plasma treatment on water diffusivity and permeability of an unsaturated polyester resin. *J. of Fluorine Chemistry*, 2001. 107: p. 199-203.

[56] Bank, L.C., T.R. Gentry, and A. Barkatt, Accelerated Test Methods to Determine the Long-Term Behavior of FRP Composite Structures: Environmental Effects. *J. of Reinforced Plastics and Composites*, 1995. 14: p. 559-587.

[57] Collings, T.A., R.J. Harvey, and A.W. Dalziel, The use if elevated temperature in the structural testing of FRP components for simulating the effects of hot and wet environmental exposure. *Composites*, 1993. 24(8).

[58] McMillan, A.R., et al., Statistical study of environmental degradation in resin transfer moulded structural composites. Composites Part A, 1998. 29A: p. 855-865.

[59] Davies, P., et al. Accelerated marine ageing of composites and composite/metal joints. in 4th International Conference on Durability Analysis of Composite Sytems - DURACOSYS 99. 1999. Brussels Belgium: A. A. Balkema.

[60] Gellert, E.P. and D.M. Turley, Seawater immersion ageing of glass-fibre reinforced polymer laminates for marine applications. Composites Part A, 1999. 30: p. 1259-1265.

[61] Feichtinger, K., R. Alvarez, and A. Lakis. 14-year sanwich panel moisute permeation study. in Composites Fabricators Association Composites 2001 Convention and trade Show. 2001. Tampa, FL USA.

[62] Chekalkin, A.A., et al., Long-term durability of glass-fiber-reinforced composites under operation in pulp and reactant pipelines. *Mechanics of Composites Materials*, 2003. 39(3): p. 273-282.

[63] Perreux, D., D. Choqueuse, and P. Davies, Anomalies in moisture absorption of glass fibre reinforced epoxy tubes. *Composites Part A*, 2002. 33: p. 147-154.

[64] Kootsookos, A. and A.P. Mouritz, Seawater durability of glass- and carbon-polymer composites. *Composites Science and Technology*, 2004. 64.

[65] Cantwell, W.J. and J. Morton, The impact resistance of composite materials - a review. *Composites*, 1991. 22(5): p. 347-362.

[66] Richardson, M.O. and M.J. Wisheart, Review of low-velocity impact properties of composite materials. *Composites Part A*, 1996. 27(A): p. 1123-1131.

[67] Abrate, S., Impact on laminated composite materials. *Applied Mechanics Reviews*, 1991. 44(4): p. 155-190

[68] Abrate, S., Impact on laminated composites: Recent advances. *Applied Mechanics Reviews*, 1994. 47(11): p. 517-544

[69] Rydin, R.W., M.B. Bushman, and V.M. Karbhari, The influence of velocity in low-velocity impact testing of composites using the drop weight impact tower. *J. of Reinforced Plastics and Composites*, 1995. 14

[70] Guillaumat, L., Influence of the shape of a striker on the mechanical responses of composite beams, in Recent developments in Durability Analysis of Composite Systems, A.H. Cardon, et al., Editors. 2000, Balkema, Rotterdam. p. 337-342

[71] Wu, E. and K. Shyu, Responce of composite laminates to contact loads and relationship to low-velocity impact. J. of Composite Materials, 1993. 27(15): p. 1443-1464

[72] Wen, H.M., Penetration and perforation of thick FRP laminates. *Composites Science and Technology*, 2001. 61: p. 1163-1172
[73] Guillaumat, L., Reliability of composite structures - impact loading. *Computers & Structures*, 2000. 76: p. 163-172
[74] Hirai, Y., H. Hamada, and J.-K. Kim, Impact responce of woven glass-fabric composites-I. Effect of fibre surface treatment. Composites Science and Technology, 1998. 58: p. 91-104
[75] Varelidis, P.C., N.P. Kominos, and C.D. Papaspyrides, Polyamide coated glass fabric in polyester resin: interlaminar shear strength versus moisture absorption studies. *Composites Part A*, 1998. 29A: p. 1489-1499
[76] Raju, B.B., D. Liu, and X. Dang. Thickness effects on impact response of composite laminates. in 13th Annual Technical Conference on Composite Materials. 1998. Baltimore, Maryland: American Society for Composites
[77] Cantwell, W.J. and J. Morton, Geometrical effects in the low velocity impact responce of CFRP. *Composite Structures*, 1989. 12: p. 39-59
[78] Sutherland, L.S. and C. Guedes Soares, Impact tests on woven-roving E-glass/polyester laminates. *Composites Science and Technology*, 1999. 59: p1553-1567
[79] Belingardi, G. and R. Vadori, Influence of the laminate thickness in low velocity impact behavior of composite material plate. *Composite Structures*, 2003. 61: p. 27-38
[80] Belingardi, G. and R. Vadori, Low velocity impact tests of laminate glass-fiber-epoxy matrix composite material plates. *International Journal of Impact Engineering*, 2002. 27: p. 213-229
[81] Zhou, G., Effect of impact damage on residual compressive strength of glass-fibre reinforced polyester (GFRP) laminates. *Composite Structures*, 1996. 35: p. 171-181.
[82] Zhou, G. and G. Davies, Impact response of thick glass fibre reinforced polyester laminates. *International Journal of Impact Engineering*, 1995. 16(3): p. 357-74.
[83] Khan, B., R.M.V.G.K. Rao, and N. Venkataraman, Low velocity impact fatigue studies on glass epoxy composite laminates with varied material and test parameters-effect of incident energy and fibre volume fraction. *J. of Reinforced Plastics and Composites*, 1995. 14.
[84] Bibo, G.A. and P.J. Hogg, Review The role of reinforcement architecture on impact damage mechanisms and post-impact compression behaviour. *J. of Materials Science*, 1996. 31: p. 1115-1137.
[85] Shyr, T.-W. and Y.-H. Pan, Impact resistance and damage characteristics of composite laminates. *Composite Structures*, 2003. 62: p. 193-203.
[86] Rydin, R.W., A. Locurcio, and V.M. Karbhari, Influence of reinforcing layer orientation on impact response of plain weave RTM composites. *J. of Reinforced Plastics and Composites*, 1995. 14.
[87] Fuoss, E., P. Straznicky, and C. Poon, Effects of stacking sequence on the impact resistance in composite laminates - Part 2: prediction. *Composite Structures*, 1998. 41: p. 177-186.
[88] Fuoss, E., P. Straznicky, and C. Poon, Effects of stacking sequence on the impact resistance in composite laminates - Part 1: parametric study. *Composite Structures*, 1998. 41: p. 67-77.

[89] Davallo, M., et al., Low energy impact behaviour of polyester-glass composites formed by resin transfer moulding. *Plastics, Rubber and Composites Processing and Applications*, 1998. 27(8): p. 384-391.

[90] Karbhari, V.M., Impact characterization of RTM composites Part I Metrics. *Journal of Materials Science*, 1997. 32: p. 4159-4166.

[91] Yang, B., et al., Bending, compression, and shear behavior of woven glass fiber-epoxy composites. *Composites Part B: engineering*, 2000. 31: p. 715-721.

[92] Hosur, M.V., M.R. Karim, and S. Jeelani, Experimental investigations on the response of stitched/unstitched woven S2-glass/SC15 epoxy composites under single and repeated low velocity loading. *Composite Structures*, 2003. 61: p. 89-102.

[93] Strait, L.H., M.L. Karasek, and M.F. Amateau, Effects of seawater immersion on the impact resistance of glass fiber reinforced epoxy composites. *J. of Composite Materials*, 1992. 26(14): p. 2118-2133.

[94] Komai, K., K. Minoshima, and H. Yamasaki. Evaluation of low-velocity impact induced delamination by scanning acoustic microscope and influence of water absorption on delamination and compression after impact of CFRP. in 1st International Conference on Mechanics of Time Dependent Materials. 1995. Ljubljana.

[95] Ishai, O., C. Hiel, and M. Luft, Long-term hygrothermal effects on damage tolerance of hybrid composite sandwich panels. *Composites*, 1995. 26(1): p. 47-55.

[96] Imielinska, K. and L. Guillaumat, The effect of water immersion ageing on low-velocity impact behaviour of woven aramid-glass fibre/epoxy composites. Composites Science and Technology, 2004. 64: p. 2271-2278.

[97] Sierakowski, R.L. and G.M. Newaz, Damage Tolerance in Advanced Composites. 1995: Technomic Publishsing Company Inc.

[98] NASA, NASA Reference publication 1092 - Standard tests for toughened resin composites. 1983, NASA.

[99] BOEING, BSS 7260 Boeing Specification Support Standard "Advanced Composite Compression Tests". 1988, The Boeing Company: Seattle, Washington.

[100] SACMA, SACMA recomended Method "Compression After Impact Properties of Oriented Fiber-Resin Composites". 1994, Suppliers of Advanced Composites Materials Association.

[101] JIS, K 7089 Testing method for compression after impact properties of carbon fibre reinforced plastics. 1996, JIS.

[102] AITM, AITM 1.0010 Airbus Industrie Test Method, "Fiber Reinforced Plastics Determination of compression strength after impact". 1994, Airbus Industrie. p. 11.

[103] EFA, EFA-CFC-TP-014C EF2000 CFC Material Test Procedure. 1998.

[104] Habib, F.A., A new method for evaluating the residual compression strength of composites after impact. *Composite Structures*, 2001. 53: p. 309-316.

[105] Hogg, P.J., J.C. Pritchard, and D.L. Stone. A miniaturized post impact compression test. in ECCM-CTS. 1992.

[106] Preuss, T.E. and G. Clark, Use of time-of-flight C-scanning for assessment of impact damage in composites. *Composites*, 1988. 19(2): p. 145-148.

[107] Hosur, M.V., et al., Estimation of impact-induced damage in CFRP laminates through ultrasonic imaging. *NDT&E International*, 1998. 31(5): p. 359-374.

[108] Scarponi, C. and G. Briotti, Ultrasonic technique for the evaluation of delaminations on CFRP, GFRP, KFRP composite materials. Composites: Part B, 2000. B31: p. 237-243.

[109] Strycek, J.O. and H. Loertscher, Ultrasonic Air-Coupled Inspection of Advanced Material. 1999.

[110] Audoin, B., Non-destructive evaluation of composite materials with ultrasonic waves generated and detected by lasers. *Ultrasonics*, 2002. 40: p. 735-740

[111] Chen, A.S., D.P. Almond, and B. Harris, Impact damage growth in composites under fatigue conditions monitored by acoustography. *International Journal of Fatigue*, 2002. 24: p. 257-261

[112] Chen, A.S., D.P. Almond, and B. Harris, In situ monitoring in real time of fatigue-induced damage growth in composite materials by acoustography. *Composites Science and Technology*, 2001. 61: p. 2437-2443

[113] Kotsikos, G., et al., Environmetally enhanced fatigue damage in glass fibre reinforced composites characterised by acoustic emission. *Composites Part A*, 2000. 31: p. 969-977

[114] Hong, G., A. Yalizis, and G.N. Frantziskonis, Hygrothermal degradation in glass/epoxy - evaluation via stress wave factors. *Composite Structures*, 1995. 30: p. 407-417

[115] Tauchert, T.R. and A.N. Guzelsu, Measurements of the elastic moduli of laminated composites using an ultrasonic technique. *J. of Composite Materials*, 1971. 5: p. 549-552

[116] Choi, N.S., et al., Influence of weathering on unreinforced and short glass fibre reinforced thermoplastic polyester. *Journal of Materials Science*, 1998. 33: p. 2529-2535

[117] Kazys, R. and L. Svilainis, Ultrasonic detection and characterization of delaminations in thin composite plates using signal processing techniques. *Ultrasonics*, 1997. 35: p. 367-383

[118] Wooh, S.-C. and C. Wei, A homomorphic deconvolution technique for improved ultrasonic imaging of thin composite laminates. *Review of Progress in Quantitative Nondestructive Evaluation*, 1998. 17: p. 807-814

[119] Nesvijski, E.G., Some aspects of ultrasonic testing of composites. *Composite Structures*, 2000. 48: p. 151-155

[120] Banks, W.M., et al., Dielectric and mechanical assessment of water ingress into carbon fibre materials. *Computers & Structures*, 2000. 76: p. 43-55

[121] Apicella, A., et al., The role of dielectric and dynamic-mechanical techniques in the process monitoring and environmental resistance control of polymeric matrices for composites. *Plastics, Rubber and Composites Processing and Applications*, 1992. 18: p. 127-137

[122] Gryzagoridis, J., D. Findeis, and D.R. Schneider, The impact of optical NDE methods in vessel fracture protection. *Int. J. Pres. Ves. & Piping*, 1995. 61: p. 457-469

[123] Hung, Y.Y., Applications of digital shearography for testing of composite structures. *Composites Part B: engineering*, 1999. 30: p. 765-733

[124] Nyongesa, H.O., A.W. Otieno, and P.L. Rosin, Neural fuzzy analysis of delaminated composites from shearography imaging. *Composite Structures*, 2001. 54: p. 313-318

[125] Hung, Y.Y., Shearography for Non-destructive Evaluation of Composite Structures. *Optics and Lasers in Engineering*, 1996. 24: p. 161-182

[126] Summerscales, J., ed. Non-destructive testing of fibre-reiforced plastics composites. Vol. 2. 1990, Elsevier Science Publishers LTD

[127] Khandetskii, V.S. and L.Y. Martynovich, Eddy-current nondestructive detection of delaminations near edges of composite materials. *Russian Journal of Nondestructive Testing*, 2001. 37(3): p. 207-213
[128] Vavilov, V.P., A.G. Klimov, and V.V. Shiryaev, Active thermal detection of water in cellular aircraft structures. *Russian Journal of Nondestructive Testing*, 2002. 38(12): p. 927-936
[129] Lin, S.S. and E.G. Henneke, Analytical and experimental Investigations of Composites using vibrothermography. ASTM STP 1059, ed. S.P. Garbo. 1990, Philadelphia
[130] Qi, B. and I. Herszberg, An engineering approach for predicting residual strength of carbon/epoxy laminates after impact and hygrothermal cycling. *Composite Structures*, 1999. 47: p. 483-490
[131] Soutis, C. and P.T. Curtis, Prediction of the post-impact compressive strength of CFRP laminated composites. *Composites Science and Technology*, 1996. 56: p. 677-684
[132] Symons, D.D. and G. Davis, Fatigue testing of impact-damaged T300/914 carbon-fibre-reinforced plastic. *Composites Science and Technology*, 2000. 60: p. 379-389
[133] Aramah, S.E., Aspect of the damage tolerance of quasi-isotropic noncrimp fabrics, in Materials Department. 2001, Queen Mary and Westfield: London, UK. p. 350
[134] Soutis, C. and N.A. Fleck, Static Compression Failure of Carbon Fibre T800/924C Composite Plate with a Single Hole. J. *of Composite Materials*, 1990. 24: p. 536-558
[135] Soutis, C., N.A. Fleck, and P.A. Smith, Failure Prediction Technique for Compression Loaded Carbon Fibre-Epoxy Laminate with Open Holes. *J. of Composite Materials*, 1991. 25: p. 1476-1498
[136] Hawyes, V.J., P.T. Curtis, and C. Soutis, Effect of impact damage on the compressive responce of composite laminates. Composites Part A, 2000. 32: p. 1263-1270
[137] Zhou, G., The use of experimentally-determined impat force as a damage measure in impact damage resistance and tolerance of composite structures. *Composite Structures*, 1998. 42: p. 375-382
[138] Whitney, J.M. and R.J. Nuismer, Stress fracture criteria for laminated composites containing stress concentrations. *Composite Materials*, 1974. 8: p. 253-265
[139] Aramah, S. and P.J. Hogg. Compression testing of glass epoxy non-crimp fabrics containing elastic inserts. in ICCM13. 2001. Beijing, China
[140] Caprino, G., Residual Strength Prediciton of Impacted CFRP Laminates. *J. of Composite Materials*, 1983. 18: p. 508-518
[141] Phani, K.K. and N.R. Bose, Temperature dependence of hygrothermal ageing of CSM-laminate during water immersion. *Composites Science and Technology*, 1986. 29: p. 79-87
[142] Upadhyay, P.C. and J. Prucz, Parametric damage modelling of Composites due to moisture absorption. *J. of Reinforced Plastics and Composites*, 1992. 11: p. 198-210
[143] Papanicolaou, G.C. and S. Giannis. Effect of hygrothermal aging on the low energy impact behaviour of fibre-reinforced polymers. in 8th International Conference on Fibre Reinforced Composites FRC2000. 2000. University of Newcastle, UK: Woodhead Publishing Ltd, Cambridge, UK
[144] Papanicolaou, G.C. and C.D. Stavropoulos, New approach for residual compressive strength prediction of impacted CFRP laminates. Composites, 1995. 26: p. 517-523
[145] Liu, D., Impact-induced delamination - A view of bending stifness mismatching. *J. of Composite Materials*, 1988. 22: p. 674-692.

[146] Guedes, R.M., et al., Prediction of long-term behaviour of composite materials. *Computers & Structures*, 2000. 76: p. 183-194.

In: Resin Composites: Properties, Production and Applications ISBN: 978-1-61209-129-7
Editor: Deborah B. Song

Chapter 2

FABRICATION AND EVALUATION OF BIOACTIVE DENTAL COMPOSITES BASED ON AMORPHOUS CALCIUM PHOSPHATE

C.H. Davis, J.N.R. O'Donnell, J. Sun, and D. Skrtic
Paffenbarger Research Center, American Dental Association Foundation,
Gaithersburg, MD 20899, USA

1. INTRODUCTION

With the exception of a small portion of the inner ear, all hard tissues of the human body are formed of calcium phosphate(s), CaPs [1]. CaPs also occur in pathological calcifications (dental and urinary calculi, atherosclerotic lesions). The atomic arrangements of CaPs are built up around an orthophosphate (PO_4) network that gives stability to their structures. It is, therefore, of no surprise that structural and chemical properties of CaPs have been extensively studied (1-6). A total of eleven crystalline and one amorphous form of CaP with different Ca/P molar ratios, solubilities and crystallographic properties have been identified, seven of which are biologically-relevant (**Table 1**; [1-3]). The general rule in CaP family is: the lower the Ca/PO_4 molar ratio, the more acidic and water-soluble the CaP. The *in vivo* presence of small peptides, proteins and inorganic additives considerably influences CaP precipitation, making it difficult to predict the possible phases that may form. The least soluble, hydroxyapatite (HAP), preferentially formed under neutral and basic conditions, is usually non-stoichiometric. In normal *in vivo* calcifications, formation of apatite is reportedly preceded by an amorphous calcium phosphate (ACP) precursor that spontaneously converts into thermodynamically favored HAP [1-3, 6].

Table 1. Chemical and thermodynamic properties of biologically relevant CaPs [1-3]

Name (acronym)	**Formula**	**Ca/P**	K_{sp} [#]
Dicalcium phosphate dihydrate (DCPD)	$CaHPO_4 \cdot 2H_2O$	1.00	6.6
Octacalcium phosphate (OCP)	$Ca_8(HPO_4)_2(PO_4)_4 \cdot 5H_2O$	1.33	96.6
Amorphous calcium phosphate (ACP)	$Ca_3(PO_4)_2 \cdot nH_2O$*	1.50[@]	N_d
α-tricalcium phosphate (α-TCP)	$Ca_3(PO_4)_2$	1.50	25.5
β-tricalcium phosphate (β-TCP)	$Ca_3(PO_4)_2$	1.50	28.9
Hydroxyapatite (HAP)	$Ca_{10}(PO_4)_6(OH)_2$	1.67	116.8
Fluoroapatite (FAP)	$Ca_{10}(PO_4)_6F_2$	1.67	120.0

*Approximate formula; n =3.0-4.5. [#]K_{sp} is the negative logarithm of the ionic product for the given formula at 25 °C (hydrate water excluded). [@]Average value for ACP synthesized in our group; reported values vary from 1.2 to 2.2 (1, 3). N_d- could not be determined precisely; reported values range from 25.7 to 32.7. Dissolution in acidic buffer decreases as follows: ACP >> α-TCP >> β-TCP > non-stoichiometric HAP > HAP > FAP.

1.1. ACP Composites: Physicochemical Considerations

All biological hard tissues are complex composites of inorganic mineral phase (provides strength) and organic phase (contributes to ductility). There is an abundance of information on a large variety of bone substituting composites made of CaPs and organic polymers of either synthetic or biological origin [1]. Because of their inherent rigidity and brittleness CaP biomaterials are primarily designed for the filling of bone defects in oral and orthopedic surgery or coatings for dental implants and metallic prosthesis [1, 2]. It is well known that almost 50 % of all dental fillings require replacement because of recurrent mineral loss in tooth enamel, dentine and/or cementum. Although there is no known method for regenerating large amounts of tooth structure, a variety of treatments are used in dental clinics to restore teeth to proper form, function, and aesthetics. Among those treatments, the proposed use of bioactive, remineralizing ACP composites is of particular importance in preventive dentistry, orthodontics and endodontics [9].

ACP composites are made up of the organic resin phase (predominantly methacrylate based monomers) and the inorganic phase (ACP filler). The composite's physicochemical performance is determined by chemical interactions between the ACP filler and the polymer matrix at their interphase and the ability of the resin to penetrate into surface cavities formed during filler particle agglomeration. In the case of dental composites based on the siliceous filler, the bonding at the filler/polymer matrix interphase is commonly enhanced by introducing alkyloxy silanes as coupling agents. The silane coupling agents strengthen this critical interphase by concomitant covalent bonding of its organic functionalities with the polymer matrix and hydrolyzable silane groups with the inorganic filler. The use of silanes in ACP composites, however, failed to significantly improve ACP/matrix bonding [12]. The main drawback of ACP composites compared with glass-reinforced composites is their mechanical inferiority caused by the uncontrolled agglomeration of ACP particles [13]. As a result of the relatively low strength and toughness, ACP composites are not able to resist cracking under masticatory stresses. To overcome this shortcoming, we have focused on developing strategies for improving the interfacial properties of the ACP/resin matrix interphase by alternative approaches, i.e., by better controlling the surface properties and the particle size distribution (PSD) of ACP fillers and by fine-tuning the chemical structure and composition of the resins. Extensive physicochemical studies performed on various ACP polymeric composites described in this Chapter are the basis for broadening ACP's utility beyond already proposed topical gels, tooth pastes, mouthrinses, chewing gums and glass ionomers [14-16]. These studies illustrate the importance of reviewing the chemistry of composites for an appreciation of the potential for interaction between ACP filler, the monomer components of these materials, and the oral environment.

ACP obtained via spontaneous precipitation from supersaturated calcium and phosphate solutions is highly agglomerated. The extent of its agglomeration is a major factor affecting the physical and chemical properties of ACP-based composites. This hard-to-control factor determines the stability of filler/polymer interphase and the composite's performance in aqueous environments (mechanical strength, water sorption, release of remineralizing ions and the release of organic components). In order to attain the desired chemical and mechanical properties of ACP-filled composites, it is essential to achieve a fairly uniform distribution of ACP particulates in the resin, i.e., minimize the uneven formation of filler-rich and filler-poor areas within the composite. As an alternative to the unsuccessful silanization of ACP filler [12], we explored the effect(s) of various cations and polyelectrolytes (polymers) on the structure, composition, morphology and PSD of the precipitating ACP. The hypothesis was that these additives, when introduced *ab initio* during the synthesis, will affect the extent of ACP's agglomeration and its stability in aqueous milieu, thus controlling the mechanical performance and the efficacy of ACP as a mineral ion-releasing component of the composites. The mechanism and the extent of additive-ACP interaction would primarily depend on the chemical nature of additives. With cations, the interactions were expected to be controlled by their ionic potential (defined by the water coordination number, multiplicity of charges and ionic radius [17]). In the case of polyelectrolytes, the multiplicity of the ionizable groups was expected to be a controlling factor.

Methacrylate networks with good solvent resistance are widely used in dentistry in a variety of restorative materials [18]. Typical dental resins are composed of at least two dimethacrylate monomers: a relatively viscous base monomer (serves to enhance the modulus of cured polymer and to reduce the polymerization shrinkage (PS) due to its relatively large molecular volume) and a low viscosity diluent monomer (provides good handling properties and improves degree of vinyl conversion (DVC) due to its smaller molecular volume and increased diffusivity [19]. The most commonly utilized copolymers are based on the base monomer 2,2-bis[p-(2-hydroxy-3-methacryloxypropoxy)phenyl]propane (Bis-GMA) and the diluent monomer tri(ethyleneglycol) dimethacrylate (TEGDMA). The hydroxyl groups of Bis-GMA and the ethylene oxide segments of TEGDMA contribute to the relatively high water sorption (WS) of Bis-GMA/TEGDMA copolymers [20]. The relatively low cure efficiency at ambient temperatures and subsequent plasticization of Bis-GMA/TEGDMA copolymers by oral fluids affect the service life of these composites. Dental polymers based on ethoxylated bisphenol A dimethacrylate (EBPADMA), a relatively hydrophobic analog of Bis-GMA with a higher molecular mass and a more flexible structure and lower viscosity, reportedly show higher DVC and lower PS than Bis-GMA/TEGDMA resins [21]. Urethane dimethacrylate (UDMA) monomer has been shown to be more reactive than Bis-GMA or EBPADMA [19]. In order to overcome some of the known shortcomings of the Bis-

GMA/TEGDMA copolymers, we have formulated a series of experimental resins with distinctively different degrees of hydrophilicity. These experimental resins comprised the alternative base monomers, various diluent monomers and multifunctional co-monomers. It was hypothesized that fine-tuned resins will have improved physicochemical properties and that their ACP composites will have improved filler/matrix coherence, thus yielding composites with satisfactory remineralization potentials without diminishing their stability in aqueous environments. The physicochemical basis for the expected interactions is a Lewis acid/ base reaction in which multifunctional co-monomer is the electron donor and the ACP is the electron acceptor [22, 23].

1.2. ACP Composites: Cellular Aspects

Equally important to the physicochemical evaluation of composites is their biodegradation at the interface with tooth structures. The susceptibility to degradation is inherent to the choice of chemistries selected for the formulation of dental composites. It is the promotion by salivary enzymes and related cofactors that results in generating defined chemical products [24] which affect the biological activity of cells and oral bacteria interfacing with the composite material.

In vivo, the interactions between the biomaterial and its "bioenvironment" are multifaceted due to non-equilibrium conditions and the undefined amount of compounds participating in these interactions. It has recently been documented that controlled release of the ionic dissolution products of bioactive materials results in regeneration of tissues [25]. Such controlled release of soluble moieties from bioactive filler/resorbable polymer composites leads to gene activation and provides the conceptual basis for the molecular design of biomaterials optimized for *in situ* tissue regeneration. However, the exact mechanism of CaPs bioactivity is not yet well understood. The majority of bioresearchers have embraced a concept introduced by Prof. Hench [26] which entails eleven successive reactions steps. The initial five steps are of chemical nature (hydrolysis, formation of Si-OH bonds, poly-condensation reaction, formation of an amorphous CaP, crystallization of HAP) and they result in the formation of bone-like apatite. In the case of the experimental ACP-based material, formation of the bone-like apatite would be spurred by spontaneous, *in situ* ACP conversion rather than the formation of an amorphous precursor via negatively charged hydroxyl, carbonyl and/or phosphate functionalities as is the case with bio-glasses and other synthetic CaP biomaterials [1, 2]. The six biological steps (adsorption of biological moieties in HAP layer, action of macrophages, cell attachment and differentiation, generation and maturation of matrix) ultimately lead to the formation of new bone. CaP-related parameters that can affect cellular activity are: dissolution-precipitation behavior [27-39], chemical composition [29, 39-42], topography [43, 44] and surface energy [42]. In particular, a clear link has been demonstrated between the levels of free Ca and PO_4 ions in culture medium and osteoblastic activities [27-34] as well as early bone formation *in vitro* and *in vivo* [27, 28]. Any change in Ca/P ratio as a consequence of compositional changes in CaP phase directly affects ion exchange mechanisms. Incorporation of Zn and Si in CaP ceramics promoted osteoblast attachment and proliferation [40, 41]. On the other hand, presence of carbonate in the apatitic network of the biomaterial had deleterious effects on osteoblast proliferation and alkaline phosphatase (ALP) production [29, 39, 42]. When in contact with both micro- and macro-porous CaP ceramic, osteoblasts can bridge pores many times larger than their full length [43]. Osteoblasts are also sensitive to the shape and size of apatite crystals (45, 46). Their initial proliferation activity can be affected by surface energy, although this factor appears to be of lesser significance at the latter stages of osteoblastic activity (15, 39).

Polymerization of dental resin composites is never complete (DVC typically reaches (60 to 70) %) and almost every component can be detected in the extracts of polymerized materials [47, 48]. Some of the released, unpolymerized resin monomers may elicit various unwanted biological effects [49, 50]. We hypothesize that the type of ACP filler, chemical structure of the constituent monomers, composition of the resins and the type of polymerization initiator systems determine the attainable DVC and regulate the kinetics of ion release and the release of leachable organic moieties. In a simplified approach to leachability issues, we assume that copolymers derived from highly converted resins (high DVC values) would generally yield polymeric ACP composites with low leachability, therefore, high DVC is taken as indicator of high biocompatibility [13, 51]. However, to properly assess how the release of calcium and phosphate ions and leachability of unreacted monomers affect osteoblast-like cells, *in vitro* cellular tests need to be performed as an integral part of the composite evaluation.

Over the last decade our group has been designing ACP/methacrylate composites intended for dental applications as pit and fissure sealants, orthodontic adhesives and endodontic sealers. The main working hypotheses of our studies were: 1) that it is feasible to formulate ACP/polymethacrylate composite pastes with different viscosities required for various dental applications by surface-modification of the ACP filler phase and fine-tuning of the resin phase, 2) that high DVC attained in these experimental resins will be only moderately reduced by the incorporation of ACP filler into polymer phase of the composites 3) that due to the overall hydrophilicty of the resin phases and ACP's affinity for water, the resulting resin composites will exhibit sufficient water sorption (WS) to generate adequate release of mineral ions for tooth mineral regeneration while not seriously diminishing the mechanical stability of the composites, 4) that relatively high PS of the composites with high DVC could be offset by the significant hygroscopic expansion (HE) occurring when composites are exposed to aqueous milieu, and 5) due to the high DVC values attained in our experimental materials, that leaching out of un-reacted monomers and/or degradation products, and cellular responses to our experimental materials will be equal to or exceed the performance of the commercial controls.

In this Chapter, we describe the fabrication of ACP remineralizing composites formulated for various dental applications, explore the structure-composition-property relationships resulting from the comprehensive physicochemical evaluation of both unfilled resins (copolymers) and composites, and assess their *in vitro* cytotoxicity. Understanding the interplay between the surface properties of ACP, chemical structure and composition of the resin matrix and the critical physicochemical and biological properties of copolymers and their corresponding composites is especially valuable when designing remineralizing ACP composites for different dental utilities. The lessons learned from these studies may also provide new insight(s) into physicochemical, molecular and cellular interactions that may be essential for the future design of CaP-based materials intended for general bone regeneration.

2. EXPERIMENTAL METHODS

The analytical techniques utilized to evaluate ACP fillers, the unfilled resins (copolymers) and their ACP composites are summarized in Table 2. Details on the experimental protocols are provided in the appropriate subsections below.

2.1. Synthesis and Characterization of ACP Fillers

ACP fillers were synthesized as originally reported in refs. [8-10]. ACP precipitated instantaneously in a closed system at 23 °C upon rapidly mixing equal volumes of 80 mmol/L $Ca(NO_3)_2$ solution and 54 mmol/L Na_2HPO_4 solution containing a molar fraction of 2 % $Na_2P_2O_7$, a well-documented ACP stabilizer [52]. Additives were introduced during *ab initio* as follows. In cationic series, an appropriate volume of a 250 mmol/L of either $AgNO_3$, $FeCl_2{\cdot}4H_20$, $ZnCl_2$, $AlCl_3{\cdot}6H_2O$, $FeCl_3{\cdot}6H_2O$, $(C_2H_5O)4Si$, or $ZrOCl_2$ aqueous solution needed to achieve a mole fraction of 10 % cation based on Ca reactant was combined simultaneously with the Ca and PO_4 reactant solutions. The precipitating cation-ACPs were designated Ag-, Fe(II)-, Al-, Fe(III)-, Si-, Zn- and Zr-ACP, respectively. Polymeric additive, i.e., poly(ethylene oxide) $H(C_2H_4)_nH$ (PEO; n=180, 2300 or 22700 corresponding to the molecular masses of 8K, 100K and 1000K, respectively) was introduced at a mass fraction of 0.25 %. The reaction pH varied between 7.9 and 9.0. The suspension was filtered, the solid phase washed subsequently with ice-cold ammoniated water and acetone, freeze-dried and then lyophilized. Dry solids were used as-synthesized (as-made or am-ACP) or were subjected to grinding (g-ACP; [53]) or milling (milled or m-ACP [54, 55]). The g-ACP was obtained by simple dispersing of am-ACP in isopropanol (common non-aqueous dispersant that prevents ACP's premature conversion into apatite) and grinding it with a porcelain mortar and pestle in a wet state for 1 min. To make m-ACP, am-ACP and high-density zirconia oxide balls (3 mm in diameter; Glen Mills Inc., Clifton, NJ, USA) were mixed at an approx. mass ratio 1:25, combined with 150 mL analytical grade isopropanol and sealed in a grinding jar. The jar was clamped into the latching brackets, counterbalanced and milled with rotation reversed every 15 min for 2 h at 42 rad/s (planetary ball mill PM 100, Retch Inc., Newton, PA, USA). The m-ACP was separated from the zirconia oxide balls by sieving. Isopropanol was evaporated in a vacuum oven (Squaroid Labline, Melrose Park, Il, USA; 70 °C; 24 h). Approximately 80 % by mass of the initial ACP was retrieved on average after the ball milling and recovery. All ACP fillers were stored under vacuum to avoid exposure to humidity and possible conversion to HAP prior to its utilization in the composite preparation and the subsequent physicochemical and cytotoxicity evaluation of composites.

The particle size distribution (PSD) of the fillers in dry and/or wet state was measured using laser obscuration concurrently with a computerized inspection system (CIS-100 particle size analyzer; Ankersmid Ltd., Yokneam, Israel) [54-56]. Morphology was examined by scanning electron microscopy (SEM; JEOL 35C instrument, JEOL, Inc., Peabody, MA, USA) after specimens were sputter-coated with gold. The overall water content and the thermal decomposition profiles of ACP fillers were determined by thermogravimetric analysis (TGA; 7 Series Thermal Analysis System, Perkin Elmer, Norwalk, CT, USA). Powdered ACP samples (initial weight (5 to 10) mg) were heated at a rate of 20 °/min (temperature range: (30 to 600) °C) in air. The amorphous state of ACP fillers was verified by powder X-ray diffraction (XRD; DMAX 2000 diffractometer, Rigaku/USA Inc., Danvers, MA, USA) and Fourier-transform spectroscopy (FTIR: Nicolet Magna-IR FTIR 550 spectrophotometer, Nicolet Instrumentation Inc., Madison, WI, USA). Detailed information on ACP filler characterization methods can be found in refs. [7-10].

Table 2. Methods and techniques for physicochemical and cellular testing of ACP fillers, copolymers and composites

Method	Property/Application/Information
Atomic emission spectroscopy	Compositional analysis of ACP fillers; kinetics of Ca and PO_4 release from composites
Colorimetry	Cell viability of the extracts from copolymers or composite specimens (Wst-1 or MTT assay)
Dilatometry	Volumetric changes of specimens as a measure of the PS of composites
Fourier-transform infrared (FTIR) spectroscopy and microspectroscopy	Short-range structural arrangements and DVC of copolymers and composites; intra-composite ACP to HAP conversion; distribution of the resin and ACP on composites' surface or along the cross-section
Gravimetry	Mass changes following exposure to relative humidity or aqueous immersion as a measure of WS in copolymer and composite specimens
Mechanical milling	Reducing the particle size of ACP fillers by wet milling
Mechanical testing	Mechanical strength of dry and wet copolymers and composites
Nuclear magnetic resonance spectroscopy (NMR)	Identification and quantification of leachables
Particle size distribution (PSD) analysis	Histograms of the volume and number particle size distribution; size range and median diameter of ACP fillers
Phase contrast microscopy	Morphology of cells cultured in specimen extracts
Scanning electron microscopy (SEM)	Morphology/topology of ACP fillers and/or composites
Thermogravimetry (TGA)	Water content and thermal stability of ACP solids
Tensometry	Shrinkage stress (PSS) developed in composites
X-ray diffraction (XRD)	Long-range crystalline order of the fillers; intra-composite ACP to HAP conversion

2.2. Formulation and Characterization of the Resins

The experimental resins were formulated from the commercially available dental monomers. Base monomers, diluent monomers and adhesive monomers as well as the polymerization initiatior systems utilized in structure/composition/property studies are listed in Table 3 (indicated acronyms are used throughout this Chapter).

Table 3. Monomers and the components of the polymerization-initiating systems utilized in the studies

Component	Chemical name	Acronym
Base monomers	2,2-Bis(p-2'-hydroxy-3'-methcryloxypropoxy)phenyl-propane Ethoxylated bisphenol A dimethacrylate Urethane dimethacrylate	Bis-GMA EBPADMA UDMA
Diluent monomers	Di(ethyleneglycol)methyl ether methacrylate Glyceryl methacrylate Glyceryl dimethacryalte 2-hydroxyethyl methacrylate Hexamethylene dimethacrylate 2-methoxyethyl methacrylate Poly(ethylene glycol) extended urethane dimethacrylate Triethylene glycol dimethacrylate	DEGMEMA GMA GDMA HEMA HmDMA MEMA PEG-U TEGDMA
Adhesive monomers	Maleic acid Methacrylic acid Methacryloyloxyethyl phthalate Mono-4-(methacryloyloxy)ethyl trimellitate Vinyl phosphonic acid	MaA MA MEP 4MET VPA
Polymerization initiators	L(+) Ascorbic acid Benzoyl peroxide Camphorquinone Diphenyl(2,4,6-trimethylbenzoyl) phosphine oxide & 2-hydroxy-2-methyl-1-phenyl-1-propanone 2,2'-Dihydroxyethyl-p-toluidine Ethyl-4-N,N-dimethylamino benzoate Bis(2,6-dimethoxybenzoyl)-2,4,4-trimethylpentyl phosphine oxide & 1-hydroxycyclohexyl phenyl ketone 2-benzyl-2-(dimethylamino)-1-(4-(4-morphollinyl)phenyl)-1-butanone Phenyl bis(2,4,6-trimethylbenzoyl) phosphine oxide t-Butyl perbenzoate	AA BPO CQ 4265 Darocur DHEPT 4EDMAB 1850 Irgacure 369 Irgacure PbTMBPO TBPB

After combining monomers in mass ratios required for the specific formulation, the mixtures were magnetically stirred (38 rad/s; in the absence of blue light) at room temperature until achieving uniform consistency. The following formulations were evaluated: A) binary and ternary Bis-GMA-, EBPADMA- or UDMA-based resins with HEMA, HmDMA or TEGDMA as co-monomers, B) ternary Bis-GMA/X/TEGDMA resins with X = hydroxyl- (DEGMEMA, GMA, GDMA, HEMA and MEMA) or carboxyl-containing monomers (MaA, MA, 4MET and VPA), C) EBPADMA/HEMA/TEGDMA resins with surface active MEP co-monomer, and D) UDMA/HEMA/MEP/PEG-U resins. Compositions of the experimental resins are indicated in Table 4.

Resins were activated for light-cure (LC), chemical-cure (CC) or dual cure (DC, i.e., LC + CC) depending on the intended clinical application. Besides the conventional camphorquinone (CQ)/tertiary amine system, the commercial acyl phosphine oxide photoinitiators (Ciba Specialty Chemicals Corporation, Tarrytown, NY, USA) were evaluated as possible alternatives to eliminate or reduce the need for a tertiary amine in the resins. These widely used co-initiators are also known for their unwanted post-cure effects [57].

Table 4. Composition (mass fraction, %) of experimental resins evaluated in studies A-D. Polymerization intiation: CC – chemical cure, DC – dual cure (CC + LC), LC – light cure

A1. Bis-GMA based resins (LC)				
Resin/monomer	Bis-GMA	HEMA	HmDMA	TEGDMA
BHm	52.44		46.56	
BT	49.50			49.50
BHHm	39.67	29.20	32.83	
BHT	35.50	28.00		35.50
A2. EBPADMA based resins (LC)				
Resin/monomer	EBPADMA	HEMA	HmDMA	TEGDMA
EHm	49.57		49.43	
ET	46.70			52.30
EHHm	34.33	30.44	34.23	
EHT	32.90	29.17		36.93
A3. UDMA based resins (LC)				
Resin/monomer	UDMA	HEMA	HmDMA	TEGDMA
UHm	51.75		47.25	
UT	48.82			50.18
UHHm	36.24	29.46	33.30	
UHT	34.48	28.32		35.85

B1. Bis-GMA/X/TEGDMA resins; X = neutral co-monomer (LC)							
Resin/monomer	Bis-GMA	DEGMEMA	GMA	GDMA	HEMA	MEMA	TEGDMA
BT	49.50						49.50

Table 4. (Continued)

BDT	31.50	36.00					31.50
BGmT	34.00		31.00				34.00
BGdT	27.50			44.00			27.50
BHT	37.00				25.00		37.00
BMT	35.65					27.70	35.65

B2. Bis-GMA/X/TEGDMA resins; X = acidic co-monomer (LC)*						
Resin/monomer	Bis-GMA	MaA	MA	4MET	VPA	TEGDMA
BT	49.03					49.03
BMaT	48.37	1.30				48.37
BMT	48.37		1.30			48.37
B4MT	46.57			4.90		46.57
BVT	48.21				1.64	48.21

C. EBPADMA/HEMA/MEP/TEGDMA Resins (LC)**				
Resin/monomer	EBPADMA	HEMA	MEP	TEGDMA
EHMT	16.80 to 62.85	10.00 to 10.36	2.51 to 5.00	23.22 to 67.20

D. UDMA/HEMA/MEP/PEG-U Resin (CC, LC, DC)***				
Resin/monomer	UDMA	HEMA	MEP	PEG-U
UHMP	47.20 to 48 70	16.80 to 17.30	2.90 to 3.00	29.10 to 30.00

LC photo-initiator system (content expressed as mass fraction %) CQ (0.2 %) and 4EDMAB (0.8 %). *PbTMBPO ((1.94 to 1.96) %). **LC: CQ (0.2 %) and 4EDMAB (0.8 %) or 1850 Irgacure (1.0 %). *** LC: CQ (0.2 %) + 4EDMAB (0.8 %), 1850 Irgacure (1.0 %) or CQ (0.4 %) + 4EDMAB (0.4 %) + 4265 Darocur (0.8 %) + 369 Irgacure (1.5); CC: BPO (2.0 %) + DHEPT (1.0 %) or AA (0.5 %) + TBPB (1.0 %); DC: 1850 Irgacure (1.0 %) + BPO (2.0 %) & DHEPT (1.0 5) or 1850 Irgacure (1.0 %) + AA (0.5 %)& TBPB (1.0 %).

In LC formulations containing acidic monomers, the diacyl phosphine oxide, PbTMBPO, was utilized as a photoinitiator because of possible storage stability problems usually encountered with the use of CQ and 4EDMAB and resins containing acidic monomers.

In addition to BPO systems, which are also known for inadequate storage stability, the alternative initiator system that employed AA and hydroxyperoxides as oxidants was assessed. After introducing the appropriate amounts of the initiating systems to the monomer blends, the activated resin mixtures were again magnetically stirred until fully homogenized. To avoid the accidental exposure of the LC-activated resin formulations to visible light, the introduction of the initiators and the storage of resins were performed in the absence of blue light.

2.3. Fabrication and Physicochemical Evaluation of Copolymers and Composites

Composite pastes were made by mixing the resin (mass fraction 60 %) and am-, g- or m-ACP filler (mass fraction 40 %) by hand spatulation. Once homogenized, pastes were kept under a moderate vacuum (2.7 kPa) overnight to eliminate the air entrained during mixing. Light-curable composite pastes were then packed into Teflon molds, each opening was covered with thin Mylar film and a glass slide, and the assembly clamped in place by spring clips. The clamped specimens were photo-polymerized by irradiating sequentially each side of the mold assembly for 120 s with visible light (Triad 2000, Dentsply International, York, PA, USA). The specimens for BFS testing were (13.0 to 15.0) mm in diameter and (1.2 to 1.5) mm in thickness. The specimens for cytotoxicity tests were (5.3 ± 0.1) mm in diameter and (3.1 ± 0.1) mm in thickness). In CC systems, BPO- or AA-containing paste and the DHEPT- or TBPB-containing paste were combined in 1:1 mass ratio before packing the mixture into the molds in the same manner as for the LC specimens. In the cases where DC was used, CC specimens were additionally irradiated and then stored for 24 h in air at 23 °C before being randomly selected for dry BFS, wet BFS, ion release or cytotoxicity testing. WS tests were typically performed with the broken disk specimens collected after dry BFS determinations. LC, CC and DC copolymer (unfilled resin) specimens were prepared following the same procedures utilized for fabrication of the composite specimens. When the composite samples were made from commercial materials (commercial controls), manufacturer-recommended curing procedures were followed. Typical ACP powder, uncured composite paste, molded uncured composite disk specimen and the composite disk specimen after cure are shown in Figure 1.

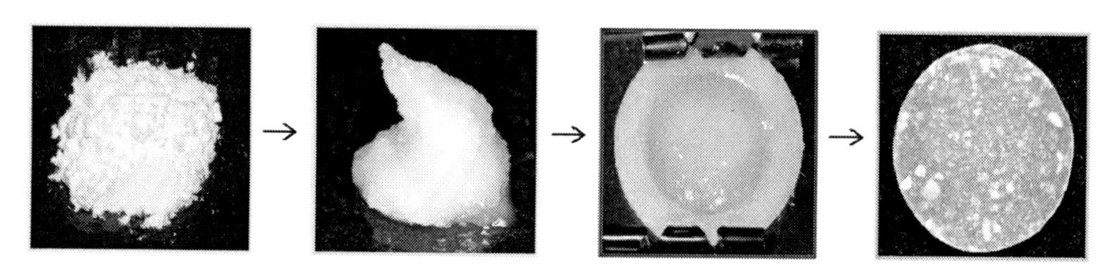

Figure 1. From left to right: ACP powder, composite paste after mixing ACP filler with the resin, Teflon mold with the uncured paste and cured composite disk specimen.

Mid-FTIR or near-IR (NIR) screening was utilized to determine DVC of the unfilled resins (copolymers) and their ACP-filled composites. Mid-FTIR measurements included monitoring the reduction in the 1637 cm-1 absorption band for the vinyl group against that of an unchanged aromatic peak (1538 cm-1; internal standard) [9, 10]. Spectra were acquired before cure and at predetermined time intervals after cure (usually immediately after cure and 24 h post-cure) by collecting 64 scans at 2 wave-number resolution. DC values determined by NIR method [19] were calculated from the % change in the integrated peak area of the 6165 cm-1 methacrylate=CH_2 absorption band between the cured specimen (polymer) and the uncured specimen (monomer). Use of an internal reference was not required for the NIR measurements, provided that the thickness of monomer and polymer sample was measured. The FTIR-m (a Nicolet Magna-IR 550 FTIR spectrophotometer equipped with a video camera, a liquid nitrogen cooled-mercury cadmium telluride detector, a computerized, motorized mapping stage and the Omnic Atlus software, Spectra-Tech Inc., Shelton, CT, USA) was utilized to analyze the intact copolymer and composite surfaces as well as cross-sections of copolymer and composite specimens in dry and wet states (after exposure to aqueous environment).

BFS values of dry (stored for 24 h in the air at 23 °C) and wet (after immersion in HEPES-buffered, pH=7.4, saline solutions at 23 °C for a minimum of one month) copolymer and composite disk specimens were determined using a piston-on-three-ball loading cell (Figure 2 a-c) and a computer-controlled Universal Testing Machine (Instron 5500R, Instron Corp., Canton, MA, USA) operated by Testworks 4 software. The BFS values were calculated according to the mathematical expressions defined in ASTM F394-78 specification [58]:

$$BFS = A \cdot L/t^2 \tag{1}$$

where $A = -[3/4\pi(X-Y)]$, $X = (1+\nu)\cdot\ln(r_1/r_s)^2 + [(1-\nu)/2]\cdot(r_1/r_s)^2$, $Y = (1 + \nu)\cdot[1 + \ln(r_{sc}/r_s)^2]$, ν is the Poisson's ratio (value of 0.24 was used in accordance with the published data on elastic properties of resin-based composites [59]), r_1 is the radius of the piston applying the load at the surface of contact, r_{sc} is the radius of the support circle, r_s is the radius of the disk specimen, L is the applied load at failure, and t is the thickness of the disk specimen.

Figure 2. Piston-on-three-ball loading cell utilized in BFS measurements (left). Close-ups of specimen positioning (right).

WS of copolymer and composite specimens was determined as follows. After initially drying the specimens over anhydrous $CaSO_4$ until a constant mass was achieved (± 0.1 mg), specimens were immersed in saline solutions as described in the BFS measurements. Gravimetric mass changes of dry-tissue padded specimens were recorded at predetermined time intervals. The degree of WS of any individual specimen at a given time interval (t), expressed as a % mass fraction, was calculated using the equation:

$$WS = [(W_t - W_o)/W_o] \times 100 \quad (2)$$

where W_t represents the sample mass at time t, and W_o is the initial mass of the dry sample.

In some systems WS was determined by exposing dry specimens to an air atmosphere of 75 % relative humidity (RH) at 23 °C (specimens were suspended over saturated aqueous NaCl slurry in closed systems). Gravimetric mass changes were recorded in the same manner as with the saline-immersed specimens and calculated according to Eq. (2). Data collected at 75 % RH provided information on the water uptake that was not affected by the ACP filler dissolution and leaching out of any water-soluble monomeric and/or polymer degradation species (processes that may take place parallel to water sorption of the specimens immersed in saline).

The PS of composite resin samples was measured by a computer-controlled mercury dilatometer (**Figure 3**; Paffenbarger Research Center (PRC); American Dental Association Foundation (ADAF), Gaithersburg, MD, USA). Composite pastes were cured using a standard 60 s plus 30 s exposure and data acquisition of 60 min + 30 min. PS of a specimen corrected for temperature fluctuations during the measurement was plotted as a function of time. The overall shrinkage (volume fraction, %) was calculated based on the known mass of the sample (50 to 100 mg) and its density. The latter was determined by means of the Archimedean displacement principle using an attachment to a microbalance (Sartorius YDK01 Density Determination Kit; Sartorius, Goettingen, Germany).

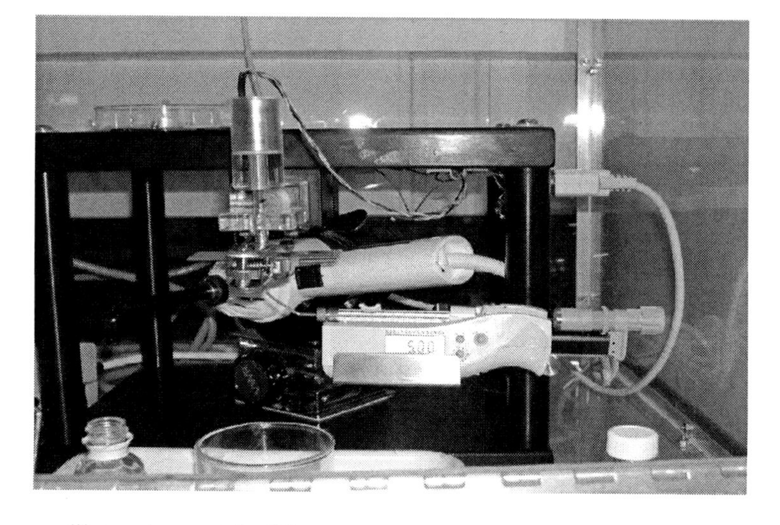

Figure 3. Mercury dilatometer used for PS measurements.

A tensometer (designed and fabricated at PRC-ADAF, Gaithersburg, MD, USA) was utilized to assess the polymerization stress (PSS) of the composites (Figure 4). The tensometer is based on the theory that a tensile force generated by the bonded polymerizing sample causes a cantilever beam to deflect. For a rectangular prismatic cantilever beam of a linearly elastic material with a small deflection which is under a concentrated normal load F, the displacement at the end of the cantilever beam is defined by the following expression [60, 61]:

$$\varepsilon/F = 2a^2(3L-a)/Ebd^3 \quad (3)$$

In Eq. (4), ε is the displacement at the beam end (μm); E is the Young's modulus of the cantilever beam (MPa); F is the load (N) needed to generate the displacement ε; L is the total beam length (cm); a is the distance from the position of the applied load to the end of the beam (cm); b is the width of the beam (cm) and d is the height of the beam (cm). The deflection of the cantilever beam was measured with a linear variable differential transformer. The force was calculated from a beam length (12.5 cm) and a calibration constant (3.9 N/μm). PSS was obtained by dividing the measured force by the cross sectional area of the sample.

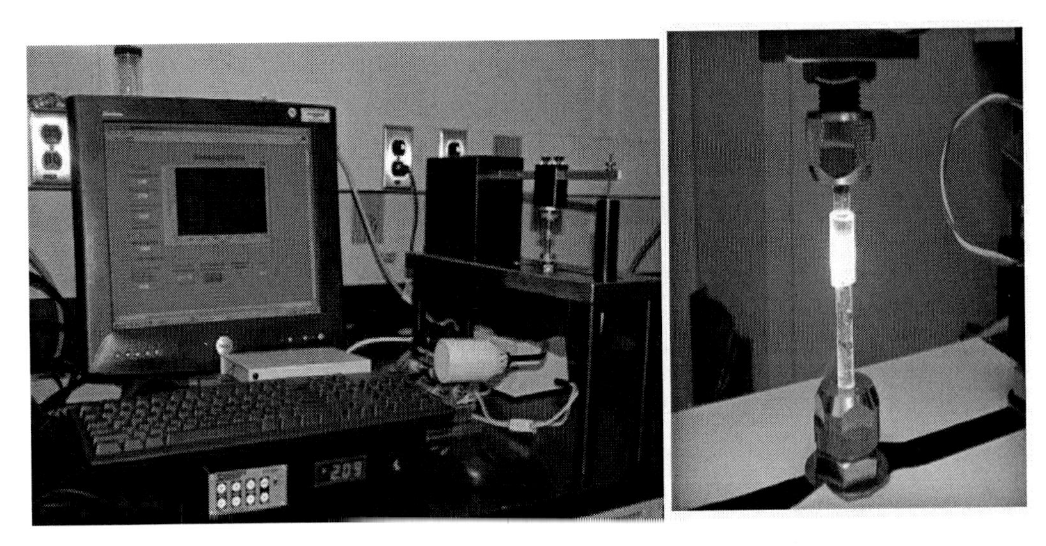

Figure 4. Tensometer used for PSS measurements (left). Close-up of the composite specimen during light-curing (right).

Mineral ion release from the individual composite disk specimens was examined at 23 °C in magnetically stirred, HEPES-buffered saline solutions (pH=7.4). Saline solution was replaced at predetermined time intervals (up to three months). Kinetic changes in the calcium and phosphate levels were determined by utilizing atomic emission spectroscopy (prodigy High Dispersion ICP-OES, Teledyne Leeman labs, Hudson, NH, USA). Ion release data were corrected for variations in the total area of the composite disk specimen exposed to the immersion solution using the simple relation for a given surface area, A: normalized value = measured value X (500/A).

2.4. Leachability Studies

The leachability of unreacted monomers and/or components of the initiator system from copolymer and composite specimens was assessed by nuclear magnetic resonance (NMR). Extraction studies that preceded the NMR experiments were performed on copolymer and composite samples with a variety of polar aprotic (acetone and dichloromethane), non-polar (cyclohexane) and polar protic solvents (ethanol and methanol). In all extraction experiments, polymerization inhibitor, butylated hydroxyl toluene (BHT) was added to organic solvents at 0.01 mass % to prevent secondary monomer polymerization upon extraction.

Typically, NMR is used to determine the chemical structure and conformation, but more recently has gained attention as a quantitative technique [62-64]. NMR has the advantage of easy sample preparation and quickly generated results that are easy to interpret, more accurate, and more consistent than chromatographic techniques [65]. NMR measurements rely on the principal that NMR signal peak intensities are directly proportional to the number of nuclei generating the resonance line. Integration of NMR signals determines the ratio of protons present in the sample, which are used to calculate the amount of sample present.

In this study, the concepts of NMR were applied to effectively identify and measure the leachable components of the UHMP copolymers and composite specimens. All ^{1}H NMR samples were run on a JEOL GSX 270 MHz Fourier transform nuclear magnetic resonance spectrometer in acetone-d_6 (99.9 atom % D, containing a volume fraction of 0.03 % tetramethyl silane; Sigma-Aldrich Co., St. Louis, MO, USA).

Initially, each UHMP comonomer (UDMA, HEMA, MEP, and PEG-U) was analyzed and integrated separately by ^{1}H NMR in acetone-d_6 at 20 mass-% to verify structure, ratio of protons, and determine differentiating peak values. The corresponding protons on the molecular structure represented were counted and recorded for each differentiating peak chosen. For example, the spectrum of HEMA displays a peak triplet at 4.22 ppm (Figure 5). These peaks represent the 2 CH_2 protons (d) and were therefore assigned an integration value of 2. All other values in the spectrum are based off this peak and correlate to representative protons in the molecular structure. In the initial integration of each comonomer these proton counts were confirmed based on molecular structure. These values will be used in the equations below to determine leachable content. Initiators (CQ and 4EDMAB) and the inhibitor BHT were also analyzed to confirm that there was no interference with comonomer peaks of interest. The UHMP comonomers were then combined in equal 5 mass-% parts (total 20 mass-%) in acetone-d_6. The sample was analyzed by ^{1}H NMR and the differentiating peaks representing each monomer were integrated at the same values previously determined. Integration values were based around the CH_2 peak value of 2 for HEMA at 4.22 ppm. The integration values and number of protons were used to determine mol-% in two steps:

$$M_x = I/P \quad (4)$$

where M_x is the mole fraction of each comonomer, I is the peak integration value, and P is the number of protons associated with that peak in the pure sample. These values were plugged into the following equation to give mol-% for each comonomer.

$$\text{Mol-\%}_x = M_x/[M_1+M_2+M_3+M_4] \times 100 \quad (5)$$

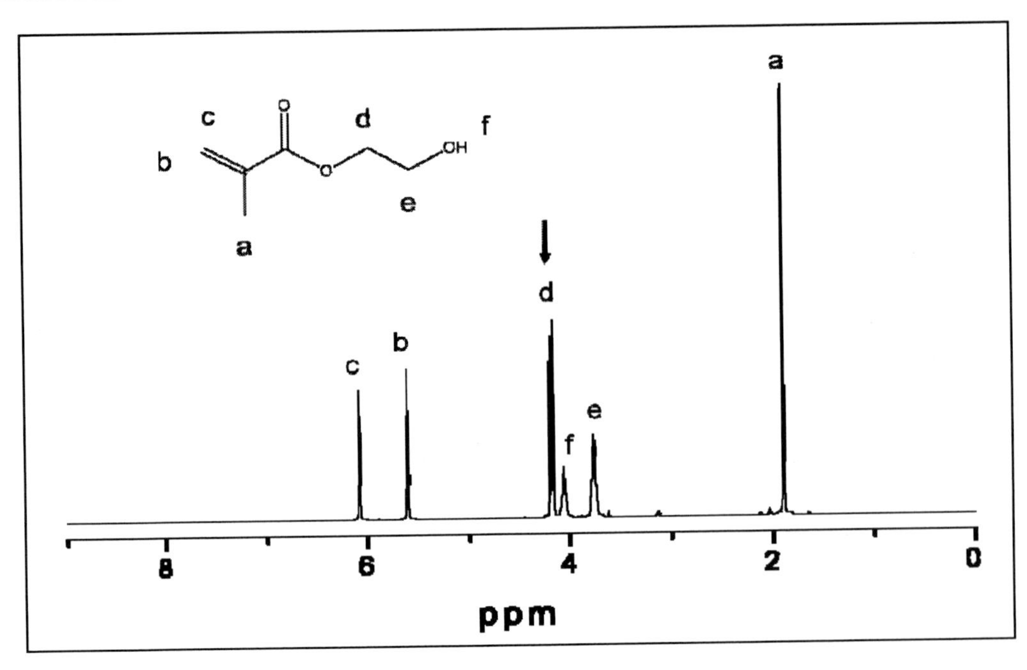

Figure 5. ^{1}H NMR spectrum of HEMA in acetone-d_6. Peak integration values based on peak d at 4.22 ppm, representing two $-CH_2$ protons.

These calculations were developed based on peak integration values of the mixed sample and proton count of the pure samples to give a mol-% of each component present in the acetone-d_6 solution. These values were compared with the calculated mol-% values based on the masses in the prepared sample. To test the equations, several comonomer samples in varying concentrations were run, integrated, and evaluated. Again the NMR mol-% values were compared with the calculated mol-% values based on sample masses. The NMR values versus calculated values in all samples showed minimal difference with a variation of (1.0 ± 0.3) %.

As mentioned previously, extraction studies preceding these experiments were performed on copolymer and composite samples in a variety of solvents. Based on the gravimetric screening study, acetone was selected as the solvent for NMR evaluation. Three copolymer and three composite samples were analyzed. In each case, the acetone was evaporated from the sample jar as it diluted the sample spectrum, making peaks difficult to detect. After evaporation the jars were weighed to confirm leachable comonomer mass and then filled with 1.0 mL acetone-d_6. The solutions were lightly agitated to ensure solvation of the comonomers and were transferred to NMR tubes. ^{1}H NMR was run on each sample and the peaks of interest for each comonomer were integrated using BHT as an internal standard. These integrations values were again plugged into the developed equations to give mol-% values of the leached comonomers. Using the measured disk mass losses and the mol-% data obtained by NMR, the overall mass-% loss of each comonomer was calculated.

2.5. Cellular Tests

The *in vitro* cytotoxicity studies were performed to evaluate early interactions between the experimental orthodontic and endodontic copolymers and their corresponding ACP composites and osteoblastic cells. Cellular studies included phase contrast microscopy and assays for dehydrogenase activity (Wst-1 and MMT tests as described in [66, 67], respectively). Osteoblast-like MC3T3-E1 cells (Riken Cell Bank, Hirosaka, Japan) were maintained in α-modification of Eagle's minimum essential medium (Biowhittaker, Walkerville, MD, USA) with a volume fraction of 10 % fetal bovine serum (Gibco-BRL-Life Technologies, Rockville, MD, USA) and 60 mg/L kanamycin sulfate (Sigma, St Louis, MO, USA) in a fully humidified atmosphere with a volume fraction of 5 % CO_2 at 37 °C. The medium was changed twice a week. Cultures were passaged with EDTA-containing (1 mmol/L) trypsin solution (mass fraction of 0.25 %; Gibco, Rockville, MD, USA) once a week. All disks were sterilized with 70 % ethanol prior to extraction experiments. Each disk was then washed with 2 mL of media for 1 h and then fresh media was placed on each disk for 1 day extraction in the cell incubator at 37 °C. In parallel, a flask of 80 % confluent MC3T3-E1 cells was passaged, cells seeded into well plates with 10,000 cells per well in 2 mL of media, and then placed in the incubator overnight. On the second day of the experiment, the medium from each "cell well" was removed and replaced with the 2 mL of extraction medium from one of the disk specimens (or with the positive or negative control media). The cells were incubated in the extracts for 3 days, photographed (digital photography using an inverted phase contrast microscope, Nikon TE300, Melville, NY, USA) and then prepared for the cytotoxicity assays. To measure cellular dehydrogenase activity, the Wst-1 colorimetric assays were performed according to the following procedure: Extract-cultured cells and the controls without cells were combined with a Wst-1 (2-(4-iodophenyl)-3-(4-nitophenyl)-5-(2,4-disulfophenyl)-2H-tetrazolium, monosodium salt; Dojindo, Gaithersburg, MD, USA) solution in HEPES buffer, individually added to wells and incubated for 2 h at 37 °C. Aliquots from each well were transferred to a well-plate and absorbance was read at 450 nm with a plate-reader (Wallac 1420 Victor^2, Perkin Elmer Life Sciences, Gaithersburg, MD, USA). MTT assays were performed as follows: cells cultured in the extracts were rinsed with 1 mL phosphate buffered saline solution (PBS; 140 mmol/L NaCl, 0.34 mmol/L Na_2HPO_4, 2.9 mmol/L KCl, 10 mmol/L HEPES, 12 mmol/L $NaHCO_3$, 5 mmol/L glucose, pH = 7.4) and 0.125 mL/well of 3-(4,5-dimethylthiazol-2-yl)-2,5-diphenyltetrazolium bromide (MTT) solution (5 mg/mL MTT in PBS). After 2h incubation at 37 °C, the MTT solution was removed and the insoluble formazan crystals were dissolved in 0.1 mL dimethylsulfoxide (DMSO). Finally, the absorbance was measured at 540 nm with a plate reader. The blank values (the well that contain only the PBS, MMT and DMSO solutions) were subtracted from each of the experimental values as background.

2.6. Statistical Analysis

The number of test specimen for each evaluation step was chosen so that there is a reasonable chance (power) to detect the minimum desired difference between the groups [68].

The analysis of variance (ANOVA) was performed to evaluate the experimental data as a function of composite makeup, storage times or any other relevant factor involved in the experimental design. When the overall statistically significant effects were found with ANOVA, multiple comparison tests (2-sided; α=0.05) were used to determine the significant differences between the specific groups. Statistical analyses of the data were done by means of Microsoft Office Excel 2007 and/or SigmaStat version 2.03 (SPSS Inc., Chicago, IL, USA).

One standard deviation (SD) is identified in this paper for comparative purposes as the estimated standard uncertainty of the measurements.

3. Results and Discussion

3.1. ACP Filler Modification

It is well known that addition of metal cations, organic additives containing hydroxyl, carboxyl or amino functionalities, polymeric molecules and polyelectrolytes to calcium and phosphate supersaturated solutions influences the kinetics of precipitation and affects the type and stability of precursor phases [69, 70]. Despite its important role in understanding the processes that control the kinetics of spontaneous precipitation from supersaturated solutions [71, 72], the role of particle agglomeration has scarcely been studied in models involving HAP [73, 74], and completely ignored with regard to ACP. It is the spontaneous and uncontrolled agglomeration of ACP particles during ACP synthesis that yields highly clustered ACP solid that disperses non-uniformly in matrix resins or polymers. As a result of ACP's heterodispersivity, ACP polymeric composites are mechanically inferior to surface-treated, glass-reinforced resin materials. Moreover, the state of ACP agglomeration also affects the ion release from these composites. The underlying premise of the filler modification studies was that cations and/or poly(ethylene oxide) introduced during the synthesis would reduce the extent of ACP agglomeration (consequently, mean particle size would decrease) without having a detrimental effect on its stability in aqueous environments, thereby, preserving ACP's bioactivity as a calcium and phosphate releasing agent.

Cation-ACPs (with the exemption of Fe(II)- and Fe(III)-ACP) and PEO-ACP generally showed two diffuse, broad bands resembling the XRD patterns of glasses and certain polymers (Figure 6a). The corresponding FTIR spectra (Figure 6b) exhibited two wide bands typical for phosphate stretching and phosphate bending in the region (1200 to 900) cm^{-1} and (630 to 500) cm^{-1}, respectively. For Fe(II)- and Fe(III)-ACP, signs of conversion to crystalline apatite were seen in their XRD and FTIR scans. In addition, significant color changes occurred in Ag-, Fe(II)- and Fe(III)-ACP samples. This unwanted color change was due to the co-precipitation of light–sensitive Ag- and colored Fe-phosphates with ACP. Because of the color instability and accelerated conversion to apatite, Ag-, Fe(II)- and Fe(III)-ACPs were not further evaluated. The median particle diameter, d_m, obtained from the PSD measurements and water content (TGA data) of cation- and PEO-ACPs (findings were independent of the PEO's molecular mass, therefore the average values for the combined 8K, 100K and 1000K molecular mass PEO are reported), and the mechanical strength of the composites based on these ACPs are summarized in Table 5. The PSD data revealed heterogeneous distribution with particles ranging from submicron sizes up to 200 μm or more in diameter. The heterogeneity of the particle sizes was confirmed by SEM observations (images not shown). In cation series, d_m of Zn-ACP and Al-ACP was significantly lower than the d_m of Si-ACP and Zr-ACP. PEO-ACPs had significantly higher d_m compared to cation-ACPs or no-additive, control ACP. Observed differences in the mean values of d_m between different cations would suggest slight modifications in the degree of ACP's agglomeration, which appear random rather than systematically related to ionic potential of cations. The apparently higher extent of PEO-ACP agglomeration could possibly be attributed to a mechanism similar to "polymer bridging", that reportedly controls the agglomeration of apatite particles in the presence of high-molecular mass polyacrylates [70]. The average water content of modified ACPs (mass fraction of 15.7 %) appeared unaffected by the type of additive used during the synthesis. It corresponds to approximately 2.6 water molecules per structural ACP unit assuming $Ca_3(PO_4)_2$ as the compositional formula. Regardless of the type of additive, roughly 2/3 of the ACP's water was surface-bound (mobile water; weight loss below 130 °C) while the remaining 1/3 was structurally incorporated (weight loss in the temperature region (130 to 600) °C)). Water content of all ACPs correlated well with the values reported in the literature (mass fraction of up to 18 % [52]). Apparently, no correlation existed between the ACP particle size and the BFS of composites in either dry or wet state. The highest dry BFS value was attained in PEO-ACP composite group (76.0 MPa). Dry BFS values of cation-ACP specimens ranging from 41.5 MPa to 54.3 MPa were not statistically different from the unmodified ACP control (56.0 MPa). After exposure to aqueous environment, the BFS of PEO-ACP composites was reduced by 69 % compared to their dry values. In cation-ACP series, the extent of BFS reduction in going from dry to wet state was nonexistent (Zn-ACP) or ranged from 10 % (Zr-ACP) to 41 % (Al-ACP). The same reduction in the control (unmodified) ACP group was 29 %.

Table 5. Effect of cations and PEO on the particle size and water content of ACP fillers and the biaxial flexure strength (BFS) of cation-ACP, PEO-ACP and control ACP/BHT composites. Indicated are mean values with one standard deviation in parentheses. Number of replicate experiments in each experimental group: n □ 5. Control: ACP synthesized in the presence of sodium pyrophosphate as a stabilizer

Additive	Median diameter, dm (μm)	Water content (mass fraction, %)	BFS (MPa)	
			Dry	Wet
Zn^{2+}	1.4 (0.5)	16. 6 (2.5)	45.2 (13.9)	47.7 (13.2)
Al^{3+}	2.2 (1.3)	16.8 (2.8)	41.5 (7.7)	24.6 (3.1)
Si^{4+}	5.8 (1.6)	13.2 (1.6)	46.8 (7.4)	31.8 (6.8)
Zr^{4+}	7.4 (2.3)	16.1 (2.0)	54.3 (11.4)	49.3 (11.6)
PEO	14.1 (4.7)	15.6 (2.0)	76.0 (13.9)	24.0 (4.1)
none (control)	7.6 (2.5)	15.8 (3.9)	56.0 (10.0)	40.0 (9.0)

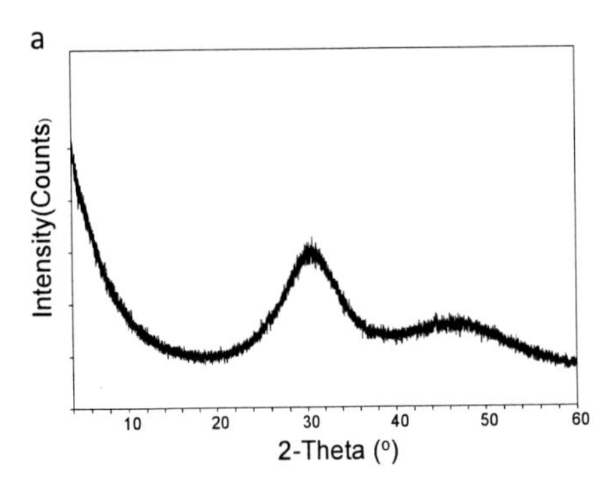

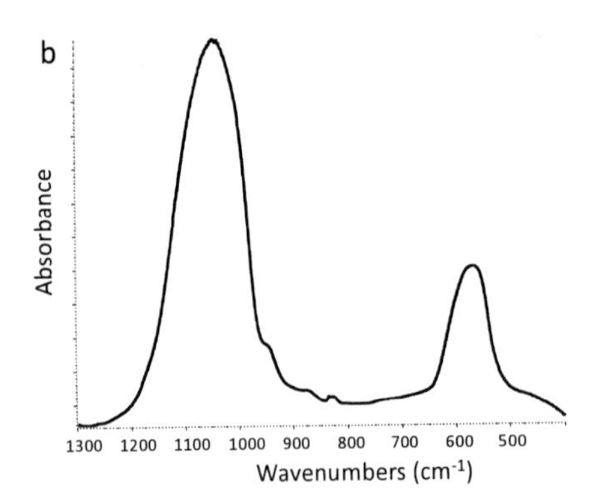

Figure 6. XRD pattern (a) and FTIR spectrum (b) representative of cation- or PEO-ACP.

Since the introduction of cations during the spontaneous precipitation of ACP from supersaturated calcium and phosphate solutions generally yielded ACPs with only marginally reduced particle sizes (the addition of PEO even had a contrary effect) and since only composites formulated with Zn- or Zr-ACP showed a minimal increase in strength after immersion compared to the composites formulated with the unmodified ACP filler, we explored alternative ways of breaking up large ACP agglomerates. Grinding and milling experiments were performed with Zr-ACP, which was shown to successfully blocks, by adsorption, potential sites for apatite nucleation and growth, and, as a result, maintains the ACP's remineralizing potential for prolonged time periods [75]. Typical volume size distribution histograms of am-, g- and m-ACP dispersed in isopropanol are shown in Figure 7.

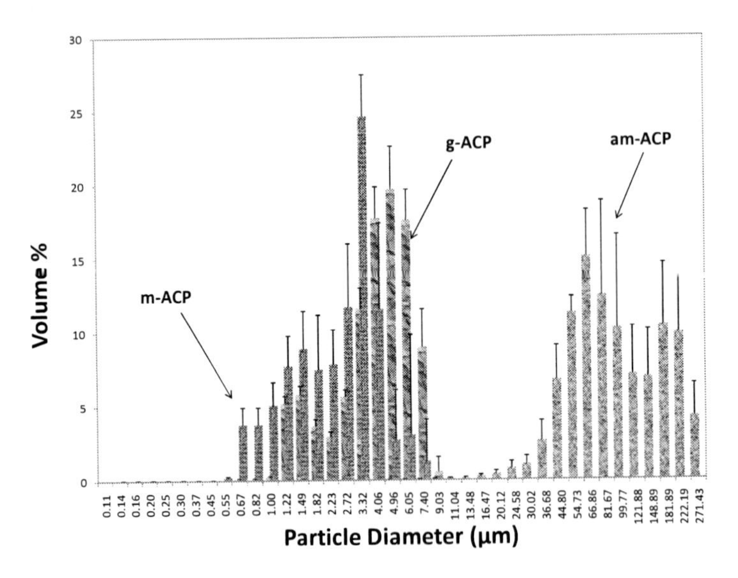

Figure 7. Particle size distribution of as made (am-), ground (g-) and milled (m-) Zr-ACP dispersed in isopropanol. Shown histograms represent the mean values + standard deviation of three repetitive runs in each group.

The issue of how closely sample handling and measuring conditions of the PSD measurements (dry vs. wet state, sonication vs. no-sonication) reflect the actual ACP distribution within composites has been extensively discussed [54]. It has been concluded that sonicating and dispersing ACP in isopropanol before the measurement is more representative of the true PSD. The PSDs shown in Figure 7 clearly show that volume fraction of fine particles increases in following order: m-ACP >> g- ACP ≥ am-ACP. The higher volume fraction of fine particles in m-ACP and, to a lesser extent, in g-ACP significantly affected the median volume diameter, d_m of these fillers (Table 6). It should be noted that d_m alone, although used as a main factor in demonstrating the effects of grinding and/or milling, provide insufficient information about the PSD of the sample. The complete characterization required both quantitative PSD and qualitative SEM screening. Additional indication of the changes in ACP particle sizes is obtained during the fabrication of composite specimens. Hand spatulation of g- and, particularly, m-ACP into the resin is much easier and takes less time than the same process using am-ACP. Moreover, at the same filler level (a mass fraction of 40 %) pastes with g- and m-ACP are more flowable compared to the am-ACP composite paste, which is regularly very viscous and not at all flowable. When fabricating composites for experimental and clinical applications, care must be taken to ensure that minimal defects are introduced into a composite's structure. Poor processing techniques such as inadequate packing of ACP composites into molds can lead to internal inclusion of air bubbles, which can lead to failures [76]. The narrower PSD obtained through grinding and, especially, milling apparently improved dispersion of these fillers within the matrices and, in turn, the mechanical properties of g- and m-ACP composites (Table 6).

Table 6. Results of particle size distribution analysis for the am-, g- and m-ACP fillers, the mechanical strength (after 1 mo aqueous immersion), maximum water sorption (WS_{max}) at 75 % relative humidity and the anti-demineralizing/remineralizing capacity (expressed as the thermodynamic stability of the solutions with respect to stoichiometric HAP; ΔG^0) of their EHT and EHMT resin composites. Values represent a group mean + one standard deviation. Number of specimens in each group: n □ 5

Parameter	am-ACP	g-ACP	m-ACP
Particle size range (μm)	(3.0 to 271.4)	(0.6 to 11.0)	(0.4 to 8.2)
Median particle diameter, d_m (μm)	80.0 ± 4.7	4.5 ± 0.8	3.3 ± 0.5
Biaxial flexure strength, BFS (MPa)	42.2 ± 6.7	50.0 ± 8.0	56.4 ± 7.7
Water sorption, WS_{max} (mass %)	3.1 ± 0.4	2.5 ± 0.5	± 0.2
Thermodynamic stability, (kJ/mol)	-(5.7 ± 0.2)	-(4.9 ± 0.6)	-(5.1 ± 0.3)

More homogeneous dispersion of g- and m-ACP fillers throughout the composites, i.e., a lesser number of voids/defects existing throughout the bodies of the composite disk specimens, resulted in reduced WS in g- and m-ACP composites. However, these composites steadily released calcium and phosphate ions into buffered saline solution. The attained levels of the mineralizing ions were adequate to create an environment favorable for regenerating mineral-deficient tooth structures (ΔG^0 values < 0). The thermodynamic stability of the immersion solutions containing the maximum concentrations of calcium and phosphate ions released from the composite disk specimens was calculated with respect to stoichiometric HAP using the Gibbs free-energy expression [8-10]:

$$\Delta G^0 = -2.303(RT/n)\ln(IAP/K_{sp}) \tag{5}$$

where IAP is the ion activity product for HAP, K_{sp} is the corresponding thermodynamic solubility product, R is the ideal gas constant, T is the absolute temperature, and n is the number of ions in the IAP (n=18).

Based on the above discussed findings, grinding and/or milling are identified as simple yet effective approaches for reducing the size and number of large agglomerates that exist in am-ACP filler. Further modifications in the milling and/or grinding protocols may be required to additionally enhance the mechanical interlocking at the filler/matrix interfaces which should boost the mechanical performance of composites.

3.2. Copolymers and ACP Composites: Structure/Property Relationships

The results of physicochemical screening and the mechanical testing of the series of the experimental resins based on Bis-GMA, EBPADMA and UDMA (systems A1- A3; Table 4) and their corresponding am-ACP composites (envisioned for applications as dental sealants and/or base/liners) are compiled in Table 7.

LC Bis-GMA- and EBPADMA-based copolymers and composites, and to a lesser extent UDMA-based copolymers and their composites, all achieved higher DVC values when hydrophilic HEMA was included as a co-monomer in the resin. Higher DVC values attained in resins with relatively high content of HEMA ($\geq$ 28 mass %) are attributed to HEMA's high diffusivity and mono-functionality. Similar findings are reported for LC Bis-GMA resins containing hydroxypropyl methacrylate as a co-monomer [77], a monomer homologous to HEMA. Regardless of the resin matrix composition, the DVCs of Bis-GMA- and EBPADMA-based ACP composites were generally lower than DVC values attained in their UDMA-based counterparts. This phenomenon can be explained by the higher reactivity of UDMA monomer in comparison with Bis-GMA or EBPADMA [19]. Furthermore, the following order of decreasing DVC existed in composites formulated with binary or ternary resins: XHT $\geq$ XHHm > XT $\geq$ XHm, with X being Bis-GMA, EBPADMA or UDMA.

As is expected, composite materials that yielded generally high DVCs (from 76.0 % to 92 %) also showed relatively high PS (ranging from 4.2 vol % to 8.7 vol % for all experimental groups). The PS showed no clear-cut correlation with the composition of the resin matrix. Practically all experimental composites exceeded the PS values typically reported for the commercial composite materials (PS values ranging from 1.9 % to 4.1 % [78, 79]). Measured PS values fell into the category of either flowable composites or adhesive resins (reported PS values (3.6 to 6.0) % and (6.7 to 13.5)%, respectively [79]). These high PS values can be attributed in part to a much lower filler level in our experimental materials (40 mass % ACP) compared to that of up to 85 % of silica-based fillers in highly filled conventional composites. Based on PS results alone, adjustments in resin composition by inclusion of bulkier but relatively low viscosity resins or ring-opening monomers [80, 81] may be necessary in designing composites with lower PS while maintaining satisfactory DVC.

Research on WS of dental materials indicates that excessive water uptake generally causes a decrease in mechanical strength, depression of the glass transition temperature due to plasticization [82], solvation, reversible rupture of weak inter-chain bonds and irreversible disruption of the polymer matrix [83]. In the case of ACP polymeric composites, water-ACP filler interactions additionally contribute to the overall WS profiles. Besides affecting the mechanical integrity of the composites, water diffusion affects the kinetics of the intra-composite ACP conversion to HAP and ultimately determines the remineralizing capacity of these bioactive materials. WS_{max} values for all resins and their ACP composites were reached within two weeks of soaking and were generally highest for ternary BHT, EHT and UHT formulations, and lowest for binary resins that contained HmDMA. The observed differences are primarily related to the presence of hydrophilic (HEMA and TEGDMA) or hydrophobic (HmDMA) monomers in the matrix.

In all systems, after one month of composite specimen soaking, calcium and phosphate ion levels attained in the immersing solution were sufficient to create an environment highly supersaturated with respect to HAP, potentially leading to HAP's re-precipitation. While no significant differences in the overall WS were observed between Bis-GMA-, EBPADMA- and UDMA-based resins, ion release from their composites appears affected by the chemical structure and the composition of the monomer system. Generally, higher supersaturations (more negative ΔG^0 values) were attained in systems containing EBPADMA as a base monomer and HEMA as a co-monomer. The most probable reason for higher ion releases obtained with EBPADMA-based composites is a more open cross-linked network structure of their resin matrix. On the other hand, hydrophilic HEMA-enriched matrices increased internal mineral saturation by allowing the uptake of more water and/or better accessibility of ACP filler to the water already entrained.

The mechanical strength of Bis-GMA- and EBPADMA-based copolymers did not deteriorate while UDMA-based copolymers failed to maintain their strength upon exposure to aqueous milieu. Generally, dry ACP composite specimens had substantially lower BFS than the corresponding copolymers regardless of the resin composition. The strength of all but binary, HmDMA-containing composites deteriorated further upon soaking. It appears that with am-ACP, the existence of numerous defects/voids (resin-rich, phosphate-depleted regions [13]), i.e., the uneven distribution of highly agglomerated filler is responsible for inadequate filler/resin interlocking and the adverse effect on the overall mechanical strength.

The effects of chemical structure and composition of the resin phase on DVC and BFS of copolymers and the corresponding am-ACP composites were additionally investigated in LC Bis-GMA/X/TEGMA ternary formulations (X being a neutral or acidic co-monomer (systems B1 and B2; Table 4)). The results of the study are summarized in Table 8.

In the neutral termonomer series (B1 group), DVC values decreased in the following order: BDT > (BHT, BMT) > (BGmT, control) > BGdT for copolymers, and BDT > (BGmT, BHT, control) > (BGdT, BMT) for composites. The DVCs of both copolymer and composites in acidic termonomer series (B2 group) showed practically no dependence on structural variations of the acidic monomer. Generally, the DVCs of composites were lower than the DVCs of the corresponding copolymers (the reduction ranged from 4.5 % to 16.4 %) regardless of the resin matrix composition (the only exception are BMaT formulations, where, surprisingly, the reverse effect was observed). The fact that BDT copolymers and composites showed the highest DVC is likely due to the highly flexible nature of DEGMEMA monomer.

Table 7. Mean value ± standard deviation of the DVC (24 h post-LC), WS_{max} (at 75 % relative humidity), PS, BFS (dry and wet specimens after 1 mo of aqueous immersion) of Bis-GMA-, EBPADMA and UDMA-based copolymers and composites, and the supersaturation (ΔG^0) corresponding to the maximum release of calcium and phosphate ions from the immersed composite specimens. Number of repetitive runs in each group, n = 4 (ΔG^0; WS), n ◘ 4 (BFS), n ◘ 8 (DVC) and n ◘ 9 (PS)

Resin/specimen	DVC (%)	WS_{max} (mass %)	PS (vol %)	BFS (MPa)		$\Delta\Gamma^0$ (κϑ/μολ)
				Dry	Wet	
A1. Bis-GMA based						
BHm copolymer	86.2 ± 2.4	1.3 ± 0.2	nd	101 ± 26	123 ± 26	n/a
composite	77.4 ± 3.0	2.2 ± 0.2	5.5 ± 0.7	53 ± 13	55 ± 11	-[6.02 ± 0.10]
BT copolymer	85.6 ± 2.6	3.2 ± 0.2	nd	132 ± 27	123 ± 22	n/a
composite	82.0 ± 5.0	2.6 ± 0.3	8.7± 2.6	62 ± 15	62 ± 13	-[5.49 ± 0.12]
BHHm copolymer	89.7 ± 3.9	3.3 ± 0.3	nd	155 ± 45	133 ± 36	n/a
composite	83.2 ± 4.2	2.8 ± 0.2	4.2 ± 1.7	71 ± 10	55 ± 11	-[6.32 ± 0.15]
BHT copolymer	92.8 ± 1.8	4.8 ± 0.6	nd	156 ± 40	144 ± 52	n/a
composite	80.5 ± 6.0	3.6 ± 0.2	7.2 ± 0.9	56 ± 10	40 ± 10	-[5.99 ± 0.11]
A2. EBPADMA based						
EHm copolymer	87.5 ± 1.5	0.8 ± 0.1	nd	95 ± 18	93 ± 24	n/a
composite	76.0 ± 4.1	1.9 ± 0.2	7.2 ± 1.1	59 ± 8	53 ± 11	-[6.91 ± 0.17]

ET copolymer	91.6 ± 1.8	[illegible].2 ± 0.3	nd	114 ± 19	125 ± 35	n/a
composite	78.0 ± 2.9	2.2 ± 0.1	6.4 ± 0.7	61 ± 6	59 ± 7	-[6.47± 0.28]
EHHm copolymer	92.8 ± 1.7	2.0 ± 0.3	nd	122 ± 13	120 ± 27	n/a
composite	79.5 ± 5.5	2.9 ± 0.1	8.2 ± 1.1	58 ± 9	51 ± 9	-[8.19 ± 0.27]
EHT copolymer	94.4 ± 1.5	3.8 ± 0.2	nd	133 ± 38	128 ± 49	n/a
composite	91.1 ± 2.0	3.4 ± 0.3	7.8 ± 1.5	57 ± 11	49 ± 8	-[7.38 ± 0.19]
A3. UDMA based						
UHm copolymer	81.8 ± 2.0	3.5 ± 0.3	nd	183 ± 22	117 ± 38	n/a
composite	81.2 ± 2.1	2.3 ± 0.2	5.8 ± 1.4	65 ± 8	60 ± 13	-[3.48 ± 0.11]
UT copolymer	85.8 ± 1.4	2.2 ± 0.1	nd	192 ± 46	93 ± 30	n/a
composite	84.4 ± 2.0	2.7 ± 0.1	6.5 ± 0.5	61 ± 11	57 ± 10	-[5.80 ± 0.20]
UHHm copolymer	91.0 ± 1.4	3.3 ± 0.4	nd	170 ± 32	123 ± 14	n/a
composite	89.1 ± 1.9	2.9 ± 0.1	6.9 ± 0.8	63 ± 7	37 ± 12	-[6.10 ± 0.22]
UHT copolymer	92.8 ± 1.8	1.2 ± 0.2	nd	124 ± 30	74 ± 27	n/a
composite	92.0 ± 2.0	3.3 ± 0.1	7.5 ± 0.8	54 ± 11	40 ± 11	-[6.00 ± 0.15]

The observed reduction in DVC in going from copolymers to ACP composites is attributed primarily to the reduction in exotherm of resin polymerization by ACP phase, although other factors such as greater air entrapment and light scattering (likely facilitated by the heterogeneous size distribution of ACP filler) may also add to this reduction. Somewhat lower reduction in DVC of acidic copolymers vs. their composites compared to DVC reduction of neutral copolymers vs. their composites could possibly be explained by the much lower molar mass of the acidic co-monomers compared to the neutral co-monomers. In addition, the greater expected affinity of the acidic co-monomers for ACP, especially those with carboxylate functional groups (MaA, MA and 4MET) may augment the interfacial conversion of methacrylate groups more than the neutral co-monomers.

In neutral co-monomer series, the BFS of dry and wet copolymer specimens (on average (68.9±8.7) MPa and (68.6±6.8) MPa, respectively) appears unaffected by the co-monomer structure. However, introduction of ACP filler into group B1 resins caused reduction in BFS of up to 23 % for dry composite specimens compared to copolymer counterparts. The BFS of composites deteriorated further upon soaking (additional reduction between 44 % and 56 % compared to dry composite specimens). Similar general trends in BFS of copolymers and composites were observed in acidic co-monomer series. In this series, a reduction in the BFS of dry vs. wet copolymers was overshadowed by the extreme data scattering. The smallest reduction in the BFS of composites (23%) was achieved in B4MT formulation, i.e., with 4MET as a co-monomer. The extreme drop in strength of BMaT copolymers (64%) and composites (62%) may possibly be attributed to the increased hydrophilicity of the resin due to the presence of two carboxylic acid groups in MaA. The favorable arrangement of these two carboxylic acid groups in the BMaT copolymer may have resulted in the increased cross-linking arising from enhanced intermolecular hydrogen bonding and, ultimately, yielding dry copolymer with very high BFS. Disruption by water of the hydrogen bond-mediated cross-links may explain the precipitous decrease in BFS values. The observed reduction in mechanical strength of composites is caused by either reduction in ACP's integrity and rigidity at the filler/matrix interface (possibly chemical reactions involving ACP and acid groups), spatial changes resulting from calcium and phosphate ion efflux as a consequence of intra-composite ACP conversion to HAP, or excessive water sorption. As a result of this study, it seems that inclusion of DEGMEMA and 4MET into polymer matrices may aid in attaining high level of DVC while preserving the mechanical integrity of composites upon aqueous immersion. However, defining the optimal levels of these co-monomers in the resin matrix would require further testing.

The effects of resin composition on the properties of the experimental EBPADMA/ HEMA/MEP/TEGDMA composites intended for orthodontic application were assessed with two series of matrices (series C; Table 4), and am- and m-ACP as bioactive remineralizing fillers. Two sets of EHMP resins were examined: a) with the molar ratio of EBPADMA/TEGDMA (0.13 to 0.50) at a constant HEMA/MEP ratio of 4.28 (assigned $EHMT_{0.13\text{-}0.50}$), and b) with the molar ratio of EBPADMA/TEGDMA (0.35 to 0.50) at a constant HEMA/MEP ratio of 8.26 (assigned $EHMT_{0.50*\text{-}1.35}$). The results of physicochemical and mechanical evaluation of EHMT composites are compiled in Table 9.

Compositional variations of the resin matrix did not significantly influence the DVC, WS or BFS of am-ACP composites in $EHMT_{0.13\text{-}0.50}$ series. However, the composition of the resin matrix, i.e., the relative content of the high molecular mass EBPADMA had an effect on the PS that developed in these composites: the PS increased with the decreasing

EBPADMA/TEGDMA molar ratio in the resin. Both am- and m-ACP/EHMT$_{0.50*-1.35}$ composites attained lower DVC and showed lesser affinity for WS compared to am-ACP/EHMT$_{0.13-0.50}$ counterparts. Apparently lower BFS (series B: on average (36.1±5.2) MPa) of am-ACP composites with higher content EBPADMA in the polymer compared to BFS (series A: on average (45.6±6.8) MPa) of am-ACP composites with lower EBPADMA content was not statistically significant.

Table 8. Mean value ± standard deviation of the DVC (24 h post-LC) and the mechanical strength (BFS; dry and wet specimens (after two weeks of aqueous immersion)) of Bis-GMA/X/TEGDMA (X = neutral (B1 series: DEGMEMA, GMA, GDMA, HEMA or MEMA) or carboxylic (B2 series: MaA, MA, 4MET or VPA) co-monomer) ternary resins and their am-ACP composites. Number of repetitive measurements per group: n □ 6 (DVC) and n □ 5 (BFS)

Resin/specimen	DVC (%)	BFS (MPa)	
		Dry	Wet
B1. Neutral co-monomers			
BDT copolymer	87.0 ± 0.8	61.3 ± 11.7	63.7 ± 11.0
composite	75.0 ± 3.2	49.3 ± 12.0	27.8 ± 5.0
BGmT copolymer	72.6 ± 1.6	60.3 ± 11.5	61.3 ± 11.3
composite	66.6 ± 2.0	62.7 ± 6.7	32.3 ± 6.3
BGdT copolymer	66.4 ± 2.4	68.7 ± 9.2	66.3 ± 14.2
composite	61.3 ± 4.5	67.0 ± 5.4	31.7 ± 10.0
BHT copolymer	80.2 ± 2.1	84.7 ± 14.8	73.0 ± 14.7
composite	67.7 ± 1.5	65.3 ± 14.0	31.6 ± 8.7
BMT copolymer	78.9 ± 1.3	69.5 ± 10.7	80.0 ± 14.7
composite	61.0 ± 2.6	57.0 ± 14.9	25.0 ± 9.7
B2. Acidic co-monomers			
BMaT copolymer	63.4 ± 3.9	132.8 ± 28.9	47.7 ± 37.2
composite	66.6 ± 3.8	70.3 ± 11.2	26.8 ± 3.7
BMT copolymer	71.8 ± 0.7	88.6 ± 27.3	57.7 ± 14.9
composite	65.9 ± 2.5	87.5 ± 15.4	37.9 ± 7.0
B4MT copolymer	74.6 ± 2.4	91.2 ± 37.4	50.3 ± 3.8
composite	69.7 ± 3.3	86.0 ± 11.7	66.7 ± 13.2
BVT copolymer	72.5 ± 2.6	85.7 ± 49.7	76.5 ± 42.2
composite	60.6 ± 4.2	66.8 ± 4.2	31.8 ± 2.5
BT (control)			
copolymer	73.6 ± 4.3	80.4 ± 21.2	70.1 ± 23.0
composite	67.4 ± 1.3	68.2 ± 8.5	50.1 ± 6.0

Table 9. DVC (24 h post-LC), PS, WS (at 75 % relative humidity), BFS (wet specimens after 1 mo immersion) and solution thermodynamic stability (ΔG^0) corresponding to plateau calcium and phosphate concentrations for am-ACP/EHMT composites (series A), and am-and m-ACP/EHMT composites (series B). Indicated are mean values ± standard deviation for n □ 6 (DVC), n □ 3 (PS), n = 5 (WS, BFS) and n = 4 (ΔG^0)

nd – not determined

Resin/composite	DVC (%)	PS (vol %)	WS (mass %)	BFS (MPa)	ΔG^0 (kJ/mol)
A. $EHMT_{0.13-0.50}$ series					
$EHMT_{0.13}$	86.6 ± 5.0	7.8 ± 0.6	3.3 ± 0.5	44.4 ± 8.2	-[5.07 ± 0.31]
$EHMT_{0.25}$	82.2 ± 6.5	7.2 ± 0.4	3.2 ± 0.5	45.9 ± 4.2	-[4.44 ± 0.30] -
$EHMT_{0.33}$	84.4 ± 5.0	6.5 ± 0.7	3.6 ± 0.5	48.5 ± 6.9	[4.61 ± 0.13] -
$EHMT_{0.50}$	85.8 ± 5.3	6.1 ± 0.5	3.3 ± 0.5	43.6 ± 7.8	[4.50 ± 0.19]
B. $EHMT_{0.50*-1.35}$ series					
$EHMT_{0.50*}$ am-ACP	76.9 ± 3.8	nd	2.5 ± 0.2	36.9 ± 4.9	-[4.35 ± 0.21] -
m-ACP	80.3 ± 1.4	nd	2.1 ± 0.2	60.9 ± 9.4	[3.42 ± 0.39] -
$EHMT_{0.85}$ am-ACP	70.5 ± 3.8	nd	2.3 ± 0.2	35.0 ± 4.2	[4.65 ± 0.22] -
m-ACP	74.8 ± 3.9	nd	1.8 ± 0.2	52.9 ± 8.9	[4.32 ± 0.36] -
$EHMT_{1.33}$ am-ACP	69.1 ± 1.2	nd	2.5 ± 0.3	36.4 ± 6.7	[4.67 ± 0.23] -
m-ACP	75.1 ± 1.2	nd	1.6 ± 0.1	55.8 ± 17.2	[4.79 ± 0.20]

However, the improved BFS of composites was achieved with the use of milled ACP (series B: on average (56.5±11.8) MPa) instead of am-ACP. This improved strength and lower WS of m-ACP composites is related to the improved dispersion of milled ACP filler throughout the composites compared to the coarse, am-ACP. This lower WS only marginally affected the overall remineralizing potential of composites (the supersaturations considerably above the theoretical minimum needed for remineralization ($\Delta G^0 < 0$) were attained in all formulations).

The results of extensive evaluation of the experimental endodontic sealer formulated with UDMA/HEMA/MEP/PEG-U resin and am- and g-ACP filler are presented in Table 10.

At 24h post-cure, all LC UHMP copolymers attained exceptionally high DVC values. Differences between various LC systems (for formulation details see subsection 2.2.) were not statistically significant. Similarly there was no difference between the two CC and two DC formulations. The DVC values obtained in CC copolymers were, on average, 35 % lower than DVC values achieved in LC formulations. Based on these findings, CC composites were excluded from further testing. DVC values obtained for LC am- and g-ACP composites were between 7.6 % and 9.7 % lower than DVC attained in the corresponding copolymer. In DC series, the attained DVCs were comparable (am-ACP composites) or even higher (g-ACP composites) than DVC of copolymers. Since there was no discernible difference between the two DC systems (1850 Irgacure + BPO & DHEPT vs. 1850 Irgacure + AA & TBPB), the remaining tests were performed with 1850 Irgacure + BPO & DHEPT due to more favorable handling properties of the pastes made with these initiators.

PS could be measured successfully only with LC composites. In DC system, hardening of the paste occurred within 10 min of mixing the chemically activated components. Some degree of contraction occurred before the sample was even placed in the instrument, thus making the material unsuitable for measurement by tensometry. The mean PS values for LC am- and g-ACP UHMP composites (7.1 vol % and 6.9 vol %, respectively) were comparable to PS measured in EBPADMA- and UDMA-based composites (Table 7: (6.4 to 8.2) vol% and (5.8 to 7.5) vol%, respectively), and EHMT composites (Table 9: (6.1 to 7.8) vol %) but have exceeded the PS values of LC am-ACP/UDMA and am-ACP/PEG-U formulated without HEMA and adhesive monomer (5.2 vol % on average [84]). The relatively high contraction of ACP/UHMP composites may be attributed to the intensified hydrogen bonding that is likely to occur in UHMP matrices that contain a relatively high amount of HEMA (16.8 to 17.3 mass %). This excessive hydrogen bonding could ultimately lead to the densification of polymerization [85]. In LC UHMP composites, the polymerization stress (PSS) decreased in going from am-ACP to g-ACP. PSS was generally lower in DC compared to LC UHMP composites. The PSS developed in LC am-ACP/UHMP formulations ((4.1 to 4.7) MPa) compares well with the PSS that developed in LC am-ACP/UDMA/HEMA composites (on average (4.5±0.1) MPa [84]. This finding would imply that the stress originating from the PS in UHMP matrices was not elevated by the simultaneous inclusion of HEMA and PEG-U into the resin matrix. In binary UDMA/HEMA and UDMA/PEG-U based composite, the differences in PSS may possibly be related to the higher relative molecular mass and more flexible character of the PEG-U oligomer compared to the less flexible poly(HEMA) segments in the matrix. It would, however, be erroneous to use the same reasoning to explain PSS that develops in more complex UHMP resins and their ACP composites.

Table 10. Properties of LC and DC UHMP copolymers and their ACP composites (mean value ± standard deviation). Number of repetitive runs ≥ 3/experimental group. am – as made ACP; g – ground ACP; nd – not determined; n/a – not applicable

Property/Specimen	LC			DC		
	Copolymer	Composite		Copolymer	Composite	
		am-ACP	g-ACP		am-ACP	g-ACP
DVC (%)	95.7±2.2	86.4±1.98	8.4±2.4	79.3±3.4	76.0±4.4	85.4±3.2
PS (vol %)	nd	7.1±0.3	6.9±0.1	n/d		
PSS (MPa)	nd	4.8±0.2	4.1±0.2	nd	3.7±0.3	3.6±0.2
WS (mass %) RH	3.2±0.3	3.2±0.2	3.2±0.1	2.5±0.5	2.9±0.1	2.6±0.2
immersion	6.7±0.6	8.7±0.4	9.4±0.4	6.2±0.5	7.4±0.5	8.6±0.4
HE (vol %)	5.4±2.4	12.5±1.8	11.8±1.3	6.7±1.7	13.0±1.1	13.6±1.7
BFS (MPa)	137.1±24.9	39.4±3.3	44.4±4.4	124.3±3.4	49.4±9.4	47.3±8.9
ΔΓ⁰ (κϑ/μολ)	na	-[7.37±0.33]	-[7.14±0.46]	na	-[7.44±0.39]	-[6.96±0.32]

Therefore, the apparent discrepancy between the PS and PSS in relation to the composition of the resin matrix has yet to be resolved. In UHMP resins and their composites, relatively high PS is likely to be compensated by the significant hygroscopic expansion (HE) of these materials ((5.4 to 6.7) vol % for copolymers and (11.8 to 13.6) vol % for composites) due to water uptake upon immersion in aqueous medium. Beneficial aspects of HE have indeed been demonstrated by other researchers [86, 87].

The results of BFS testing of ACP/UHMP composites have re-confirmed the well known fact that ACP composites are generally too weak to be considered for use as direct filling materials [7-10]. However, for the intended endodontic application, a material's mechanical strength is not the most critical property. It is, though, important that with respect to BFS, DC formulations were not inferior to LC counterparts. On average, wet specimens across the four ACP/UHMP composite groups compare well with the BFS values of the experimental orthodontic adhesive (Table 9), which efficiently restored subsurface carious lesions in human specimens [11].

Compared to other experimental formulations (Tables 7 & 9), all ACP/UHMP composites exhibited equal or higher remineralization potential, thus confirming their strong potential to inhibit or possibly even reverse root caries in endodontic applications. Although described physicochemical and mechanical tests provide insufficient evidence to distinguish between the DC and LC formulations, or between the am- and g-ACP composites, DC g-ACP/UHMP formulation should be favored due to the nature of the intended application and the fact that g- and/or m-ACP composites are generally mechanically superior to coarse (am) ACP composites (Table 6) while not significantly diminishing their bioactivity, i.e., their remineralizing potential [7-10, 54, 56, 59].

Although the LC methacrylate monomer systems have been studied extensively [88-95], the kinetics of both PS and PSS in these systems is still not fully understood. The polymerization process is affected by the type and concentration of initiators which determine reaction kinetics and DVC attained upon polymerization. In addition to material factors (filler type and/or content, resin type and composition, polymerization mode), processing factors such as cavity configuration (C-factor [96-98]) were identified as factors that control the performance of bonded composite materials. In our study, an attempt was made to mimic constrained PS and PSS that occurs in composites bonded to tooth structure. To investigate whether larger surface area (lower C-factor) yields lower PSS values by allowing greater plastic deformation to occur during polymerization before the gel point is reached (as proposed in [98]), we have compared the effect of variations in C-factor on PSS in a typical ACP/Bis-GMA/TEGDMA remineralizing composite and a typical commercial glass-filled composite. The measured PSS values (PSS_{meas}) obtained by cantilever beam tensometry for specimens with variable C-factors were normalized for mass to specimens with a C-factor of 1.33 (height = 2.25 mm) as controls to give calculated PSS values (PSS_{calc}). DVC attained in the experimental ACP composite and in the commercial control were measured by NIR spectroscopy.

In both ACP/BT and control composite, PSS_{calc} increased with the increasing C-factor, confirming the hypothesis that cavity configuration affects PSS values. Other studies of PSS as a function of specimen thickness in applications simulating the cementation of inlays, i.e., very thin layer of adhesive or composite [99, 100], have shown a substantial disparity in stress with specimen thickness. For the range of specimen thickness examined in our study ((0.50 to 3.75) mm), no correlation existed between the PSS_{meas} and the specimen thickness for both types of composites. One could possibly attribute such results to the insufficient sensitivity of tensometer to detect differences in PSS for composite specimens over the range of C-factors studied. However, for a specimen thickness more akin to the resin-composite direct restorations (thickness (0.8 to 1.50) mm), it was also found that the variations in PSS are minimal and not affected by the instrument compliance [101]. Higher PSS values for the experimental am-ACP/Bis-GMA/TEGDMA composite compared to the commercial control are not unexpected. The higher DVC attained in ACP composites lead to higher PS and PSS compared to a less converted and more highly filled control material. The greater translucency of the ACP composite may have enhanced its degree of radiance and contributed to its higher DVC. Additional critical factor is the composition of resin phase: Bis-GMA/TEGDMA (ACP composite) vs. UDMA-modified Bis-GMA/TEGDMA (control). Urethane-modified Bis-GMA oligomer would be expected to shrink less than Bis-GMA/TEGDMA. With respect to PSS, high levels of filler in composite are desirable since their contribution to stress is usually minimal [102], but filler may also contribute to PSS by increasing elastic modulus. Other material factors being equal, the most rigid material, i.e., the one with highest modulus will show the highest PSS. In view of lack of correlation between PSS_{meas} and C-factor for composites, it may be prudent to investigate the temperature of the composites' exothermic polymerization, and cooling that follows, along with the PSS measurements [103]. Nevertheless, results of this study suggest that processing factors need to be considered when assessing PSS development in composites.

3.3. Biocompatibility of ACP Composites: Leachability and Cellular Responses

The chemical structure/property relations of the constituent monomers, compositional differences involving polymers and initiator systems, and the attainable DVC, especially as it relates to the leachable monomers, are important contributing factors that control the biocompatibility of ACP composites. Cellular responses to these materials are likely to depend on leachable residual monomers and other leachable organic species. We have conveniently used the DVC attained upon polymerization to indirectly assess the potential leachability of unreacted monomeric species from bioactive ACP composites [9, 51, 53, 59]. However, to better understand the correlation between the cytotoxicity and the extent of vinyl conversion it seems necessary to identify and quantify organic moieties that leach-out from the composites.

Leachability studies were performed with UHMP copolymers and composites specimen using ^{1}H NMR. This study is unique in that NMR is not usually used to quantify leachables in polymeric systems. Typically, chromatographic techniques, such as high-performance liquid chromatography, gas chromatography and liquid chromatography-mass spectrometry are used for that purpose. However, these techniques have longer sample preparation and results can be difficult to interpret. We have demonstrated that ^{1}H NMR is a valuable technique that provides both qualitative and quantitative information on leachables without the burden of either sample preparation and/or data interpretation. The preliminary gravimetric extraction study that preceded the NMR experiments entailed a variety of solvents. Protic solvents are expected to solvate negatively charged solutes via hydrogen bonding while aprotic ones tend to have large dipole moments and solvate positively charged species via their negative dipole.

In all extraction experiments, polymerization inhibitor butylated hydrohytoluene was added to organic solvents at 0.01 mass % to prevent secondary monomer polymerization upon extraction. At given experimental conditions (7 days of extraction in hermetically sealed containers at room temperature with continuous magnetic stirring), the extractable portion of the copolymer and composite specimens ranged from (0.5 to 13.9) mass % and (1.6 to 6.2) mass %, respectively. The fact that higher extraction values were obtained with aprotic polar solvents compared to protic polar solvents would suggest that leachables from the UHMP copolymers and composites are most likely positively charged leading to the enhanced salvation via negative dipoles of acetone and/or dichloromethane. Non-polar cyclohexane did not appear to be as effective as polar solvents (mass losses in cyclohexane extracts were minimal). Normalizing the mass losses in composite specimens to a 100 % resin for acetone and dichloromethane extractions revealed that leachability of composite specimens was (25 to 44) % lower compared to the corresponding copolymer (unfilled resin) specimens. Based on the gravimetric screening study, acetone was selected as a solvent for the quantitative NMR evaluation. Initial testing showed that evaporation of the acetone solvent before running ^{1}H NMR was needed for signal detection. An aliquot of leached monomers in acetone from a test sample was added to acetone-d_6 in an NMR tube and thoroughly mixed. The resulting spectrum showed peaks that were barely visible and difficult to integrate. Evaporating off the acetone and adding acetone-d_6 produced a spectrum with clearly defined peaks that were easy to integrate. Therefore, the acetone in all experimental samples was evaporated before NMR analysis. Based on weighed masses and calculations, this did not affect results.

Signal peaks for each comonomer were chosen that were easily distinguished and separate from each other to facilitate precise integration. Spectral evaluation of the first experimental sample had shown that the initiator 4EDMAB was present in the leachable content. A representative peak was integrated and equation (5) (see Section 2.4.) was manipulated to include the initiator component as follows:

$$\text{Mol-}\%_x = M_x/[M_1+M_2+M_3+M_4+M_5] \times 100 \quad (6)$$

The peaks were integrated at the same values in each sample analyzed for consistency. Using molecular weight and disk mass loss data, the overall mass-% loss from the original composition for each comonomer and initiator was determined (Figure 8).

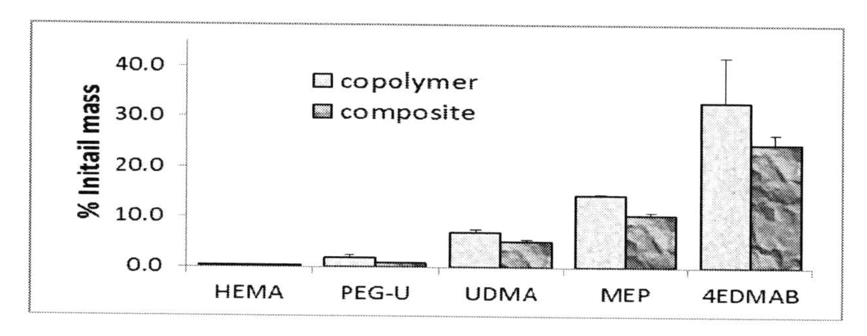

Figure 8. Unreacted monomers and photo-reductant that leached out from UHMP copolymers and their ACP composites (as detected by ^{1}H NMR) expressed as % of the initial amounts (mean value + SD). Extraction medium: acetone.

The levels of the unreacted HEMA, MEP, PEG-U and UDMA detected in the copolymer and composite extracts ranged from 0.30 % to 14.29 % and 0.12 % to 10.39 % of the initial content, respectively. Photo-reductant, (ethyl-4-N,N-dimethylamino benzoate, 4EDMAB) showed the highest leachability (33.06 % and 24.66 % in copolymer and composite extracts, respectively). However, when the composite leachability data are normalized with respect to the initial amount of the resins, differences between the copolymer and composite values become marginal indicating that introduction of ACP into UPHM resins has no significant effect on the leachability of non-polymerized monomeric species from the experimental sealer. This can be explained by correlating the leachability data with the DVC results shown in Table 10. Although there is a 9 % difference in the degree of conversion between the copolymer and composite samples, the conversion is so high that in terms of a cross-linked system, this difference is not significant. In systems this highly cross-linked, the degree of mobility in the polymer chains is small and there are not many pathways for free monomer to leach out of the system. The copolymer and composites systems are above some DVC threshold over which mobility is very low and leachability has become practically constant.

In order to shed light on the interactions between the ACP filler and/or ACP composites, as well as the corresponding copolymers, *in vitro* cytotoxicity studies were performed with EBPADMA/UDMA/TEGDMA/HEMA (EUHT) [62] and UHMP formulations. EUHT copolymer, m-ACP powder and the corresponding m-ACP/EUHT composite were extracted in media overnight and the osteoblast-like cells were then cultured in extracts for 3 d. Phase contrast images of MC3T3-E1 cells cultured in extracts from ACP powder and different resin composites for 3 d (not shown) indicated a normal, spread, polygonal morphology. Only cell remnants were seen in a positive control, detergent-containing samples, indicating that 0.1 mass % detergent in the medium was cytotoxic. Qualitatively, an approximately equivalent amount of cells was found in each experimental system suggesting no adverse cellular response to ACP powder, composite, copolymer or the commercial orthodontic adhesive (COA). However, based on a colorimetric assay of cellular dehydrogenase activity Wst-1 (Figure 9), the extracts from the EUHT copolymer, m-ACP powder, m-ACP/EUHT and COA caused a mild drop in the viability of cells compared to the negative control. This mild reduction in cell viability was attributed to leachable residual monomers and other species. No attempt was made to correlate the cytotoxicity results with DVC attained in these systems.

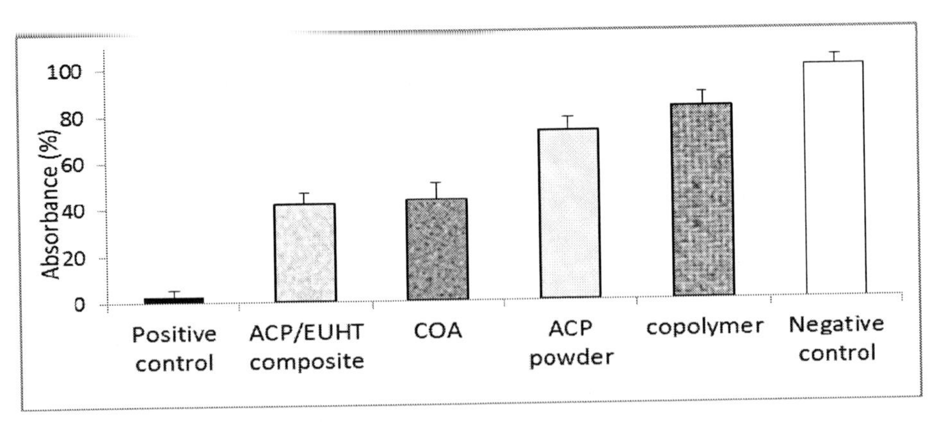

Figure 9. Dehydrogenase activity (Wst-1 assay; mean value + SD) as a measure of cell viability for EUHT copolymer, m-ACP powder and m-ACP/EUHT composite compared to commercial orthodontic adhesive (COA) and positive control (0.1 mass % detergent).

Cellular responses to UHMP copolymers and their ACP composites were assessed in terms of cellular morphology and cell proliferation. UHMP copolymer, ACP pellets and the corresponding ACP/UHMP composites were extracted in media for 24 h, and murine pre-osteoblasts (MC3T3-E1) were then cultured in the extracts for 24 h. Extracts from a commercial endodontic sealer (CES) were used as a reference, and media without any extracts were applied as a control. The cell morphology was examined *in situ* at 24 h using optical microscopy. Cells in the control media showed the spread, polygonal morphology described above (Figure 10a). However, the cells cultured in extracts of composites including CES exhibited a contracted, spherical morphology (Figures 10 b, c). As for the cause of the morphology change, the copolymer and ACP powder did not induce the change individually because cells showed the polygonal morphology in their extracts. The spherical cell morphology and slow cell proliferation rate of this MC3T3-E1 cell line has been reported in hydrogel scaffolds designed for bone regeneration, in which the preosteoblasts generated bone-like minerals [104]. It is also speculated that the morphology changes alternate the cytoskeletal tension on nucleus and nucleus organization and hence influence the mineralization of the osteoprogenitors [105, 106].

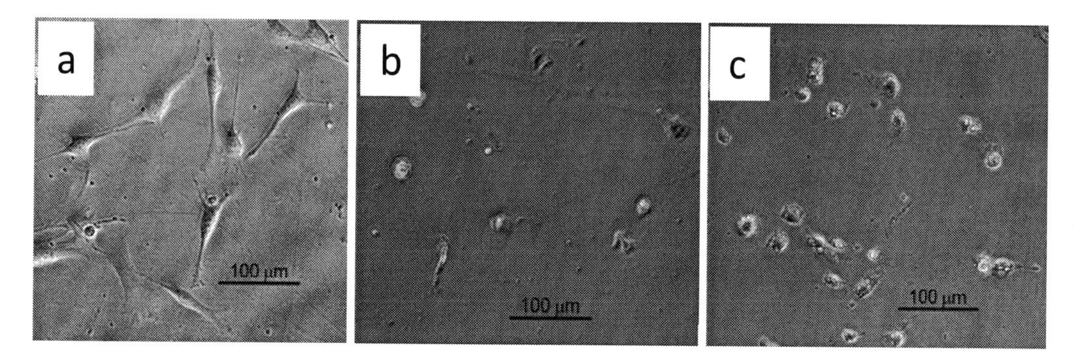

Figure 10. *In situ* cell morphology under optical microscopy. a) spread and polygonal cells in media and extracts of copolymer and ACP pellets. b) and c) contracted and spherical cells in extract of copolymer/ACP composites and CES composites, respectively.

In addition to the morphological changes, cells exposed to the extracts from ACP/UHMP composites and CES also showed slow proliferation. Cells with the polygonal shape proliferated approximately 2.5 times faster than those contracted and sphere-shaped cells in 24 h according to cell viability tests using MTT methods. (Figure 11, SD indicated by bars). Further testing (possibly modified cell viability tests and cell proliferation experiments) will be required to better understand cellular responses to both the experimental ACP composite intended for endodontic application and the commercial control sealer. To-date performed *in vitro* cytotoxicity tests nevertheless suggest that our bioactive, remineralizing ACP sealer is as good a candidate for the envisioned endodontic application as the chosen commercial control which was selected for testing as a representative of the contemporary endodontic materials with a resin matrix similar to the resin phase of the ACP/UPHM composite.

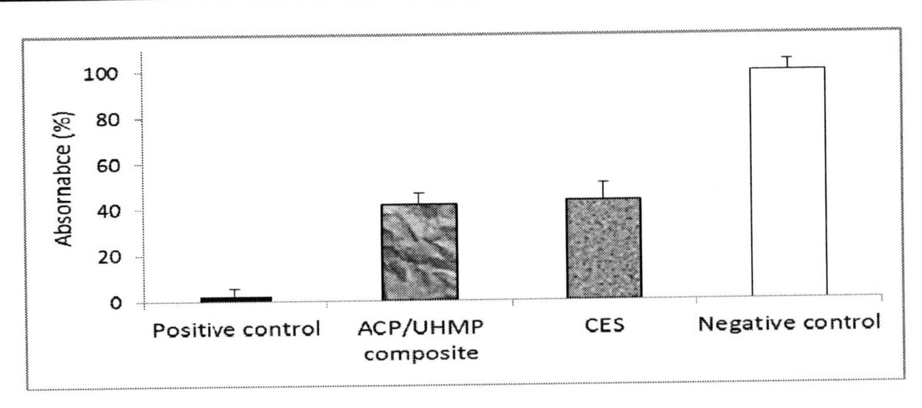

Figure 11. Cell viability (mean value + SD) for ACP/UHMP composite, commercial endodontic sealer (CES), negative control (medium only) and positive control (medium + detergent)..

Conclusion

The comprehensive physicochemical testing of the unfilled resins (copolymers) and their composites, quantitative leachability studies and the *in vitro* cellular responses to these materials are integral parts of our design of bioactive ACP-based composites.

The spontaneous and uncontrolled agglomeration of particles during the ACP's synthesis typically yields highly clustered ACP solid that disperses non-uniformly in matrix resins or cured composites. The more favorable, narrower size distribution of ACP particles is routinely obtained through the mechanical treatment of ACP filler. ACP's grinding and, especially, milling rather than ACP's surface modification with various additives, typically results in better dispersion of these mechanically homogenized ACP fillers within polymer matrices and yield remineralizing ACP composites with improved mechanical stability.

Light-cure Bis-GMA-, EBPADMA- and UDMA-based copolymers and their ACP composites regularly achieve high degrees of vinyl conversion (DVC) when the hydrophilic HEMA is included as a co-monomer in the resin at relatively high content (HEMA $\geq$ 28 mass %). These higher DVC values are attributed to HEMA's high diffusivity and mono-functionality. As a consequence of high DVC, the experimental ACP composites undergo high shrinkage upon polymerization (PS). Their PS values typically exceed values reported for the commercial composite materials. This phenomenon can be attributed in part to a much lower filler level in our experimental materials (40 mass % ACP) compared to highly filled (up to 85 mass %) in conventional glass composites. A possible ways to fabricate ACP composites with lower PS while maintaining satisfactory DVC would be the preparation of bulkier but relatively low viscosity experimental resins or inclusion of ring-opening monomers. In some formulations, such as UDMA/HEMA/MEP/PEG-U resins and their composites, relatively high PS is likely to be compensated by the significant hygroscopic expansion (HE) of these materials upon immersion in aqueous medium. Beneficial aspects of HE have indeed been demonstrated by other researchers for different composite materials.

In the case of ACP composite materials, besides the water-polymer interactions, strong water-ACP filler interactions additionally contribute to the overall water sorption (WS) profiles. Besides affecting the mechanical integrity of the composites, water diffusion affects the kinetics of the intra-composite ACP conversion to HAP and ultimately determines the remineralizing capacity of these materials. Generally, higher remineralizing potential is attained when the polymer phase contains EBPADMA or UDMA as base monomers and HEMA or HEMA plus PEG-U as co-monomers. The most probable reason for higher ion releases obtained with these base monomers is a more open cross-linked network structure of their resin matrix. Hydrophilic HEMA-enriched matrices generally increase internal mineral saturation by allowing the uptake of more water and/or better accessibility of ACP filler to the water already entrained.

In addition to material factors (filler type and/or content, resin type and composition, polymerization mode), processing factors such as cavity configuration (C-factor) need to be considered when assessing polymerization shrinkage stress (PSS) development in composites. However, the apparent discrepancies between the PS and PSS in relation to the composition of the resin matrix have yet to be resolved.

The chemical structure/property relationships of the monomers, compositional differences involving polymers and polymerization initiation systems, and the attainable DVC values are important factors that determine the cytotoxicity of the polymeric ACP composites. *In vitro* cytotoxicity tests comparing the experimental material with the representative commercial control are a good predictor of the new material's suitability for the intended applications and should be one of the main parameters when considering its recommendation for clinical trials.

The results of the above discussed studies are useful guidelines in designing remineralizing ACP composites for different dental utilities. The lessons learned from these studies have the potential of providing new insight(s) into physicochemical, molecular and cellular interactions that may be essential in the future development of calcium phosphate-based biomaterials intended for general bone regeneration.

Disclaimer

Certain commercial materials and equipment are identified in this article for adequate definition of the experimental procedures. In no instance does such identification imply recommendation or endorsement by the American Dental Association Foundation or the National Institute of Standards and Technology or that the material and the equipment identified are necessarily the best available for the purpose.

Acknowledgment

The reported studies were supported by the National Institute of Dental and Craniofacial Research through research grant DE 13169, by the American Dental Association Foundation and by the National Institute of Standards and Technology. Authors would like to thank Esstech, Essington, PA, USA, for donation of Bis-GMA, EBPADMA, UDMA, TEGDMA, HEMA and PEG-U monomers.

REFERENCES

[1] Dorozhkin, S.V. (2009). Calcium orthophosphate cements and concretes. *Materials*, 2, 221-291.

[2] Dorozhkin, S.V. (2009). Calcium orthophosphate-based biocomposites and hybrid materials. *J. Mater. Sci.*, 44, 2343-2387.

[3] Wang, L. & Nancollas. G.H. (2009). Pathways to biomineralization and biodemineralization of calcium phosphates: the thermodynamic and kinetic controls. *Dalton Trans.*, 15, 2665-2672.

[4] Elliot, J.C. Structure and chemistry of the apatites and other calcium orthophosphates. Amsterdam: Elsevier; 1994.

[5] LeGeros, R.Z. & LeGeros, J.P. (1996). Calcium phosphate biomaterials in biomedical applications. *Bioceramics*, 9, 7-10.

[6] Putlyaev, V.I. & Safronova, T.V. (2006). A new generation of calcium phosphate biomaterials: the role of phase and chemical compositions. *Glass & Ceramics*, 63(3-4), 99-102.

[7] Skrtic, D., Antonucci, J.M. & Eanes, E.D. (2003). Amorphous calcium phosphate-based bioactive polymeric composites for mineralized tissue regeneration. *J. Res. Natl. Inst. Stand. Technol.*, 108(3), 167-182.

[8] Antonucci, J.M. & Skrtic, D. Physicochemical properties of bioactive polymeric composites: Effects of resin matrix and the type of amorphous calcium phosphate filler. In: Shalaby SW, Salz U, editors. Polymers for dental and orthopedic applications. Boca Raton: CRC Press; 2007; 217-242.

[9] Skrtic, D. & Antonucci, J.M. Design, characterization and evaluation of biomimetic polymeric dental composites with remineralization potential. In: Lechov M, Prandzheva S, editors. Encyclopedia of polymer composites: Properties, performance and applications. Hauppauge: Nova Science Publishers; 2010; 281-318.

[10] JM Antonucci, J.M., Skrtic, D., Hailer, A.W. & Eanes, E.D. Bioactive polymeric composites based on hybrid amorphous calcium phosphate. In: Ottenbrite RM, Kim SW, editors. Polymeric drugs & drug delivery systems. Lancaster: Technomics Publ. Co., Inc.; 2000; 301-310.

[11] Langhorst, S.E., O'Donnell, J.N.R. & Skrtic, D. (2009). *In vitro* remineralization effectiveness of polymeric ACP composites: Quantitative micro-radiographic study. *Dent. Mater.*, 25, 884-891.

[12] Skrtic, D. & Antonucci, J.M. (2003). Effect of bifunctional co-monomers on mechanical strength and water sorption of amorphous calcium phosphate- and silanized glass-filled Bis-GMA-based composites. *Biomaterials*, 24(17), 2881-2888.

[13] Skrtic, D., Antonucci, J.M., Eanes, E.D. & Eidelman, N. (2004). Dental composites based on hybrid and surface-modified amorphous calcium phosphates – A FTIR microspectroscopic study. Biomaterials, 25, 1141-1150.

[14] Tung, M.S. & Eichmiller, F.C. (1999). Dental applications of amorphous calcium phosphates. *J. Clin. Dent.*, 10, 1-6.

[15] Reynolds, E.C., Cai, F., Shen, P. & Walker, G.D. (2003). Retention in plaque and remineralization of enamel lesions by various forms of calcium in mouthrinse or sugar-free chewing gum. *J. Dent. Res.*, 82(3), 206-211.

[16] Mazzaoui, S.A., Burrow, M.F., Tyas, M.J., Dashper, S.G., Eakins, D. & Reynolds, E.C. (2003). Incorporation of casein phosphopeptide-amorphous calcium phosphate into a glass ionomer cement. *J. Dent. Res.*, 82(110, 914-918.

[17] Wilson, A.D. & Nicholson, J.W. Chemistry of solid state materials. 3. Acid-base cements. Their biomedical and industrial applications. Cambridge: Cambridge Univ. Press; 1993; 56-102.

[18] Morgan, D.R., Kalachandra, S., Shobha, H.K., Gunduz, N. & Stejskal, E.O. (2000). Analysis of a dimethacryalte copolymers (Bis-GMA and TEGDMA) network by DSC and ^{13}C solution and solid-state NMR spectroscopy. *Biomaterials*, 21, 1897-1903.

[19] Stansbury, J.W. & Dickens, S.H. (2001): Network formation and compositional drift during photo-initiated copolymerization of dimethacrylate monomers. *Polymer*, 42, 6363-6369.

[20] Antonucci, J.M. & Stansbury, J.W. Molecularly designed dental polymers. In: Arshady, R, editor. Desk reference of functional polymers syntheses and applications. Washington DC: *American Chemical Society*; 1997; 719-738.

[21] Antonucci, J.M., Liu, D.W. & Stansbury, J.W. (1993). Synthesis of hydrophobic oligomeric monomers for dental applications. *J. Dent. Res.*, *72*, 369.

[22] 22. Asmussen, E. & Peutzfeldt, A. (1988). Influence of UEDMA, Bis-GMA and TEGDMA on selected mechanical properties of experimental resin composites. *Dent. Mater.*, 14, 51-56.

[23] 23. Morra, M. (1993). Acid-base properties of adhesive dental polymers. *Dent. Mater.*, 19, 375-378.

[24] Santerre, J.P., Shajii, L. & Leung, B.W. (2001). Retention of dental composite formulations to their degradation and the release of hydrolyzed polymeric-resin-derived products. *Crit. Rev. Oral Biol. Med.*, 12(2), 136-151.

[25] Vallet-Regi, M. & Gonzales-Calbert, J.M. (2004): Calcium phosphates as substitution of bone tissues. *Prog. Solid State Chem.*, 32, 1-31.

[26] Hench, L.L., Hynos, I.D. & Polak, J.M. (2004): Bioactive glasses for in situ tissue regeneration. *J. Biomater. Sci: Polym.* Ed., 15(4), 543-562.

[27] de Bruijn, J.D., Davies, J.E. & Klein, C.P.A.T. Biological responses to calcium phosphate ceramics. In: Ducheyne, P, Kokubo, T, van Blitterswijk, CA, editors. Bone bonding biomaterials. Leiderdorp: Reed Healthcare Com.; 1993; 57-72.

[28] deBruijn, J.D., Bovell, Y.P., Davies, J.E. & van Blitterswijk, C.A. (1994). Osteoclastic resorption of calcium phosphates is potentiated in postosteogenic culture conditions. *J. Biomed. Mater. Res.*, 28(1), 105-112.

[29] Midy, V., Dard, M. & Hollande, E. (2001). Evaluation of the effect of three calcium phosphate powders on osteoblast cells. *J. Mater. Sci.: Mater. Med.*, 12(3), 259-265.

[30] Knabe, C., Berger, G., Gildenhaar, R., Meyer, J., Howlett, C.R., Markovic, B. & Zreiqat, H. (2004). Effect of rapidly resorbable calcium phosphates and a calcium phosphate bone cement on the expression of bone-related genes and proteins *in vitro*. *J. Biomed. Mater. Res.*, 69A(1), 145-154.

[31] Siebers, M.C., Walboomers, X.F., Leeuwenburgh, S.C.G., Wolke, J.C.G. & Jansen, J.A. (2004). Electrostatic spray deposition (ESD) of calcium phosphate coatings, an *in vitro* study with osteoblast-like cells. *Biomaterials*, 25(11), 2019-20127.

[32] Wang, C., Duan, Y., Markovic, B., Barbara, J., Howlett, C.R., Zhang, X. & Zreiqat, H. (2004). Phenotypic expression of bone-related genes in osteoblasts grown on calcium

phosphate ceramics with different phase compositions. *Biomaterials*, 25(13), 2507-2514.

[33] Arinzeh, T.L., Tran, T., Mcalary, J., & Daculsi, G. (2005). A comparative study of biphasic calcium phosphate ceramics for human mesenchymal stem-cell induced bone formation. *Biomaterials*, 26(17), 3631-3638.

[34] Berube, P, Yang, Y., Carnes, D.L., Stover, R.E., Boland, E.J. & Ong, J.L. (2005). The effect of sputtered calcium phosphate coatings of different crystallinity on osteoblast differentiation. J. Periodontol., 76, 1697-1709.

[35] Bellows, C.G. (1992). Inorganic phosphate added exogenously or released from beta-glycerophosphate initiates mineralization of osteoid nodules *in vitro*. *Bone Miner*., 17, 15-29.

[36] Meleti, Z., Shapiro, I.M. & Adams, C.S. (2000). Inorganic phosphate induces apoptosis of osteoblast-like cells in culture. *Bone*, 27(3), 359-366.

[37] Adams, C.S., Mansfield, K., Perlot, R.L. & Shapiro, I.M. (2001). Matrix regulation of skeletal cell apoptosis – Role of calcium and phosphate ions. *J. Biol. Chem*., 276, 20316-20322.

[38] Dvorak, M.M., Siddiqua, A., Ward, D.T., Carter, D.H., Dallas, S.L., Nemeth, E.F. & Riccardi, D. (2004). Physiological changes in extracellular calcium concentration directly control osteoblast function in the absence of claciotropic hormones. *Proc. Natl. Acad. Sci. USA*, 101(14), 5140-5145.

[39] Anselme, K., Sharrock, P., Hardouin, P & Dard, M.(1997). *In vitro* growth of human adult bone-derived cells on hydroxyapatite plasma-sprayed coatings. *J. Biomed. Mater. Res*., 34(2), 247-259.

[40] Ishikawa, K., Miyamoto, Y., Yukasa, T., Ito, A., Nagayama, M. & Suzuki, K. (2002). Fabrication of Zn containing apatite cement and its initial evaluation using human osteoblastic cells. *Biomaterials*, 23(2), 423-428.

[41] Thian, E.S., Huang, J., Best, S.M., Barber, Z.H. & Bonfield, W. (2005). Magnetron co-sputtered silicon-containing hydroxyapatite thin films – an *in vitro* study. *Biomaterials*, 26(16), 2947-2956.

[42] Redey, S.A., Nardin, M., Bernache-Assolant, D., Rey, C., Delannoy, P., Sedel, L. & Marie P.J. (2000). Behavior of human osteoblastic cells on stoichiometric hydroxyapatite and type A carbonate apatite: role of surface energy. *J. Biomed. Mater. Res*., 50(3), 353-364.

[43] Annaz, B., Hing, K.A., Kaiser, M., Buckland, T. & Di Silvio L. (2004). An ultrastructural study of cellular response to variation in porosity in phase-pure hydroxyapatite. J. Microsc., 216(2), 97-109.

[44] Lu, X & Leng, Y (2003). Quantitative analysis of osteoblast behavior on microgrooved hydroxyapatite and titanium substrata. *J. Biomed. Mater. Res*., 66(3), 677-687.

[45] Chou, Y.F., Dunn, J.C.Y. & Wu, B.M. (2005). In vitro response of MC3T3-E1 preosteoblast within three-dimensional apatite-coated PLGA scaffolds. *J. Biomed. Mater. Res*., 75B(1), 81-90.

[46] Chou, Y.F., Huang, W., Dunn, J.C.Y., Miller, T.A. & Wu, B.M. (2005). The effect of biomimetic apatite structure on osteoblast viability, proliferation and gene expression. *Biomaterials*, 26(3), 285-295.

[47] Pelka, M., Distle, R.W. & Petshelt, A. (1999). Elution parameters and HPLC-detection of single components from resin composite. *Clin. Oral Invest*., 3, 194-200.

[48] Spahl, W., Budzikiewicz, H. & Geurtsen, W. (1998). Determination of leachable components from four commercial dental composites by gas and liquid chromatography/mass spectrometry. J. Dent., 26, 137-145.

[49] Schweikl, H., Schmalz, G. & Spruss, T. (2001). The induction of micronuclei *in vitro* by unpolymerized resin monomers. *J. Dent. Res.* 80(7), 1615-1620.

[50] Ratanasthien, S., Wataha, J., Hanks, C.T. & Dennison, J.B. (1995). Cytotoxic interactive effects of dentin bonding components on mouse fibroblasts. *J. Dent. Res.* 74, 1602-1606.

[51] Antonucci, J.M. & Skrtic, D. (2005). Matrix resin effects on selected physicohemical properties of amorphous calcium phosphate composites. *J. Bioact. Comp. Polym.*, 20, 29-49.

[52] Eanes, E.D. Amorphous calcium phosphate: Thermodynamic and kinetic considerations. In: Amjad Z., editor. Calcium phosphates in biological and industrial systems. Boston: Kluwer Academic Publ.; 1998; 21-39.

[53] O'Donnell, J.N.R. & Skrtic, D. (2009). Degree of vinyl conversion, polymerization shrinkage and stress development in experimental endodontic composite. *J. Biomim. Biomater. Tissue Eng.*, 4, 1-12.

[54] O'Donnell, J.N.R., Antonucci, J.M., & Skrtic, D. (2006): Amorphous calcium phosphate composites with improved mechanical properties. *J. Bioact. Compat. Polym.*, 21(3), 169-184.

[55] O'Donnell, J.N.R., Antonucci, J.M. & Skrtic, D. (2008): Illuminating the role of agglomerates on critical physicochemical properties of amorphous calcium phosphate composites. *J. Comp. Mater.*, 42(21), 2231-2246.

[56] Lee, S.Y., Regnault, W.F., Antonucci, J.M. & Skrtic, D. (2007). Effect of particle size of an amorphous calcium phosphate filler on the mechanical strength and ion release of polymeric composites. J. Biomed. Mater. Res., 80(B), 11-17.

[57] Nguyen, C.K., Kuang, W. & Brady, C.A. (2003). A new class of tertiary amines. RadTech Report, July/August, 32-38.

[58] ASTM F394-78 (re-approved 1991): Standard test method for biaxial strength (modulus of rupture) of ceramic substrates.

[59] O'Donnell, J.N.R., Langhorst, S.E., Fow, M.D., Skrtic, D. & Antonucci, J.M. (2008). Light-cured dimethacrylate-based resins and their composites: Comparative study of mechanical strength, water sorption and ion release. *J. Bioact. Compat. Polym.*, 23(5):207-226.

[60] Lu, H., Stansbury, J.W., Dickens, S.H., Eichmiller, F.C. & Bowman, C.N. (2004). Probing the origins and control of shrinkage stress in dental resin-composites: I. Shrinkage stress characterization technique. *J. Mater. Sci.: Mater. Med.*, 15, 1097-1103.

[61] Lu, H., Stansbury, J.W., Dickens, S.H., Eichmiller, F.C. & Bowman, C.N. (2004). Probing the origins and control of shrinkage stress in dental resin-composites: II. Novel method of simultaneous measurement of polymerization shrinkage stress and conversion. *J. Biomed. Mater. Res.: Appl. Biomater.*, 71B, 206-214.

[62] Malz, F. & Jancke, H. (2005). Validation of quantitative NMR. *J. Pharm. & Biomed. Anal.*, 38, 813-823.

[63] Pauli, G. F., Jaki, B. U. & Lankin, D. C. (2005). Quantitative ^{1}H NMR: Development and potential of a method for natural products analysis. *J. Nat. Prod.*, 68, 133-149.

[64] Diehl, B. W. K., Malz, F. & Holzgrabe, U. (2007). Quantitative NMR spectroscopy in the quality evaluation of active pharmaceutical ingredients and excipients. *Spectroscopy Europe*, 19(5), 15-19.

[65] Wells, R. J., Hook, J. M., Al-Deen, T. S. & Hibbert, D. B. (2002). Quantitative nuclear magnetic resonance (qNMR) spectroscopy for assessing the purity of technical grade agrochemicals: 2,4-dichlorophenoxyacetic acid (2,4-D) and sodium 2,2-dichloropropionate (dalapon sodium). *J. Agric. Food Chem.*, 50, 3366-3374.

[66] Simon, C.G. Jr., Antonucci, J.M., Liu, D.W. & Skrtic, D. (2005): *In vitro* cytotoxicity of amorphous calcium phosphate composites. *J. Bioact. Compat. Polym.*, 20(5), 279-295.

[67] Ishiyama, M., Shiga, M., Sasamoto, K., Mizoguchi, H. & He, P.G. (1993). A new sulfonated tetrazolium salt that produces a highly water-soluble formazan dye. *Chem. Pharm. Bull.*, 41, 1405-1412.

[68] Neter, G., Kutner, M.H., Nachtsheim, C.J. & Wasserman, W. Applied linear statistical models. 4th edition. Boston: WCB/McGraw-Hill; 1996.

[69] Amjad, Z. (2004). Inhibition of the amorphous calcium phosphate phase transformation reaction by polymeric and non-polymeric inhibitors. *Phosphorus Res. Bull.*, 7, 45-54.

[70] Ofir, P.B.Y., Govrin-Lipman, R., Garti, N. & Furedi-Milhofer, H. (2004). The influence of polyelectrolytes on the formation and phase transformation of amorphous calcium phosphate. *Cryst. Growth Design*, 4, 177-183.

[71] Ryal, R.L., Ryal, R.G. & Marshall, V.R. (1981). Interpretation of particle growth and aggregation patterns obtained from the Coulter counter. A simple theoretical model. *Inv. Urol.* 18, 396-399.

[72] Skrtic, D., Markovic, M., Komunjer, Lj. & Furedi-Milhofer H. (1986). Precipitation of calcium oxalates from high ionic strength solutions. IV. Testing of kinetic models. *J. Cryst. Growth*, 79, 791-796.

[73] Hansen, N.M., felix, R., Bisaz, S. & Fleish, H. (1976). Aggregation of hydroxyapatite. Biochim. Biophys. Acta, 451, 549-559.

[74] Nancolas, G.H. & Budz, J.A. (1990). Analysis of particle size distribution of hydroxyapatite crystallites in the presence of synthetic and natural polymers. *J. Dent. Res.*, 69, 1678-1685.

[75] Skrtic, D; Antonucci, JM; Eanes, ED; Brunworth, RT. (2002). Silica- and zirconia-hybridized amorphous calcium phosphate. Effect on transformation to hydroxyapatite. *J. Biomed. Mater.* Res. 59(4), 597-604.

[76] Quinn, J.B., Regnault, W.F., Antonucci, J.M. & Skrtic, D. (2005). Fractographic analysis of three ACP-filled resin systems. J. Dent. Res., 84 (Special Issue A), AADR Abstract No. 0585.

[77] Venhoven, B.A.M., de Gee, A.J. & Davidson, C.L. (1993). Polymerization contraction and conversion of light-curing Bis-GMA based methacrylate resins. *Biomaterials*, 14(11), 871-875.

[78] Labella, R., Lambrecths, P., Van Meerbeck, B. &Vanherle, G. (1999). Polymerization shrinkage and elasticity of flowable composites and filled adhesives. *Dent. Mater.*, 15, 128-137.

[79] Price, R.B., Rizkalla, A.S. & Hall, G.C. (2000). Effect of stepped light exposure on the volumetric shrinkage and bulk modulus of dental composites and unfilled resin. *Am. J. Dent.*, 13, 176-180.

[80] Guggenbarger, R. & Weinmann, W. (2002). Exploring beyond methacrylates. *Am. J. Dent.*, 13, 82D-84D.

[81] Tilbrook, D.A., Clarke, R.L., Howle, N.E. & Braden, M. (2000). Photocurable epoxy-polyol matrices for use in dental composites. *Biomaterials*, 21, 1743-1753.

[82] Arima, T., Hamada, T. &Mccabe, J.F. (1995). The effects of cross-linking agents on some properties of HEMA-based resins. *J. Dent. Res.*, 74(9), 1597-1601.

[83] Garcia-Fiero, J.L. & Aleman, J.V. (1982). Sorption of water by epoxide prepolymers. *Macromolecules*, 15, 1145-1149.

[84] Antonucci, J.M., Regnault, W.F. & Skrtic, D. (2010). Polymerization shrinkage and polymerization stress development in amorphous calcium phosphate/urethane dimethacrylate polymeric composites. *J. Comp. Mater.*, 44(3), 355-367.

[85] Skrtic, D & Antonucci, J.M. (2007). Dental composites based on amorphous calcium phosphate – Resin composition/physicochemical properties study. *J. Biomater. Appl.*, 21(4), 375-393.

[86] Momoi, Y. & McCabe, J.F. (1994). Hygroscopic expansion of resin based composites during 6 months water storage. Br. Dent. J., 176, 91-96.

[87] Huang, C., Tay, F.R., Cheung, G.S.P., Kei, l.H., Wei, S.H.Y. & Pashley, D.H. (2002). Hygroscopic expansion of a compomer and a composite on artificial gap reduction. *J. Dent.*, 30, 11-19.

[88] Labella, R., Lambrechts, P., VanMeerbeek, B. & Vanherle, G. (1999). Polymerization shrinkage and elasticity of flowable composites and filled adhesives. *Dent. Mater.*, 15, 28-137.

[89] Ferracane JL. (2005). Developing a more complete understanding of stresses produced in dental composites during polymerization. *Dent. Mater.*, 21, 36-42.

[90] Kinomoto, Y. & Torii, M. (1998). Photoelastic analysis of polymerization contraction stresses between self- and light-cured composites. *J. Dent.*, 26, 165-171.

[91] Venhoven, B.A.M., de Gee, A.J. & Davidson, C.L. (1996). Light initiation of dental resins: dynamics of the polymerization. *Biomaterials*, 176, 2313-2318.

[92] Braga, R.R. & Ferracane, J.L. (2002). Contraction stress related to degree of conversion and reaction kinetics. *J. Dent. Res.*, 81, 114-118.

[93] Calheiros, F.C., Braga, R.R., Kawano, Y. & Ballester, R.Y. (2004). Relationship between contraction stress and degree of conversion in restorative composites. *Dent. Mater.*, 20, 939-946.

[94] Stansbury, J.W., Trujillo-Lemon, M., Lu, H., Ding, X., Lin, Y. & Ge, J. (2005). Conversion-dependent shrinkage stress and strain in dental resins and composites. *Dent. Mater.*, 21, 56-67.

[95] Kleverlaan, C.J. & Feilzer, A.J. (2005). Polymerization shrinkage and contraction stress of dental resin composites. *Dent. Mater.*, 21, 1150-1157.

[96] Choi, K.K., Ruy, G.J., Choi, S.M., Lee, M.J., Park, S.J. & Ferracane, J.L. (2004). Effects of cavity configuration on composite restoration. *Oper. Dent.*, 29, 462-469.

[97] Uno, S., Tanaka, T., Inoue, S. & Sano, S. (1999). The influence of configuration factors on cavity adaptation in compomer restorations. *Dent. Mater.*, 18, 19-31.

[98] Feilzer, A.J., de Gee, A.J. & Davidson, C.L. (1990). Quantitative determination of stress reduction by flow in composite restorations. *Dent. Mater.*, 6, 167-171.

[99] Alster, D., Feilzer, A.J., de Gee, A.J. & Davidson, C.L. (1997). Polymerization contraction stress in thin resin composite layers as a function of layer thickness. *Dent. Mater.*, 13, 146-150.

[100] Choi, K.K., Condon, J.R. & Ferracane, J.L. (2000). The effects of adhesive thickness on polymerization contraction stress of composite. *J. Dent. Res.*, 79, 812-817.

[101] Watts, D.C. & Satterthwaite, J.D. (2008). Axial shrinkage stress depends upon both c-factor and composite mass. *J. Dent. Res.*, 24, 1-8.

[102] Munksgaard, E.C., Hansen, E.K. & Kato, H. (1987). Wall-to-wall polymerization contraction of composite resins versus filler content. *Scand. J. Dent.*, 95, 526-531.

[103] Antonucci, J.M., Giuseppetti, A.A., O'Donnell, J.N.R., Schumacher, G.E. & Skrtic, D. (2009). Polymerization stress development in dental composites: Effect of cavity design factor. *Materials*, 2, 169-180.

[104] Chatterjee, K., Lin-Gibson, S., Wallace, W.E., Parekh, S.H., Lee, Y.J., Marcus T. Cicerone, M.T., Young, M.F. & Simon, C.G. (2010). The effect of 3D hydrogel scaffold modulus on osteoblast differentiation and mineralization revealed by combinatorial screening. *Biomaterials*, 39(19), 5051-5062.

[105] Dalby, M.J., Gadegaard, N., Tare, R., Andar, A., Riehle, M.O., Herzyk, P., Wilkinson, C.D.W. & Oreffo, R.O.C. (2007). The control of human mesenchymal cell differentiation using nanoscale symmetry and disorder. *Nature Materials*, 6, 997-1003.

[106] Dalby, M.J., Biggs, M.J.P., Gadegaard, N., Kalna, G., Wilkinson, C.D.W., Curtis, A.S.G. (2007). Nanotopographical stimulation of mechanotransduction and changes in interphase centromere positioning. *J. Cell. Biochem.*, 100(2), 326-338.

Appendix 1. List of Acronyms Used throughout the Chapter

AA	ascorbic acid
ACP	amorphous calcium phosphate
ADAF	American Dental Association Foundation
am-ACP	as made ACP
ANOVA	analysis of variance
APTMS	3-aminopropyltrimethoxysilane
ASTM	American Society for Testing and Materials
BDT	Bis-GMA/DEGMEMA/TEGDMA resin
BFS	biaxial flexural strength
BHHm	Bis-GMA/HEMA/HmDMA resin
BHm	Bis-GMA/HmDMA resin
BHT	Bis-GMA/HEMA/TEGDMA resin
BGdT	Bis-GMA/GDMA/TEGDMA resin
BGmT	Bis-GMA/GMA/TEGDMA resin
Bis-GMA	2,2-bis[p-(2-hydroxy-3-methacryloxypropoxy)phenyl]propane
BMT	Bis-GMA/MEMA/TEGDMA resin
BmaT	Bis-GMA/MaA/TEGDMA resin
BMT	Bis-GMA/MA/TEGDMA resin
B4MT	Bis-GMA/4MET/TEGDMA resin

BPO	benzyl peroxide
BHTZ	Bis-GMA/HEMA/TEGDMA/ZrDMA resin
BT	Bis-GMA/TEGDMA resin
BVT	Bis-GMA/VPA/TEGDMA resin
C factor	cavity configuration factor
CaP	calcium phosphate
CC	chemical cure
CES	commercial endodontic sealer
COA	commercial orthodontic adhesive
CQ	camphorquinone
4265 Darocur	commercial photo-initiator system
DCPA	dicalcium phosphate anhydrous
DCPD	dicalcium phosphate dehydrate
DEGMEMA	di(ethyleneglycol)methyl ether methacrylate
DHEPT	2,2'-dihydroxyethyl-p-toluidine
d_m	median particle diameter
DVC	degree of vinyl conversion
EBPADMA	ethoxylated bisphenol A dimethacrylate
EDMAB	ethyl-4-N,N-dimethylamino benzoate
EDTA	ethylenediamine tetraacetic acid
EHHm	EBPADMA/HEMA/HmDMA resin
EHm	EBPADMA/HmDMA resin
EHMT	EBPADMA/HEMA/MEP/TEGDMA resin
EHT	EBPADMA/HEMA/TEGDMA resin
ET	EBPADMA/TEGDMA resin
EUHT	EBPADMA/UDMA/HEMA/TEGDMA resin
FAP	fluorapatite
FTIR	Fourier transform infrared spectroscopy
FTIR-m	FTIR micro-spectroscopy
ΔG^o	Gibbs free energy
g-ACP	ground ACP
GDMA	glyceryl dimethacrylate
GMA	glyceryl methacrylate
HAP	hydroxyapatite
HE	hygroscopic expansion
HEMA	2-hydroxyethyl methacrylate
HEPES	4-(2-hydroxyethyl)-1-piperazineethane sulfonic acid
HmDMA	hexamethylene dimethacrylate
IAP	ion activity product
1850 Irgacure	commercial photo-initiator system
369 Irgacure	commercial photo-initiator system
K_{sp}	thermodynamic solubility product
LC	light cure
MaA	maleic acid
MA	methacrylic acid
m-ACP	milled ACP

MCPA	monocalcium phosphate anhydrous
MEMA	2-methoxyethyl methacryalte
MEP	methcryloyloxyethyl phthalate
4MET	mono-4-(methacryloyloxy)ethyl trimellitate
MPTMS	methacryloxypropyltrimethoxysilane
NIR	near infrared spectroscopy
OCP	octacalcium phosphate pentahydrate
PAA	poly(acrylic acid)
PbTMBPO	phenyl bis(2,4,6-trimethylbenzoyl) phosphine oxide
PEG-U	poly(ethylene glycol) extended urethane dimethacrylate
PEO	poly(ethylene oxide)
PMGDMA	pyromellitic glycerol dimethacrylate
PRC	Paffenbarger Research Center
PS	polymerization shrinkage
PSD	particle size distribution
PSS	polymerization shrinkage stress
R	ideal gas constant
RH	relative humidity
SEM	scanning electron microscopy
SD	standard deviation
T	absolute temperature
TBPB	t-butyl perbenzoate
TCP	tricalcium phosphate
TEGDMA	triethylene glycol dimethacrylate
TGA	thermogravimetric analysis
TTCP	tetracalcium phosphate
UDMA	urethane dimethacrylate
UHHm	UDMA/HEMA/HmDMA resin
UHm	UDMA/HmDMA resin
UHT	UDMA/HEMA/TEGDMA resin
UHMP	UDMA/HEMA/MEP/PEG-U resin
UT	UDMA/TEGDMA resin
VPA	vinyl phosphonic acid
WS	water sorption
Wst1	mitochondrial dehydrogenase activity assay
XRD	X-ray diffraction
ZrDMA	zirconyl dimethacrylate

In: Resin Composites: Properties, Production and Applications ISBN: 978-1-61209-129-7
Editor: Deborah B. Song

Chapter 3

NANOSTRUCTURED ORGANOSILICATE COMPOSITES: PRODUCTION, PROPERTIES AND APPLICATION

E.F. Kudina **and G.G. Pechersky***
V.A. Belyi Metal-Polymer Research Institute of NASB,
32A Kirov Str., 246050 Gomel, Belarus

ANNOTATION

The present paper is devoted to organosilicate nanocomposites, representing interest as a new class of materials, with a high chemical resistance and mechanical durability in combination with optimal thermophysical properties, which provides effective increase of service characteristics of composite materials. Modern representations on obtaining and study of structure, chemical and physico-mechanical properties of organosilicates composites are considered.

The authors discuss technological problems of modifying soluble alkaline sodium silicate by organic reagents, physico-chemical processes of obtaining stable in aggregate sols based on organosilicate precursors, methods of realizing directed and controllable in time sol-gel transition of stabilized combined systems with the purpose of management of structure and properties of new multifunctional organosilikate nanocomposites. The technological features of formation of high-dispersed hybrid products from organosilicate solutions and homogeneous highly elastic gels are considered.

The possibility to control nanocomposite properties under thermal effect allowing to form highly elastic gels or highly dispersed xerogels is shown. This also concerns microwave heating that results in polymerization of silicate matrix. On the basis of the received experimental data the mechanisms of formation of organosilicate nanostructured materials and correlation of dependences are considered: structure and properties of the initial reagents → reaction technology of combining components → physical and chemical conditions of sol-gel transition → nanophase structure of a product → property of a material. The features of the nanophase structure of organosilicate composites depending on structure and technology of the formation product are considered. The properties of the initial components of combined systems and received hybrid nanocomposites are compared.

---

* E-mail: kudina_mpri@tut.by

The promises of using the developed hybrid fillers for modifying thermoplastic and lubricant materials, and homogeneous highly elastic gels as waterproofing screens, grouting solutions and sealants are shown. The efficiency of using hybrid fillers in thermoplastics and greases depending on their structure and composition is estimated. The influence of dispersed organosilicate fillers on polyolefins, polyamide 6, phenilone and PTFE operation properties is considered.

INTRODUCTION

One of the important trends in modern materials science is the development of composite materials for mechanical engineering as well as chemical industries. Such materials must have enhanced characteristics along with multifunctional capabilities. They must be functionally active, corrosion resistant, highly elastic, superstrong, etc. To make such materials, carbon compounds, metals, ceramics, glass-fiber-reinforced plastics and polymers are employed. Recently, much attention has been shown to the use of aqueous solutions of alkaline silicates (ASS) owing to their low cost, availability, incombustibility, no toxicity, and the fact that they meet the requirement for ecological safety.

Of a particular interest are organosilicate nanocomposites based on a silicate matrix. Modification of silicates by organic reagents allows to produce hybrid products, which show high chemical stability and mechanical strength in combination with optimum thermal properties [1-5].

The sol-gel method which can compatibilize starting reagents in solution is a most efficient means for making hybrid nanocomposites with a particular molecular structure and physicochemical properties [4,5]. The advantage of the sol-gel synthesis is that fragments of both organic and inorganic components can be varied on the molecular level [4,6,7] thus realizing a directional synthesis of organosilicate nanocomposites that can combine properties of ceramics and polymers.

While varying the nature of the organic components, as well as the thermodynamic parameters of the sol-gel process, the structure of the synthesized nanocomposites can be changed directionally [4,5,8-11].

There is no a single generally accepted classification of sol-gel processes at present. According to the known classification of J.D. Mackenzie [12], the sol-gel systems are divided into two groups: 1) processes of the first generation which lead to oxide composites; 2) processes of the second generation which lead to hybrid organo-inorganic nanocomposites.

The hybrid organo-inorganic nanocomposites formed by chemical bonding can be divided, according to U. Shubert's classification [13], into two classes with the following differences:

- organic and inorganic fragments are joined by stable chemical bonds;
- organic molecules are trapped into inorganic matrix networks of the gel formed, or on the contrary – inorganic molecules are trapped into interstructural zones formed by chains of the organic molecules. The inorganic and organic components can be only linked by weak physical bonds.

It should be mentioned that all products of the sol-gel synthesis are nanomaterials [4]. The process of gel formation provides for systems transition from loosely dispersed state to bound-dispersed, which results either in precipitation from solution of the ultra dispersed product, or in formation of a hybrid gel. This mode of nanocomposite production is called condensation [5]. A nanocomposite is made from inorganic and organic molecules, fragments, etc. in the course of phase transformation. While a new product is formed, chemical reactions usually take place and form a new phase, i.e. chemical transformations lead to a physical process – phase transition.

The functional application of the nanocomposite obtained depends on the properties of the modifying reagents which, during formation of sterically arranged structures, bonds (e.g., if sodium silicate solution is used as a precursor) sodium polysilicates, and simultaneously intercalates polymeric or oligomeric molecules and forms fine particles.

The sol-to-gel transition has been found advantageous for producing materials with a wide range of physicochemical properties: insulating nanocomposites [14-16]; highly-porous materials with nanopores, and materials with controllable pore size [17, 18]; block-type and dispersed organosilicate materials that contain nanosized oxide inclusions [5, 19-23]; specialty additives like pigments and chromophores [21, 24-27]; luminofores [7, 28-30], etc.

The present work deals with the technological particularities of producing, from ASS, homogeneous high-elasticity gels and hybrid high-dispersion products; their structure and basic properties; fields of application.

RESEARCH TECHNIQUE

The object of research were organosilicate nanocomposites produced by sol-gel process from compatibilized systems based on a colloidal alkaline silicate solution. The alkaline silicate solution was 53% aqueous solution of sodium silicate (ASS, n = 2.9; ρ = 1.49 g/cm^3). The ASS was modified by functionally reactive organic reagents (OR) of different structures: monomers – acrylic acid (Acr.A), acetic acid (Ac.A), glycerin (Gc), N,N'-meta(orto)-phenylene-bis-maleimide (PBMI), acryl amide (AA), and ε-caprolactam (CL); oligomer – epoxydiane resin ED-20 (ER); polymers – phenoloformaldehyde resin (PFR), polyamide (PA).

The compatibilized systems were prepared by mixing starting reagents under standard conditions and various proportions. Several hybrid compositions were first combined and then microwave-heated (MWH, frequency υ = 2463 MHz, time – 5 min, or until a constant weight was reached) with the purpose of activating the physicochemical interaction.

In order to obtain gels the composition sdution was subjected to isothermal influences at T = 60-90 °C until a homogeneous gel was formed throughout the volume. The time of gel formation (TGF) was determined as a time of full loss of flow ability of the system.

To obtain a dispersed precipitate, the mixed sol was kept in air for 2-24 h. The sols modified by monomers were matured, filtered and rinsed to produce inert filtrate. The sol modified by an oligomer was matured, subjected to coagulation in aqueous solution of salt of a polyvalent metal (e.g. Co^{2+}, Ni^{2+}, Fe^{3+} or Cu^{2+}). To complete the processes of physicochemical interaction, the nanocomposites formed were heat-treated at T = 120-160 °C. The xerogels prepared were dispersed and fractionated.

The prepared products were tested as fillers to polyolefins high-density polyethylene, HDPE; ultra-high molecular weight polyethylene, UHMWPE), polyamides (PA6 and phenylon S-2, PS, polytetrafluoroethylene, PTFE). The UHMWPE- and PS-based compositions were prepared by mixing dispersed components in a rotating electromagnetic field [31-33]. The UHMWPE- and PS-based materials were produced by compression moulding; those based on PA6 – by injection moulding; PTFE – by cold moulding followed by caking at T = 367÷377 °C.

The processes of physicochemical interaction in the mixed systems were studied using the IRS, DTA and TGA techniques. IR-spectra of the compositions prepared were recorded using FTIR-spectrophotometer NICOLET 5700 at 500-4000 cm^{-1} frequency range.

The thermal analysis of the samples was run on derivatograph Q-1500 D, type Paulik-Paulik-Erdey. The tests were run in ceramic crucibles in air at 25-1000 °C and heating rate of 5.0 °C/min. The size, condition and phase contents of the powders were determined by the transmission electron microscopy (TEM) using TEM 125. The phase contents were determined from electron-diffraction photomicrographs and results of X-ray phase analysis. Roentgenografic studies were performed the diffractometer DRON-3.8 and Cu-K_{α} radiation.

The degree of chemical cross-linking and physicomechanical properties of dispersed organosilicate composites were found following standard procedures. The true density was determined with a pycnometer, the apparent density – in bulk after shaking.

The strength characteristics (ultimate tensile stress at 0.2% ($\sigma_{0.2}$) strain and elastic modulus in compression) of the thermoplastics-based composite materials were determined on the tensile-testing machine FP-100.

The triboengineering characteristics of the composite materials were studied on friction-testing machine SMTs-2 at a sliding speed υ = 0.5 m/s and load 10 MPa. To estimate the application possibilities of dispersed organosilicate products as additives to lubricants, the triboengineering properties of lubricating compositions (LC) based on industrial oil I-40 thickened by ceresin (5 wt%) and filled with prepared nanocomposites (2.5 wt%). The triboengineering characteristics of LC were tested on friction machine SMT-1, using the shaft-sector arrangement at a sliding speed υ = 1.5 m/s. The lubrication was done by immersing the roller into the vessel containing the lubricating composition.

Preparation of Organosilicate Composites From ASS

The reactivity of ASS toward different organic substances, as well preparation of materials with a set of properties peculiar to both the organic and inorganic components, depends on the content of hydroxyl groups and coordination-unsaturated silicon atoms in the silicon-oxygen structures [7, 34]. It has been agreed-upon in the silicate chemistry, as well as in ASS chemistry, that the result of chemical interaction is difficult, and sometimes even impossible, to represent as chemical reactions in formulas; the product composition is sometimes so complicated that it is difficult to determine exactly. Interaction of ASS and numerous substances is a multistage process. This is explained, on the one hand, by complex and often unknown polymer content in the solution and its colloidal nature which have great influence on polymer transformations in the course of interaction and formation of reactive products usually of non-stoichiometric composition. On the other hand, this is mostly associated with the amorphous state of reactive products which have developed surface. In this case, the adsorptive phenomena play an important role.

The SiO_2 gel produced is a binding substance for modifier particles and influences cross-linking as well as sol-gel transition in silicate compositions [6, 35].

The ASS becomes stronger owing to dehydration [36] either after hydrolysable substances have been added or under the effect of heat. Under these conditions ASS gradually acquires a slightly stable state characterized by onset of structurization.

Numerous substances easily mix with ASS: cane sugar, glucose, glycerin, alkaline resinates, salts of fatly acid, etc. [7, 34, 37]. Animal and vegetable fats cause formation of stable emulsions. Many organic substances (e.g. acetone, aldehyde, phenol, chloralhydrate, creosote, gelatin, esters, fluorescin, rhodamin, etc.) cause the system to coagulate immediately [34, 38]. In order that modified ASS find practical application, the multi component solutions must be aggregation-stable and retain the stability with time.

It is not simple to study experimentally physicochemical processes in ASS-based systems because of their physical-and-chemical structure, which respond to even insignificant changes in the test conditions (e.g. environmental temperature, concentration of carbonic acid in the air, etc.). Understanding of the processes occurring during mixing organic reagents with ASS is also complicated by hydrolysis. The reaction equation for alkaline-silicate solutions should be treated, therefore, as an approximate scheme and not as an exact reproduction of quantitative proportions of components reacting with ASS.

Production of organosilicate nanostructurized composites is associated with the difficulty of realizing an intended process of sol-gel synthesis which implies compatilization of organic and inorganic components in solution with subsequent formation of an aggregation-stable system, because colloidal alkaline-silicate sols are thermodynamically unstable and capable of uncontrollable aggregation and coagulation; these must be accounted for when hybrid products are being formed.

The chemistry of silicate solutions differs in that the result of reagents interaction depends on both their chemical origin and reaction conditions along with numerous technological factors such as concentration of reagents and sequence of their mixing, mixing rate, time of exposure in the air, etc. [7, 37, 38]. Gel formation on the interface of the interacting or mixed phases is common to these phenomena [37]. This leads to complications when a reaction system undergoes homogenization; the role of diffusion processes, that prevents chemical interaction of reagents, becomes more important. Therefore, different technological procedures, used to ensure components interaction, may play a decisive role in the development of systems with required properties. Introduction of an organic modifier into an alkaline-silicate solution and realization of the sol-gel process under certain conditions allow to prepare high-dispersion hybrid products and monolithic structurized gels. The growth process of particles can be controlled by metal salts added to the system in a particular concentration. All materials prepared by the sol-gel process are nanostructurized ones.

Hybrid composites are prepared from ASS in the following sequence (Figure 1): preparation of starting solution → introduction of a modifying reagent → preparation of a stable compatibilized solution → maturing of a binary sol → gel formation → drying → heat treatment of the xerogel → dispersion → hybrid nanocomposite [31, 39].

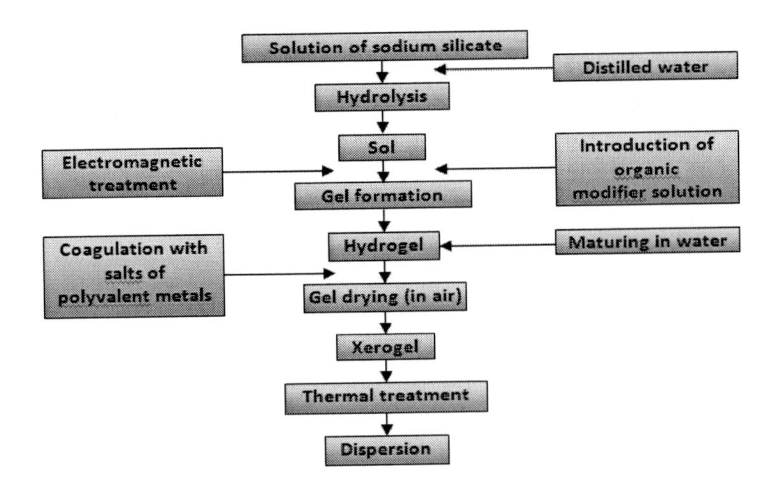

Figure 1. Diagram of sol-gel process to produce organosilicate nanocomposites.

An organic reagent can be components of various structures: AcrA, CL, PBMI, ER, PFR, etc.

Physicochemical method of analysis confirmed interaction between inorganic and organic components [31, 39, 40]. Probable reactions of producing silicates modified by organic reagents (e.g., oligomers, polymers, etc.) [39] are shown in Figure 2.

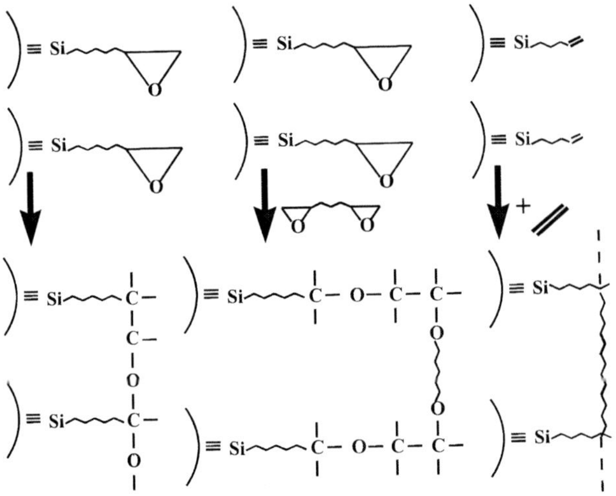

Figure 2. General route of reactions for organosilicate cross-linking.

Synthesis of organosilicate nanocomposites performed by this scheme ensures an opportunity to realize a series of physicochemical interactions between mixed reagent leading to bulky organosilicate and polysilicate matrixes or a high-dispersion hybrid product. All of the processes may take place simultaneously in the organosilicate solution, but may, more than likely, differ considerably in the rate of phase transition.

Fabrication of a hybrid particle from sodium silicate solution is represented schematically in Figure 3.

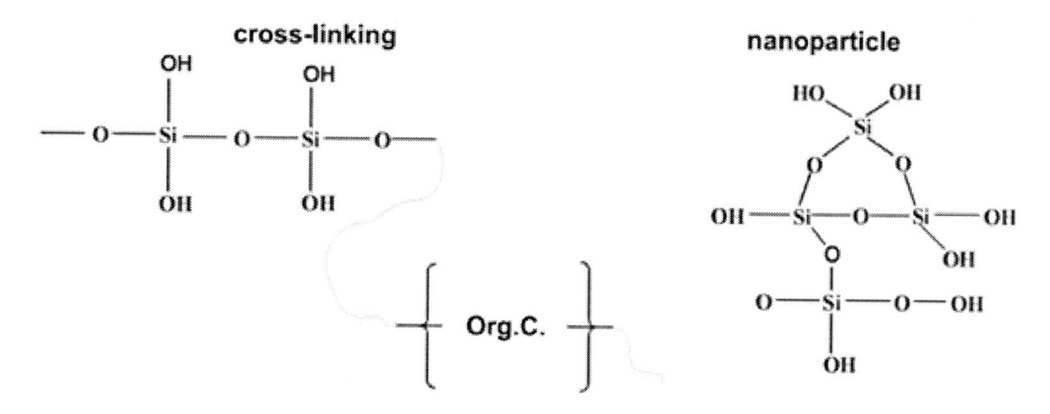

Figure 3. Schematic representation of structural arrangement organosilicate composites.

Physicochemical methods of testing ASS-organic component systems were used to establish that during the sol-gel process several interrelated physicochemical processes take place in organosilicate systems [40, 41]:

- formation of silicate matrix – as a result of hydrolysis and condensation of the silicate precursor – the polymerization degree of which depends on the conditions of sol-gel transition:
- formation of nanodispersed silica or hybrid particles;
- interaction of the organic and inorganic phases to form chemical bonds between the silicon-oxygen matrix and the functional group of the organic reagent.

Besides, in all of the cases it was found that with the introduction of polyvalent-metal-compounds into a multicomponent blend, a metal ion interacts with the silicon-oxygen matrix formed and intercalates in situ.

The sol-gel process allows to make bimetal (tri-, etc) silicas. Hybrid composites containing transition metals in the oxide networks are of particular interest because they can have a wide spectrum of special properties: magnetic, luminescent, chromophore, etc. In addition, the introduction of metals into the hybrid product matrix allows to use subsequently the electromagnetic effects to distribute homogeneously the prepared dispersed filler in a polymeric matrix.

ORGANOSILICATE GELS

The research showed that addition in certain quantifies of ASS-reagents of the acid group makes possible production of aggregation-stable solutions; these solutions, if the sol-gel process is run at elevated temperatures, form homogeneous bulky gels. It was established experimentally, that reduction of time for gel formation in binary systems leads to gels of higher strength (Figure 4)

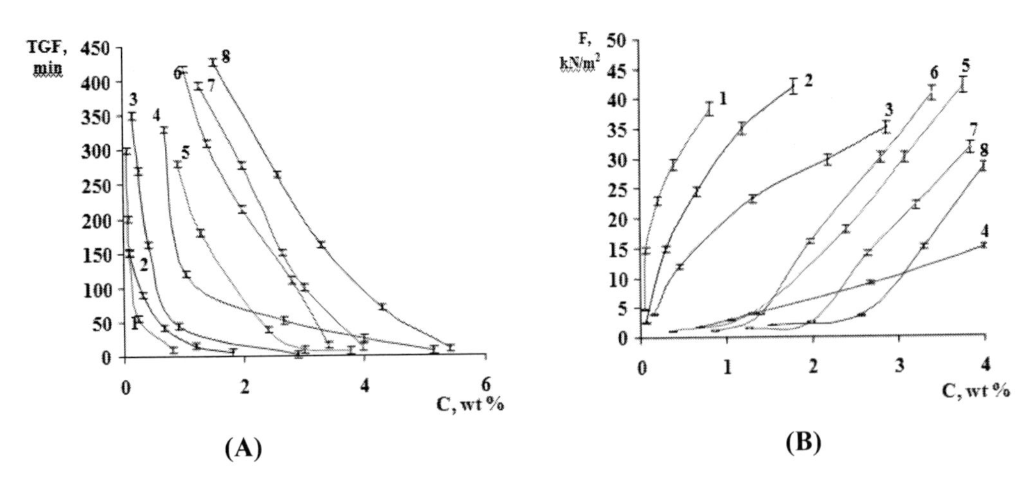

Figure 4. Dependence of time of gel formation (TGF) at 70 °C (a) and gel strength (b) on acid concentration in an ASS\acid system: 1 – sulphuric acid, 2 – hydrochloric, 3 – ortho-phosphoric, 4 – succinic, 5 – acetic, 6 – oxalic, 7 – maleic, 8 – acrylic acids.

The analysis of data obtained show that it is practically impossible to ensure a long for solution gel-formation and prepare strong gels using two-component systems. Higher concentrations of ASS, that provide for formation of the polysilicate phase and strengthen the gel, do not give a desired result, because the total reduction of an acid in the system prevents structurization and gel formation all through the volume.

This problem can be solved, on the one hand, by adding stabilizing substance to the mixed system; on the other hand, by polymerization of the organic modifier to form a polymeric matrix [42-45].

The introduction of glycerin, as a stabilizer, into an ASS/acetic base composition (Figure 5) appeared to increase much TGF (from 70 to 105 min) and obtain gels of strength up to 15.5 kN/m^2.

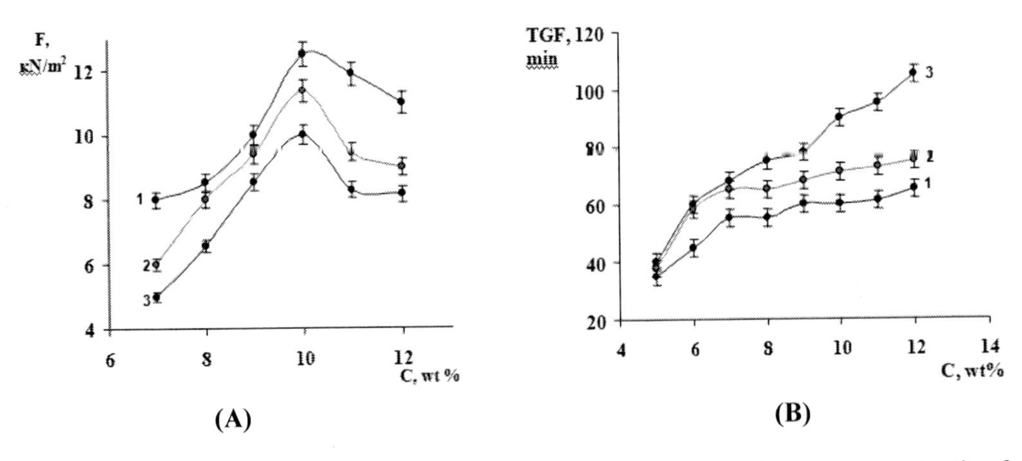

Figure 5. Dependence of gel strength (a) and TGF (b) on GL concentration. Component ratios for dehydrated ASS/acetic acid systems: 1 – 2.0/3.0; 2 – 2.0/2.9; 3 – 2.0/2.8.

In addition to all this, glycerin lowered the freezing range from -2 up to -10 °C for the composition, thus making it possible to use ASS/acetic acid/GL composition at lower temperatures.

To prepare compositions and gels with higher service characteristics, unsaturated carboxylic acid – acrylic acid (CH_2=CHCOOH) and its derived compound – acrylamide (CH_2=$CHONH_2$) were tested. These reagents cause a weak cross-linking effect on ASS, they mix well with water; form polymers: polyacrylic acid [-CH_2-CH(COOH)-]$_n$ or polyacrylamide [-CH_2CH($CONH_2$)-]$_n$, when the temperature is raised and a polymerization initiator (PI) is present.

For such systems stabilizers may be mineral acids (MA) in small concentrations. It was learned that highest values of TGF for the compositions were reached with HCl or H_3PO_4 acid (Figure6) with no decrease in the gel strength.

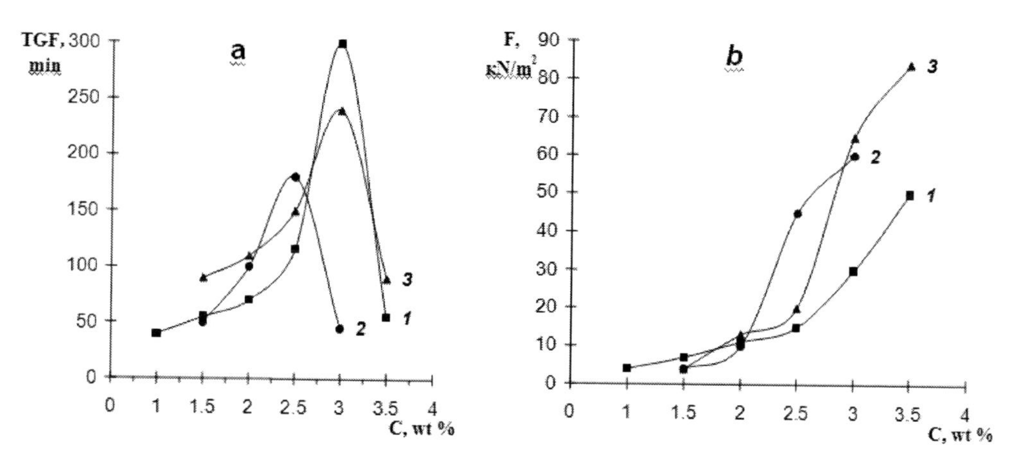

Figure 6. Dependence of TGF (a) for ASS/acrylic acid/MA/PI compositions and gel strength (b) on concentration of mineral acid (in % of dehydrated ASS in the systems) 1 – HCl; 2 – H_2SO_4; 3 – H_3PO_4.

An optimum component ratio in a composition ensures obtaining an aggregation-stable solution at elevated temperatures (60-90 °C) up to ~ 300 min; phase transition causes formation of homogeneous high-elastic strong (up to ~ 400 kN/m^2) gel capable of shape recovering after mechanical effects. Similar results were obtained with addition of AA to ASS.

The investigation of processes in blended get-forming compositions using the IR-spectroscopy technique showed that acid-base interaction of ASS and acids results in a silicon-oxygen matrix. Simultaneously, Acr. A. or AA undergoes polymerization to give polyacrylic acid or polyacrylamide. As a result, gel is formed that combines, first of all, properties of polymer flexible chains-which make the gel elastic, flexible and capable of shape recovery after application of mechanical effects; secondly, properties of silicon-oxygen matrix that uniformly strengthens the gel through the volume and increases the mechanical properties.

Microstructural studies revealed that the prepared gels have a homogenous structure (Figure 7) that depends on the modifier used.

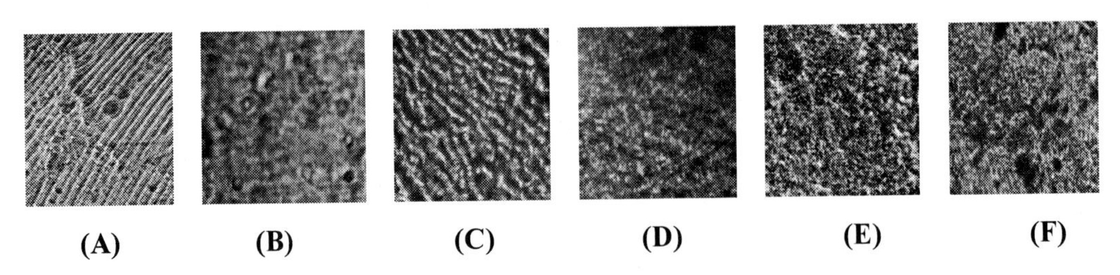

Figure 7. Structures of gels based on ASS modified by: acetic acid (a), acetic acid/GC (b); acrylic acid (c); acrylic acid/MA/PI (d), AA/MA/PI (e) in comparison with AKOR-BN102 (f).

The prepared gals are water-insoluble and have a high water-absorptive capacity (Figure 8). After being dried, the gel samples recover their initial shape, elastic properties and water-absorptive capacity while retaining this capacity up to 10 drying-saturation cycles.

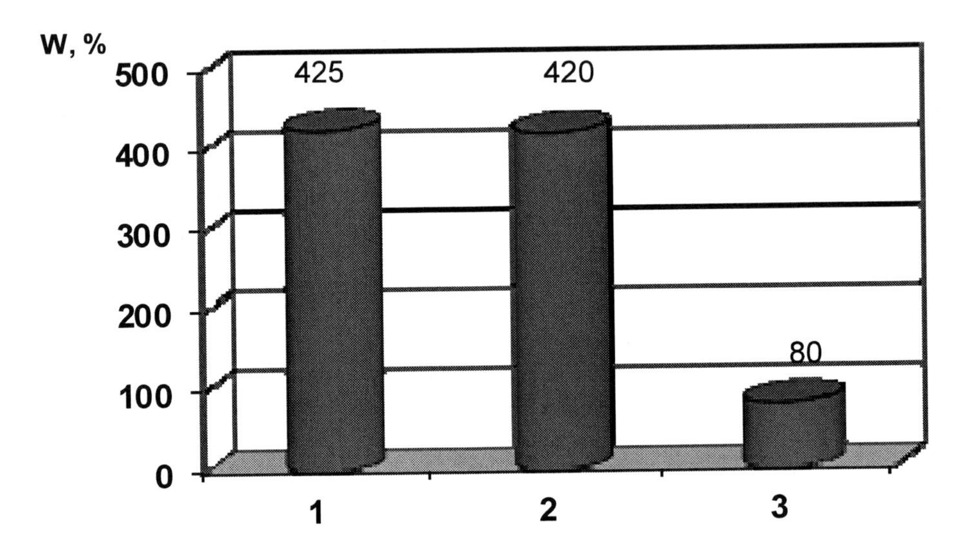

Figure 8. Water-absorptive rate of organosilicate gels: 1 – ASS/Acr. A./MA/PI; 2 – ASS/AA/MA/PI; 3 – AKOR-BN102.

Gel-forming organosilicate materials can be efficiently used in repair-insulating work when supplying water to production wells of oil-and-gas industry, for hydro insulation of underground constructions (tunnels, main pipelines, etc.) [43, 45].

Organosilicates used to make gel compositions are not corrosion-active substances; their corroding activity, toward steel, is 1.2 to 17.5 times as low as that of gel-forming insulation solution based on tetraethoxysilane AKOR-BN102 (Russia), which is widely used in oil industry.

The developed get-forming compositions do not degrade construction materials (e.g., cement mixes, bricks, plaster), but on the contrary, they favor their strength (Figure 9)/

When a construction materials gets in contact with a gel-forming composition, the strength of grouting cement increases 2.6 times, that of red bricks – 2.0 times, silica bricks – 1.2 times, laying cement – 1.3 times against AKOR-BN102 material.

Besides, gel-forming organosilicate compositions can increase the adhesion strength of material-gel-material bonds (Table 1).

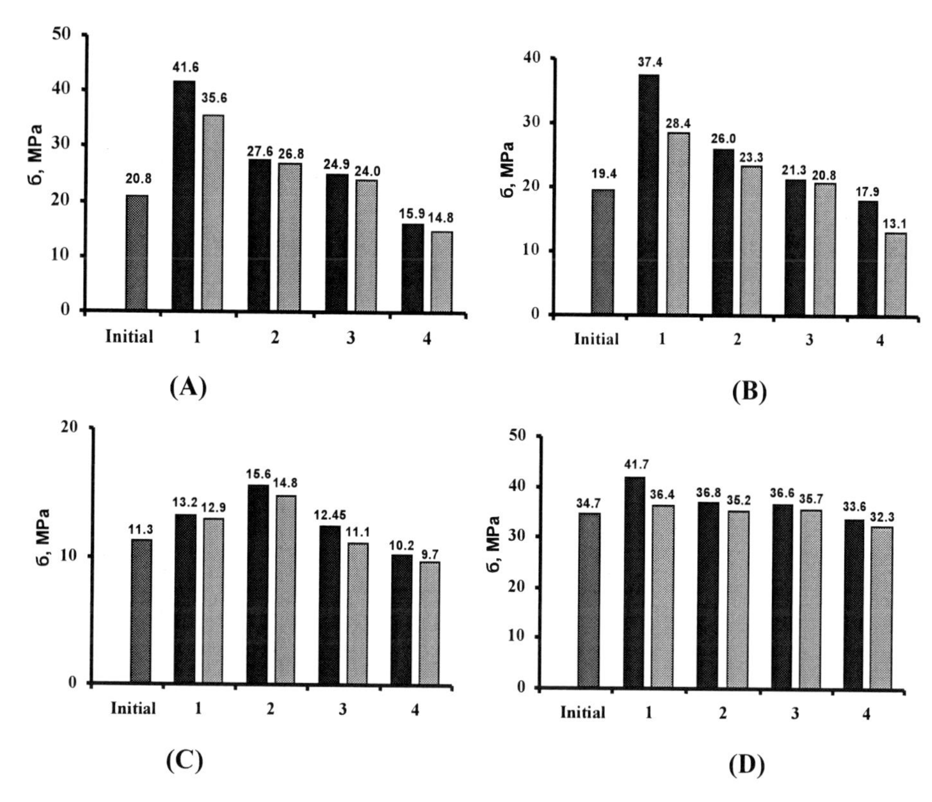

Figure 9. Effect of gel-forming compositions (1 – WSS/acrylic acid/MA/PI; 2 – ASS/AA/MA/PI; 3 – ASS/acetic acid/Gl; 4 – AKOR-BN102) on the strength of mineral construction materials: a – grouting cement; b – ceramic bricks; c – laying cement; d – silicate bricks.

Table 1. Adhesion strength* of material-gel-material bonds

Adhesive Composition	Contact Materials					
	Aluminum	Titan	PE	PP	PVC	Wood
ASS/acryl acid/MA/PI	3.89*	128.80	0.01	0.34	0.01	3.86
ASS/AA/MA/PI	4.40	134.20	0.01	0.17	0.01	3.20
Neutral silicon	2.54	105.40	0.01	0.30	0.01	3.10

Notes: PE – polyethylene; PP – polypropylene;
* Adhesion strength (N/m^2) of a material-gel-material bond.

The organosilicate gels prepared are on a par with the silicon hermetic in the adhesion strength when used in bonds with polymer materials; the former increase contact interaction to aluminum (by 53-73%), titan (22-27%), wood (up to 24%). The compositions developed combine high adhesion and low corrosion activity to metals, which prompts their application as sealing materials for metallic structures.

Dispersed Organosilicate Products

Table 2 lists experimental data on properly changes in dispersed organosilicate composites depending on modifiers and forming conditions.

Table 2. Property evolution for dispersed organosilicate composites

Nos	OrgC	Ratio of WSS/ OrgC***, %	Heat treatment, °C	Gel fraction, %	Density****		Porosity, %
					bulk, g/cm^3	true, g/cm^3	
1	–	100/0	120	0	0.78	2.2	66
2***	–	100/0	MWH	0	0.79	2.14	63
3****	–	100/0	MWH	0	0.86	2.07	58
4	Acr.A	10/1	25	44.0	0.71	2.15	67
5	Acr.A	10/1	120	17.7	0.67	2.06	67
6	Acr.A	10/1	160	25.7	1.0	1.94	48
7**	Acr.A	10/1	160	4.3	0.94	1.75	46
8****	Acr.A	10/1	160	0	0.63	1.61	61
9	CL	10/1	н.у.	0	0.71	2.19	68
10****	CL	10/1	MWH	0	0.54	2.45	78
11	CL	5/1	25	46.0	1.1	2.07	78
12	CL	2/1	25	68.3	0.13	1.96	93
13	ER	4/1	25	42.7	0.35	1.72	80
14	ER	4/1	120	48.7	0.45	1.79	75
15	ER	4/1	160	65.7	0.61	2.07	71
16**	ER	4/1	MWH, 120	14.0	0.45	1.82	75
17****	ER	4/1	MWH, 120	62.0	0.36	1.71	79
18**	ER	4/1	MWH, 160	26.5	0.52	1.90	73
19****	ER	4/1	MWH, 160	68.0	0.50	1.70	71
20	PBMI	25/1	25	10.0	0.53	2.11	75
21****	PBMI	25/1	MWH	57.0	0.45	2.1	79
21	PBMI	25/2	25	33.0	0.46	1.94	76

* Components ratio is for dehydrated systems;

** Dispersed fraction size d<50 μm;

*** Microwave-treated system (t=5 min);

**** Microwave-treated system heated to complete dehydration.

The experimental results show that the introduction of a modifier in WSS and formation of dispersed hybrid particles at certain time-temperature regimes result in a wide variation of powder properties [31, 39, 40]. The products obtained easily undergo dispersion and have low hydroscopicity. A high proportion of voids between particles almost completely excludes particle agglutination and is supported by low caking of pocoders in storage. The gel-fraction of the prepared products varies between 0 and 68%. High values of chemical cross-linking in the parent reagents led to products of high thermal stability (Table 3).

Table 3. Thermal properties of organosilicate nanocomposites

Composition	T_0*, °C	T_5, °C	T_{10}, °C	T_{20}, °C
WSS–AcrA	30	40	50	110
WSS –CL	27	76	117	703
WSS –ER	75	242	365	640
WSS –PBMI	82	164	289	647

* Onset temperature and weight loss, %: 5, 10, 20.

The IR-spectra of xerogels have wide intensive bands at 900-1200 cm^{-1} resulted from internal vibrations of atoms in tetrahedrons $[SiO_4]^{4-}$ and implies a polymerization process in the system [39, 40]. The widening in the spectra at 900-1200 cm^{-1}, in comparison with the parent silica spectrum, is explained by structural groupings with similar vibration frequencies, i. e. a longer silicon-oxygen chain along with transition to a more complex structure. The MWH gives a band with a maximum at ~ 1020 cm^{-1} that is indicative of cyclic polysilicates, while at ~ 850 and 866 cm^{-1} it is indicative of a greater proportion of Si-O-Si bonds; this is explained by the fact that silicon-oxygen chains of different lengths participate in formation of hybrid-product structure.

The IR-spectroscopy allowed to establish that between inorganic and organic components polycondensation processes take place in which carboxyl –COOH, siloxyl –SiOOH or silanol ≡Si–OH groups participate to form Si–O–C bonds [15-18]. IR-spectral analysis of MWH-treated products showed that the intensity of absorption bands at 900-1200 cm^{-1} grows, indicating a stronger chemical interaction.

Structural studies of the composites by the TEM technique revealed that the sol-gel transition in the parent WSS under standard regimes result in polymeric phases: α-$Na_2Si_2O_5$ and β-$Na_2Si_2O_5$, with SiO_2 inclusions, the particle size being 20-25 nm. In formation of WSS-xerogel during MWH-treatment, the composition of the base polymeric phase remains, in fact, unchanged; SiO_2 nanocrystals are formed followed by a reduction in the crystal lattice parameters [46].

The hybrid nanocomposite structure can be described as a mixed hybrid structure (Figure 10), where regions with dispersed amorphous and crystalline particles (β-$Na_2Si_2O_5$, SiO_2) of size between 5 and 50 nm are present, along with intercalated polymer-silicate nanoclusters [39, 40].

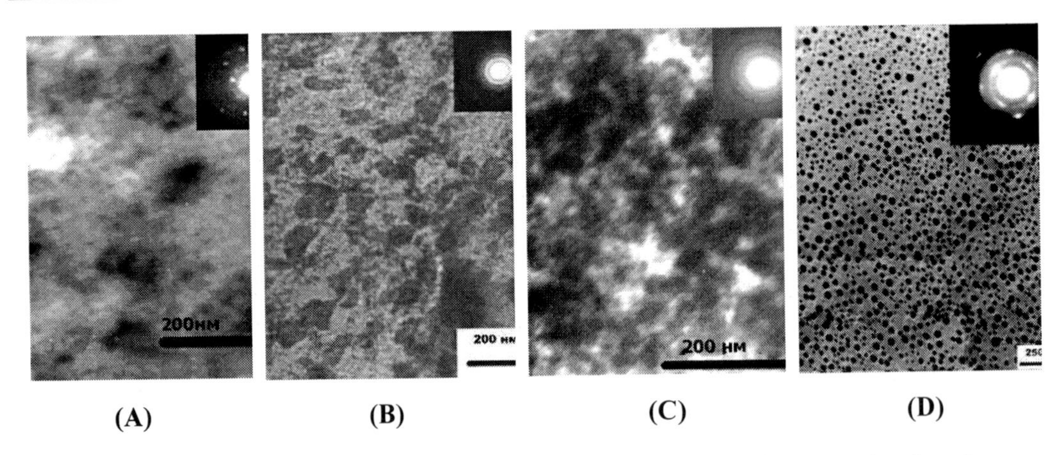

Figure 10. TEM-photomicrographs and micro diffraction patterns of respective xerogel regions from systems: a – WSS/Acr.A; b – WSS/CL; c – WSS/ER; d – WSS/PBMI. Modifier concentration = 10 wt%.

It was learned experimentally that additional MWH-treatment of systems during sol-gel transition causes the silicon-oxygen matrix to polymerize ($Na_2Si_3O_7$ and $Na_6Si_8O_{19}$ phases are formed); the degree of crystallization rises; the dispersed-phase inclusions become much smaller in size, below 15 nm. Figure 11 compares structures of composites produced from organosilicate systems subjected to additional MWH-treatment.

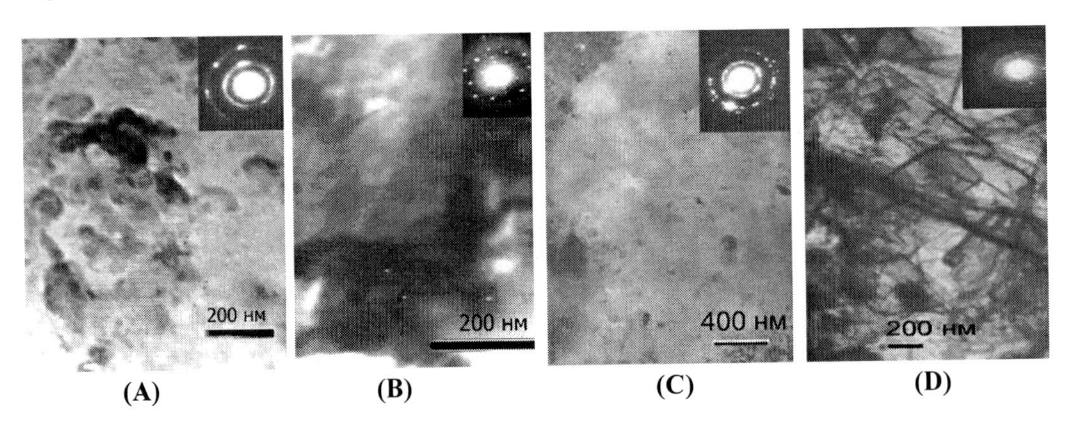

Figure 11. TEM-photomicrographs and microdiffraction patterns of respective xerogel regions; xerogels were prepared under MWH-conditions from systems: a – WSS/Acr.A; b – WSS/CL; c – WSS/ER; d – WSS/PBMI.

The MWH-treatment promotes formation of composites with a more homogeneous structure; exeption is WSS/PBMI composition in which fibers and spherulites are formed.

Changes in the structure and properties of nanostructurized nanocomposites caused by polyvalent metal compounds were investigated in WSS/ER. IR-spectra of xerogels prepared from WSS/ER with different salts are shown in Figure 12.

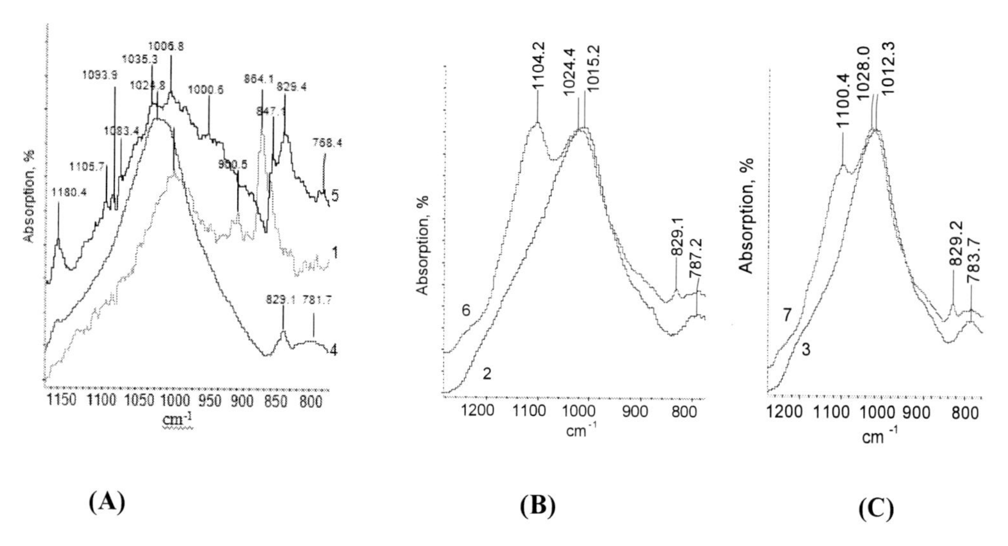

Figure 12. IR-spectra of xerogels: WSS (1), WSS/$CoSO_4$ (2), WSS/$NiSO_4$ (3) and WSS/ER, modified: 4 – $FeCl_2$; 5 – $Cu(HCOO)_2$; 6 – $CoSO_4$; 7 – $NiSO_4$.

The IR-spectra show that the salt added to the WSS/ER system influences considerably the formed product structure; cation's influence is the greatest. The spectra of all xerogels have a wide band at 900-1200 cm^{-1} caused by valent asymmetrical vibrations of silica and oxygen atoms in Si–O–Si bridge bonds. The higher band intensity of the composites, compared with WSS xerogel (Figure 12a, curve 1), can be explained by a higher proportion of Si–O–Si bonds, as the process of structure-formation proceeds along with formation of inorganic networks. The shift of the main peak to the higher-frequency spectral region is indicative of silica polymerization. The shift of the main band in all spectra of compositions in which the metal has been intercalated into an epoxy-silicate matrix (Me→ESM) composition results, probably, from the intercalated metal: 24.2; 23.8;27.4; 6.2 (as compared with IR-spectrum of WSS-xerogel) and 8.4; 9.2; 15.7 and 0 cm^{-1} (as compared with IR-spectrum of Me-WSS composition), with Fe, Co, Ni and Cu, respectively. Such band shifts can be explained by metal participation in composite formation and incorporation of a silicate matrix as fragment of Si-O-Me into its structure [47, 48].

The diffraction patterns of WSS-xerogel and cured ER (amorphous halo at d=18.9 A) indicate their amorphous structure. The X-ray analysis showed that in all samples of Me-ESM composition the crystallinity increased independently of parent metal salts, or formation of respective silicates. The marked changes in the structure of the three-component systems obtained can be explained as follows. The ER oligomers contain π-conjugated electrons in the aromic ring of molecules that are separated by maleic bridges to break bonds in the molecules. When a system is supplied with ions of transition metals containing unpaired electrons, an interaction takes place with π-electrons of aromatic rings, which strengthens the bond and favors denser packing of molecules. In that way, a self-arranging process of nanophases occurs in the systems under the affect of added metals, which is confirmed by a noticeable dependence of the structures formed on the type of added metal (i.e., its electron configuration).

Electronic photos of xerogel microstructures in accord with intercalated metal are shown in Figure 13 [49].

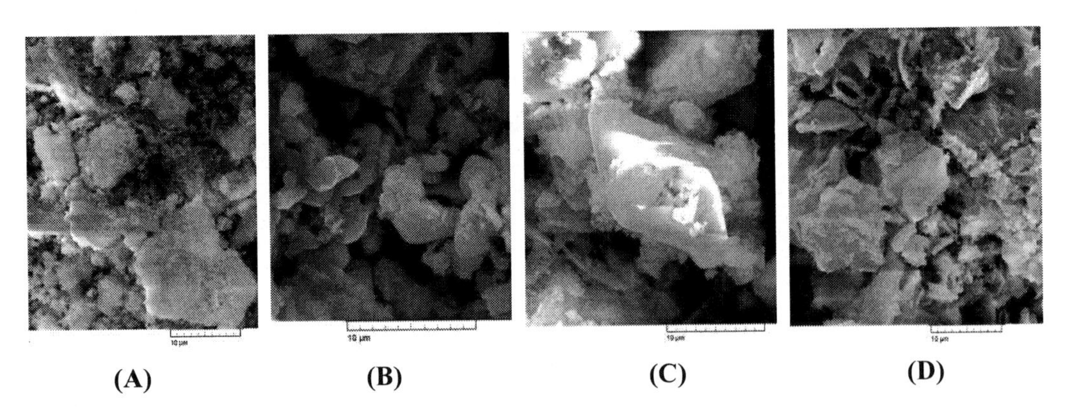

Figure 13. Microstructures of xerogels based on epoxysilicate matrix intercalated by metal (from respective salt of metal sulphate): a – iron; b – cobalt; c – nickel; d – copper.

The microstructural analysis showed that during phase transition – if a metal sulphate is added to an organosilicate gel – a hybrid matrix results being layered spherulites of quite a regular form; their size depends on the incorporated metal.

Substitution of acid anion causes a sharp change in the dispersed particles formed. With addition of chlorides (Figure 14 a and b), bulky homogeneous rounded agglomerates are formed without phase interfaces; formiates lead to elongated "fibrouse" formations (Figure 14 a and b).

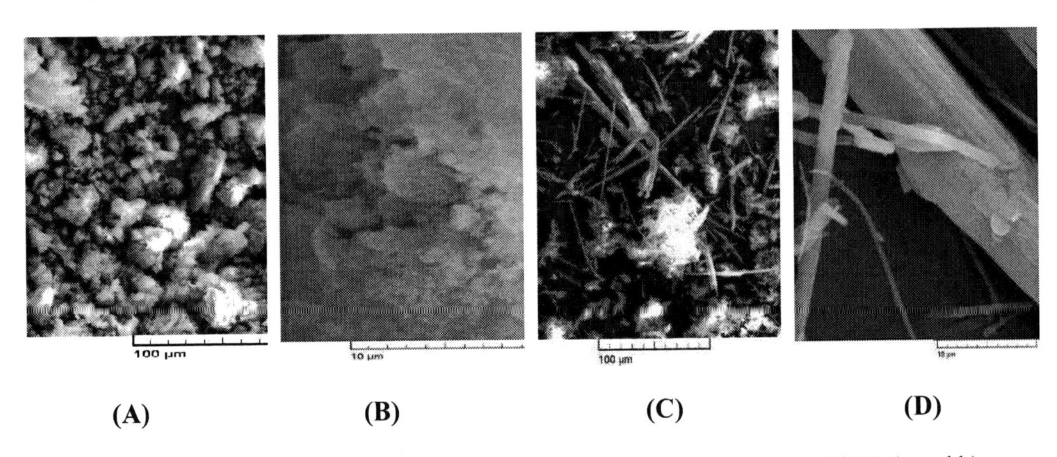

Figure 14. Microstructures of xerogels produced from compositions: $FeCl_2 \rightarrow$(WSS/ER) (a and b); $Cu(HCOO)_2 \rightarrow$(WSS/ER) (c and d).

The properties of high-dispersion hybrid products according to transition metal ions are listed in Table 4.

The analysis of data in Table 4 shows that composites made by the sol-gel procedure have high values of gel-fraction, which confirms components interaction in the systems. The physicochemical interaction of components that leads to products of networked structure with intercalated metallosilicate and organic fragments explains the high values of thermal stability shown by prepared nanocomposites (Table 5).

Table 4. Properties of metallo-organosilicate nanocomposites

Composition*	Density**		Porosity, %	Gel fraction, %
	in bulk, g/cm^3	true, g/cm^3		
Cr→ ESM	0.28	1.94	85.6	59.8
Fe→ ESM	0.29	1.53	81.1	65.7
Co→ ESM	0.31	1.48	79.1	69.2
Ni→ ESM	0.33	1.43	76.9	72.5
Cu→ ESM	0.41	1.41	70.9	83.5

* metal incorporated in epoxysilicate matrix; organic component concentration in system = 6 wt%
** dispersed fraction size d<50 μm.

Table 5. Thermal properties of organosilicate nanocomposites

Composition	T_0^*, °C	T_5, °C	T_{10}, °C	T_{20}, °C
Mn**→(WSS + 12 wt % ER)	50	330	483	931
Fe→(WSS + 12 wt % ER)	50	317	398	722
Co→(WSS + 12 wt % ER)	50	301	390	640
Cu→(WSS + 12 wt % ER)	75	242	365	650
Co→(WSS + 6 wt % ER)	50	288	322	630
Cu→(WSS + 6 wt % ER)	60	140	288	637

* Temperatures of: onset and 5, 10 and 20 % weight loss;
** Metals added from salts: $MeSO_4$ (where Me = Mn, Fe, Co) and $Cu(CHOO)_2$

A wide variation in structural features, along with physicochemical properties, of metallo-organosilicate and organosilicate nanostructurized composites allowed to anticipate efficiency of their use as functionally active fillers for thermoplastics as well as additives to lubricating materials. As fillers, the products are used as produced by WSS modification with acrylic acid (acrylosilicate filler, AcrSF), acetic acid (acetic-silicate filler, AcSF), imide PBMI (imidosilicate filler, ISF), ε-caprolactam (caprolactamosilicate filler, CSF), epoxydiane resin (epoxysilicate filler, ESF), phenoloformaldehyde resin (phenoloformaldehyde-silicate filler, PSF), polyamide (polyamide-silicate filler, PASF).

The investigation results on the effect of hybrid organosilicate fillers, added to the lubricating composition (LC) [50, 51], on coefficient of friction for steel surfaces lubricated by LC under testing are depicted in Figure 15.

The tests showed that triboengineering properties of LC depend, first of all, on the organic phase in the organosilicate filler [50]. The incorporation of organosilicates in LC (curves 3 to 5) causes a substantial reduction in the friction coefficient (at loads up to 30 MPa) in comparison with LC containing ultradispersed sialon (curve 1). The lowest coefficient of friction was reached with ESF. Under moderate loading regimes (<10 MPa), the lowest friction coefficient was observed with AcSF, the load-carrying capacity of the separating layer, however, was not high; with rising load – the friction coefficient becomes higher. AcrSF leads to a somewhat higher coefficient of friction for LC as compared with ESF or AcSF. This can probably be explained by end polar carbonyl groups and their interaction with polar groups of I-40-oil components. Even under severe loading (≥10 MPa) AcrSF is capable of preventing surface sticking, and the friction process remains quite steady [50].

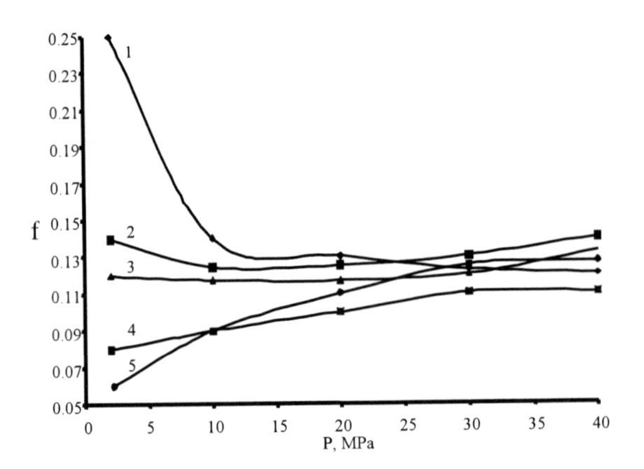

Figure 15. LC friction coefficient (v=0.5 m/s; S_f=0.5 cm^2) versus load and filler (2.5 wt%): 1 – sialon; 2 – WSS-xerogel; 3 – AcrSF (AcrA – 5 wt%); 4 – ESF (Cu→ESM, ES-concentration – 12 wt%); 5 – AcSF (AcA-concentration – 6.5 wt%). Initial composition I-40 at P>40 MPa is inefficient.

Variations in the friction coefficient of LC with load are explained by filler behavior. Sialon [52], for example, behaves as follows: ultradispersed spherical particles prevent a direct contact between the metallic surfaces; they act similar to microbearings; as a result, friction mode mixed (sliding and rolling) and ensures steady operation of the friction pair, especially under high contact loads. Some workers [53] explain the behavior of nanodispersed fillers, particularly sialon, by the fact that particles possess inherent or acquired uncompensated charge and form "charge clusters that promote lubricant structurization. This results in a lubricating layer of higher load-carrying capacity with simultaneous achievement of the thixotropic effect giving a low friction coefficient.

The performance of designed fillers can probably be explained by step-like decay of hybrid additives, as well as formation of a stable separating layer on the friction surface that provides for high performance capability of LC under severe regimes with a lower friction coefficient. The decay of a hybrid product in friction is associated with higher contact pressures. Because of filler degradation, the nanoparticles dispersed in the organosilicate matrix become distributed in the LC and start to act as ball bearings. Higher triboengineering properties of LC with added ESF is, evidently, associated with the incorporated metal (copper) that leads to a separating layer of higher load-carrying capacity as a result of copper compounds present in the system, as well as a plating action of the metal, That weakens the intensity of oxidizing reactions in the friction zone.

Test result for LC filled with ESF versus organic phase concentration and hybrid product thermal treatment are shown in Figure 16.

The experimental data show the lowest friction coefficient was reached with ESC containing ES-12 wt% and treated at 120 °C [50].

The study of AcrSF and ESF as fillers to lubricating composition (the formers were formed with additional electric current- or MWH-treatment (Figure 17) and subsequent thermal treatment) revealed that a more efficient way of treating binary sols is the electric current procedure (as a more severe effect) in comparison with MWH used to make a filler; the incorporation of this filler in LC causes a greater reduction in the friction coefficient (especially with ESF). This is probably associated with the fact that an additional electromagnetic treatment of the composite on the mixing stage of components, ensures most favorable conditions for making products with uniform distribution of both organic and inorganic phases.

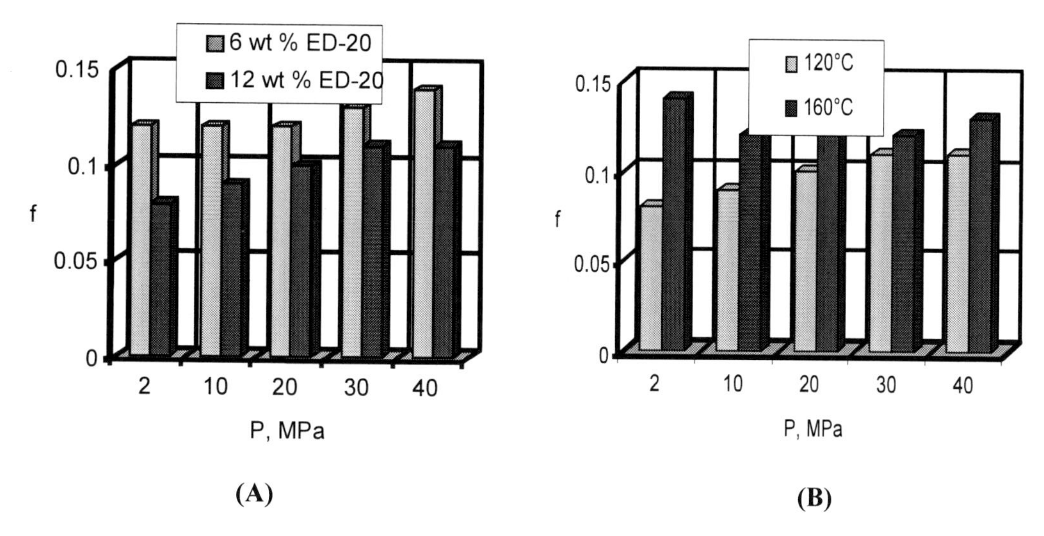

Figure 16. Coefficient of friction versus load for LC filler with Cu→ESM: a – WR concentration in the filler (treatment at T=120 °C); b – treatment temperature for filler (ER concentration in ESF = 12 wt%).

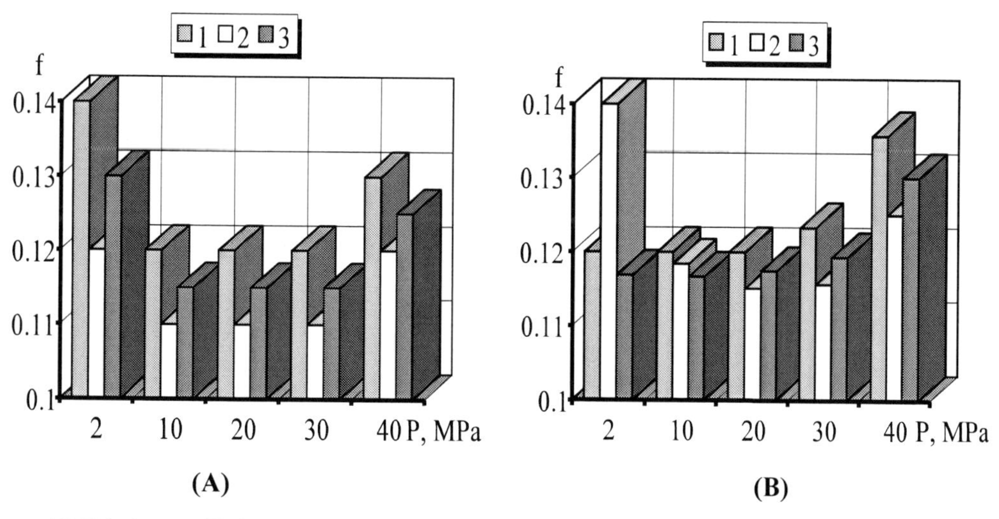

Figure 17. Friction coefficient of LC versus fillers: ESF (composition Cu→ESM, ER-concentration = 12 wt%) (a) and AcrSF (b). Fillers were obtained: 1 – without additional treatment; 2 – treated by electric current (V = 130 V; j = 3-4 A/dm^2; t = 5 min); 3 – MWH (t = 5 min). The fillers obtained were heat-treated at T = 160 °C.

Table 6. Comparative effect of designed fillers

Properties	LDPE			PTFE		PA6		
	Fillers*							
	–	ISF	AcrSF	–	ESF	–	ESH	PASF
Breaking stress, MPa: – at compression – at tension	 – 24	 – 22	 – 24	 – 20	 – 17	 85 65	 145 125	 120 115
Unnotched impact strength, kJ/m^2	unbr.**	unbr.	unbr.	unbr.	unbr.	80	70	55
Relative elongation at break, %	260	150	195	90	48	120	80	70
Vicat softening point, ^oK	353	393	383	393	473	473	488	478
Coefficient of friction (при $P = 1.5$ MPa, $v = 0.5$ m/s)	–	–	–	0.15	0.10	0.30	0.25	0.35
Wear rate, $I \cdot 10^{-8}$	55	–	–	3500	2.5	9	3	3

* Filler concentrations: 5 wt% for HDPE; 20 wt% for PTFE;

** Unbroken.

The designed fillers added to a lubricating base also enhance the lubricant temperature characteristics [51]. The result of comparative tests with widely industrially used lubricants Uniol-1 and Litol-24 (Russia) are graphically depicted in Figure 18.

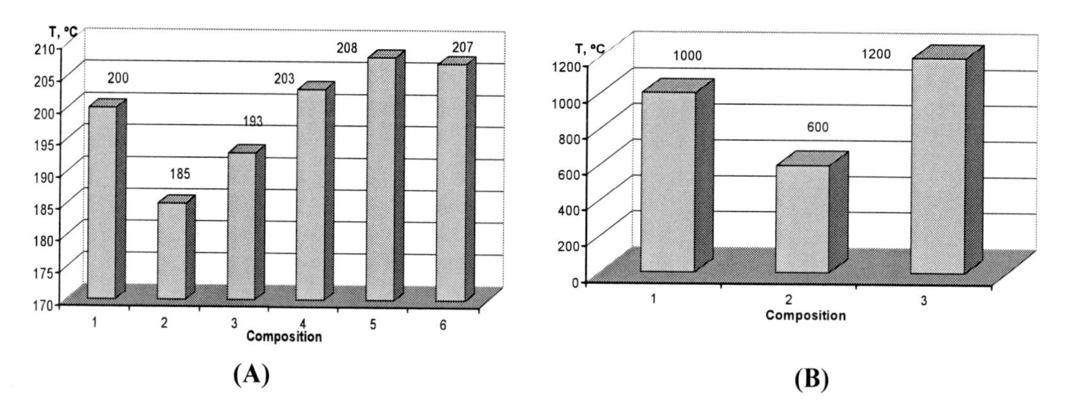

Figure 18. Dropping point (a) and critical sticking load (b) versus LC-composition: 1 – Uniol-1 lubricant; 2 – Litol-24; 3 to 6 – I-40 oil modified by WSS xerogel (3); Cu→WSS (4); Cu→ESM (5); ISF (6).

The dropping point, as well as separation temperature of modified lubricating compositions rise considerably (up to 24 %) after preliminary treatment at 210 °C [51]. This allows to expect that at a high temperature the physicochemical processes become intensive between the filler and parent lubricating composition, and result in a stronger soap structure. Thus, hybrid fillers can upgrade the temperature characteristics and LS antiscore qualities; they can be used in lubricating materials to serve in heavy-duty friction units. The triboengineering properties of the designed fillers much depend on the composition and production of the final product.

Table 6 compares data on the influence of organosilicate fillers on properties of various thermoplastics [31, 39, 54]

The analysis of data chows that organosilicate products can be used as functional fillers to improve thermoplastics: with PA6 – ultimate compression stress by 18 to 71 %: ultimate tensile stress by 38 to 92 %; Vicat softening point by 20 % and 11 %, respectively. The friction coefficient drops by 10-20 % and 50 % with wear rate reduction 3 and 1400 times for PA6 and PTFE, respectively.

Table 7 shows properties of PA6, filled with hybrid fillers (5 wt%) as a function of organic phase in the composite [54].

Table 7. Comparative properties of modified polyamide 6

Properties	$PA6_{par.}$	Fillers				
		ISF	PSF	CSF	ESF	TiO_2
Ultimate stress, MPa						
– at compression	85	145	136	120	125	100
– at tension	65	125	120	115	120	90
Impact strength, kJ/m^2	80	70	65	56	60	50
Vicat softening point, °K	200	215	210	205	210	205

The results obtained show the organic phase in the organosilicate matrix of a nanostructurized filler to influence substantially its functional properties.

The effect of intercalated in an organosilicate matrix on hybrid filler properties was investigated while preparing materials based on UHMWPE and phenylone modified by epoxy-silicates of metals.

The results showed the crystallization degree of UHMWPE-based materials to vary negligibly (Table 8) but to depend on the incorporated metal [49, 55].

Table 8. Crystallinity degree for UHMWPE and CM based on that

Indexes*	UHMW PE	Filler Composition				
		Cr → ESM**	Fe → ESM	Co → ESM	Ni → ESM	Cu → ESM
I_1	0.079	0.104	0.096	0.089	0.097	0.08
I_2	0.082	0.085	0.083	0.078	0.092	0.077
I_1/I_2	0.96	1.22	1.16	1.14	1.05	1.04

* Absorption band intensity at 1894 (I_1) and 1303 cm^{-1} (I_2);
**Epoxy-silicate matrix (ESM)

Crystallinity reduction in UHMWPE modified by hybrid fillers enhances strength characteristics, especially with addition of Ni→ESM and Cu→ESM (Table 9).

Table 9. Strength characteristics of UHMWPE-based materials

Properties	UHMWPE	Filler Composition				
		Cr → ESM	Fe → ESM	Co → ESM	Ni → ESM	Cu → ESM
$\sigma_{0,2}$, MPa	16.4	16.0	16.6	18.6	20.1	20.9
E, MPa	1023.7	1137.7	1196.4	1224.7	1246.2	1248.1

Hybrid fillers added to UHMWPE increase compressive strength by 27 %, elastic modulus by 22 % as a result of interaction between the polymer matrix and filler particles. The analysis of physicomecanical properties of materials show that property improvements depend on the polyvalent metal intercalated in the organosilicate matrix. The strength enhancement correlates (Table 9) with nucleus charge changes of the metal in the filler matrix in the sequence: $_{24}Cr \rightarrow _{26}Fe \rightarrow _{27}Co \rightarrow _{28}Ni$ (Figure 19). The filler composition affects the yield point and elastic modulus of the formed composites.

Reinforcement of the composites causes changes in the failure mode of the materials (Figure 20).

The second period of failure – with crack development – starts from the state that developed after the loss of stability. In addition to stability loss, the parent UHMWPE samples contained cracks typical of brittle failure.

So, synthesized nanocomposites added to UHMWPE improve service characteristics of the materials, which is explained, first of all, by the filler composition and mechanism of making thermoplastic materials.

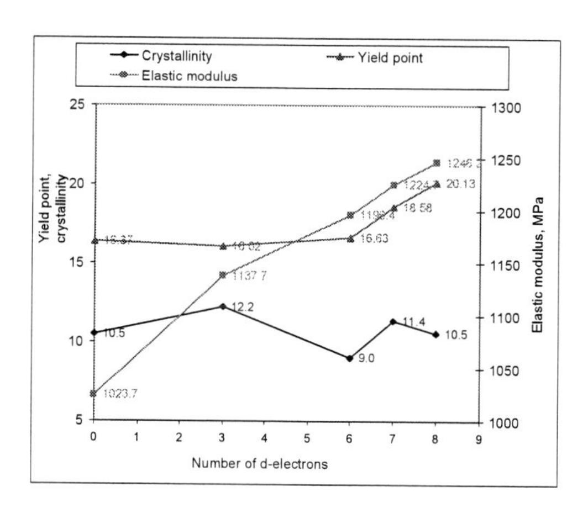

Figure 19. UHMWPE-based material properties versus metal structure (d-electron quantity on 3d-sublevel) in the filler (Me→ESM): 0 – parent UHMWPE; 3 – Cr; 6 – Fe; 7 – Co; 8 – Ni. Crystallinity rates are: Ax10.

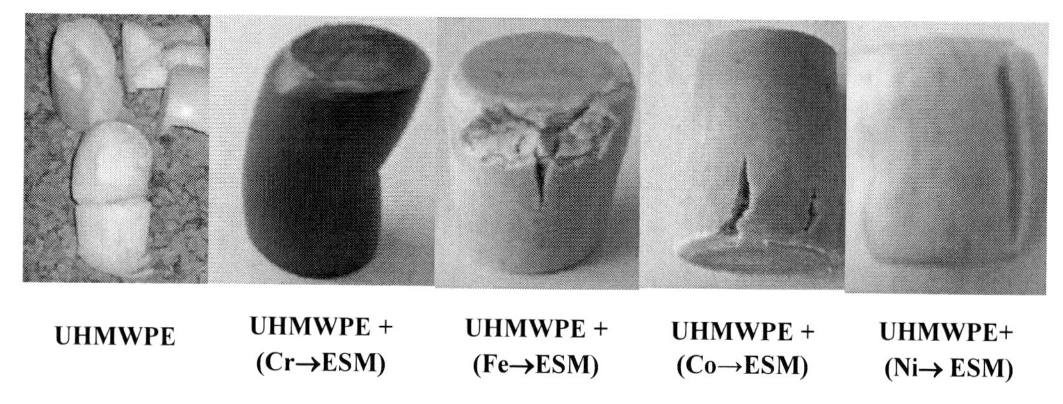

UHMWPE	UHMWPE + (Cr→ESM)	UHMWPE + (Fe→ESM)	UHMWPE + (Co→ESM)	UHMWPE+ (Ni→ ESM)

Figure 20. Appearance of samples after compression testing.

The study of hybrid fillers as fillers to phenylone showed that their introduction in a thermoplastic matrix allows the following: to increase thermal conduction by 14.7 to 20.8 %; elastic modulus by 12.3 %, and compression strength by 16.2 %; to decrease the friction coefficient 1.2 times [39, 40, 49]. The estimated efficiency of the fillers against electron structure the metal intercalated in the epoxy-silicate matrix indicated that Cr→ESM increases the mechanical properties of the phenilone-based material; Co→ESM and Ni→ESM improve the triboengineering properties, while Fe→ESM and Co→ESM increase thermal stability. It was established that changes in the thermophysical properties of the materials correlates with the nucleus charge and atom's radius of the intercalated metal. The values of heat conduction examined over the test temperature range show that organosilica fillers raise heat conduction of base polymer by 14.7-20.8 %.

The efficiency of hybrid fillers grows with introduction of thermoexpanded graphite (TEG) in the composite; this is explained by synergistic effect that allows to vary widely the thermophysical, physicomechanical, as well as triboengineering properties of the material [56]. It was found that hybrid fillers are most efficient in increasing the performance characteristics of phenylone when used in combination with thermoexpanded graphite (Figure 21).

The analysis of operating conditions (pressure and sliding velocity) effect on triboengineering characteristics of the material (Figure 22) with an optimum filler (TEG/(Co→ESM)=1/1) concentration, that ensures a set of highest thermophysical and mechanical characteristics [49, 56] depicted in Figure 21, allows the following conclusion: wear and temperature in the contact zone vary proportionally; the friction coefficient vary inversely proportionally with pressure and sliding velocity; at similar PV-values, the contact temperature is lower at a higher velocity, i. e. when the friction contact time is shorter.

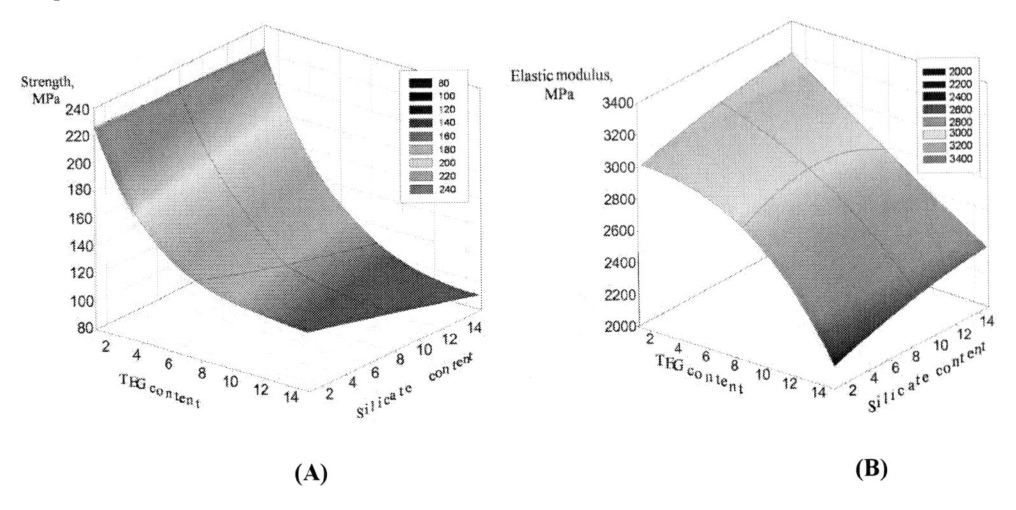

Figure 21. Strength (a) and elastic modulus (b) dependence at compression versus composition of phenylone-based composite modified by TEG and (Co→ESM).

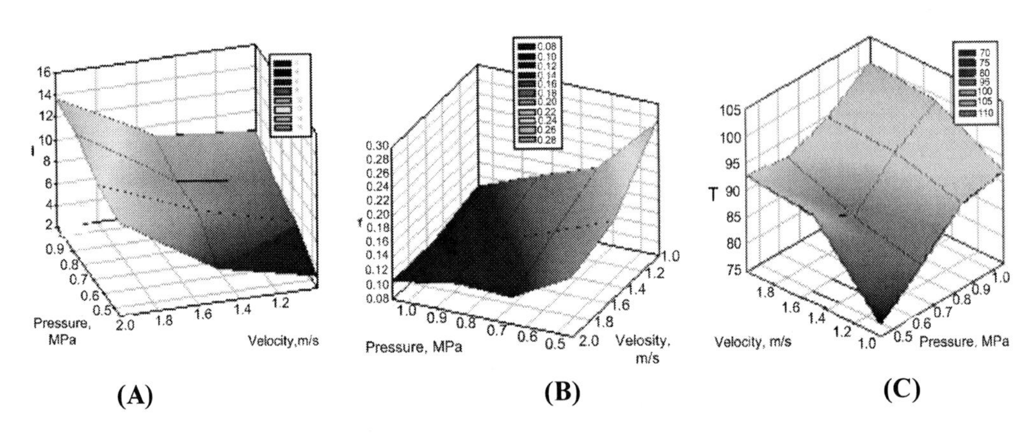

Figure 22. Linear wear rate (a), friction coefficient (b) and temperature (c) in the contact zone versus sliding velocity and pressure (steel 45, HRC 52, $R_a = 0.17$) for phenylone-based composite with a hybrid filler (TEG/(Co→ESM) = 1/1).

The surface of composition samples is glassy within the test load range, which is typical of fatigue wear.

The results obtained showed that joint use of fillers causes mutual activation and synergistic effect that allows a wide variation in the thermophysical, physicomechanical and triboengineering properties of the materials.

Conclusion

The investigation was conducted to show that homogeneous high-elastic gels or high-dispersed products can be produced depending on the intermixed solution composition and regimes of the sol-gel process in the case of multicomponent systems based on alkaline-silicate solutions. Nanostructurized high-elastic gels have high strength and water absorption, they can recover shape after mechanical effects and can be efficiently applied as water-insulating shields, sorbents or hermetics.

The structural arrangement and properties of organosilicate, as well as metallo-epoxsilicate, dispersed nanocomposites depend on the organic phase and transition metal contained in the hybrid matrix; they ensure functional activity of the formed products; these products can be employed as functionally-active fillers to thermoplastics, or additives to lubricants, thus enhancing the physicomechanical and performance properties.

References

[1] Inorganic and Organometallic Polymers with Special Properties Ed. by Laine R.M. NATO ASI Ser. New York: Kluwer, 1992. 206 ps.

[2] Andriyevsky R. A. *Progress of chemistry*. 2002. V. 71, No. 10. p. 967-981.

[3] Kudina E.F., Kushnerov D.N., Tyurina S.I., Chmykhova T.G. *Friction and wear*. 2003. V. 24, No. 5. p. 547-553.

[4] Shilova O. A., Shilov V. V. Nanosystems, nanomaterials, nanotechnologies: collection of scientific works. Kiev: Akademperiodika, 2003. V. 1. p. 9-83.

[5] Pomogailo A.D., Rosenberg A.S., Ufland I.E., Metallic Nanoparticles in Polymers. Khimia, M., 2000, 672 ps.

[6] Melnikhov I.V., Vestnik of Russian Academy. 2002. V. 72, No 10, p. 900-909.

[7] Shabanova N.A., Popov V.V., Sarkisov P.D. Chemistry and Technology of Nanodispersed Oxides: Textbook. Akademkniga, M., 2006. 309 ps.

[8] Khimich N.N., Koptelova L.A., Khimich G.N. *J. Appl. Chem.* 2003, V. 76, No 3, p. 457-462.

[9] Poddenezhny E.N., Boiko A.A. Sol-Gel Synthesis of Optical Quartz Glass. P. Sukhoi University, Gomel, 2002, 210 ps.

[10] Bronshtein L.M., Sidorov S.N., Valetsky P.M. Uspekhi khimii (Progress in Chemistry). 2004, V. 73, No 5, p. 542-557.

[11] Lomakin S.M., Zaikov G.E., Vysokomolekularnye soedinenia (High-Molecular Weight Compounds) Series B. 2005, V. 47, No 1, p. 104-120.

[12] J.D. Mackenzie *Sol-Gel Sci. Tech.* 2003. V. 26, No. 1/3. p.23.

[13] U. Schubert *Sol-Gel Sci. Tech.* 2003. V. 26, No. 1/3. p.47.

[14] Yankovsky Yu. N. et al. Neftianoe khoziaistvo (Petroleum Industry), 1984, No 8, p. 52-55.

[15] Stroganov V.M., Stroganov A.M. et al. Collection of Papers of Petroleum Research Institute: Problems of securing and finishing oil wells. PRI Publisher, Krasnodar, 1991, p. 140-145.

[16] Vagner G.R. Development of Structures in Silicate Dispersions. Naukova dumka, Kiev, 1989, 169 ps.
[17] C. Song, G. Villemure *J.Microporous and Mesoporous Materials*. 2001. V.44-45. p.675-680.
[18] Z. Deng, J. Wang, A. Wu et al. *J. Non-cryst. Solids*. 1998. V. 225. p.98-102.
[19] Kudina E.F., Tyurina S.I. 4th International Conference "Chemistry of highly ordered substances and the scientific bases of nanotechnology": Abstracts of reports. St. Petersburg, 2004. p. 239-240.
[20] Kudina E.F., Tyurina S.I. Composite Materials in Industry. 25th International Conference and Exibition. Abstracts. Yalta, 2005. p. 84 – 87.
[21] Kudina E.F., Pleskachevsky Yu.M. Composite Materials in Industry. 27th International Conference and Exibition. Abstracts. Yalta, 2007. p. 44-48.
[22] Qing Bing Gua, Min Zhi Rong, Guo Liang Jia et. al. *Wear*. 2009. V.266. p.658-665.
[23] Z. Changsheng, Z. Xiaodong, L. Shikai, S. Yaogang. *Plast. Sci. and Technol.* 2005. No. 1. p. 45-49.
[24] Boiko A.A., Poddenezhny E.N., Stotskaya O.A., Kudina E.F., Tyurina S.I. Poland-Belarus Scientific Seminar: Abstracts. Olsztyn, 2004. p. 145-146.
[25] Poddenezhny E.N., Boiko A.A., Stotskaya O.A., Kudina E.F., Tyurina S.I. Cooperation in Solving Problems of Wastes. 2nd International Conference. Abstracts. Kharkov, 2005. p. 50-51.
[26] Boiko A.A., Poddenezhny E.N., Kudina E.F., Tyurina S.I., Stotskaya O.A. Polymer composites and Tribology. International Scientific Conference. Abstracts. Gomel, 2005. p. 203-204.
[27] Stotskaya O.A., Poddenezhny E.N., Boiko A.A.,Drobyshevskaya N.E., Kudina E.F., Tyurina S.I. Reactive – 2005. XVII International Sciencific Conference. Abstracts. BG.TU-Publisher, Minsk-Ufa, 2005. p.93.
[28] Gurin V.S., Prokopenko V.B., Alexeenko A.A. et. al. *J.Inorg. Mater*. 2001. V.3. p. 493-496.
[29] Strek W., Deren P., Maruszewski K. e.a. *J. Fluoresc*. 1999. V. 9, No. 4. p. 343-345.
[30] Malashkevich G.E., Makhanek A.P., Semchenco A.V. et al. *Physics of Solids J.*, 1999. V.41, No. 2. p.229-234.
[31] Kudina E.F., Pleskachevsky Yu.M., Burya A.I. *Composite materials J.* 2007. V.1, No. 1. p.8-18.
[32] Fomichev I.A., Burya A.I., Gubenkov M.G. *Electron Trreatment of Materials J.* 1978. No. 4. p. 26-27.
[33] UA Patent 42843. Polymer Composition, Burya A.I., Gayun N.S., Kudina E.F., Yaremko Yu.O., 2009
[34] Eiler R. Silica Chemistry. Russian transl. Mir, M. 1982. Vs. 1-2. 1127 ps.
[35] French Patent 1476636. Organic Gel-forming Substances for Alkaline Silicates. Fillet P., Bonnel B. / Заявл. 28.12.65; опубл. 6.03.67.
[36] Skala M., Stepan P., Mojzisek S. Hardeners for Liquid-Glass Moulding Mixtures. Slevarenstvi. 1971. V.19, No. 5. p. 189-192.
[37] Korneev V.I., Danilov V.V. Production and Use of Soluble Glass. Liquid Glass. Stroiizdat, L., 1991. 176 p.
[38] Grigoriev P.N., Matveev M.A. Soluble Glass. Promstroiizdat, M, 1956. 443 ps.

[39] Kudina E.F., Pleskachevsky Yu.M., Burya A.I. Organosilicate nanocamposites: production, structure, properties. *Vestnik of Fundamental Research Foundation J.* 2008. No. 3. p.16-28.

[40] Kudina E.F., Pleskachevskii Yu.M. Modification of Alkali Silicate Solutions by Organic Reagents and Investigation of the Properties of the Final Products Glass Physics and Chemistry. 2009. V. 35, No.4. p. 442-448.

[41] Kudina E.F. Composite Materials in Industry. 30th International Conference and Exibition. Abstracts. Yalta 2010. p. 19-22.

[42] Kudina E.F., Pechersky G.G., Ermolovich O.A. Vesti of Belarus Academy. *Chemistry Ser.* 2010. No. 1. p.30-34.

[43] Kudina E.F., Pechersky G.G., Ermolovich O.A., Makarevich A.V., Gulevich V.V. *Chemistry and Chemical Technology*. 2009. No. 2. p.125-130.

[44] Kudina E.F. *Composite Material J.* 2008. V.2, No. 2. p.41-49.

[45] Kudina E.F., Ermolovich O.A., Pechersky G.G. *J. Appl. Chem.* 2009. V.82, No. 12. p.1963-1970.

[46] Kudina H.F. III Int.Conference on Colloid Chemistry and Physicochemical Mechanics. IC-CCPCM 2008. Moscow, 2008. 1 el. opt. disk (CD-ROM). 2 ps.

[47] Svidersky V.A., Voronkov M.G., Klimenko S.V., Bystrov D.N. *J. Appl. Chem.* 2003. V.76, No. 5. p.810-813.

[48] Svidersky V.A., Voronkov M.G., Klimenko S.V., Bystrov D.N. et. al. *J. Appl. Chem.* 2005. V.78, No.3. p.380-383.

[49] Kudina H.F., Burya A.I., Shapovalov V.M., Gayun N.S. Materials of 9th International Conference «Research and Development in Mechanical Industry» (RaDMI): Abstracts of reports. Vrnjaska Banja, Serbia, 2009. V. 2. p.1118-1123.

[50] Kudina E.F., Kushnerov D.N., Tyurina S.I., Chmykhova T.G. *Friction and wear*. 2003. V. 24, No.5. p. 547-553.

[51] Volnianko E.N., Kudina E.F., Chmykhova T.G., Smurugov V.A. *Friction and wear*. 2006. V. 27, No. 2. p. 232-235.

[52] Volnianko E., Chmykhova T., Smurugov V. J. of Balkan Tribological Association, 1998. V.4, No. 3-4. p. 232-236.

[53] Liuty M., Kostiukovich G.A., Skaskevich A.A., Struk V.A., Kholodilov O.V. *Friction and wear*. 2002. V.23, No. 4. p. 411-424.

[54] Kudina E.F., Burya A.I. Materials of 6th International Conference «Research and Development in Mechanical Industry» (RaDMI): Abstracts of reports. Budva - Montenegro, 2006. 1 el. opt. disk (CD-ROM): XXII, 1200, xxvi str. ISBN 86-83803-21-X (HTMS). Session B (B-5). 4 p.

[55] Burya A.I., Redchuk A.S, Kudina E.F., Pleskachevsky Yu.M. V Polish - Ukrainian conference «Polymers of special applications»: abstracts of reports. Radom - Swieta Katarzyna, 2008. Poland. p. 28.

[56] Kudina E.F., Burya A.I., Pleskachevsky Yu.M., Yaremko Yu.A. and Kuznecova O.Yu. *Chemistry and Chemical Technology*. 2008. No.6. p.66-71.

In: Resin Composites: Properties, Production and Applications ISBN: 978-1-61209-129-7
Editor: Deborah B. Song

Chapter 4

ELECTROMAGNETIC PROPERTIES OF A COMPOSITE MADE OF METAL PARTICLES DISPERSED IN RESIN

Kenji Sakai, Norizumi Asano, Yoichi Wada, Yuuki Sato and Shinzo Yoshikado

Department of Electronics, Doshisha University, Tatara Miyakodani, Kyotanabe, Kyoto, Japan

ABSTRACT

The frequency dependences of the relative complex permeability μ_r^* and relative complex permittivity ε_r^* for the composite made of metal particles (aluminum particles) dispersed in polystyrene resin were measured in the frequency range from 1 MHz to 40 GHz. The volume mixture ratio and particle size of aluminum were varied and the dependences of the volume mixture ratio and particle size on the frequency dependences of μ_r^* and ε_r^* were investigated. In addition, theoretical values of the real and imaginary part of μ_r^*, μ_r' and μ_r", for the composite made of metal particles dispersed in polystyrene resin were calculated using Maxwell's equations.

The measured value of the real part of ε_r^* was independent of frequency and both real and imaginary part of ε_r^* increased with increasing the volume mixture ratio of aluminum particle. The measured value of μ_r' was found to decrease with increasing frequency in the low frequency range and became constant in the high frequency range. Meanwhile, the measured value of μ_r" increased with frequency, had a maximum and decreased with increasing frequency. Moreover, at high frequencies where the skin depth is much smaller than the radius of aluminum, μ_r' was found to depend only on the volume mixture ratio of aluminum, whereas μ_r" was determined by both the volume mixture ratio and the aluminum particle size. These results almost agreed with the calculated values. Thus, the frequency dependences of μ_r' and μ_r" were found to be predicted by the theoretical calculation and to be controlled by the volume mixture ratio and particle size of aluminum.

For the application of this composite to an electromagnetic wave absorber, the return loss of this composite was calculated from the measured values of μ_r^* and ε_r^*. The return loss of the composite made of aluminum particles dispersed in polystyrene resin was less than -20 dB (absorption of 99% of an electromagnetic wave power) in the frequency range from 1 to 40 GHz when a suitable volume mixture ratio, particle size, and sample thickness were selected. Therefore, this composite can be used as an electromagnetic wave absorber in the gigahertz range. Furthermore, the absorption characteristics of such composite can be tailored based on the ability to control the values of μ_r' and μ_r" independently by adjusting the volume mixture ratio and aluminum particle size.

1. INTRODUCTION

This chapter deals with a composite made of metal particles dispersed in resin, and electromagnetic properties, such as frequency dependences of the relative complex permeability μ_r^* and the relative complex permittivity ε_r^*, are discussed to develop a new electromagnetic wave absorber using this composite.

Recently, electromagnetic waves with frequencies higher than 1 GHz are being increasingly used for telecommunication devices such as wireless local area network (LAN), and the frequencies used by these devices is expected to shift to the high frequency range of above 10 GHz in the future. For this reason, the development of an electromagnetic wave absorber suitable for these frequency bands is required.

To design such an absorber, control of the frequency dependence of μ_r^* and ε_r^* is important factor because the absorption of electromagnetic waves is determined by these parameters. The composite made of magnetic material, such as ferrite, dispersed in resin or rubber has the frequency dependence of μ_r^* and ε_r^* that can satisfy the absorption condition of electromagnetic wave in the GHz range. Thus, the frequency dependence of μ_r^* and ε_r^* for the composite made of magnetic material has been investigated [1-5].

However, according to Snoek, composite electromagnetic wave absorbers made of magnetic materials dispersed in resin or rubber exhibit an upper frequency limit of the electromagnetic wave absorption center frequency [6]. Moreover, μ_r', the real part of μ_r^*, must be less than unity to satisfy the absorption condition of electromagnetic wave at frequencies above 10 GHz and a flexible control of μ_r^* and ε_r^* is required to design an absorber at any frequency range. The composite made of magnetic materials dispersed in resin or rubber is difficult to meet these requirements. Therefore, a new absorbent material is required. As one of the new absorbent materials, the design of an absorber using metamaterials, in which μ_r^* and ε_r^* can be artificially controlled, has been reported [7-9]. However, the proposed absorber using metamaterial is unsuitable for practical applications because a metamaterial needs precise patterning or an array of a conductive material, and its mass production and cost reduction are difficult.

According to Nishikata et al., it has been reported that the magnetic moment is induced by an eddy current flowing on the surface of an aluminum particle and this magnetic moment contributes to the absorption of an electromagnetic wave [10-11]. This phenomenon indicates that a composite made of aluminum particles dispersed in an insulating matrix can artificially exhibit magnetic properties although both aluminum and insulating matrix is non-magnetic material, and it is possible to absorb electromagnetic waves in a wide frequency range because there is no practical upper limit on the absorbing frequency. The principal of an absorber using the composite made of aluminum particles dispersed in polystyrene resin is explained as follows. Aluminum is a paramagnetic substance, its nondimensional magnetic susceptibility is approximately 2.4×10^{-4}, and its μ_r' is almost 1. When an electromagnetic wave of a high frequency enters an aluminum particle, an eddy current flows whose strength depends on the skin effect of the incident magnetic field on the particle surface [13-16], and a reverse magnetic moment appears. Then, the energy of the electromagnetic wave is converted into thermal energy by the eddy current, which flows on the surface of the aluminum particles. As mentioned above, since μ_r' must be smaller than unity for absorption of electromagnetic waves in the high-frequency range, the conditions under which the magnetic moment appears is an important factor when absorbent materials are developed.

In the case of this composite, the artificial magnetic properties are expected to be realized not by a periodic array but random dispersion of aluminum particles because the magnetic moments induced in each particle contribute to the macroscopic magnetic properties of the composite. Therefore, this artificial medium enables the mass production of an absorber with a low cost because complicated processes such as precise patterning are not required. Moreover aluminum is a light-weight and low-cost material. These advantages are suitable for preparing a practical absorber, because materials used for absorbers should be low-cost and abundant in order to avoid the depletion of global resources.

In this chapter, the frequency dependence of μ_r^* for the composite made of aluminum particles dispersed in polystyrene resin is discussed in the frequency range from 1 MHz to 40 GHz. The volume mixture ratio and particle size of aluminum were varied and the dependences of volume mixture ratio and particle size of aluminum were investigated. Then, μ_r' and μ_r", the real and imaginary components of μ_r^*, were analytically-calculated and were compared with the measured values in order to control these parameters artificially. In the calculations, the eddy currents which flow inside the aluminum particles, were calculated using Maxwell's equations and used to determine μ_r' and μ_r". After that, the frequency dependence of ε_r^* is shown. Finally, the return loss of a metal backed single layer absorber is calculated using the values of μ_r^* and ε_r^* for the composite made of aluminum particles dispersed in polystyrene resin and the absorption characteristics of this composite are discussed.

2. Experiment

Commercially available aluminum particles, with average diameters of approximately 8, 30, and 50 μm were used. Chips of polystyrene resin were dissolved in acetone and the particles were mixed in until they were uniformly dispersed within the resin. Aluminum particle volume ratios of 16.4, 33.8 and 50 vol% were used for 30 and 50 μm aluminum particle and those of 16.4, 33.8 40 50 and 60 vol% were used for 8 μm aluminum particle. The mixture was then heated to melt the polystyrene resin and was hot-pressed at a pressure of 5 MPa to form a pellet. This was allowed to cool naturally to room temperature and was processed into a toroidal-core shape (outer diameter of approximately 7 mm, inner diameter of approximately 3 mm) for use in a 7 mm coaxial line in the frequency range 1 MHz to 12.4 GHz, or into a rectangular shape (P-band: 12.4-18 GHz, 15.80 mm × 7.90 mm, K-band: 18-26.5 GHz, 10.67 mm × 4.32 mm, R-band: 26.5-40 GHz, 7.11 mm × 3.56 mm) for use in a waveguide. The sample was mounted inside the coaxial line or waveguide using silver past to ensure that no gap existed between the sample and the walls of the line/waveguide. The complex scattering matrix elements S_{11}^* (reflection coefficient) and S_{21}^* (transmission coefficient) for the TEM mode (coaxial line) or TE_{10} mode (rectangular waveguide) were measured using a vector network analyzer (Agilent Technology, 8722ES) by the full-two-port method in the frequency range from 1 to 40 GHz. The values of μ_r^* ($\mu_r^* = \mu_r' - j\mu_r''$, $j = \sqrt{-1}$) and ε_r^* ($\varepsilon_r^* = \varepsilon_r' - j\varepsilon_r''$) were calculated from the data of both S_{11}^* and S_{21}^*. In the frequency range from 1 MHz to 1 GHz, the values of μ_r^* and ε_r'' were calculated from the data of the impedance measured by an impedance analyzer (Agilent Technology, 4291A). The complex reflection coefficient Γ^* for a metal backed single layer absorber was then determined from the values of μ_r^* and ε_r^*. The return loss R for each sample thickness was calculated from Γ^* using the relation $R = 20 \log_{10}|\Gamma^*|$. R was calculated at 0.1 mm intervals in the sample thickness range 0.1 to 30 mm.

3. Results and Discussions

3.1. Dispersion State of Aluminum Particles in Polystyrene Resin

The effect of magnetic moment generated by an eddy current on the frequency dependences of μ_r' and μ_r'' is speculated to increase with increasing the amount of aluminum particles, because the surface area where the eddy current flows increase with increasing the amount of aluminum particles.

However, if the amount of aluminum particles increases, it is considered that the effect of magnetic moment is saturated or decreases, because aluminum particles are in mutual contact, the total surface area of the particles decreases, and the total surface current of the particles decreases. Furthermore, as the aluminum particles come in contact, the conductivity of composite increases. This increase in conductivity results in an increase of the reflection coefficient of the composite, and it becomes difficult for an electromagnetic wave to pass through the composite. Eventually, the wave cannot be effectively absorbed because the Joule loss becomes small. Therefore, the dispersion state of aluminum particles is very important factor to develop a composite made of metal particles dispersed in resin or rubber. In this section, the surface of the composite are observed using an optical microscope (Nikon, ECLIPSE ME600) and the dispersion state of aluminum particles are evaluated.

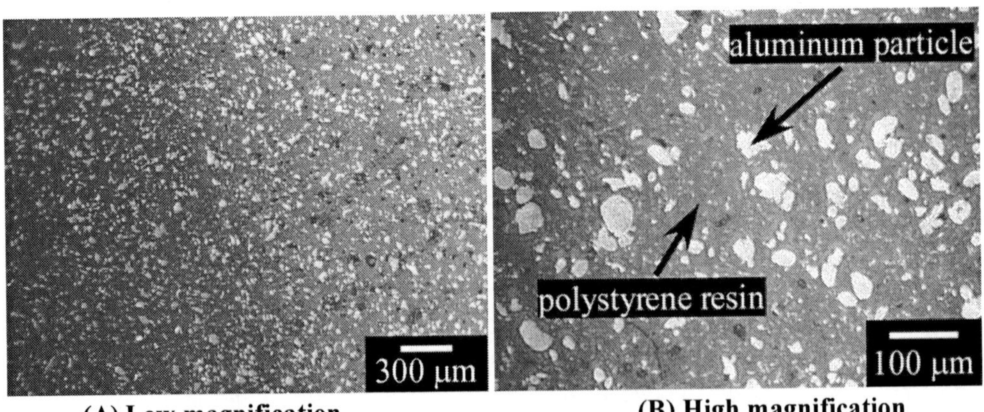

Figure 1. Optical microphotographs of composites made of polystyrene resin and aluminum particles with sizes of 30 μm. Volume mixture ratio of aluminum 16.4 vol%.

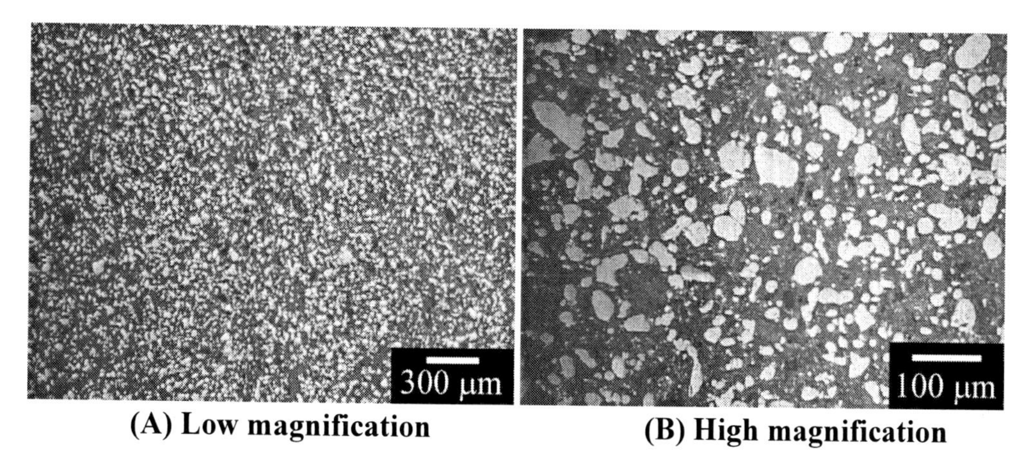

Figure 2. Optical microphotographs of composites made of polystyrene resin and aluminum particles with sizes of 30 μm. Volume mixture ratio of aluminum 33.8 vol%.

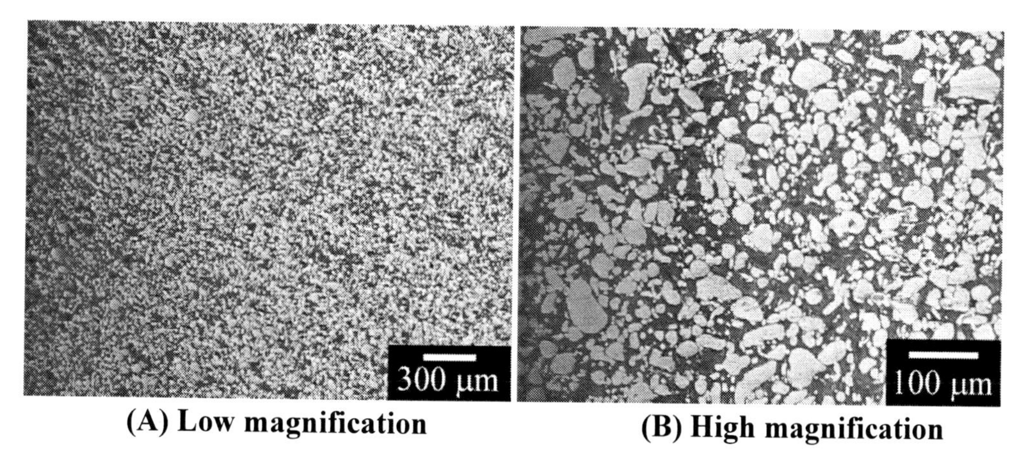

Figure 3. Optical microphotographs of composites made of polystyrene resin and aluminum particles with sizes of 30 μm. Volume mixture ratio of aluminum 50.0 vol%.

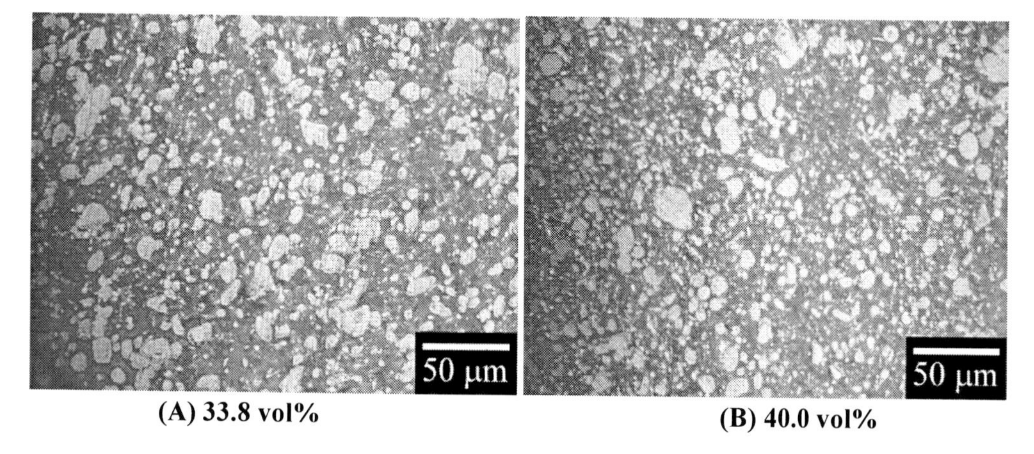

Figure 4. Optical microphotographs of composites made of polystyrene resin and aluminum particles with sizes of 8 μm. Volume mixture ratio of aluminum 33.8 and 40.0 vol%.

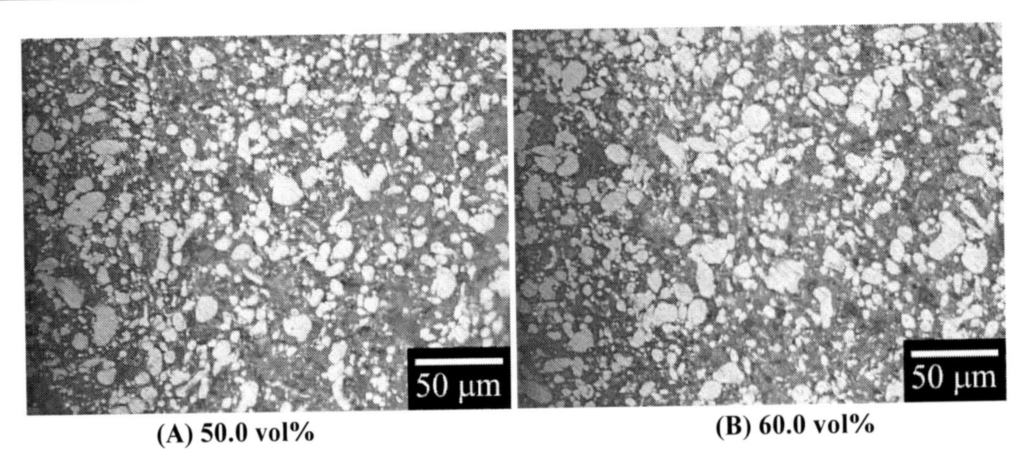

Figure 5. Optical microphotographs of composites made of polystyrene resin and aluminum particles with sizes of 8 μm. Volume mixture ratio of aluminum 50.0 and 60.0 vol%.

Figures 1, 2, and 3 show surface optical microphotographs of the composite made of aluminum particles of 30-μm diameter with volume mixture ratios of 16.4, 33.8, and 50.0 vol%, respectively. It was found from Figure 1(a), 2(a), and 3(a) that aluminum particles were uniformly dispersed in polystyrene resin. As shown in Figures 1(b) and 2(b), aluminum particles were not in contact with each other and isolated in the polystyrene medium in the case of the sample with volume mixture ratios of 16.4 and 33.8 vol% for the aluminum particles. However, the aluminum particles were in mutual contact in the sample with a volume mixture ratio of 50.0 vol%. This result indicates that aluminum particles were uniformly dispersed in polystyrene resin when the amount of the aluminum particles was low. This result is also in agreement with the percolation theory (the particles are not in mutual contact up to the percolation threshold of 33.3 vol%) [15].

Figures 4 and 5 show surface optical microphotographs of the composite made of aluminum particles of 8-μm diameter with volume mixture ratios of 33.8, 40.0, 50.0 and 60.0 vol%, respectively. When the particle size of aluminum is small, the particles tend to be isolated with each other even if the amount of aluminum particles increased. As shown in Figure 4(b), the aluminum particles were isolated in the polystyrene medium in the sample with a volume mixture ratio of 40.0 vol%, which exceeded the percolation threshold. From above result, it is concluded that the dispersion state of aluminum particle depended on particles size and particles with a volume exceeding the percolation threshold could be dispersed in the resin without mutual contact using aluminum particles of small size.

3.2. Frequency dependences of μ_r' and μ_r" for the Composite Made of Polystyrene Resin and Aluminum Particles of Various Particle Sizes and Volume Mixture Ratios

Figure 6, 7, and 8 show the frequency dependences of μ_r' and μ_r" for the composite made of polystyrene resin and aluminum particles with various volume mixture ratios. The aluminum particle size is 8, 30, and 50 μm, respectively. Regardless of the particle size and volume mixture ratio, all composites showed similar frequency dependences of μ_r' and μ_r". Although the values of μ_r' was almost unity in the low frequency range, μ_r' decreased with increasing frequency and became almost constant in the high frequency range, as shown in Figures 6(a), 7(a), and 8(a). On the other hand, the values of μ_r" increased with increasing frequency, had a maximum, and decreased with increasing frequency, as shown in Figures 5(b), 6(b), and 7(b). This phenomenon is explained as follows.

When an electromagnetic wave of high frequency incidents to a metal particle, the eddy current flows in the layer of a thickness δ on a metal particle and generates a magnetic moment antiparallel to the incident magnetic field.

The skin depth δ of a metal for an electromagnetic wave is given by

$$\delta = \sqrt{\frac{2}{\omega\sigma\mu_0\mu_{Mr}'}}. \quad (1)$$

Here, ω represents the angular frequency of the electromagnetic wave; σ, the conductivity of the metal; μ_0, the permeability of free space; and μ_{Mr}'; the real part of the relative complex permeability of the metal. For example, the skin depth δ of aluminum ranges from 2.6 to 0.9 μm in the frequency range from 1 to 10 GHz and δ is shorter than the diameter of the aluminum particles used for experiments. Therefore, the eddy current flows in the region from the surface of the aluminum particles to the skin depth and generates a magnetic field in each aluminum particle. The dimensionless magnetic susceptibility of aluminum is very small (2.1×10^{-5}), and μ_{Mr}' is approximately 1. However, in the high-frequency range, μ_r' of an aluminum particle is smaller than unity because the magnetic field inside the aluminum particle is canceled by the magnetic field generated by the eddy current. Therefore, μ_r' for the composite made of aluminum particles dispersed in polystyrene resin becomes less than unity. Because of this phenomenon, the magnetic loss occurs and μ_r" increases.

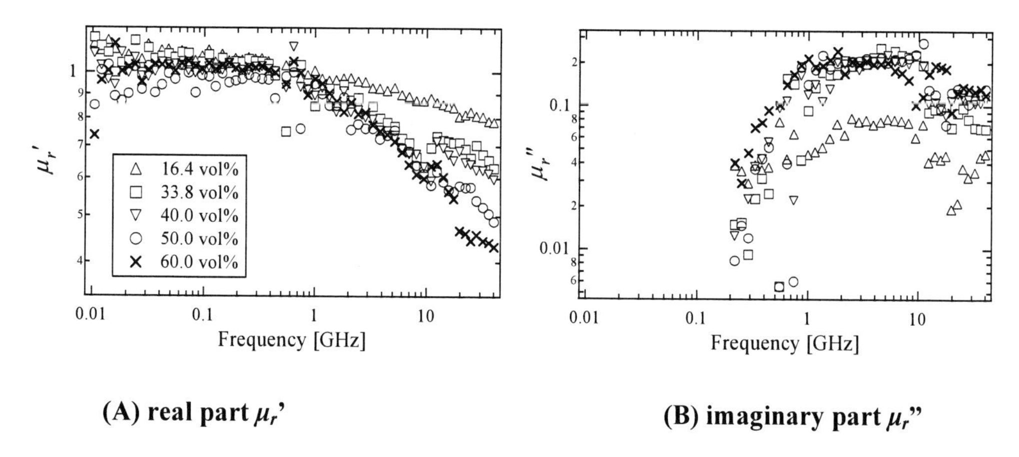

(A) real part μ_r' **(B) imaginary part μ_r"**

Figure 6. Frequency dependence of μ_r' and μ_r" for composites made of polystyrene resin and aluminum particles with sizes of 8 μm. The aluminum volume mixture ratios are 16.4, 33.8 40.0 50.0 and 60.0 vol%.

Although the shape of frequency dependences of μ_r' and μ_r" were similar, the values of μ_r' and μ_r" depended on the volume mixture ratio and particle size of aluminum. When the volume mixture ratio of aluminum increased, the values of μ_r' in the high frequency range were smaller and the values of μ_r" larger in the measured frequency range, compared with the composite made with low volume mixture ratio of aluminum particle. When the aluminum particle size became small, the frequency where μ_r' begins to decrease shifted to a high frequency range and the frequency where μ_r" is maximum increased. The frequency f ' at which the radius of aluminum particle is equal to δ can be obtained from Equation (1).

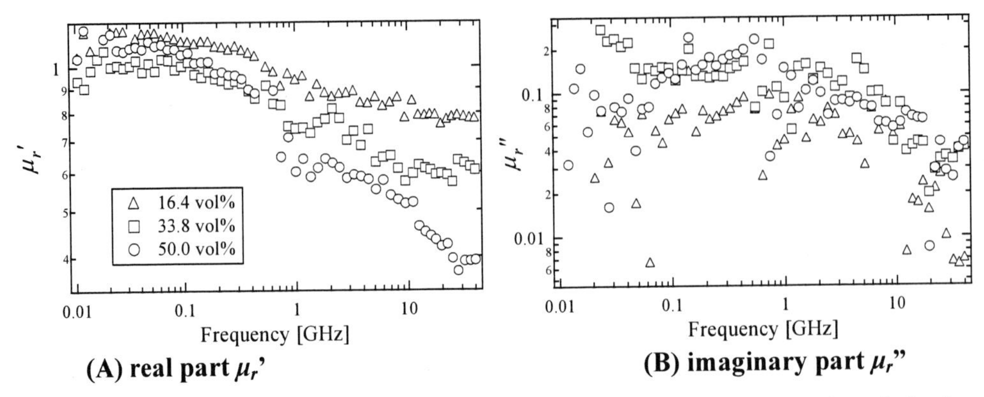

Figure 7. Frequency dependence of μ_r' and μ_r"for composites made of polystyrene resin and aluminum particles with sizes of 30 μm. The aluminum volume mixture ratios are 16.4, 33.8, and 50.0 vol%.

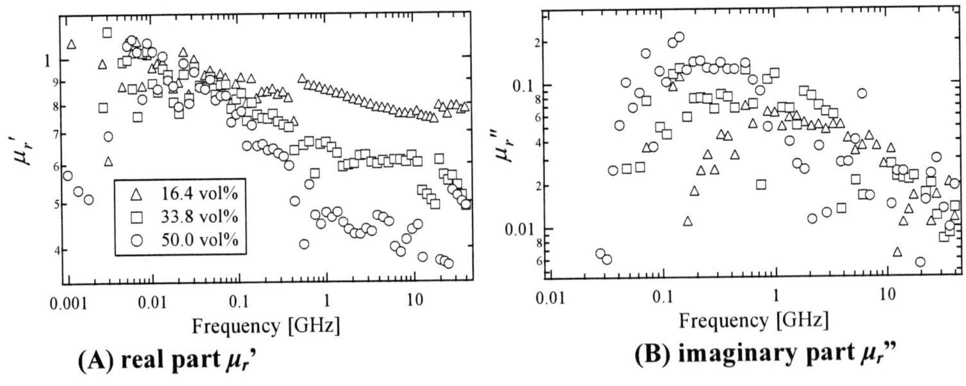

Figure 8. Frequency dependence of μ_r' and μ_r" for composites made of polystyrene resin and aluminum particles with sizes of 50 μm. The aluminum volume mixture ratios are 16.4, 33.8, and 50.0 vol%.

$$f'=\frac{1}{a^2\pi\sigma\mu_r'}. \quad (2)$$

f'values for aluminum particles for 8, 30, and 50 μm are approximately 420, 30, and 11 MHz, respectively. This result shows that the frequency where the effect of the magnetic moment generated by the eddy current becomes apparent increases with a decrease in the particle size. Therefore, it is speculated that the measured values of μ_r' and μ_r" depended on the particle size.

These results indicate that the values of μ_r' and μ_r" are expected to be controlled by simply changing the volume mixture ratio and particle size of aluminum. To clarify the mechanism of the frequency dependences of μ_r' and μ_r", the values of μ_r' and μ_r" are calculated theoretically based on the effect of magnetic moment generated by the eddy current on μ_r' and μ_r" and these values are compared with measured values in the next sections.

3.3. Theoretical Calculation of Relative Complex Permeability for the Composite Made of Metal Particles Dispersed in Resin or Rubber

To simplify the discussion, the shape of an aluminum particle is approximated as a cylinder of a radius a and length $2a$. As shown in Figure 9, a model in which the eddy current flows in the outer skin of a cylindrical metallic shell of a thickness δ (equivalent to the skin depth) is used [16]. However, the eddy currents actually flow to a depth of more than δ. Therefore, in the theoretical calculation of this chapter, the eddy currents flowing inside the cylindrical metallic shell are calculated from Maxwell's equations, and μ_r' and μ_r" are subsequently determined.

For an incident magnetic field strength of H_0 parallel to the central axis of the cylinder, the eddy current density $J_\varphi(x)$ [A/m^2] may be defined as

$$J_\varphi(x) = Ae^{(-\frac{a-x}{\delta})} \qquad (3)$$

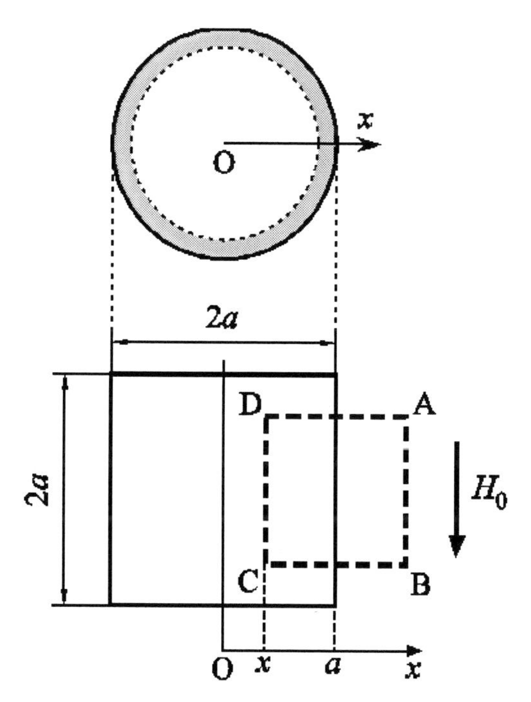

Figure 9. Model of magnetic moment with eddy current in thin cylindrical metallic shell and the integral rout of thin metallic cylindrical shell used for calculation of μ_r' and μ_r".

Here, A is the proportional coefficient and x is the distance from the center of the cylinder. Although the length of the cylinder is finite, it is assumed that a uniform magnetic field vector $\vec{H}'$ parallel to the central axis of the cylinder is generated by $J_\varphi(x)$. When Ampere's circuital law is applied to the integral route of ABCD shown in Figure 9, the following equation is obtained, because $\vec{H}'$ is $\vec{0}$ along the route AB, $\vec{H}' \perp \mathrm{BC}$ and $\vec{H}' \perp \mathrm{AD}$.

$$\oint_{\mathrm{ABCD}} \vec{H}'\cdot d\vec{s} = H'(x)\overline{\mathrm{DC}} = \int_x^a J_\varphi(x)dx\,\overline{\mathrm{DC}} \tag{4}$$

Therefore, $H'(x)$ is obtained from Equations (3) and (4).

$$H'(x) = \int_x^a Ae^{(-\frac{a-x}{\delta})}dx = Ae^{-\frac{a}{\delta}}\delta(e^{\frac{a}{\delta}} - e^{\frac{x}{\delta}}) \tag{5}$$

For an electric field vector $\vec{E}$ and a magnetic flux density vector $\vec{B}$, Maxwell's equation and Stokes' theorem give the following equations.

$$\iint_{\mathrm{S}} (\vec{\nabla}\times\vec{E})\cdot d\vec{S} = -\frac{\partial}{\partial t}\iint_{\mathrm{S}} \vec{B}\cdot d\vec{S}, \tag{6}$$

$$\iint_{\mathrm{S}} (\vec{\nabla}\times\vec{E})\cdot d\vec{S} = \oint_{\mathrm{C}} \vec{E}\cdot d\vec{s}. \tag{7}$$

Here, C is a circle of radius x, whose center is O, and S is the area inside the integral route C. When the radius of C is a ($x = a$), the integral in the right side of Equation (6) is given by

$$\iint_{\mathrm{S}} \vec{B}\cdot d\vec{S} = \iint_{\mathrm{S}} (\vec{B}_0 + \vec{B}')\cdot d\vec{S} = \pi a^2\mu_0 H_0 + \iint_{\mathrm{S}} \vec{B}'\cdot d\vec{S}, \tag{8}$$

$$\begin{aligned}\iint_{\mathrm{S}} \vec{B}'\cdot d\vec{S} &= \int_0^a 2\pi x B'(x)dx \\ &= 2\pi\mu_0 \int_0^a xH'(x)dx = 2\pi\mu_0 \int_0^a Ae^{-\frac{a}{\delta}}\delta x(e^{\frac{a}{\delta}} - e^{\frac{x}{\delta}})dx \\ &= 2\pi\mu_0 A\delta\left(\frac{1}{2}a^2 + \delta^2 - \delta^2 e^{-\frac{a}{\delta}} - \delta a\right).\end{aligned} \tag{9}$$

Here, $\vec{B}_0$ is the external magnetic flux density vector and $\vec{B}'$ is the magnetic flux density vector generated by $J_\varphi(x)$. Also, the integral in the right side of Equation (7) is given by

$$\oint_{\mathrm{C}} \vec{E}\cdot d\vec{s} = \oint_{\mathrm{C}} E_\varphi(x)ds = 2\pi a E_\varphi(a) = \frac{2\pi a J_\varphi(a)}{\sigma} = \frac{2\pi a A}{\sigma}. \tag{10}$$

Here, $E_\varphi(x)$ ($= J_\varphi(x)/\sigma$) is the electric field at the circumference of the cylinder. Thus, the following equation is obtained from Equations (8), (9) and (10).

$$\frac{2\pi aA}{\sigma} = -\frac{\partial}{\partial t}\left[\pi a^2\mu_0 H_0 + 2\pi\mu_0 A\delta\left(\frac{1}{2}a^2 + \delta^2 - \delta^2 e^{-\frac{a}{\delta}} - \delta a\right)\right]$$
$$= -j\omega\left[\pi a^2\mu_0 H_0 + 2\pi\mu_0 A\delta\left(\frac{1}{2}a^2 + \delta^2 - \delta^2 e^{-\frac{a}{\delta}} - \delta a\right)\right] \tag{11}$$

A is obtained from Equation (11).

$$A = \frac{-j\omega a\mu_0\sigma/2}{1 + j\omega\mu_0\delta\frac{\sigma}{a}\left(\frac{1}{2}a^2 + \delta^2 - \delta^2 e^{-\frac{a}{\delta}} - \delta a\right)}H_0 = \frac{-j\omega\alpha}{1 + j\omega\beta}H_0 \tag{12}$$

Here, α and β are given by

$$\alpha = \frac{a\mu_0\sigma}{2}, \tag{13}$$

$$\beta = \mu_0\delta\frac{\sigma}{a}\left(\frac{1}{2}a^2 + \delta^2 - \delta^2 e^{-\frac{a}{\delta}} - \delta a\right). \tag{14}$$

When $a >> \delta$, β becomes

$$\beta = \mu_0\delta\frac{\sigma}{a}\frac{1}{2}a^2 = \frac{\mu_0}{2}\delta\sigma a. \tag{15}$$

Therefore, $J_\varphi(x)$ is obtained from Equations (3) and (12). Here, two kinds of $J_\varphi(x)$ are considered when substituting Equation (3) into (12). One is the real part of A and $J_\varphi(x)$ becomes

$$J_\varphi(x) = \mathrm{Re}\left(\frac{-j\omega\alpha}{1 + j\omega\beta}\right)e^{(-\frac{a-x}{\delta})}H_0 = -\frac{\omega^2\alpha\beta}{1 + \omega^2\beta^2}e^{(-\frac{a-x}{\delta})}H_0 \tag{16}$$

The other is the absolute value of A and $J_\varphi(x)$ is given by

$$J_\varphi(x) = \left|\frac{-j\omega\alpha}{1 + j\omega\beta}\right|e^{(-\frac{a-x}{\delta})}H_0 = \frac{\omega\alpha\sqrt{1 + \omega^2\beta^2}}{1 + \omega^2\beta^2}e^{(-\frac{a-x}{\delta})}H_0 \tag{17}$$

Hereafter, to simplify the deviation, $J_\varphi(x)$ is defined as follows.

$$J_{\varphi}(x) = A' e^{(-\frac{a-x}{\delta})} H_0 \tag{18}$$

Here, A' is given by

$$A' = -\frac{\omega^2 \alpha \beta}{1+\omega^2 \beta^2} \text{ (real part of } A\text{)} \tag{19}$$

$$A' = \frac{\omega \alpha \sqrt{1+\omega^2 \beta^2}}{1+\omega^2 \beta^2} \text{ (absolute part of } A\text{)} \tag{20}$$

The magnetic moment m generated by $J_{\varphi}(x)$ is given by

$$m = 2a \int_0^a \pi x^2 J_{\varphi}(x) dx = 2\pi a A' e^{-\frac{a}{\delta}} H_0 \int_0^a x^2 e^{\frac{x}{\delta}} dx . \tag{21}$$

If $2a$ is constant, the number N of cylindrical particles per unit volume of the composite is given by

$$N = \frac{V}{2\pi a^3} . \tag{22}$$

Here, V is the volume mixture ratio of the particles in the composite. If it is assumed that the direction of all magnetic moments is the same and that the eddy current loss is zero, the magnetization M is given by

$$M = Nm = -\frac{V A' K_0 e^{-\frac{a}{\delta}}}{a^2} H_0 . \tag{23}$$

Here,

$$K_0 = \int_0^a x^2 e^{\frac{x}{\delta}} dx = \delta(2\delta^2 - 2a\delta + a^2)(e^{\frac{a}{\delta}} - 1) - a\delta(2\delta - a) . \tag{24}$$

Also, the following relation holds between the average magnetic flux density B and the magnetization M in the composite when M is assumed to be proportional to H_0.

$$M = \frac{B}{\mu_0} - H_0 = (\mu_r^* - 1) H_0 = (\mu_r' - 1 - j\mu_r'') H_0 \tag{25}$$

Therefore, the following equation is obtained from Equations (23) and (25).

$$\mu_r' = 1 - \frac{VA' e^{-\frac{a}{\delta}}}{a^2}\left[\delta(2\delta^2 - 2a\delta + a^2)(e^{\frac{a}{\delta}} - 1) - a\delta(2\delta - a)\right] \quad (26)$$

The Joule loss P, caused by the eddy current loss, per unit volume of the composite is

$$P = N\frac{1}{2}2a\int_0^a 2\pi x \frac{J_\varphi(x)^2}{\sigma}dx\,. \quad (27)$$

Because $J_\varphi\,(x)$ is given by Equation (16), P is given by

$$P = \frac{VA'^2 e^{-2\frac{a}{\delta}}}{\sigma a^2}\left[-\frac{\delta^2}{4}\left(e^{2\frac{a}{\delta}} - 1\right) + \frac{a\delta}{2}e^{2\frac{a}{\delta}}\right]H_0^{\,2}\,. \quad (28)$$

μ_r" is defined as the ratio of the magnetic energy lost in one cycle and the magnetic energy accumulated in the cylinder. Therefore, μ_r" is given by

$$\mu_r'' = \mu_r' \frac{P}{\frac{\omega}{2}\mu_r'\mu_0 \iiint H^2 dv}\,. \quad (29)$$

Here, H is the magnetic field in the cylinder. The integral in the denominator of Equation (29) is obtained as follow.

$$\begin{aligned}
\frac{\omega}{2}\mu_r'\mu_0 \iiint H^2 dv &= \frac{\omega}{2}\mu_r'\mu_0 \iiint_{\substack{outside\\cylinder}} H_0^{\,2} dv + \frac{\omega}{2}\mu_r'\mu_0 N \iiint_{\substack{inside\\cylinder}} \left(H_0 + H'(x)\right)^2 dv \\
&= \frac{\omega}{2}\mu_r'\mu_0(1-V)H_0^{\,2} + \frac{\omega}{2}\mu_r'\mu_0 N\int_0^a 2a\cdot 2\pi x\left(H_0^{\,2} + 2H_0H'(x) + H'(x)^2\right) \\
&= \frac{\omega}{2}\mu_r'\mu_0(1-V)H_0^{\,2} + \frac{\omega\mu_r'\mu_0 V}{a^2}\left[\frac{1}{2}a^2H_0^{\,2} + 2A'\delta\left(\frac{1}{2}a^2 - \delta a + \delta^2 - \delta^2 e^{-\frac{a}{\delta}}\right)H_0^{\,2}\right] \\
&\quad + \frac{\omega\mu_r'\mu_0 VA'^2\delta^2}{a^2}\left(\frac{1}{2}a^2 - \frac{3}{2}\delta a + \frac{7}{4}\delta^2 - 2\delta^2 e^{-\frac{a}{\delta}} + \frac{1}{4}e^{-2\frac{a}{\delta}}\delta^2\right)H_0^{\,2}
\end{aligned} \quad (30)$$

It is found from Equation (26) that μ_r' is proportional to V and depends on a and δ in the low-frequency range. It is also found from Equation (29) that μ_r" depends not only on V and a but also on δ. Further, in the high frequency range, the values of μ_r' and μ_r" given by Equations (26) and (29) can be approximated as follows.

$$\mu_r' = 1 - V \quad (31)$$

$$\mu_r'' = \frac{2V\delta}{a} \quad . \tag{32}$$

These two equations indicate that the values of μ_r' is determined by only V and that of μ_r" is determined by V and a in the high frequency range.

The above qualitative results may also be applicable to spherical aluminum particles. In the calculation of μ_r' and μ_r" using Equations (26) and (29), a was modified by $\sqrt[3]{2/3}\,a$ assuming that the volumes of the cylindrical and spherical particles are the same.

3.4. Measured and Calculated Values of μ_r' and μ_r" for a Composite Made of Aluminum Particles Dispersed in Polystyrene Resin

Figures 10 shows the measure and calculated values of μ_r' and μ_r" for the composite made of aluminum particles dispersed in polystyrene resin. The data points show the measured values for the composite made of 8 μm aluminum particles with the volume mixture ratio of 33.8 vol%, and the lines represent the values calculated using Equations (19), (20), (26), and (29). As shown in Figure 10, the measured values of both μ_r' and μ_r" were in qualitative agreement with the calculated curves. Thus, the theoretical calculation discussed in 3.3 can be applicable for determining the frequency dependences of μ_r' and μ_r" for the composite made of metal particles dispersed in resin. When the calculated values of solid and dotted lines are compared with the measured values of μ_r' and μ_r", solid line is similar to the measured values, as shown in Figure 10. This result indicate that the same phase of J_φ (x) as the incident magnetic field contributes the magnetic moment in the composite made of metal particles dispersed in resin, and that the real part of A is suitable for calculating the eddy current given by Equation (3). Therefore, the calculated values using Equations (19), (26), and (29) were compared with the measured values of μ_r' and μ_r" for the composites made of polystyrene resin and aluminum particles of various sizes and volume mixture ratios, as shown in Figure 11. The frequency dependences of the measured values of μ_r' and μ_r" qualitatively agreed with those calculated using Equations (19), (26), and (29) and the dependences of particle size and volume mixture ratio of aluminum could be explained by the theoretical calculation. From these results, it can be concluded that the frequency dependences of μ_r' and μ_r" can be predicted using Equations (26) and (29). However, the calculated values did not agree quantitatively with the measured values, as shown in Figure 11. One reason for this is considered to be the distribution of aluminum particle sizes around the average size. In addition, the low purity of the aluminum must be considered, since the conductivity of aluminum depends on its purity.

3.5. Effect of Particle Size Distribution on the Frequency Dependences of μ_r' and μ_r"

As mentioned in 3.4, the measured values of μ_r' and μ_r" did not precisely agreed with the calculated values. The reason for this difference is speculated to be the particle size distribution of aluminum. Therefore, the values of μ_r' and μ_r" were calculated by taking the particle size distribution into consideration. The values of real and imaginary parts of the relative complex permeability μ_{dr}' and μ_{dr}" obtained by taking the particle size distribution into consideration are given by

$$\mu_{dr}'=\int_0^\infty \rho(a)\mu_r'(a)da,\quad \mu_{dr}''=\int_0^\infty \rho(a)\mu_r''(a)da \tag{33}$$

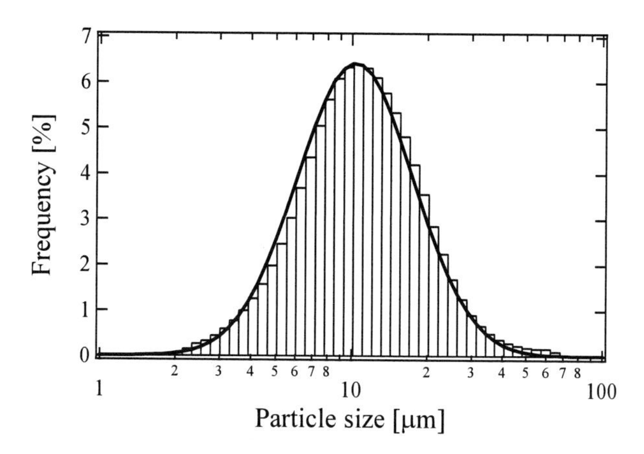

Figure 12. An example of a particle size distribution and fitted line obtained using density function of Gaussian distribution. Bars show the particle size distribution and the line shows the fitted values.

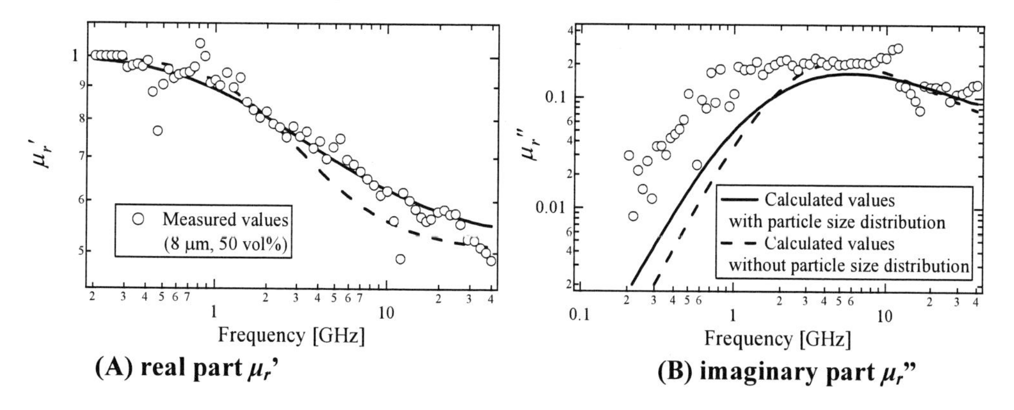

Figure 13. Values of μ_r' and μ_r" calculated by taking the particle size distribution into consideration.

Here, μ_r'(a) and μ_r"(a) are given by Equations (26) and (29), respectively; $\rho(a)$ denotes the density function of a Gaussian distribution and is given by

$$\rho(a) = A_0 \exp\left[-\left(\frac{a-a_0}{\Delta a}\right)^2\right] \tag{34}$$

The values of A_0, a_0, and Δa are determined by the particle size distribution function of aluminum using the least square method. Figure 12 shows an example of a particle size distribution and fitted line obtained using the density function of a Gaussian distribution.

Figure 13 shows the measured and values of μ_r' and μ_r" calculated using Equations (26), (29), (33), and (34). The plots show the measured values for 8-μm aluminum particles having a volume mixture ratio of 50 vol%, the solid lines show the values calculated considering the particle size distribution using Equations (33) and (34), and the dotted lines show the values calculated using only Equations (26), (29). As shown in this figure, when the particle size distribution is considered, the lines showing the frequency dependences of μ_r' and μ_r" tend to become almost horizontal after a particular frequency. This tendency is similar to that exhibited by the measured values of μ_r' and μ_r". In particular, the measured values of μ_r' almost agreed with the values calculated considering the particle size distribution. This result indicates that the frequency dependences of μ_r' and μ_r" are determined by the particle size distribution and that the values of μ_r' and μ_r" can be controlled by not only the volume mixture ratio and particle size but also the particle size distribution.

3.6. Frequency Dependences of μ_r' and μ_r" for the Composite Made of Nickel Particles Dispersed in Polystyrene Resin

To investigate the frequency dependences of μ_r' and μ_r" for the composite made of metal particles with magnetism, the values of μ_r' and μ_r" for the composite made of nickel particles dispersed in polystyrene resin were measured and compared with those for the composite made of aluminum particles. The average size of the nickel particles was between 10 and 20 μm, and the volume mixture ratio of nickel was 50 vol%.

Figures 14 and 15 respectively show the frequency dependence of μ_r' and μ_r" for the composites made of aluminum or nickel particles. The line in Figure 14 shows the value of μ_r' calculated from Equations (19) and (26) for $V = 0.5$ and that in Figure 15 shows the value of μ_r" calculated from Equations (19) and (29) for the composite made of aluminum ($V = 0.5$, μ_{Mr}' = 1, $a = 4$ μm). The calculated values of μ_r' and μ_r" for the composites made of nickel particles is not shown in Figures 14 and 15, because the value of μ_{Mr}' for pure nickel has not been obtained. The skin depths δ of nickel particles is estimated to be approximately 1.3 μm at 1 GHz using Equation (1), because the resistivity of nickel is and 6.84×10^{-8} Ωm. Thus, the eddy current also flows on the surface of nickel particles like aluminum particles. Thus, the effects of both the magnetism and the magnetic moments generated by the eddy current are observed in the composite made of nickel because nickel is a magnetic material and is conductive.

However, the value of μ_r' for the composite made of nickel decreased at frequencies above 20 GHz and approached a value of 0.5. This decrease resembles the decrease in μ_r' for the composite made of aluminum, and the value to which it approached almost agreed with that obtained from Equation (26). In addition, at frequencies above 20 GHz, the value of μ_r" for the composite made of nickel roughly agreed with that for the composite made of aluminum at frequencies above 20 GHz. These results indicate that the effect of the magnetic moment generated by the eddy current is dominant in the high frequency range because the effect of natural magnetic resonance is reduced as the frequency increases far from the resonance frequency. It is concluded from above results that the frequency dependences of μ_r' and μ_r" for the composite made of metal particles with magnetism is explained by the superposition of two effects: the natural magnetic resonance and the magnetic moment generated by the eddy current.

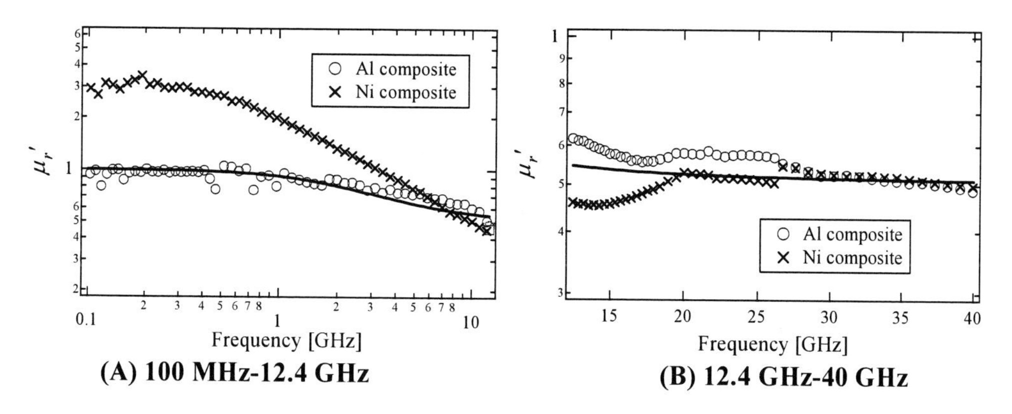

Figure 14. Frequency dependence of μ_r' for composites made of aluminum or nickel particles dispersed in polystyrene resin. The line shows the values calculated from Equations (19) and (26) for $V = 0.5$ and $a = 8$ μm.

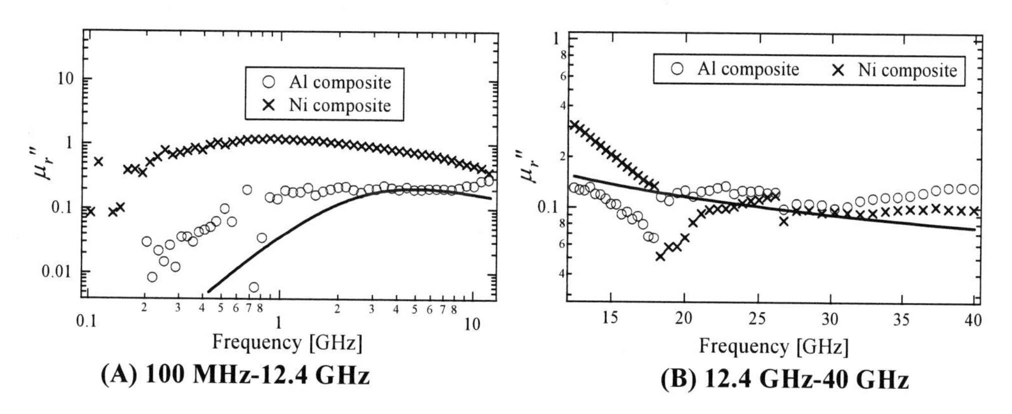

Figure 15. Frequency dependence of μ_r" for composites made of aluminum or nickel particles dispersed in polystyrene resin. The line shows the values calculated from Equations (19) and (29) for $V = 0.5$ and $a = 8$ μm.

3.7. Frequency Dependences of ε_r' and ε_r" for the Composite Made of Polystyrene Resin and Aluminum Particles

Figures 16 and 17 show the frequency dependences of ε_r' and ε_r" for the composite made of polystyrene resin and aluminum particles with sizes of 8 and 50 μm, respectively. The values of ε_r' for all composites were almost constant to the frequency, and the values of both ε_r' and ε_r" increased with increasing the volume mixture ratio of aluminum in spite of the particles size. In particular, the values of ε_r' and ε_r" increased rapidly when the volume mixture ratio of alumininum exceed 50 vol%. This rapid increase is explained by the percolation theory; the generation of clusters of aluminum particles. When the volume mixture ratio of the aluminum particles increases, the capacitance between the particles increases because the distance between the particles in the composite decreases. These capacitance elements form a 3-D network in the sample, the value of ε_r' of the composite is inverse proportional to the distance between the aluminum particles. If the volume mixture ratio exceeds the percolation threshold, the particles are partially in mutual contact and the total capacitance rapidly increases. Therefore, it is estimated that the value of ε_r' rapidly increases when the aluminum particles come in mutual contact. For the same reason, the values of ε_r" increased rapidly with a high volume mixture ratio of aluminum particles, because ε_r" is proportional to the conductivity of a material and conductivity of the composite increased owing to the mutual contact of aluminum particles.

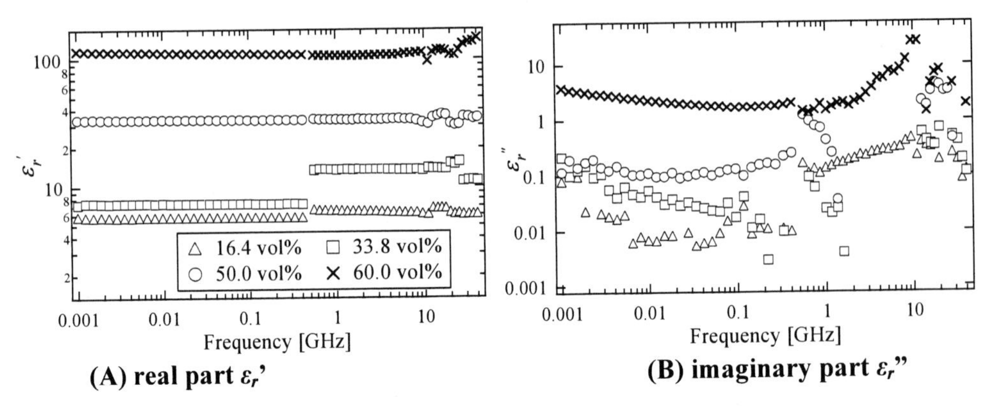

Figure 16. Frequency dependence of ε_r' and ε_r" for composites made of polystyrene resin and aluminum particles with sizes of 8 μm. The aluminum volume mixture ratios are 16.4, 33.8, 50.0 and 60.0 vol%.

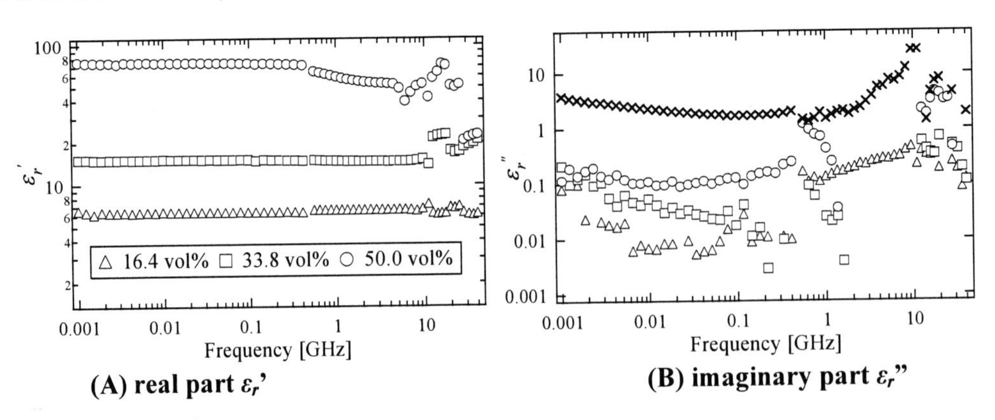

Figure 17. Frequency dependence of ε_r' and ε_r" for composites made of polystyrene resin and aluminum particles with sizes of 50 μm. The aluminum volume mixture ratios are 16.4, 33.8, and 50.0 vol%.

When the volume mixture ratio of aluminum is the same and the amount of aluminum particles is large, it is found from Figure 16 and 17 that the values of ε_r' and ε_r" for the composite made of large aluminum particles were larger than those made of small aluminum particles. This is because the large particles are more likely to contact with each other in the composite with a high volume mixture ratio of particles. However, the values of ε_r' and ε_r" were almost the same when the volume mixture ratio of aluminum is the same and the amount of aluminum particles is small. This result indicates that the frequency dependences of μ_r' and μ_r" can be controlled without changing ε_r' and ε_r" by simply changing the particle size of metal particle when the volume mixture ratio of metal particles is low.

3.8. Frequency Dependences of the Return Loss for the Composite Made of Polystyrene Resin and Aluminum Particles

The necessary condition to design an electromagnetic wave absorber is low values of the reflection and transmission coefficients. Therefore, an absorbent material is generally backed by a metal plate so that the transmission coefficient is zero and the magnitude of the magnetic field generated in the absorbent material is maximum. The absorber discussed in this chapter is a metal-backed single-layer absorber that has a low cost and is easy to fabricate.

The ideal absorption condition of a metal-backed single-layer absorber is given by the following equation [17].

$$\sqrt{\frac{\mu_r^*}{\varepsilon_r^*}}\tanh\left(\gamma_0 d\sqrt{\mu_r^*\varepsilon_r^*}\right)=1. \quad (35)$$

Equation (35) is well known as an important condition to design a metal-backed single-layer absorber. Hereafter, Equation (35) is referred to as the nonreflective condition. When this equation is satisfied, no reflection of an electromagnetic wave occurs, and the electromagnetic wave is completely absorbed in the absorbent material.

It is found from Equation (35) that to design a metal-backed single-layer absorber, an absorbent material with the values of μ_r^* and ε_r^* that satisfy the nonreflective condition should be selected or synthesized for absorbing the electromagnetic waves of the desired frequency. In addition, μ_r^* and ε_r^* of the absorbent material should have the frequency dependence that satisfies the nonreflective condition of Equation (35) because γ_0 depends on the frequency. However, for the absorbent material discussed in this chapter, only the frequency dependence of μ_r^* is the important factor because ε_r' is independent of frequency and ε_r'' can be assumed to be zero in the microwave range. In the case of a material with a constant value of ε_r' and zero value of ε_r'', a large amount of electromagnetic wave power is absorbed at a frequency where both μ_r' and μ_r'' of the absorbent material agree with or are close to the values that satisfy the nonreflective condition.

Moreover, the value of μ_r' must be less than unity to satisfy the nonreflective condition in a certain frequency range for a material with a constant value of ε_r'. Figure 18 shows the frequency dependences of μ_r' and μ_r'' that satisfy the nonreflective condition calculated using the least square method. The values of ε_r' used for the calculation are independent of the frequency and ε_r'' is assumed to be zero. From Figure 18, it is found that the values of μ_r' and μ_r'' must decrease with an increase in the frequency to satisfy the nonreflective condition. Moreover, the values of μ_r' must be less than unity in a high frequency range, even if the values of ε_r' is different. Therefore, the phenomenon that μ_r' becomes less than unity is an important factor in designing an absorber suitable for a high frequency range. The values of μ_r' for the composite made of aluminum particles dispersed in polystyrene resin were less than unity as shown in 3.2. Therefore, this composite can be used as an electromagnetic wave absorber in a high frequency range.

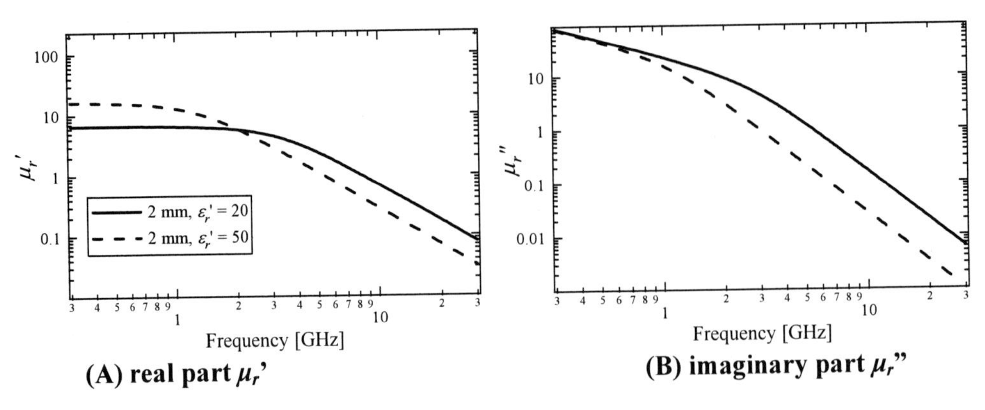

(A) real part μ_r' **(B) imaginary part μ_r"**

Figure 18. The values of μ_r' and μ_r" that satisfy the nonreflective condition when ε_r' is constant at 20 and 50 and ε_r" is assumed to be zero. The thickness of an absorber is 2 mm.

To examine the absorption characteristics of the composite made of aluminum particles dispersed in polystyrene resin, the frequency dependence of the return loss in free space was calculated from the measured values of μ_r^* and ε_r^* for all samples. The absorber used for the calculation was a metal-backed single layer absorber and the incident electromagnetic wave was perpendicular to the surface. As a result, good absorption characteristics were obtained for the composite with 33.8 vol%-aluminum and 40.0 vol%-aluminum in both cases with a 8 μm particle size, and for the composite with 33.8 vol%-aluminum particles of 30 μm. The frequency dependences of these three composite are shown in Figures 19, 20, and 21. The percentages shown in the graphs represent the normalized −20 dB bandwidth (the bandwidth Δf corresponding to a return loss of less than −20 dB divided by the absorption center frequency f_0). A value of −20 dB corresponds to the absorption of 99% of the electromagnetic wave power.

The return loss for three composites was less than −20 dB for several frequencies in the range 1 to 40 GHz if a suitable sample thickness was selected. These results indicate that practical high frequency absorbers can be realized by a suitable combination of aluminum volume mixture ratio and particle size. In particular, as shown in Figure 19(a), the composite with 33.8 vol%-aluminum particles of 8 μm had the large values of normalized −20 dB bandwidth in the range of several GHz and the sample thickness where the return loss was less than –20 dB was thinner than that of other two samples. This is because the frequency where μ_r' and μ_r" of composite with 33.8 vol%-aluminum particles of 8 μm begin to increase or decrease was the range of several GHz, as shown in Figure 6. The nonreflective condition is satisfied in the frequency range where μ_r' and μ_r" changes. Thus, the composite with 33.8 vol%-aluminum particles of 8 μm absorbed a large amount of electromagnetic wave power with a wide bandwidth at frequencies of several GHz. However, the values of normalized −20 dB for all composites were small at frequencies above 10 GHz, as shown in Figures 19(b), 20(b), and 21(b). This result is speculated to be that the values of μ_r' were almost constant in the high frequency range. If small sized aluminum particles is used, the values of μ_r' and μ_r" begin to change in the high frequency of above 10 GHz and the absorption with bandwidth is expected in the high frequency range.

The composite with high volume mixture ratio did not exhibit a return loss of less than −20 dB at frequencies above 1 GHz. This is because the values of ε_r' and ε_r" were large, as shown in Figures 16 and 17, and the nonreflective condition is difficult to be satisfied.

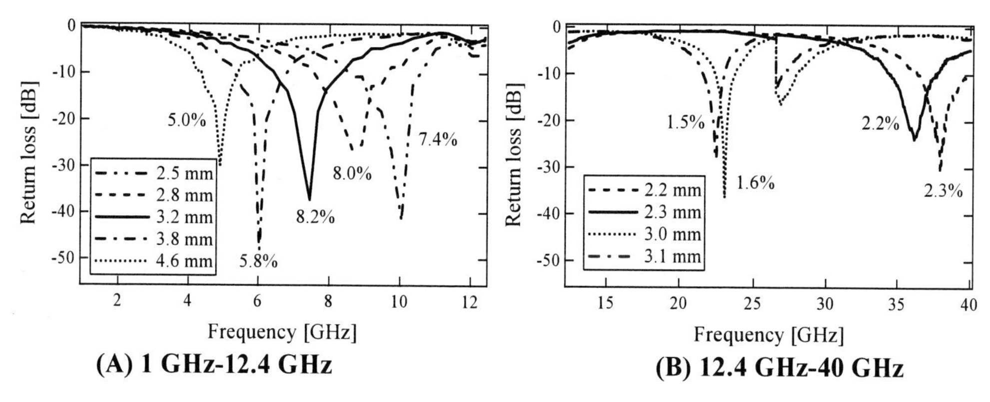

Figure 19. Frequency dependence of return loss for composites made of polystyrene resin and 8-μm aluminum particles. The volume mixture ratios of aluminum is 33.8 vol%.

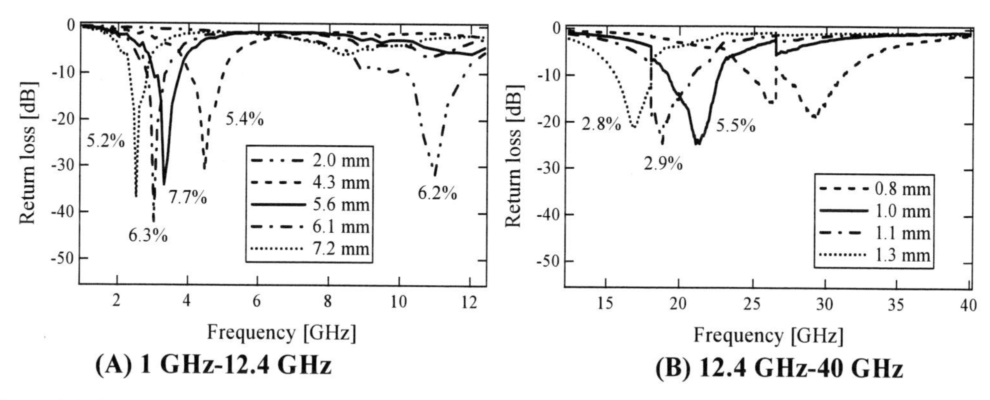

Figure 20. Frequency dependence of return loss for composites made of polystyrene resin and 8-μm aluminum particles. The volume mixture ratios of aluminum is 40.0 vol%.

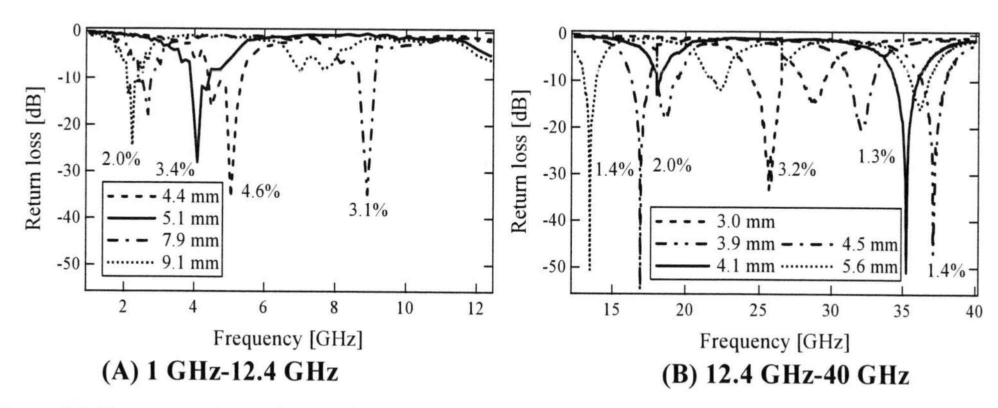

Figure 21. Frequency dependence of return loss for composites made of polystyrene resin and 30-μm aluminum particles. The volume mixture ratios of aluminum is 33.8 vol%.

To investigate the possibility of controlling the absorption characteristics, the values of μ_r' and μ_r" that satisfy the nonreflective condition given by Equation (35) were calculated using the least squares method. The values of ε_r' used for the calculation were independent of frequency and were the same as the measured values of the composite with 33.8 vol%-aluminum particles of 8 μm (ε_r' = 13). In the calculation, ε_r" was assumed to be zero and the sample thickness was 2.5 mm. Figure 22 shows the measured values of μ_r' and μ_r" for the composite with 33.8 vol%-aluminum particles of 8 μm and those calculated by Equation (35) under the above condition. As shown in Figure 22, the measured values of both μ_r' and μ_r" agreed with the calculated values at a frequency of approximately 10 GHz. Thus, a large amount of electromagnetic power could be absorbed at approximately 10 GHz, as shown in Figure 19(a). In this case, the value of μ_r' was less than unity when the nonreflective condition was satisfied. The absorbent material that the value of μ_r' becomes less than unity is limited. Therefore, the composite made of aluminum particles dispersed in polystyrene resin is suitable for fabricating an electromagnetic wave absorber of high frequency range. In addition this composite absorber is suitable for practical use, because this composite requires no scarce resource and is easy to fabricate.

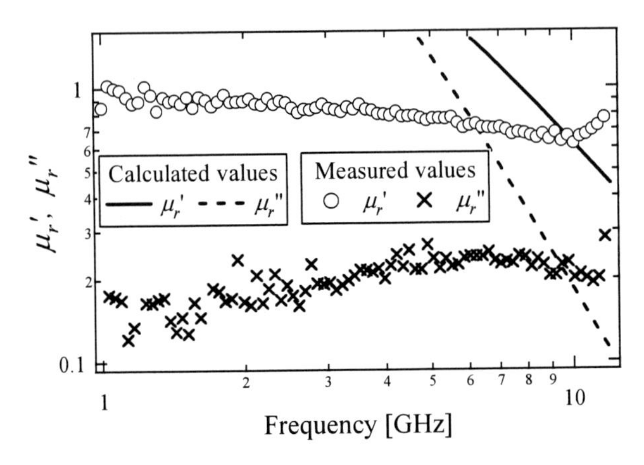

Figure 22 Measured values of μ_r' and μ_r" for the composite made of polystyrene resin and 8-μm aluminum particles with a volume mixture ratio of 33.8 vol%. Lines show the values calculated using Equation (35).

Moreover, this composite can control the absorption characteristics easily because μ_r' and μ_r" can be changed by adjusting the volume mixture ratio and particle size of aluminum, as discussed in 3.5, although the resulting increase or decrease in ε_r^* should be considered. Further, in the high frequency range where δ is much smaller than the radius of the aluminum particles, μ_r' depends only on the volume mixture ratio of aluminum, as given by Equation (33). Meanwhile, μ_r" depends on V and a at high frequencies, as given by Equation (34). Thus, μ_r' and μ_r" can be controlled independently by adjusting the volume mixture ratio and particle size of aluminum. This result suggests that the frequency at which the absorption of electromagnetic waves occurs can be selected. Moreover, it can be expected that absorption at frequencies exceeding 40 GHz is possible by modifying the values of the volume mixture ratio and particle size of aluminum.

CONCLUSION

Composites made of aluminum particles dispersed in polystyrene resin were prepared for various volume mixture ratios and particle sizes of aluminum and its electromagnetic properties were evaluated. The values of μ_r' for all samples were less than unity and the frequency dependences of μ_r' and μ_r" depended on the volume mixture ratios and particle sizes of aluminum. Moreover, μ_r' and μ_r" for the composite made of metal particles dispersed in resin were calculated theoretically and the calculated values qualitatively agreed with the measured values of μ_r' and μ_r". From this qualitative agreement, the prediction of the frequency dependences of μ_r' and μ_r" and the control of absorption characteristics of this composite was proposed. When a suitable volume mixture ratio and particle size of aluminum was selected, the composite made of aluminum particles dispersed in polystyrene resin was found to exhibit a return loss of less than −20 dB in the frequency range from 1 to 40 GHz. Therefore, it was found that this type of composite can be used as a practical absorber in the GHz range. Moreover, it was proposed that the absorption characteristics of such composite materials can be tailored based on the ability to control μ_r' and μ_r" by adjusting volume mixture ratio and the particle size of aluminum.

Acknowledgment

This research was supported by the Japan Society for the Promotion of Science (JSPS) and Grant-in-Aid for JSPS Fellows.

References

[1] T. Kasagi, T. Tsutaoka, and K. Hatakeyama, "Particle Size Effect on the Complex Permeability for Permalloy Composite Materials" *IEEE Trans. Magn.*, Vol. 35, No. 5, pp. 3424-3426, (1999).

[2] T. Tsutaoka, "Frequency Dispersion of Complex Permeability in Mn-Zn and Ni-Zn Spinel Ferrites and Their Composite Materials", *J. Magn. Magn. Mater.*, Vol. 93, pp. 2789-2796, (2003).

[3] S.-S. Kim, S.-T. Kim, Y.-C. Yoon, and K.-S. Lee, "Magnetic, Dielectric, and Microwave Absorbing Properties of Iron Particles Dispersed in Rubber Matrix in Gigahertz Frequencies", *J. Appl. Phys.*, Vol. 97, 10F905, (2005).

[4] T. Kasagi, T. Tsutaoka, and K. Hatakeyama, "Negative Permeability Spectra in Permalloy Granular Composite Materials", *Appl. Phys. Lett.*, Vol. 88, 17502, (2006).

[5] J. R. Liu, M. Itoh, T. Horikawa, E. Taguchi, H. Mori, and K. Machida, "Iron Based Carbon Nanocomposites for Electromagnetic Wave Absorber with Wide Bandwidth in GHz Range", *Appl. Phys. A*, Vol. 82, pp. 509-513, (2006).

[6] J. L. Snoek, "Dispersion and Absorption in Magnetic Ferrites at Frequencies Above One Megacycle," *Physca XIV*, Vol. 14, pp. 207-217, (1948).

[7] Y. Kotsuka and C. Kawamura, "Novel Computer Controllable Metamaterial Beyond Conventional Configurations and its Microwave Absorber Application," *IEEE MTT-S Int. Microwave Symp. Dig.*, 4264159, pp. 1627-1630, (2007).

[8] A. N. Lagarkov, V. N. Kisel, and V. N. Semenenko, "Wide-angle Absorption by the Use of a Metamaterial Plate", *PIERS Proceedings* (*PIERS 2008 in Hangzhou*), pp. 869-874, (2008).

[9] N. I. Landy, S. Sajuyigbe, J. J. Mock, D. R. Smith, and W. J. Padilla, "Perfect Metamaterial Absorber", *Phys. Rev. Lett.*, Vol. 100, Issue 20, 207402, (2008).

[10] A. Nishikata, "New Radiowave Absorbers Using Magnetic Loss Caused by Metal Particle' Internal Eddy Current", *Proceedings of EMC EUROPE 2002 International Symposium on Electromagnetic Compatibility*, pp. 697-702, (2002).

[11] T. Yamane, A. Nishikata, and Y. Shimizu, "Resonance suppression of a spherical electromagnetic shielding enclosure by conductive dielectrics", *IEEE Trans. on Electromagnetic Compatibility*, Vol. 42, pp. 441-448, (2000).

[12] Y. Wada, N. Asano, K. Sakai, and S. Yoshikado, "Preparation and Evaluation of Composite Electromagnetic Wave Absorbers Made of Fine Aluminum Particles Dispersed in Polystyrene Medium", *PIERS Online*, Vol. 4, pp. 838-845, (2008).

[13] P. Robert, *Electrical and Magnetic Properties of Materials*, Artech House, Inc., Norwood, MA, pp. 250-255, (1988).

[14] D. K. Misra, *Practical Electromagnetics*, John Wiley & Sons, Inc., New York, pp. 125-127, (2007).

[15] D. Stauffer and A. Aharony, *Introduction to percolation theory*, Taylor & Francis, London, pp. 1-7, 162-169, (1994).

[16] D. K. Misra, *Practical Electromagnetics*, John Wiley & Sons, Inc., New York, pp. 125-127, (2007).

[17] Y. Naito, and K. Suetake, "Application of ferrite to electromagnetic wave absorber and its characteristics," *IEEE Trans. Microwave Theory Tech.*, Vol. 19, No. 1, pp. 65–72, (1971).

In: Resin Composites: Properties, Production and Applications ISBN: 978-1-61209-129-7
Editor: Deborah B. Song

Chapter 5

ADVANCES ON RIGID CONDUCTING COMPOSITES FOR ELECTROANALYTICAL APPLICATIONS

Mireia Baeza, Rosa Olivé-Monllau, María José Esplandiu, Francisco Céspedes and Jordi Bartrolí
Universitat Autònoma de Barcelona, Spain

ABSTRACT

The development of composites based on conductive phases dispersed in polymeric matrices has led to important advances in analytical electrochemistry. Composite materials based on different forms of carbon as conductive phase have played a leading role in the analytical electrochemistry field, particularly in sensor devices. These materials combine the electrical properties of graphite and carbon nanotubes with the ease of processing of plastics (epoxy, methacrylate, Teflon, etc.). They show attractive electrochemical, physical, mechanical and economical features compared to the classic conductors (gold, platinum, graphite, etc.). Considering carbon composite in generally, their electrochemical properties present improvements over conventional solid carbon electrode, such as glassy carbon. The properties of these composites based are described, along with their application to the construction of electrochemical sensors.

The carbon-based composites exhibit interesting advantages, such as easy surface renewal, as well as low background current. Depending also on the conductive load, composites can behave as microelectrode arrays which are known to provide efficient mass transport of the electroactive species due to radial diffusion on the spaced carbon particles. Such improved mass transport favors the sensitive electroanalysis of a variety of reagents, including electrocatalysts, enzymes and chemical recognition agents. Moreover, the carbon surface chemistry also influences significantly the electron transfer processes at these electrodes.

During the past few decades, the electrochemical properties of different graphite powder composite materials based on different kinds of polymeric matrices were studied in detail. Nowadays, high interest is focused on composites based on carbon nanotubes (CNTs). They are attractive materials due to their remarkable mechanical and electrical properties. They have a highly accessible surface area, low resistance, high mechanical and chemical stability and their performance had been found to be superior to the other kinds of carbon material. The main drawback in CNT composite materials reside in the lack of homogeneity of the different commercial CNT lots due to different amount of impurities in the nanotubes, as well as dispersion in their diameter/length and state of aggregation (isolated, ropes, bundles). These variations are difficult to quantify and make mandatory a previous electrochemical characterization of the composite, before being used as a chemical sensor.

Another important point of consideration is the optimization of the conducting material (graphite or CNTs) loading in the composite materials for improving their electrochemical properties and analytical applications. Therefore, in this chapter we will describe the strategy to find the optimum composite proportions for obtaining high electrode sensitivity, low limit of detection and fast response. Compositions of composites can be characterized by percolation theory, electrochemical impedance spectroscopy, cyclic voltammetry, scanning electron microscopy, atomic force microscopy and chronoamperometry.

Moreover, the optimized carbon-based composite electrodes can be integrated in a continuous flow analytical system, as a flow injection analysis (FIA) or a miniaturized device, to take advantage of all the benefits provided by these automatization techniques.

1. Introduction

Environmental, clinical and industrial samples show great diversity and analytical complexity. At the same time, competition in the industrial arena has heightened considerably in the 1990s. Under these circumstances, electrochemical sensors have enormous potential as reliable, sensitive and robust devices. Although a wide range of chemical sensors and biosensors has been reported in the literature, very few of them have found their way to the market. Instrumentation prototypes are designed and built in research laboratories, but even when they show excellent analytical qualities, the devices and systems are often not suitable for industrial fabrication. Recent efforts are being made in the research of new electrochemical sensors in areas such as microelectrodes, chemically and biologically bulk- and surface-modified electrodes, and on devices based on new materials.

Among the electrode materials, carbon continues being a widely used and practical electrode material due to its desirable properties for electrochemical applications. Available in a variety of structures, carbon electrodes provide, in general, good electrical conductivity, high thermal and mechanical stability, a wide operable potential window with slow oxidation kinetics and in many cases electrocatalytical activity. Apart from that, they are recognized as versatile and easy handling materials. Thus, carbon in the form of glassy carbon, black carbon, carbon fibers, powdered graphite, pyrolytic graphite and highly ordered pyrolytic graphite (HOPG), has played for a long time an important role in solid electrode development, also favored by its rich surface chemistry which has been exploited to influence surface reactivity. However, the emergence carbon nanotubes (CNT) with their high surface area and nanometer size has exacerbated these properties and boosted in an unprecedented way its electrochemical and electroanalytical applications. For instance it has been widely praised the remarkable benefits and performance of CNT-modified electrodes in terms of low detection limits, increased sensitivity, decreased overpotentials, resistance to surface fouling and their interesting covalent/non-covalent chemical functionalization possibilities. The latter attributes can be used for selective chemical modification in the development of electrochemical (bio)sensors.

One of the more practical uses of carbon based electrodes is in a form of a resin composite. In general carbon composite materials present improvements in the electrochemical response over the ones we mentioned above of conventional solid carbon electrodes. They can exhibit the interesting advantages of high mechanical stability, easy surface renewal without extensive polishing, lower background current and in some cases microelectrode array behavior which brings about even more remarkable high signal/noise ratio, fast electrode response and low detection limits. However the fabrication of high performance carbon composites still results challenging and needs to be optimized before any practical use.

Another interesting issue under this context is related with the integration of the analytical processes into a simple device. Research fields with focus on chemical production, pharmaceutical screening, drug investigation, medical diagnostics and environmental analysis have increased their interest in rapid and on-line determinations of low sample concentrations [1, 2] that complex samples usually contain. Recently, and with the emergence of the new nano-materials it has been possible to develop analytical sensors and detection systems capable of addressing these needs. After the optimization of these sensors, they can be integrated as detectors into automatic analytical system based on continuous flow methodologies in order to obtain an automatic analytical system.

In recent years new technology has been developed that allowed great automation of the analytical process (sampling, sample pre-treatment, measurement, etc). Flow techniques such as flow injection analysis (FIA) and sequential injection analysis (SIA) have facilitated the integration of all the analytical steps in one system. Finally, with the implementation of microelectronic and computerized systems automatic control, data acquisition, treatment and interpretation could be achieved, completely closing the cycle that leads to autonomous automatic analyzers.

Indeed there is an increasing demand of in-situ, real-time, and on-line chemical information in different areas, such as biotechnology, environmental or industrial processes. This fact has accelerated the investigation and development of automatic analytical methods. Continuous flow methods, such as FIA, have a number of advantages, such as shorter analysis times, ease automation of sample pretreatment, versatility and real-time continuous monitoring. The miniaturization of these continuous flow systems adds to these advantages some others such as robustness, increase in the effectiveness of the sample manipulation and pretreatment, reduction of reagents and sample consumption, which increases the equipment autonomy, the possibility of mass and low cost production and, mainly, the portability of the systems, which permits to obtain not only real-time, but also in-situ information.

Thus, the integration of electrochemical transducers based on optimized carbon nanostructures with high-tech microfluidic systems can result in efficient chemical microanalyzers with a high analytical potential in a reduced space.

Therefore, in this chapter we describe the preparation, optimization, characterization and electroanalytical application of optimal composite electrodes based on conducting composite materials of the different allotropic carbon-polymer. Especially focus will be put in their use as improved sensors in automatic analytical systems based on different flow methodologies.

2. Conducting Composite Materials: Definition, Classification And Properties

A composite is defined as the combination of two or more dissimilar materials. Each individual component keeps its original nature while giving the composite distinctive chemical, mechanical and physical qualities, different from those shown by the individual components [3]. If any of its parts are conductor, the resultant composite become conductor and such property confer it a suitable feature in order to be used as electrochemical sensor.

The conducting composite can be classified by the nature of the conducting material. Under this classification, they can be a metal (silver, gold, etc. [4, 5]), a non-metallic conducting material (e.g., graphite, carbon nanotubes [6,7] or their mixture [4, 8, 5). On the other hand, they can be also classified taking into account the polymer matrix. Such classification is divided in insulator polymers (such as epoxy, silicone, polyurethane, polystyrene, methacrylate, or Teflon composites) and conducting polymers (such as polyanilines)[9]

At the same time, conducting composite can be also classified depending on the arrangement of such conducting material in the electrode surface. A first classification could be given depending on whether the conducting material confers an ordered or a random distribution. For a random distribution a secondary division can be given depending on whether the conducting particles are dispersed in the polymer matrix or are grouped or consolidated in clearly defined conducting zones. A composite is said to be a *dispersed composite* if the conductive material is distributed randomly and its particles have the same probability of occupying any given space in the matrix. On the other hand, c*onsolidated composites* exist when the conductor is dominant in clearly defined areas and where the insulator is the primary material (see Fig. 1).

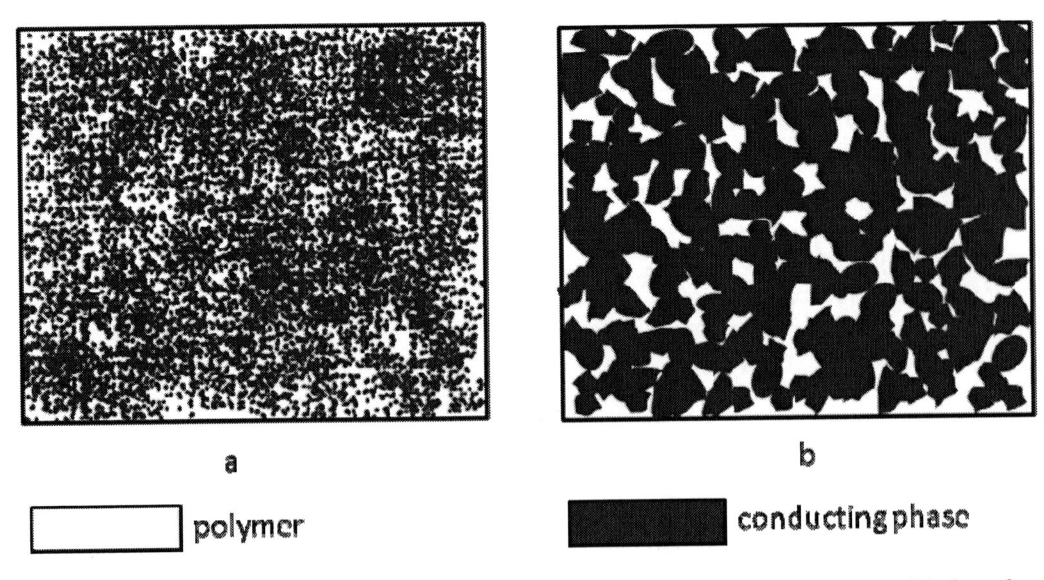

Figure 1. Morphology of the conducting composite formed by (a) dispersion and (b) consolidation of the conducting particles. Reproduced from ref. [12] with permission. Copyright (1996) Elsevier.

In this chapter we will focus on composite based on different allotropic forms of carbon (non-metallic conducting material) and polymeric matrices. Particularly, under this context, a composite formed from carbon fibers and an adhesive polymer matrix is an example where the carbon has an orderly distribution [10]. Such ordered material is technologically difficult to achieve. Indeed, if the carbon is distributed randomly, the composite is easier to prepare (see Fig. 1). An example of random distribution is that formed by a graphite powder in a soft matrix based on a water-immiscible insulating organic liquid (Nujol, hexadecane, etc), producing the well know as 'carbon paste' [11]. Other example includes epoxy resin which provides a rigid matrix by thermal curing.

Actually, the polymer should be chosen according to the final sensor application. For instance, if a biological modification is part of the construction procedure, a polymer that needs high curing temperatures should not be used. The same is true if the polymer contains additives which are harmful to the biological material. When a chemical modification is necessary (addition of redox mediators, catalysts, etc.) the modifying substance should not interfere with the curing of the polymer. Finally, consideration should be given to the environment in which the sensor is going to be applied, such as a living organism, an aqueous or organic solution, a biofluid, etc. This is important because biocompatibility or corrosion effects may have to be considered. Composite materials may be used in consonance with the polymer used. The curing conditions have to be optimized: they include the curing time, temperature and type of atmosphere (air, nitrogen, etc.)

2.1. Percolation Theory

The electrical properties of the conducting composite depend on the nature of each different component, their relative quantities and their distribution. The conductivity behavior of the composite can be explained by means of the percolation theory. This theory was developed by Broadbent and Hammersley in 1957 [13]. Initially, the original concept treated about a statistical study of probability of spread of fluid, such as liquid, gas, electrical current, infection, etc., in a random medium, such as a porous stone, the forest, etc. However, it was not until years later, when Scher and Zallen (1970) [14] applied this theory to describe the conductivity behavior in a random conductive composite material.

The simplest theoretical tridimensional model to define a composite system consists of a random compact spheres network. The spheres represent the *filler* component and the spaces among spheres correspond to the *matrix* component (see Fig. 2).

The most important parameter that this theory contributes is called percolation threshold. Such parameter is defined as the concentration of a component which provides the formation of an infinite *cluster* of this component. Such *cluster* is defined as a group of spheres from the same component, which forms a long-range conductive network in a random system [15,16].

Since its first application in conductive composite systems, several researchers have also applied it to explain the characteristics of various types of composite systems [17-19].

Navarro-Laboulais *et. al.* applied the percolation theory to interpret analytically several aspects of the composite behavior. They explained the impedance behavior of a composite based on graphite/polyethylene [20]. In such work, they presented the second percolation threshold, which corresponds to the highest load of graphite from which the composite loses its physic integrity.

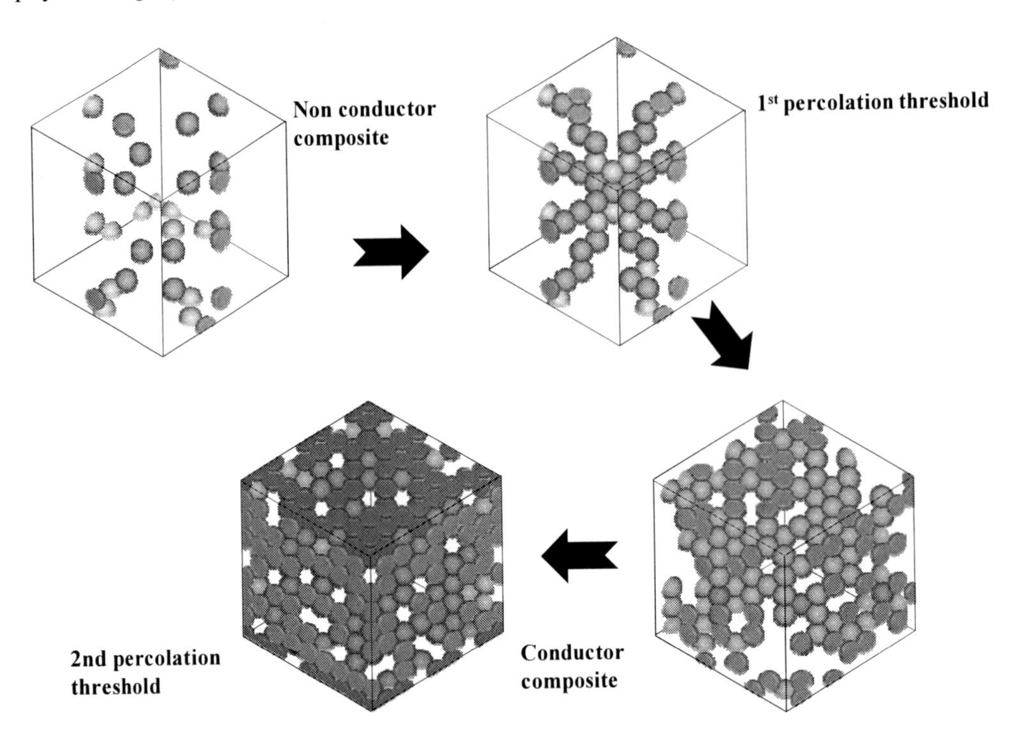

Figure 2. Percolation process of a ramdom composite material.

Recent examples of its application are the studies provided by Kovaks et al. [19]. Using a composite based on MWCNT and epoxy resin, the percolation theory allowed them to measure the shifts in the conductivity as the dispersion degree of the MWCNT in the matrix was varied. Their results showed a decrease of the percolation threshold value as the dispersion of the particles was increased. The Percolation Theory has also applied to compare different fabrication processes of composite materials [21], as well as, to optimize different parameters, such as cured temperature, mix rate, etc., in different steps of the fabrication process [17].

Thereby, the percolation theory has become an important tool to characterize, compare and evaluate conductive composite systems by the shift of the conductivity or percolation threshold value as varying one or more parameters, such as shape, size, dispersion, etc. of the particles, as well as others parameters involved in the fabrication process.

A standard percolation curve [22] is depicted in figure 3 which takes as an example a carbon based composite. It gives us information about the minimum conductor content that ensures certain conductivity, this point being known as the percolation threshold. The construction of such curve represents the first step to optimize the composite composition.

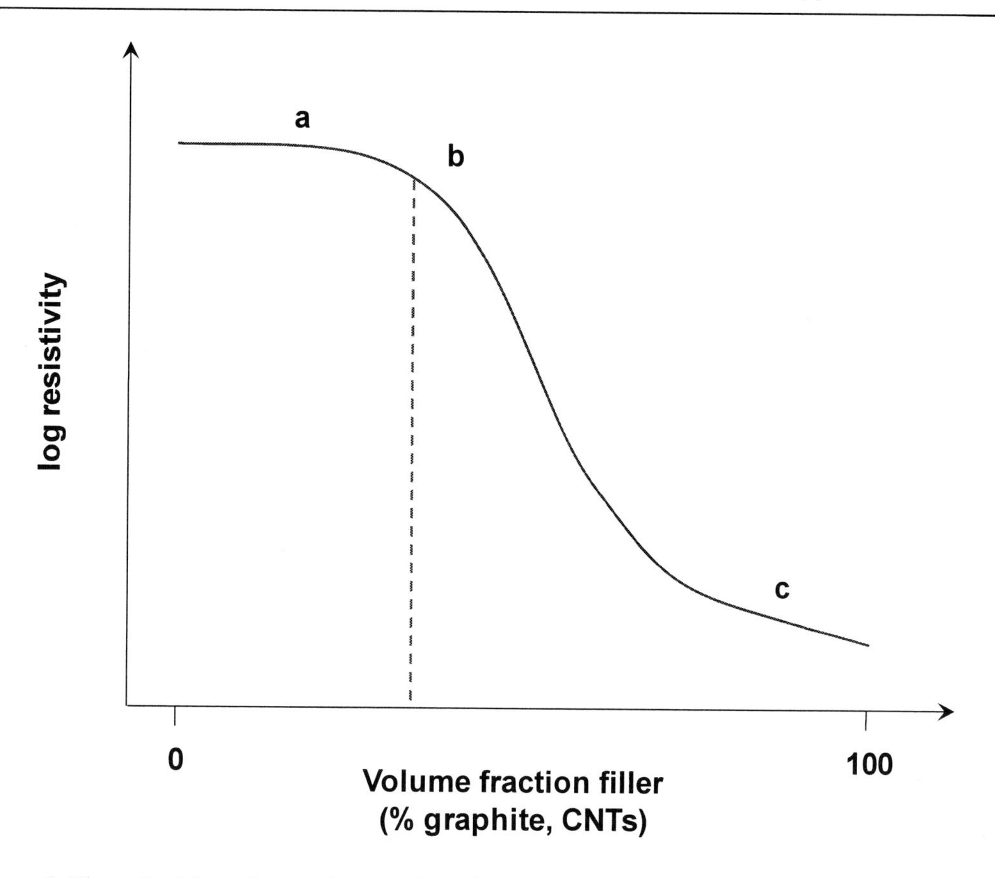

Figure 3. Theoretical dependence of composite resistivity on conductive filler content. (a) The electrical resistance of the composite is similar to the resistance of the polymer. (b) Percolation fraction: a critical conductive filler content permits the formation of the first conducting filament consisting of particle to particle contact. (c) The resistance of the composite is similar to that of pure conductive filler. Reproduced from ref. [12] with permission. Copyright (1996) Elsevier.

3. Conducting Electrodes Based on Carbon-Based Composites

In 1958 Adams proposed, for the first time, the use of carbon paste, a soft conducting composite, as a material for the construction of electrochemical transducers [11]. These composites have a conducting material, such as graphite powder, mixed with an inert, insulating, viscous liquid such as paraffin oil or nujol. Since such a first prototype, several reviews on carbon paste electrodes (CPEs) have appeared in the literature [23-25]. Such devices are cheap, easy to make, are easily miniaturised and show good biocompatibility [26]. However, they also show some undesirable features such as low chemical and physical surface stability. Such composite may soften in presence of certain electrolyte solvents and may produce undesirable residual currents [27] due to the lack of their purity. Indeed, their fast degradation has impeded their use outside the research laboratory.

However, at the beginning of the 1990s, new carbon resin composite electrodes became the low cost alternatives for amperometric sensing. They were based on new rigid composites with advantageous physical, mechanical, economical and electrochemical features. Unlike the carbon paste, such rigid composites do not show stability problems and keep the physical and cost advantages of the CPEs. Additionally, the construction method can be translated to industrial production. They can easily be bulk modified with stabilized enzymes [28, 29] to make biofuel cells or used in electro-analysis to detect acids, [30] glucose [31] and enzymes [32]. Both the conducting material and the binder can be modified with redox catalysts or selective receptors. Indeed, modified composite electrodes have found widespread biomedical applications ranging from the in vivo detection of nitric oxide [33] to bulk-modified enzyme based biosensors [34, 35]. Other important applications include electrochemical detection in Flow Injection Analysis (FIA) and Low pressure ionic Chromatography LC [36] where the relatively weak dependence of current on flow rate is an advantage [10].

Currently, carbon nanotubes (CNTs) have revolutionized the electroanalytical field. Along with the advent of CNTs, the use of nanocomposites has also been boosted.

Below we will briefly discuss the structure and properties of graphite and carbon nanotubes as the typical conductive fillers of the carbon-based composites discussed along this chapter.

Graphite

Graphite is one of the most known allotropic forms of carbon. The carbon atoms are all sp^2 hybridized, with three covalent bonds forming 120° in a same plane (hexagonal structure) and with a Π orbital perpendicular. Such delocalized orbital are important to define the electrical behavior of the graphite (see Fig. 4).

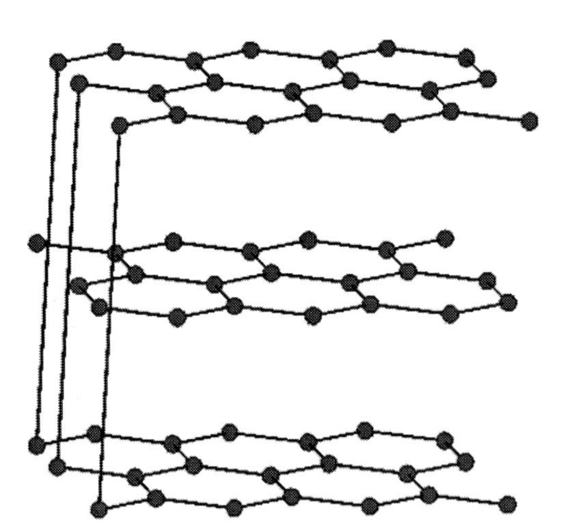

Figure 4. Graphite structure.

Its structure is laminar. The simplest graphite material is a two-dimensional graphene sheets, which is essentially a very large polyaromatic hydrocarbon. An intraplanar C=C bond is extremely strong with a length of 1.421 Å [37], whereas interplanar spacing is not so strong with a length of 3.354 Å [38]. Since its laminar structure, the graphite is deeply anisotropic, showing a higher conductive behavior in the x-plane and y-plane than z-plane.

In order to give a perspective, there are more than 10000 papers related with graphite composite electrodes and in the vast majority of them, the conventional electrode configuration consists of mixing homogeneously the graphite with the hardener and then to introduce the composite in to the body electrode, which contains the interface with the analytical instrument.

Carbon Nanotubes

Carbon nanotube structure is defined as a single or multiple layers of graphene sheets "rolled up" to form tubes of varying diameter, length and termination. Since the discovery of multiwall carbon nanotubes (MWCNT) and their homologous with single wall (SWCNT, single-walled carbon nanotubes) by Ijima [39,40], a big amount of studies focused on the development of electrochemical sensors based on these kind of materials have increased daily thanks to their remarkable electrical properties [41-45].

Such properties together with their mechanical advantages have amplified their application field. Actually they are also applied as field transmitters [46], supercapacitors [47], nanometric circuits [48], tips for scan microscopy [49], actuators [50], transistors [51], and photovoltaic devices [52].

The CNT structure can be defined using a rolled-up vector *Ch* [53] and a chiral angle θ. According the equation 1, *Ch* is a linear combination of base vectors a_1 and a_2 of the Bravais network of the graphene sheet (figure 5 A), where m and n are integers and they allow us to classify the different types of CNTs according to the different ways of rolling up the graphene sheet.

$$C_h = n\vec{a}_1 + m\vec{a}_2 \quad (1)$$

The angle and index (n,m) determine the chirality and helicity of the CNTs (figure 5 B). The electrical properties of the CNTs depend directly on such parameters. Thus, they can form metallic conductive structures such as the armchair state (10,10) or zig-zag state (10,0); or a semiconducting structures (e.g. chiral state (10,5)).

Varying different parameters in the CNT growing up process, such as dimension and nature of the catalyst, temperature, carbon source, etc., one can form CNT of a single rolled up graphene shell (SWCNTs) or multiwalled carbon nanotubes consisting of several concentrically arranged of SWCNTs nested into each other.

Generally, the CNTs are considered as unidimensional systems due to their high length to diameter ratio. SWCNTs tend to have a diameter in the range of 0.4 -3 nm, whereas that range for the MWCNTs is between 2 to 100 nm, with a layer spacing of 3.44 Å [54].

There are several methods to fabricate CNTs, but the most used are mainly the arc discharge, laser ablation and the catalytic chemical vapor deposition (CVD). Since all of these techniques produce the formation of by-products containing impurities such as metal particles and amorphous carbon, extensive research has been devoted to the purification of such CNTs. Among all of them, it is important to highlight the acid treatment [55, 56], which remove the metal particles that modify the physico-chemical properties of the CNTs; thermal oxidation [57, 58] to remove the amorphous carbon impurities, as well as the thermal curing [59, 60] to remove the partial structural defects and the chromatography [61, 62] to separate small quantities of SWCNTs in fractions of length and diameter distribution.

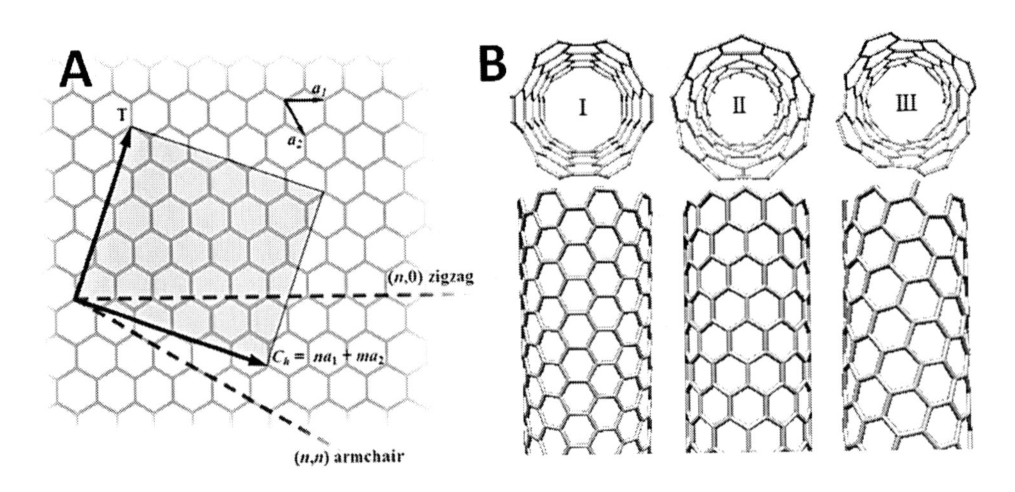

Figura 5. A) Scheme of a graphene layer showing the rolling-up vectors for SWNT formation. B) Scheme depicting three different structures of SWNTs (n,m): I) metallic nanotube (10,10), armchair; II) metallic nanotube (15,0) zig-zag; III) semiconductor nanotube (12,7), chiral.

Some important characteristics of CNT are: their insolubility in aqueous media causing CNT bundles, which provides a losing of their high aspect ratio. Usually, such bundles are avoided with the use of surfactants. Contrarily, this behavior is less observed using some organic solvents. Moreover they can be functionalized by covalent modification. They present an excellent thermal stability in inert atmosphere, being stable up to 1200 °C. Their elasticity module is 1000 GPa, five times higher than that obtained by the steal [63]

In order to construct CNT electrodes and taking into account their insolubility with some habitual solvents, some methods have been developed such as the immobilization of CNT above glassy carbon or gold electrodes by solvent dispersion [64] and polyelectrolyte [65, 66], as well as their integration in composites [67].

3.1. Fabrication of the Carbon-Based Composite Electrodes and Devices

Carbon-based composites (graphite and CTNs) represent low cost systems which makes them ideal for mass produced sensors, especially when one compares them with other materials such as gold and platinum. Allotropic carbon forms and polymer materials are readily available, and the easy preparation of the composites does not require sophisticated equipment or highly skilled personnel.

The fabrication approach is simple, cheap and the high malleability of the developed composite allows easy incorporation in a flow cell to perform electrochemical measurements in flow conditions at automatic analyzers.

As an illustrative example, here we will describe the conventional methodology used by our group to fabricate the carbon-based composites [68]. Handmade epoxy-graphite and epoxy-CTNs composites are similarly prepared mixing different amounts of conducting material and polymer.Graphite powder (particle size 50μm) is received from BDH (BDH Laboratory Supplies, Poole, UK). Epoxy resin Epotek H77A and hardener Epotek H77B are obtained from Epoxy Technology (Epoxy Technology, Billerica, MA, USA). Purified multiwall carbon nanotubes fabricated by chemical vapor deposition (MWCNTs purity > 95%, length 5-15 μm, outer diameter 10-30nm) are purchased from SES Research (Houston, TX, USA).

Epoxy-carbon composites were prepared as follows: polymer (EpoTek H77A) and its corresponding hardener (EpoTek H77B) are mixed in a proportion 20:3 (w/w) before loading different amounts of carbon. Then the resin mixture and the different carbon proportions are homogenized. The homogenization time depend on the composite system fabricated. For instance the homogenization time for epoxy-CNT composites (60 min) is higher than for epoxy-graphite composites (30 min).These differences can be related with the lack of homogeneity in the lot of CNTs and difficulty of mixing manually the CNTs and epoxy resin. Moreover, CNTs have difficulties to disperse and dissolve in any organic and aqueous medium due to the strong Van der Waals interactions which drive their aggregation.

For the electrode construction the composite is placed into a cylindrical PVC tube (i.e. 6mm i.e., 20mm length) containing a copper disk soldered to an electrical connector end. The final paste-filled cavity is 3mm long inside the PVC tube. Afterwards the composite paste is allowed to harden. As previous mentioned, the curing conditions, such as curing time and temperature are specific for each composite system and they require to be optimized. Following with our carbon-based composites, for epoxy-CNT composites, the curing conditions are during 24 hours at 80°C [69] whereas for epoxy-graphite are 24 hours at 60°C[70]. Completed process is shown in figure 6.

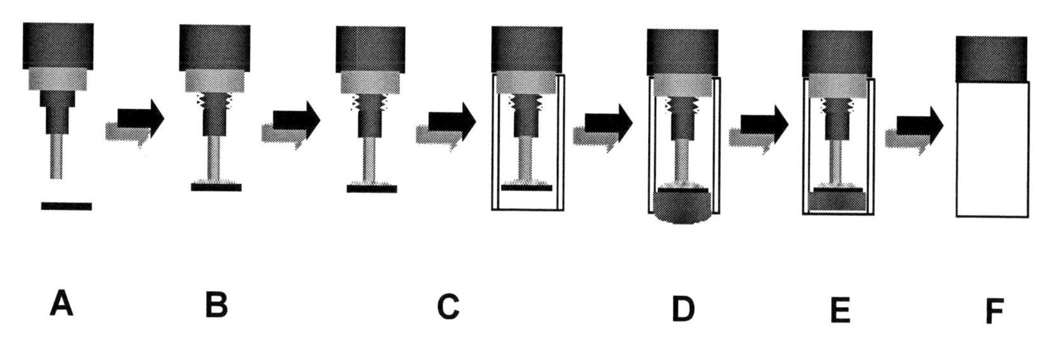

Figure 6. Different construction steps of composites electrodes for batch measurements. The carbon paste was well mixed during half or one hour (graphite and CNTs, respectively) and put in the cavity of the PVC body and thermally cured. A) Electrical connector; B) copper disk fixture; C) mount of PVC body; D) introduction of carbon-epoxy paste; E) polishing process and F) final sensor.

Finally, before electroanalytical measurements electrode surface is polished with different sandpapers of decreasing grain size. Final electrode dimensions were 28 mm^2 and 3mm for its geometric area and thickness, respectively.

We also show as illustrative example, the fabrication of the carbon-based composites as sensors in a flow system. For such purpose different strategies can be implemented but we will describe the flow cell used for our experiments in a conventional FIA system. The flow cell consists of two methacrylate blocks (Goodfellow, Oakdale, PA, USA). As seen in inset figure 7, the lower block contains a cavity filled up with transducer material, which corresponds to the working electrode whereas the upper block contains a stainless steel piece, which corresponds to the counter electrode. The upper block has two channels (45° respect to the block surface), for inlet and outlet flows (see figure 7). A mechanized silicone rubber piece with an orifice that matches with the sensor channel are used as a meeting point for both blocks, defining a chamber of 50 μL and a sensitive area of 10 mm^2 for the continuous flow sensor.

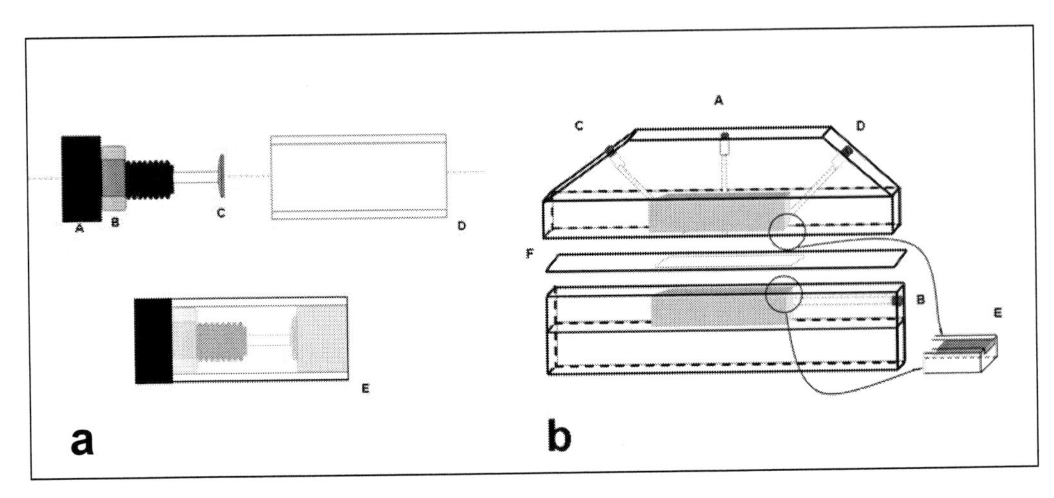

Figure 7. a) Sensor used in batch measurements for composite optimization (A: electrical connector; B: metallic nut; C: piece of copper; D: PVC tube; E: conductive resin (lower block) and stainless steel (upper block)). b) Flow cell for continuous measurements (scheme of the methacrylate cell for the integration of the sensor in a flow system. A: working electrode; B: auxiliary electrode; C/D: Inlet/outlet; E: amplification of the flow channel; F: meeting point of the perforated silicone). Adapted from ref. [71] with permission. Copyright (2009).Wiley-VCH.

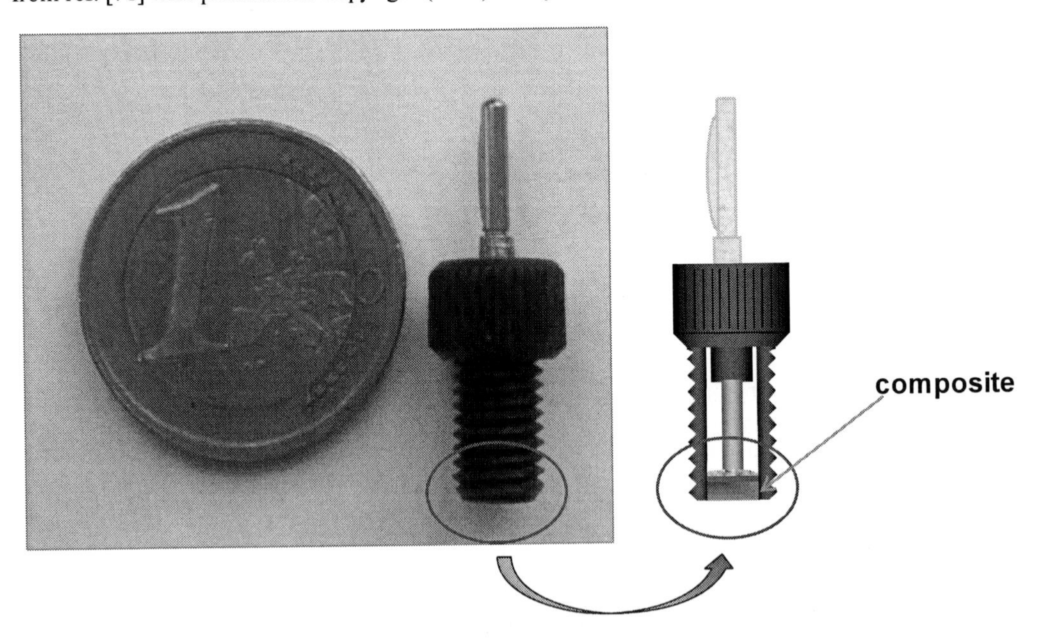

Figure 8. CNT composite-based working electrode for measurements in microfluidic systems.

On the other hand, for our miniaturized flow system a special support for composite sensor is needed. In this case, for the electrode construction, the composite mixture is placed into a PVC fitting (Cavro, 4mm i.d, 15mm length), containing a cooper disk (2 mm diameter) soldered to a 2 mm banana male end connector.

The male connector is used as electrical contact with the external electronic set up. The final dimensions of the cavity to be filled by the CNT composite are 2mm in depth and 4mm in diameter. The working electrode is integrated to the microanalyzer by means of a PVC nut previously glued to the microdevice body over its main flow channel, in a cavity specifically designed for this purpose. The final sensor is showed in figure 8.

4. Carbon-Based Composites for Amperometric Sensing

In the introductory part we mentioned the electrochemical benefits of using carbon-based composites. The polymeric matrix serves as a support for the carbon material. This support provides mechanical and chemical stability to the composite. However, the distribution of the conductor in the bulk and in the surface of the polymer will determine some electrochemical properties of the composite. As already said, carbon-based materials such as graphite or glassy carbon are ideal conductors for rigid composites. They show a high chemical inertia which provides a wide range of anodic working potentials. Moreover, they also have a highly pure crystal structure, low electrical resistance (10^{-4} Ω cm, approximately) and relatively low background currents as compared to other conventional metal electrodes.

A key feature in any electrochemical sensor is a good detection limit that is associated with a high signal-to-noise ratio. The background current or noise in a macrosensor made of a pure conducting phase depends on the sensing area. This means that a contraction of the sensing surface produces a reduction in noise. It is evident that the reduction of the sensing area implies a curtailment of the sensitivity as well. However, in accordance to the equation derived by Oldham [72], if the area is very small, as in microelectrodes, the perimeter of the surface has a more significant influence on the mass transport than the area itself. This is translated to a better sensing function, since a non-linear diffusion is established which generates a steady-state current that increases the signal-to-noise ratio. This is known as the edge effect: along with the reduced size of the electrode it has interesting and useful results. However, the generated currents are very low, calling for sophisticated and expensive voltammetric equipment. Hence, whenever small size is not required, the use of arrays or ensembles of microelectrodes is desirable. These ensembles can be seen, for example, as a macro electrode formed by a great number of carbon fiber microelectrodes. These microelectrodes are separated by an insulator and connected in parallel. The signal is the sum of the individual currents generated by each microelectrode. The end product is a sensor with a signal as strong as that of a macroelectrode but showing the signal-to-noise ratio of a microelectrode.

For a pre-design surface of a carbon microelectrodes array, Weber et al. [73] showed that the layout of such an array has a maximum efficiency when the separation between the microelectrodes is around 0.1 μm. However, the implicit difficulty of constructing these electrodes is very high. On the other hand, electrodes based on carbon based composites with a random distribution are easier to prepare. Indeed, in certain condition, their electrochemical characteristics can resemble very well to a random ensemble microelectrode, all depends on the filler composition, its size, its layout, etc.

In spite of all these advantages for sensor applications, there are some issues that need to be addressed before their use as sensors. One of these issues is the composition optimization in order to obtain the best electrochemical response of the sensor electrode. In the case of the MWCNTs one has also to overcome the difficulties associated with the lack of homogeneity of the CNT material. Such lack is due to different amount of impurities in the nanotubes, as well as dispersion in their diameter/length and state of aggregation (isolated, ropes, bundles). These variations are difficult to quantify and make mandatory a previous electrochemical characterization of the composite [74-77], before being used as a chemical sensor.

5. Modification of Carbon-Based Composites for Electroanalytical Improvement: Tuning Material

A good sensor shows high sensitivity and selectivity. However, selectivity is often lacking in amperometric sensors. This calls for chemical modification of the sensing material in order to enhance the quality of the analytical signal. The ease of modification is a further advantage of composites over conventional solid electrodes. The surface modification of electrochemical sensors based on pure conductors is generally difficult and costly, needing complex surface treatments to bond the modifying species. In contrast, the plastic nature of the composites permits easy modification of the surface by adding different species to the composite matrix [78, 9]. Such modifiers may perform several functions. They may act on the analyte by pre-concentrating it on the surface, of the electrode, they can function as catalysts, or immobilize some of the molecules involved in the electrochemical reactions, or alter the physical properties of the electrode surface.

The pre-concentration of some analytes is achieved by ligands or chemical recognition agents and ion exchangers. These species act selectively on the analyte, separating it from the initial sample and eliminating some interferents. The concentration of the analyte is determined by oxidation or reduction at the optimal potentialThe incorporation of (bio)chemical recognition agents into carbon-based composited has been widely exploited specially in the field of genosensors, immunosensors and enzymatic sensors [79-81]. Rigid carbon based composites are amenable to this type of application, since the surface can be regenerated by simple polishing. Additionally, the homogeneous addition of the modifier to the bulk of the composite guarantees that, a reproducible surface is obtained after each polishing [82, 12].

Some metals or oxides have catalytic effects on a large number of chemical species. Such metallic or oxide catalysts can be easily mixed on the bulk, allowing us to obtain a modified carbon based composite with a tailored sensing surface. Photoelectric studies have shown that the metal donates electrons to the carbon particles, favoring electron transfer [83]. Thus, Rh and Pt [84], and metal combinations such as Au-Pd [85], have been applied to the catalytic oxidation of diverse species. Improved electrochemical performance has also been observed with carbon nanotube composites modified with gold, Pt, Ag, Cu, FeCo, PtRu nanoparticles, transition metal hexacyanoferrato nanoparticles, etc. [86, 87, 9].

Gold and magnetic particles have also been added to carbon composites for improving the strategies of bioreceptor immobilization in the composite matrix. For instance it has been demonstrated that the integration of magnetic beads to graphite epoxy composite provides further advantages in terms of separation of the analyte from complex matrix and enhancement of the biological reactions, with promising applications in food safety and environmental monitoring [88].

Recently it has been reported the preparation of CNT composites with ionic liquids incorporated in the matrix. Apart from having being demonstrated synergetic enhancement of the amperometric signal to different redox analytes [89], such composites have emerged as favorable platforms for protein or enzyme immobilization preventing the deactivation of the protein functions and favoring direct electron transfer between the proteins and the carbon electrode [90].

Redox mediators represent another type of modifiers which consist of a metal atom whose oxidation states are stabilized by the π orbitals of aromatic structures [91-93]. Examples of these modifiers are cobalt phthalocyanine (CoPhC) and ferrocenes. These aromatic molecules act as electron-exchange chains mediating in the electrochemical reaction. Although the signal is related to the measured analyte, it is generated by the oxidation or reduction of the mediating species. An important drawback of these sensors is the poor stability of the mediator when one or both of its oxidation states is a soluble ionic species. Nonetheless, regeneration of the electrode is a simple maneuver, involving a simple polishing of the surface.

Another class of modifiers comprises species that confer certain physical properties on the sensor. A problem encountered in composite-based electrodes is the high hydrophobicity provided by the polymer itself. Wang and Liu [94] have proposed the addition of fumed silica to the composite to make the material more hydrophillic, thus improving the sensitivity and the stability of the sensor.

Summarizing the vast studies with (bio)chemically modified carbon composites we can remark that modifiers (chemical recognition agents, catalysts, redox mediators, inert materials, etc.) can mainly function by either changing the reactivity of the composite electrode (e.g. decreasing the working potential and enhancing the analytical sensitivity) or by providing the desired selectivity for the sensor applications.

5.1. Types of Modification

Modification can be carried out in a number of ways, depending on the type of composite. Creasy and Shaw [91] studied the advantages and disadvantages of modification of carbon fiber ensemble electrodes on the bulk (polymeric material) or on a ring around the carbon fiber (see figure 9). If the redox mediator content is raised, it is natural to expect a corresponding raise in sensitivity. Therefore, the ring modification produces stronger analytical signals because the redox modifier is in close contact with the conducting species. In bulk modification, the modifiers are embedded or trapped in the polymer matrix and therefore a fraction of the redox modifier could be inactive because it is surrounded by the insulating polymer phase of the composite. However, ring modification can have an important flaw, since the modification of the conducting phase can separate it from the polymer, reducing the adhesion between the two phases and thus the stability of the composite.

Bulk modification can be sometimes more preferred in composites based on graphite particles dispersed in a polymer matrix. It results quite attractive since the composite serves itself as a reservoir of the (bio)chemical species. Furthermore, bulk modification can be simpler and allow the regeneration of a deteriorated surface by a simple polishing procedure.

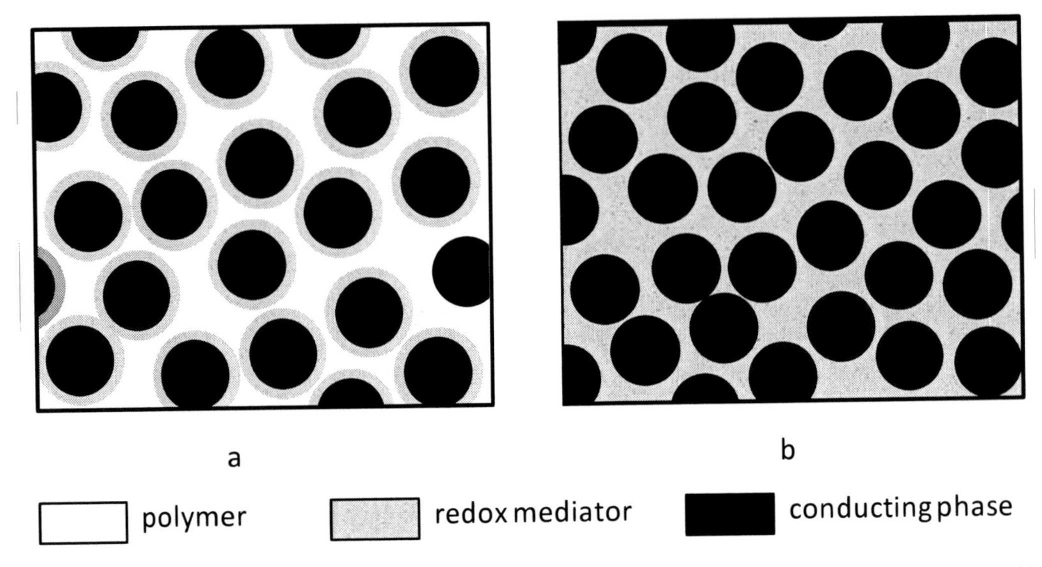

Figure 9. Schematic representation of an ensemble composite electrode array: (a) ring-modified and (b) bulk-modified. Reproduced from ref. [12] with permission. Copyright (1996) Elsevier.

With the advent of the carbon nanotube composites another modification strategy became widely used. Such approach is the surface chemical modification by anchoring covalently (bio)molecules just at the surface edges of the carbon nanotubes once the carbon composite is prepared. This kind of functionalization has resulted quite robust and at the same time very promising for wiring DNA or proteins. Such approach has been quite profited for studying electroactive proteins or redox enzymatic sensors where direct electron transfer between the biomolecule and the CNT has been observed to take place [95].

6. Optimization of Amperometric Sensors Based on Composites: Electrical and Electrochemical Properties

In general, an important feature of composite electrodes is that their overall analytical performance is strongly influenced by the carbon loading within polymeric matrix. It is due to that carbon loading influences directly on the electrochemical surface and inner structure (bulk resistance) of the composite electrode. Both parameters strongly affects on the overall electroanalytical performance of such composite electrodes, as Zhao et. al. [96] shows in their studies. The volume fraction of carbon has to be high enough in order for the carbon particles to be in contact for providing conducting pathways and sufficient low bulk resistivity. On the other hand, it is known that the electrochemical surface of composite resembles very well to a random assembly of microelectrodes. Many works have been focused on corroborating that the electrochemical behavior of different carbon composite systems electrodes resemble very well to those provided by a microelectrodes array. Some examples, is the study performed by O'Hare et al [97] for a random graphite/epoxy composite system.

Like the microelectrodes array, the composite electrodes present, the possibility to generate higher current intensity. Such improvement is due, among others things, to the way that the electroactives diffuse on the electrode surface. Two different kinds of diffusion, radial and linear, are known. The planar one is usually given in solid electrode surfaces and the diffusion of the electroactives species is linear from bulk solution to electrode surface. On the other hand, the radial diffusion is usually given in array systems where alternating zones of conductive and non-conductive zones are present at certain interdistance. Such radial diffusion provides an increase of the mass transfer (see figure 10). On the practical side, focusing on the voltammetric consequences, such increase in the mass transfer favors positively to the electroanalytical signal.

Nevertheless, the diffusional behavior in a microelectrode array strongly depends on the degree to which each electrode element in the array is diffusionally independent from the other elements. This in turn depends on both the separation of each electrode in the array relative to the diameter of the individual electrodes, and also the timescale of the voltammetric experiment [98-101].

If the electrodes within the array are sufficiently separated, and the timescale of the experiment is short enough that the diffusion layers around each electrode do not overlap, then each electrode within the array can be considered to be diffusionally independent from its neighbors[102-105]. The limit of extreme overlapping of diffusion zones occurs when the distance of separation between electrodes is small or when the timescale of the experiment is longer. In this limit a diffusion layer is formed which has uniform thickness over the whole array (planar difusion) [106].

Under this context and assuming a proper timescale of the experiment, in order for a composite electrode to maintain its analytical advantages (independent radial diffusion at each electrode), it is necessary to characterize not only its bulk conductivity but also its layout electrode surface. The layout has to contain the electrodes within the array sufficiently separated to avoid exclusively the linear diffusion and without affecting negatively the sensitivity of the system. Such decrease in the sensitivity probably could be produced by an excessive separation of the electrodes and by a higher bulk resistivity.

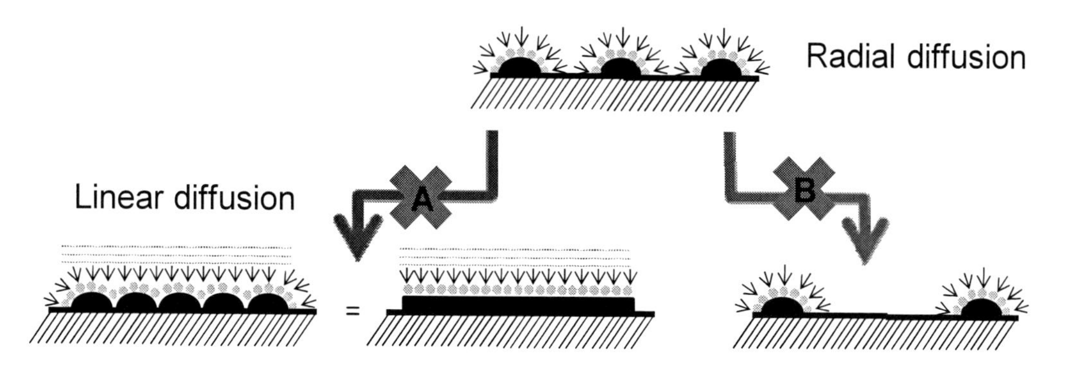

Figure 10. Scheme of the different diffusional modes in an array system. In the case A) the microelectrodes are too close, providing a linear diffusion due to the overlapping of each diffusion layer of each electrode in the array. In the case B) the microelectrodes are too far, providing a decrease of the sensitivity and higher bulk resistivity.

From an electroanalytical point of view, a proper carbon loading in the composite electrode provides some advantages respect to a solid electrode with the same active area, such as a higher sensitivity due to the radial diffusion, which provides an increase of the mass transfer. Further, it also provides a lower limit of detection caused by a decrease in the background current resulting of a decrease of the electrode capacitance, which is directly related to a decrease of the exposed carbon area.

Many theoretical or experimental studies are focused on providing an advance in understanding this kind of random systems. Compton's group [99-101] has performed interesting theoretical and experimental studies focused on the electrode surface, evaluating the voltammetric consequences of a dispersed electrode-electrolyte interface. Others studies are focused on the inner structure of the composite and the relationship between the composition of the composite and its conductivity. Conductivity behavior of the composite is well understood by the Percolation Theory. Since Navarro-Laboulais et al. [107-109] applied it, for first time, to interpret and evaluate electrochemical process in composite systems, many researchers have taken the same approach and use as a tool to characterize the electrical properties of a great variety of carbon composite materials.

Taking one step further in the understanding of carbon composites, Zhao and O'Hare [96] have reported numerical and experimental approaches considering the inner structure of a glassy carbon/ epoxy composite electrodes and its consequences for the surface electrochemistry. Some of their key findings were, on one hand, that in a random system, different conducting pathway lengths produce a shift in the composite resistance between the hook-up contact and the surface. Such shift provides, at the same time, a shift in $E_{1/2}$ (half-wave potential) and a sloping i_l (diffusion limiting current) in the voltammetric records. These results were contrasted by a three-dimensional numerical model based on the percolation theory, being both results in a good agreement. On the other hand, they also evaluated how chemical pretreatments of the carbon material before mixing could affect the voltammetric behavior. They observed, respect to the untreated carbon, that reducing agent pretreatment and acid pretreatment both created more functional groups and increased the current response of cyclic voltammetric experiments, though the effects of acid pretreatment were greater.

More recently, Faiella et.al [110] described the manufacturing of multiwalled carbon nanotubes epoxy based composite, where the electrical properties (percolation threshold) were tuned by varying different process parameters, such as the sonication time and curing temperature, and maintaining the global concentration of MWCNT constant. They assumed that such tuning is possible because the electrical connectedness varies according to the development of different nanotubes networks. This was in agreement with the study performed by Zhao and O'Hare [96].

Basically, such studies again confirm, on one hand, that the electrical properties depend on several parameters, such as nature, shape and size of the filler, the filler loading, matrix properties, fabrication process and dispersion of the filler in the matrix and electrode surface, and on the other hand, that the electrical properties affect directly on the electrochemical properties of the carbon based composite materials.

Following, we will discuss several techniques, such as electrical, morphological and electrochemical ones, which provide useful information for the characterization of the composite materials and optimization of their composition for voltammetric sensor purposes.

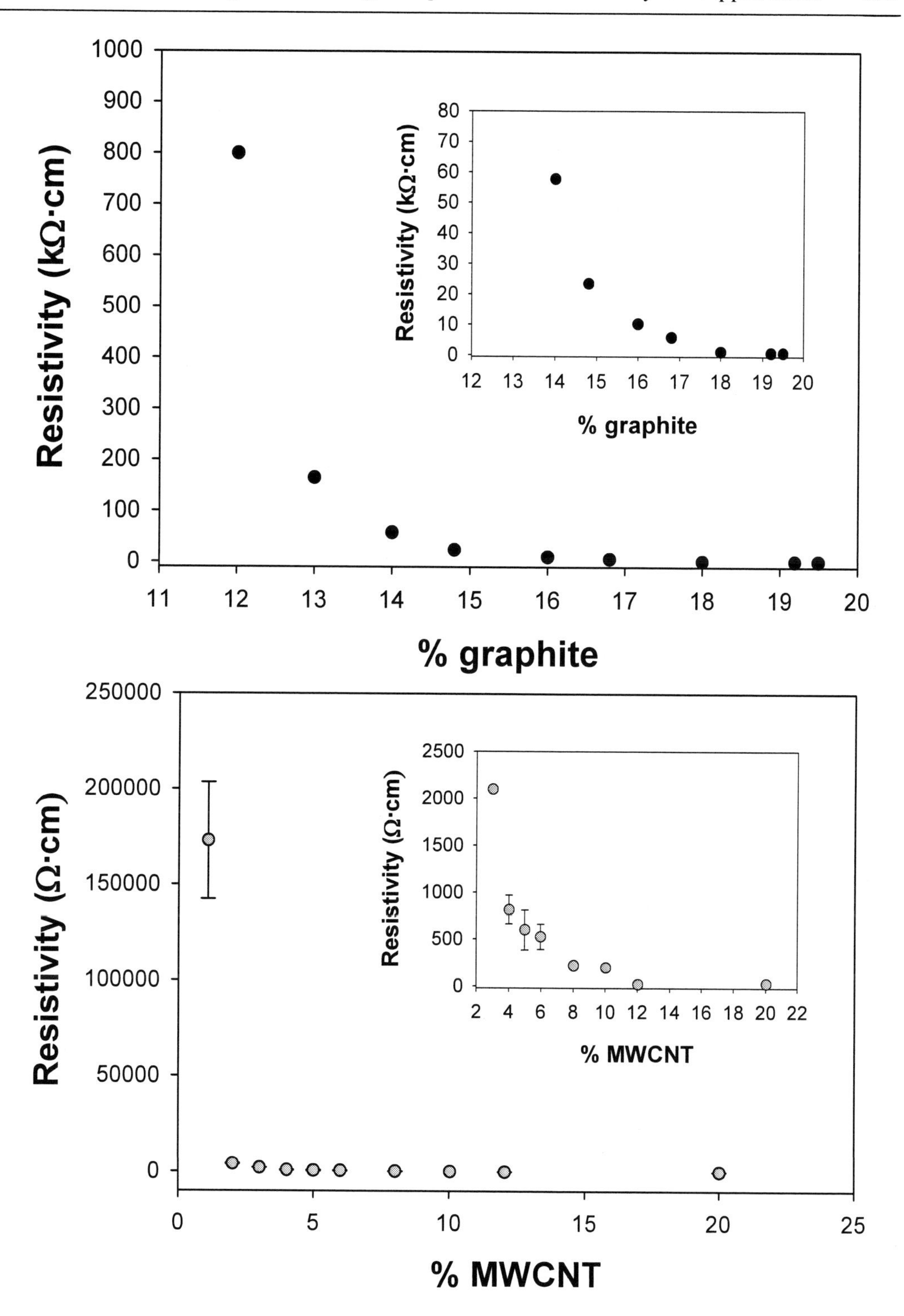

Figure 11. A) Percolation curve obtained for the rigid conducting composite based on the polymer epoxy and graphite power. B) and using epotek H77 resin and MWCNT. The trend follows the percolation theory. Adapted from ref [71] with permission. Copyright (2009) Wiley-VCH Verlag GmbH & Co. KGaA. And adapted from ref. [67] with permission. Copyright (2010) Elsevier.

6.1. Electrical Attributes of Composite Materials Trough the Percolation Theory

As mentioned before and under an electroanalytical context, it is important that the volume fraction of carbon is high enough in order for the carbon particles to be in contact for providing conducting pathways and sufficient low bulk resistivity.

As already discussed, percolation curve gives us information about the electrical properties of the composite electrodes, providing an overview of the electrical behavior by varying the carbon loading [71].

Here we will describe the percolation curves for two different composite systems, graphite/epoxy (figure 11 A) and CNT/epoxy composites (figure 11 B). In order to evaluate the electrical behavior of such composites, the electrical resistance was tested by varying the carbon loading for every composite composition. We can observe that the percolation threshold (PT) for a MWCNT composite is achieved at 1%, whereas for graphite composite is at 12%. The MWCNT composite has a lower PT value, due to the high surface ratio characteristics of the carbon nanotubes. For both, composites containing more than 20 % of carbon exhibited a poor mechanical stability and those containing less carbon loading than PT presented insulating properties with R~∞. Because of this fact such study was based on the zone above PT zone.

Analytically, a lower resistivity value produces a higher sensitivity [71]. We can observe a low resistivity zone for composites with a load between 20% and 8% of MWCNT, whereas such zone is shorter, between 20% and 15%, for graphite composite. It can be concluded that in terms of resistivity, the composites lying in such interval are suitable for electroanalytical measurements.

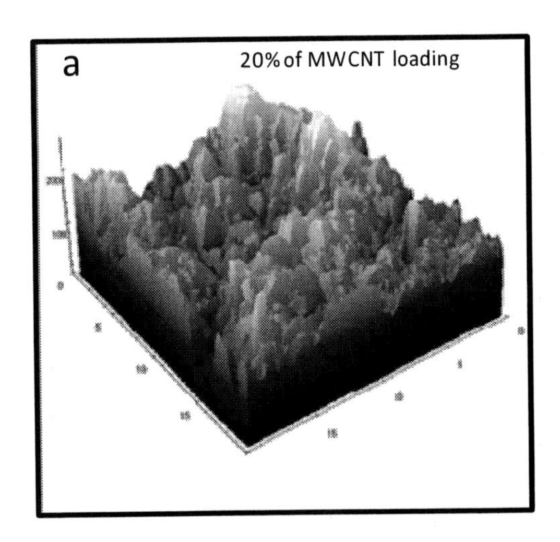

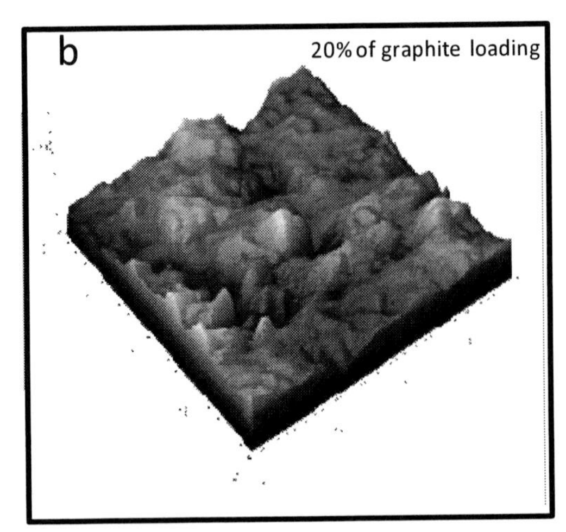

Figure 12, Topographic images obtained by means of AFM technique for the qualitative characterization of the electrodes surfaces. Image (a) corresponds to the composite electrode surface (20X20μm) for a 20% of MWCNT loading. The image (b) corresponds to the composite electrode surface (20X20μm) for a 20% of graphite loading. The x-axis, y-axis and z-axis units are μm, μm and nm, respectively. Adapted from ref. [67] with permission. Copyright (2010) Elsevier. And adapted from ref. [71] with permission. Copyright (2009) Wiley-VCH Verlag GmbH & Co. KGaA.

6.2. Morphological Characterization

Conductive Atomic Force Microscopy (C-AFM) is another way to characterize the electrical properties of the carbon composites before its analytical application. This technique consists in the use of a conductive tip operating in standard contact mode with the sample. As a voltage is applied between tip and sample, a current is generated, the intensity of which will depend on the sample nature. C-AFM technique was used in order to obtain qualitative information about the size, shape and distribution of the conducting material in the composite structures. On the other hand, the conductance mapping images obtained with AFM allowed us to observe the conductive microzones where the electrochemical charge transfer can take place (light regions), and the non conductive insulating microzones where such processes are suppressed (dark regions) for both composites.

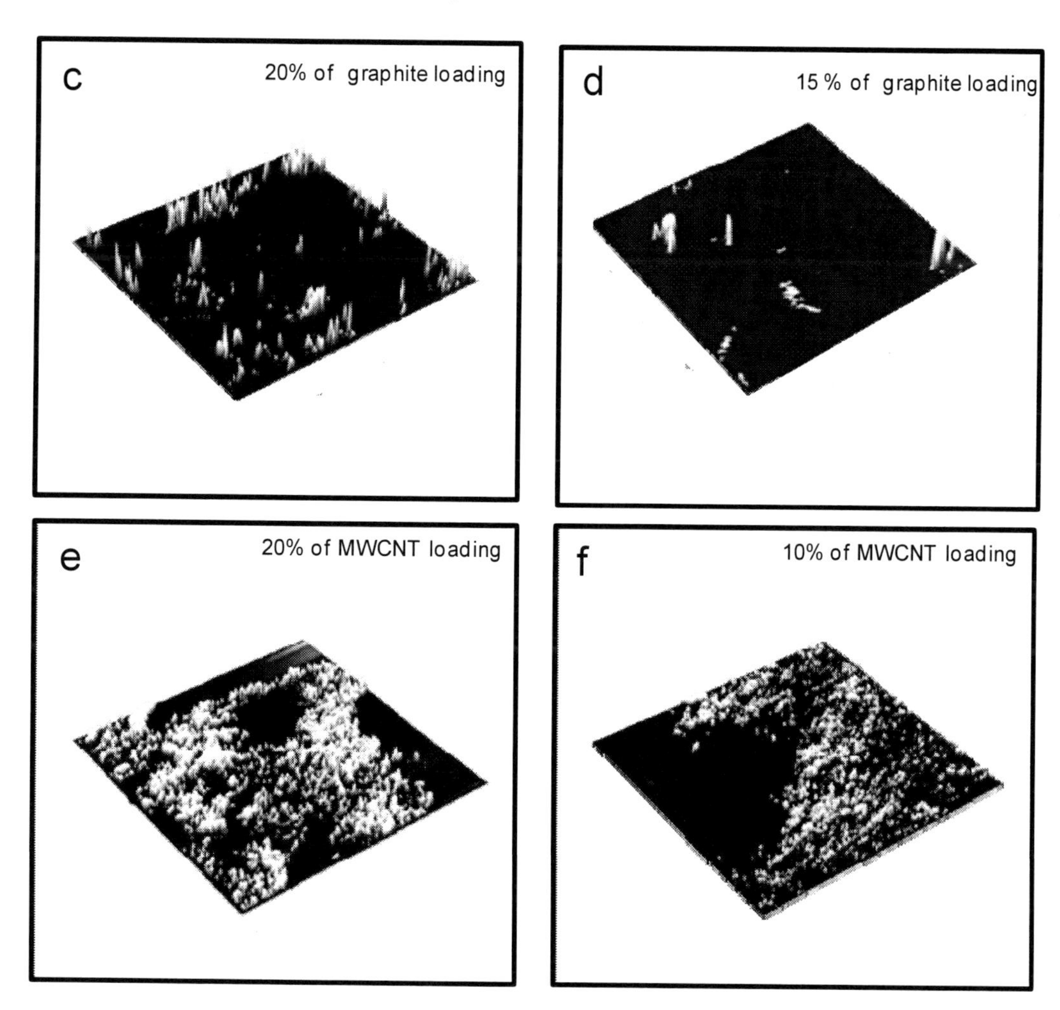

Figure 13. Conductivity mappings with CAFM technique: c) 20% of graphite loading; d) 15% of graphite loading; e) 20% of MWCNT loading. f) 10% of MWCNT loading. The x-axis, y-axis and z-axis units are μm, μm and nm, respectively. Adapted from ref. [67] with permission. Copyright (2010) Elsevier. And adapted from ref. [71] with permission. Copyright (2009) Wiley-VCH Verlag GmbH & Co. KGaA.

Figure 12 presents the most significant topographic images obtained during the electrode surface study for both graphite and carbon nanotube composites. In general, a low percentage of carbon in the material creates composites with larger spaces between the conductive channels, while a higher carbon percentage produces composites with closer conductive channels. These results allow the observation that even though the sensors have a macroelectrode appearance, their surface includes conductive microzones where the electronic transfer is produced, and non-conductive insulating microzones where electronic transfer is not possible. In order to achieve an electrochemical behavior of a microelectrode array it is important to optimize the distance between microelectrodes (conductive microzone). However and due to the randomly structure of the composites, the distance between conductive microzones are not so easy to control. Nevertheless, the distance between the conductive microzones can be slightly controlled by a decrease or an increase of the carbon loading.

Focus on the MWCNT composite, the composite with 20% of MWCNT loading showed slightly more conductive areas (32% of conductive area in front of 20x20 μm^2 total area) than in the case of the composite with 10% of MWCNT loading (17% of conductive area in front of 20x20 μm^2 total area), as expected. (Figure 13 c and d).

To gain more insights about the influence of others factors on the electrochemical behavior of the composite electrodes, we have evaluated the reproducibility of the composite construction. Such parameter is a key parameter for evaluating the methodology used in the development of handmade composites before their analytical application.

6.3. Reproducibility in the Composite Fabrication

An estimation of the electrode reproducibility due to the handmade electrode fabrication process is necessary before its analytical applications. As an illustrative example we have taken graphite-epoxy composites of variable composition and evaluated between the percolation zone composition and the low resistance zone composition. Specifically, 20%, 15% and 13% of graphite loading composites were chosen to represent the low resistivity zone composite (LRC), the near-percolation zone composite (NPC) and the percolation zone composite (PC), respectively and their electrical and electrochemical properties were analyzed (see figure 11, graphite). For each different graphite composition, five fabrication batches were produced. Each batch consisted of three identical devices for electrical measurements and three identical electrodes. Hereby, for each composition 15 electrodes and 15 devices for electrical measurements were constructed. The electrical resistance measurements were carried out according to the percolation theory explained before. The electroanalytical response was evaluated by cyclic voltammetry measurements with a benchmark redox species $Fe(CN)_6^{3-/4-}$ which is fast and reversible. In order to quantify the reproducibility in the electroanalytical response the cathodic current value was registered at a fixed potential of -0.250 V in each voltammogram (see figure 14). The statistical results as well as the average of graphite loading in the successive electrode performance for each composition are shown in table 1. We have obtained that the lowest reproducibility in the electrode production is located in the percolation zone (13% graphite loading), where smaller variation of graphite loading (RSD of 0.05%) produced a decreased in the reproducibility in terms of electrical resistance (RSD of 13%) and electroanalytical response (RSD of 40%). On the other hand, the reproducibility increased in the low resistivity (20% graphite loading) and near percolation zone (15% graphite loading). In these zone, lower changes in the reproducibility of the electrical resistance (RSD of 45% and 35%, for 20% and 15% of graphite loading, respectively) and electroanalytical response (RSD of 10%, 12%, for 20% and 15% of graphite loading, respectively) properties were produced. Moreover, it is important to remark that, in low resistance zones as well as in near percolation zone, the reproducibility value in terms of electroanalytical response was better in one same batch (three electrodes). Results showed that the reproducibility in the electrode production was similar for low resistivity zone composite and near-percolation zone composite and this value was different and worse for the percolation zone composite. In this way, in terms of reproducibility in the fabrication process, composites include in the low resistivity zone composites and near-percolation zone composites provided better features.

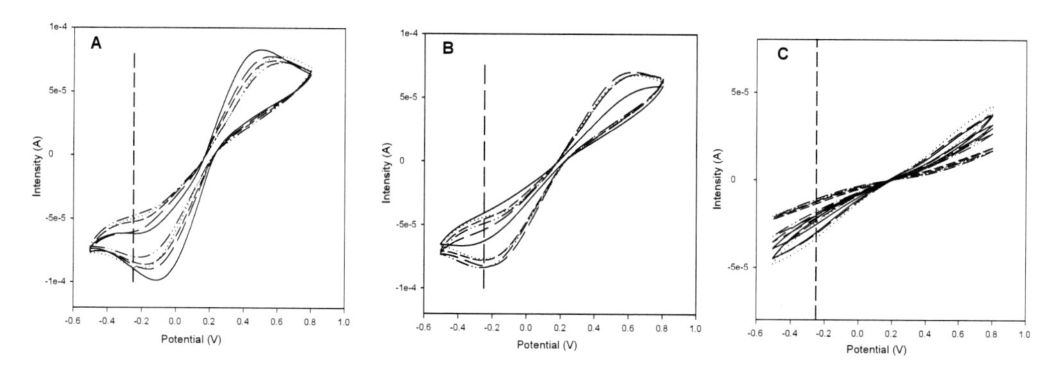

Figure 14. Representative voltammograms plots of the 0.01M $Fe(CN)_6^{3-/4-}$ in 0.1M KCl. Electroanalytical response for the electrodes fabricated, in different batches, using: A) 20% graphite loading, B) 15% graphite loading, C) 13% graphite loading composition. Experimental conditions: 100 mV s-1 scan rate; a dash line at -0.250 V (vs Ag/AgCl reference electrode) is included. Adapted from ref [71] with permission. Copyright (2009) Wiley-VCH Verlag GmbH & Co. KGaA.

Table 1. Statistical parameters obtained with electrical and electroanalytical measurements Adapted ref [71] with permission. Copyright (2009) Wiley-VCH Verlag GmbH & Co. KGaA

Zone	Batches	n	% Graphite (a)	RDS (%) (%Graphite)	R (kΩ) (b)	RDS (%) (R (KΩ))	I (A) (c)	RSD (%) (I (A))
LRC	5	3	19.97	0.2	0.4	45	$-10.4\ 10^{-5}$	10
NPC	5	3	14.98	0.4	47	35	-8.8510^{-5}	12
PC	5	3	12.95	0.05	430	130	-2.6210^{-5}	40

(a) This value is the average percentage of graphite loading in five batches of equal composition
(b) This value is the average of the electrical resistance values for five batches of equal composition
(c) This value corresponds to the analytical signal in the reduction from 0.01 M $Fe(CN)_6^{3-/4-}$ (E= -0.250V)

7. NEWS STRATEGIES FOR THE IMPROVEMENT OF THE CARBON COMPOSITE ELECTROCHEMICAL PROPERTIES

7.1. Electrochemical Impedance Spectroscopy Technique

As previously mentioned, the carbon loading affects directly the rate of electron transfer, the material stability and the background capacitance current. Analytically, these parameters influence the levels of sensitivity, limit of detection and response time. One way to optimize the carbon loading of the composite is under the criterion of maximizing the carbon loading, without losing the physical and mechanical properties of the composite [12, 111, 69] The goal of this criterion is to confer the lowest bulk resistivity value. Nevertheless, at the same time, high carbon loadings can increase the background current and smear the faradaic signal response, especially when the electroactive species are present in low concentration.

The Electrochemical Impedance Spectroscopy (EIS) technique is a useful tool to characterize composite electrodes and to follow the composite optimization strategy. This technique is capable of high precision and is frequently used for the evaluation of heterogeneous charge-transfer parameter and for studies of double-layer structure. These physical parameters are directly related with the system sensitivity, response time and limit of detection, each one being important in the overall electroanalytical performance of composite electrode.

In general, this technique involves an excitation signal of very low amplitude in order to obtain a pseudo-linear current-potential relation. In a pseudo-linear system the current obtained ($I(t)= I_0 \cos(\omega t- \theta)$)as a response to a sinusoidal potential ($E(t)= E_0 \cos(\omega t- \theta)$) is a sinusoid at the same frequency (ω) but shifted in phase (θ) (see figure 15 A).

The impedance of the system is calculated by means of an expression analogous to Ohm's Law ($Z= E(t)/I(t)$). Using the formula of Euler ($e^{j\varphi} = \cos\varphi + j\sin\varphi$), it is possible to express Z in terms of complex numbers ($Z = Z^{\circ}(\cos\emptyset + \sin\emptyset) = Z_{real} + j\,Z_{imag}$). There are many ways to plot impedance. Among them, the Nyquist plots are the most often used in the electrochemical literature when studying electron transfer kinetics because they allow for an easy prediction of the circuits elements. Nyquist plots consist of plotting the Zimag as a function of Zreal, usually showing a semicircle profile. Every experimental point in the plot corresponds to a different frequency.

EIS data is commonly analyzed by fitting it to an equivalent electrical circuit model. Most of the circuit elements in the model are common electrical elements such as resistors, capacitors, and inductors, which are combined to simulate the real impedance spectra. The utility of this technique resides on that each element in the model has a basis in the physical electrochemistry of the system, such as, charge-transfer resistance (R_{ct}), double-layer capacitance (C_{dl}) or solution resistance (R_s) and contact resistance (R_c). By means of this technique we can obtain general trends of such electrochemical parameters.

The figure 15 B shows the simplest equivalent circuit to fit the impedance data obtained from an electrochemical system with kinetic-control process. R_{ct} corresponds to the charge-transfer resistance, which is inversely proportional to the electron transfer rate, C_{dl} is the double layer capacitance and R_{Ω} corresponds to the electrolyte resistance. The figure 15 C represents the ideal impedance spectra from an electrochemical system with mixed kinetic and diffusion-control process. The element W is the Warburg Impedance parameter and arises from mass transfer limitation. Such phenomenon appears at the low frequency range of the impedance spectra.

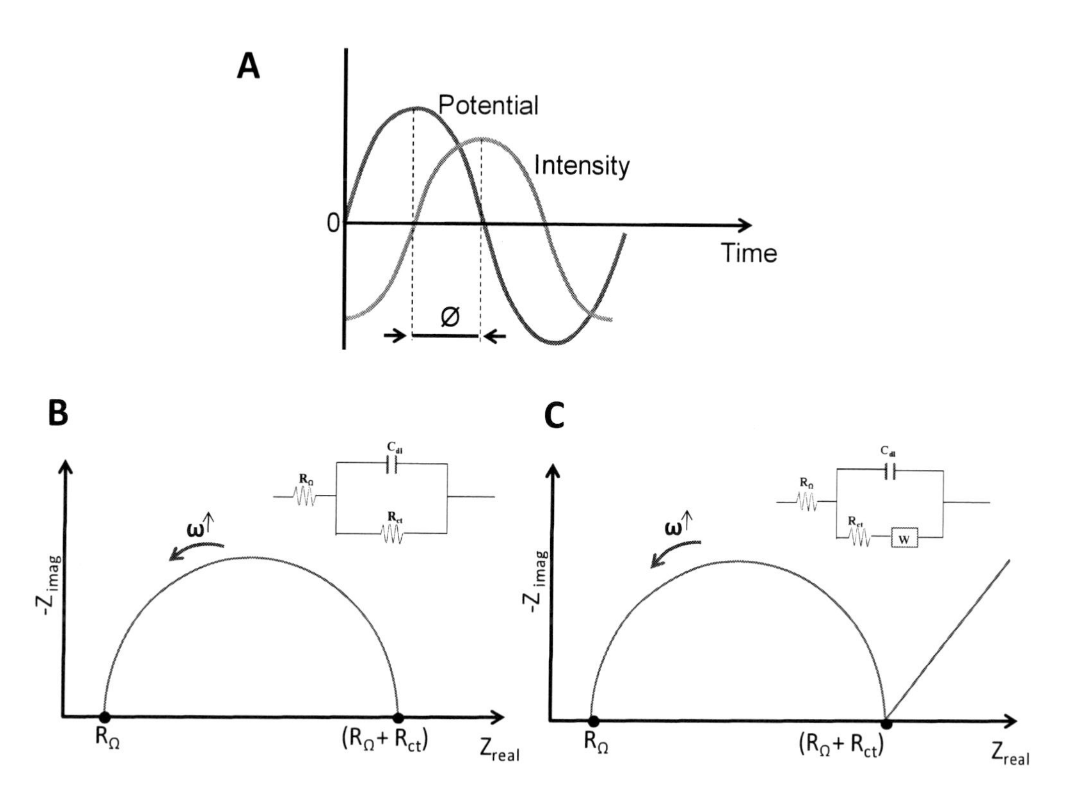

Figure 15. A) Excitation sinusoidal signal applied on the system and the sinusoidal current registered shifted in phase. B and C) depict the Nyquist plot for a electrochemical system with kinetic-control process and with mixed kinetic and diffusion-control process, respectively.

As an example of its possibilities let's focus on the EIS results obtained for a graphite/epoxy and MWCNT/epoxy composites. The purpose of this part is to demonstrate the versatility of EIS technique for the optimization of the carbon loading of different composite materials. It is important to mention that such optimum composition is specific of each composite material

Compositions were studied by varying the carbon loading from 13% to 20%, for graphite epoxy and from 3% to 20% for MWCNT epoxy composite. These ranges have been chosen according to the electrical properties (percolation curve) of each composite system showed in the section 6.1. For each carbon composition five equal electrodes were fabricated and evaluated. To perform such electrochemical characterization, the $Fe(CN)_6{}^{3-}$ / $Fe(CN)_6{}^{4-}$ was used as a benchmark redox couple since it is very sensitive to the electrode surface characteristics [26]. The measurements were carried out in a frequency range between 0.1 Hz and 100 KHz, with an amplitude around 10mV.

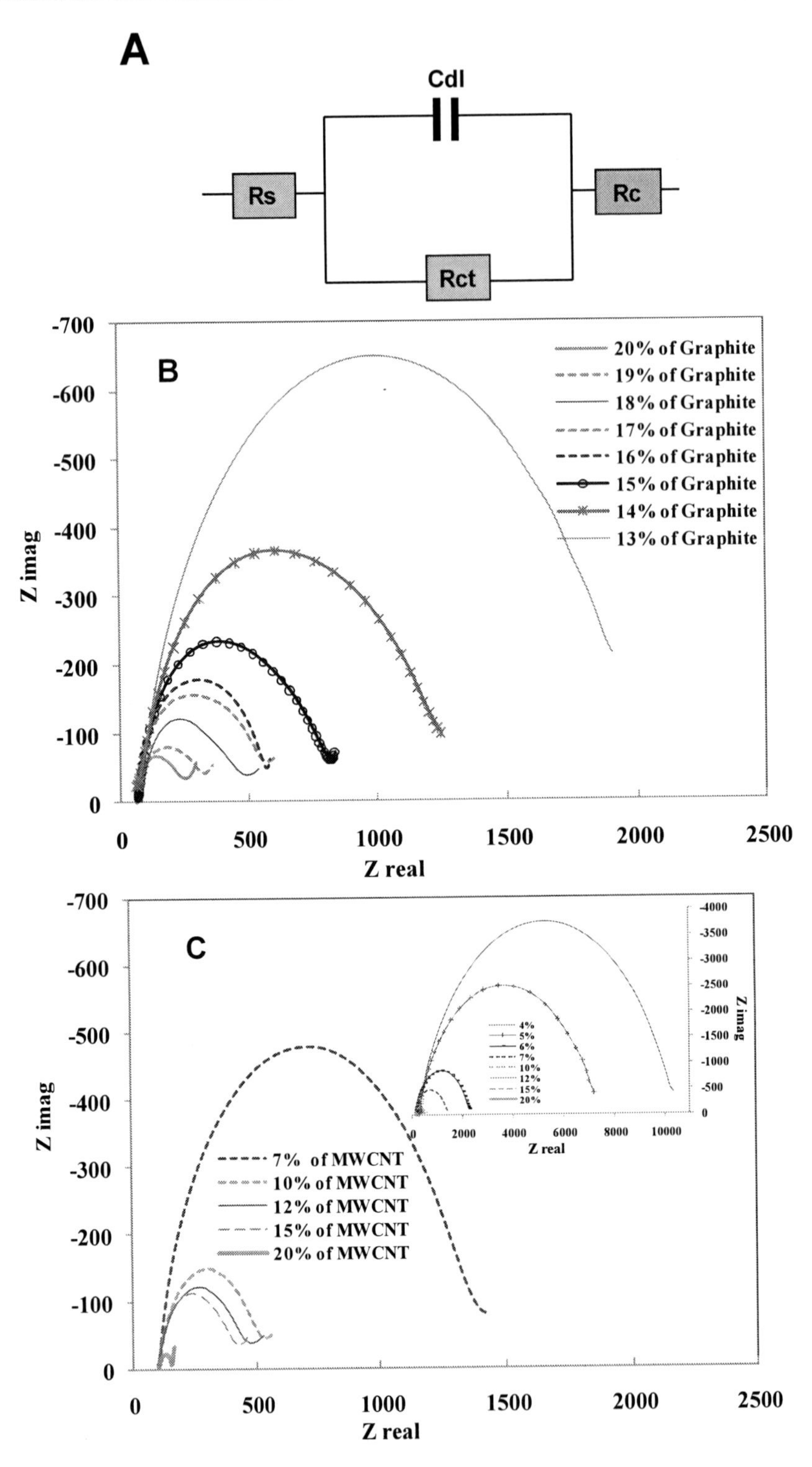

Figure 16. A) The figure shows the equivalent circuit used for the impedance spectra fitting. B and C) Nyquist plots for different carbon loadings of graphite and MWCNT, respectively, in presence of $Fe(CN)_6^{3-}$/ $Fe(CN)_6^{4-}$. The inset in C) shows a zoom of the low impedance composite electrodes. Adapted from ref. [67] with permission. Copyright (2010) Elsevier.

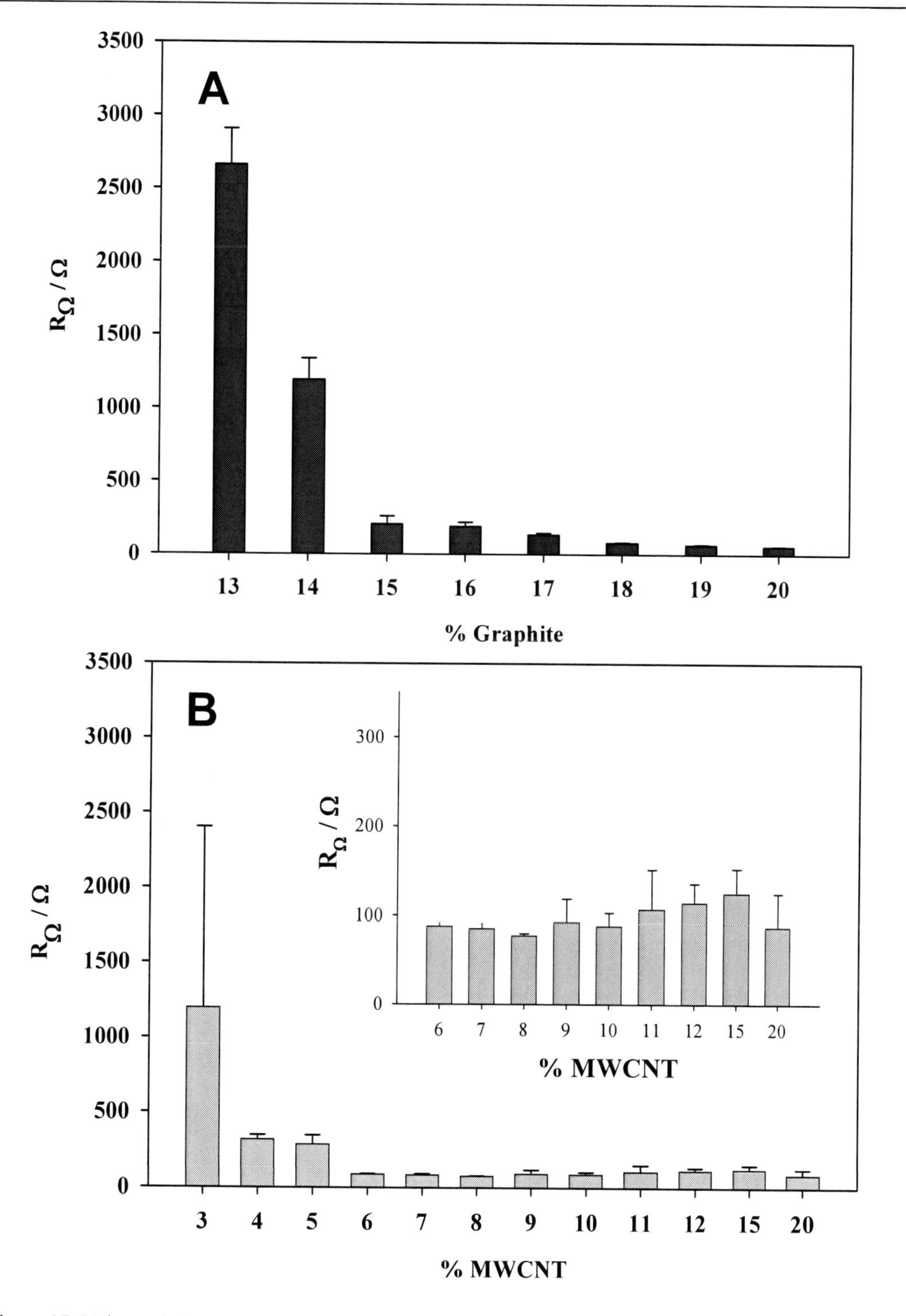

Figure 17. Values of ohmic resistance with their corresponding standard deviation for the different A) graphite loading electrodes and B) MWCNT loading electrodes, using the redox probe $Fe(CN)_6^{3-}$/ $Fe(CN)_6^{4-}$.The insets in B shows values at the lower scale for each evaluated physical parameter. Adapted from ref. [67] with permission. Copyright (2010) Elsevier.

The parameters R_{ct}, C_{dl} and R_{Ω} are obtained by fitting the impedance spectra to a simple equivalent circuit (see figure 16 A). Figure 16 B and C shows the Nyquist plot for each composite system. In general, impedance behaviors of both composites have similar trends. Composites with high resistivity (7% - 4% of MWNT loading and 14%-13% of graphite loading), appears to be dominated by a big diameter semicircle representing kinetic-controlled electrode process in all the recorded frequency range. On the other hand, the impedance plot for composites with lower resistivity (20%-10% of MWCNT loading and 20% -15% of graphite loading) is dominated by a small diameter semicircle representing again kinetic-controlled electrode process, though in some cases, the diffusion-controlled process starts to be discerned at low frequencies (linear region after the semicircle).

The ohmic resistance (R_{Ω}) parameter consists of the solution resistance (which is dependent on the ionic concentration, the type of ions and also the electrode area) in series with the contact or the ohmic composite resistance. The latter resistance is the one that has a direct relation with the dry resistance taken for the percolation plot of figure 17. Since the solution and contact resistance appear in series, they cannot be independently resolved in the impedance data. Figure 17 shows the variations of the ohmic resistance as a function of graphite or MWCNT composition. At low carbon loads ($<$ 15 % of graphite loading and $<$ 6% of MWCNT loading), the ohmic resistance is dominated by the contact or composite resistance whereas at higher carbon loads ($\geq$ 15 % of graphite loading and $\geq$ 6% of MWCNT loading), the ohmic resistance is more dominated by the solution resistance reaching an value between 100 and 200 ohms. This behavior is in agreement with the obtained results in the percolation curve. Low bulk resistance is suitable for electroanalytical materials since it favorably affects the response time and sensitivity of the electrode.

The quantitative values of R_{ct} are depicted in figure 18. The R_{ct} parameter is inversely proportional to the heterogeneous charge transfer rate and also affects the sensitivity and response time of the electrode. This parameter normalized by the electrochemical active area should be constant for conventional metal electrode interfaces. However in the case of carbon composite materials that could not be expected. Carbon structures based on graphene sheets (ex. CNT, graphite) are known to exhibit electrochemical anisotropy with higher electrode kinetics on edges and lower electron transfer rates on the walls or basal planes. That makes the electrode kinetics of such electrodes to be very dependent on the carbon nature and structure. Indeed, we have normalized R_{ct} with respect to the electrochemical active area (see next section for electroactive area evaluation) and listed its variation with the carbon composition in table 2. A decrease of the R_{ct} parameter with carbon proportion has been obtained which could strongly speak of the strong relation between electrochemical reactivity and the surface characteristics of the conducting material. As the carbon load increases, the probability of having more electroactive sites increases and hence the electrode kinetics.

Therefore, composites with low charge transfer resistances are appropriated to be used in electrochemical measurements. According to the results, composites between 20% and 9% of MWCNT loading and between 20% and 15% of graphite loading, presented lower charge transfer resistance values. The use of composites in such range of proportions will guarantee fast electron exchange. However and in spite of the enhanced kinetics, high load of conducting material can increase the background current and smear the faradaic signal response, especially when the electroactive species are present in low concentration.

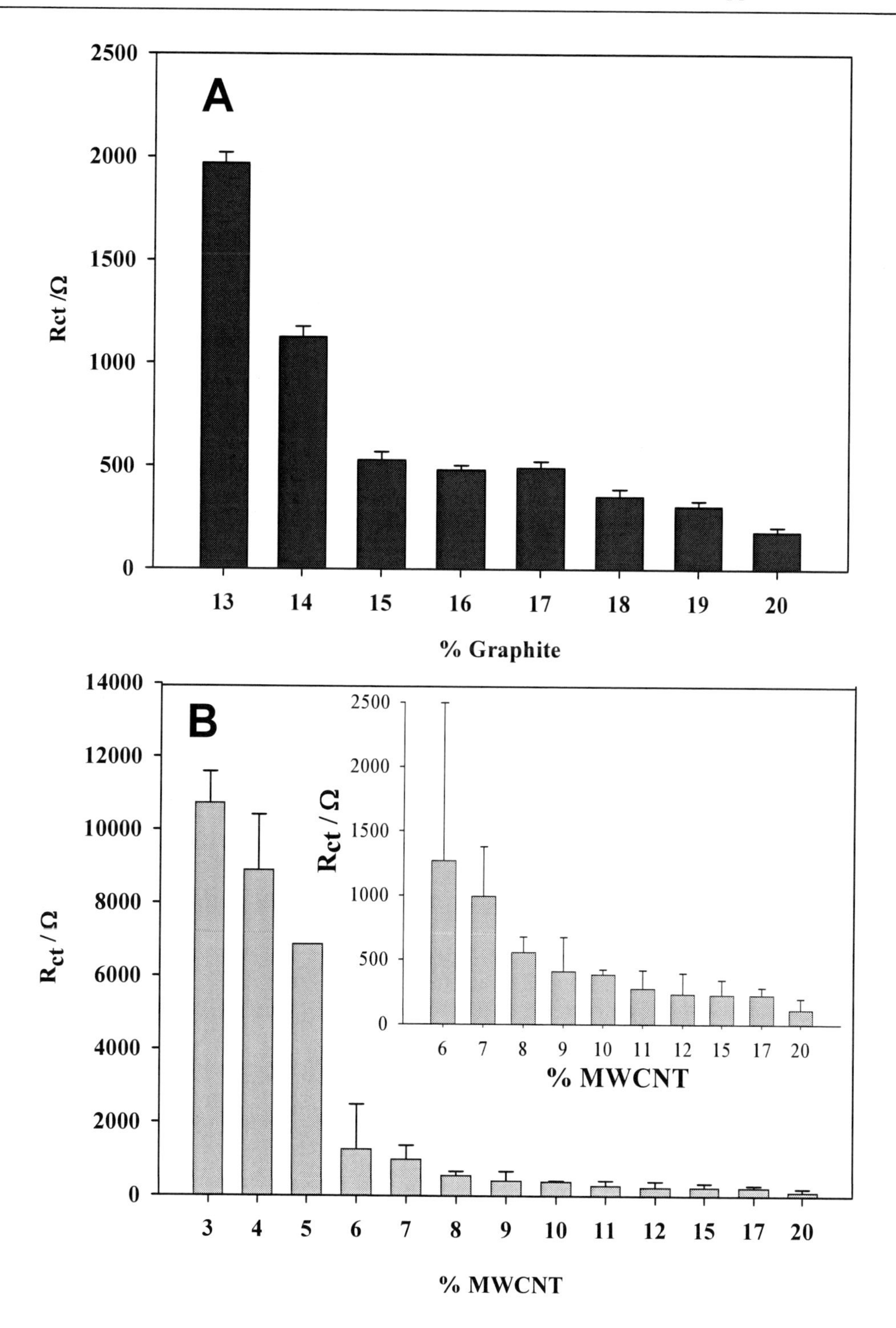

Figure 18. Values of charge transfer resistance with their corresponding standard deviation for the different A) graphite loading electrodes and B) MWCNT loading electrodes, using the redox probe $Fe(CN)_6^{3-}$/ $Fe(CN)_6^{4-}$.The insets in B show values at the lower scale for each evaluated physical parameter. Adapted from ref. [67] with permission. Copyright (2010) Elsevier.

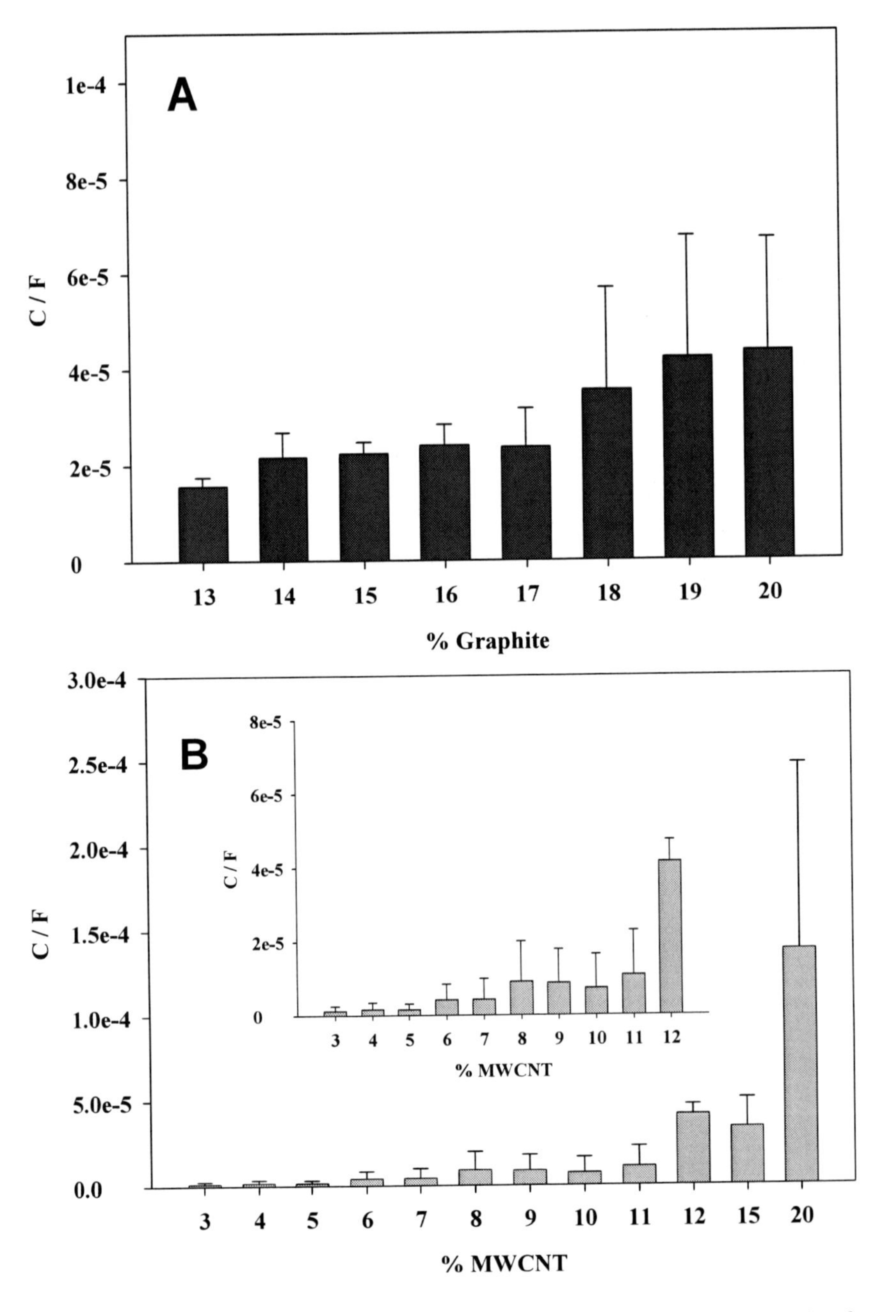

Figure 19. Values of double-layer capacitances, with their corresponding standard deviation for the different A) graphite loading electrodes and B) MWCNT loading electrodes, using the redox probe $Fe(CN)_6^{3-}$/ $Fe(CN)_6^{4-}$.The insets in B show values at the lower scale for each evaluated physical parameter. Adapted from ref. [67] with permission. Copyright (2010) Elsevier.

Therefore it is important to consider the remaining impedance parameter represented by the double-layer capacitance which is directly related to the charging or background current. This parameter exhibits increased values with electrodes comprising high surface area of conducting material. In general composites contain only a fraction of conductive area exposed to the solution with the remainder occupied by the insulating polymer. The electrode capacitance, which is determined nearly exclusively by the exposed carbon becomes low, and in turn decreases the background current. That enhances the signal to noise ratio and consequently decreases the analyte detection limits.

Figure 19 depicts, for both composite materials, the decrease of the double-layer capacitance values with the decrease of the carbon loadings. Note the remarkable lower values of the capacitance double layer between 3% and 11% of MWCNT loading whereas in the graphite composite the lowest values of capacitance were achieved between 13% and 17%.

According to the impedance results and taking into account the properties required by an electrode for electroanalytical purposes, such as rapid response time, low limit of detection and high sensitivity, the intervals between 9% and 11% of MWCNT loading and between 15% and 17 % of graphite loading seem to fulfill all these requirements. It is important to highlight that the fabricated composites in such intervals present similar electrochemical characteristics and small variations in the composite composition (due to, for example, the hand-made fabrication process, as well as the modification with (bio)catalysts) produce a small change in the electrochemical composite behavior. Indeed, under these conditions the electrochemical reproducibility is increased.

In summary we can say that EIS technique allows us to choose the suitable carbon composition, specific for each composite system, for being used as voltammetric sensor.

7.2. Carbon Composite Electrodes Characterization by Means of Cyclic Voltammetry

We have also performed cyclic voltammetry in order to compare the results obtained with the EIS technique. Cyclic voltammograms were taken for the different composite composition electrodes in presence of the benchmark $Fe(CN)_6^{3-}$ / $Fe(CN)_6^{4-}$ redox couple and compensated from any ohmic resistance.

Focusing on MWCNT composite system, it is important to point out the abrupt change in the current/potential shape between 5% and 6% of the MWCNT proportion. At 5% MWCNT proportion, the current potential profile has a more sigmoidal shape as can be appreciated in more detail in the figure 20 A, which can be ascribed to an electrochemical behavior more related with a microelectrode array. Beyond 6% of MWCNT proportion, the composite electrodes exhibit the typical peak-shaped profile corresponding to more massive electrodes with planar diffusion characteristics. The sigmoidal shape could not be captured with the graphite composite system at least with the studied carbon proportions.

For both, different parameters were extracted from the cyclic voltammograms such as the peak separation potential (ΔE) and peak current (I_p) as shown in table 2. One can observe an increase of the peak current with the carbon loading, due to an increase of the electroactive area, together with a decrease of peak separation related to an enhancement of the electron transfer rate. The relative electroactive area (table 2) was estimated from the peak-shaped voltammograms (above the proportions of 5% for MWCNT composite and 13% for the graphite one) by quantifying the peak current with the use of this relationship, $I_p = 3.01 \times 10^5 \cdot n^{3/2}(\alpha D_{red}\upsilon)^{1/2} A C^*_{red}$ [111], which is appropriate for electron transfer-controlled processes.

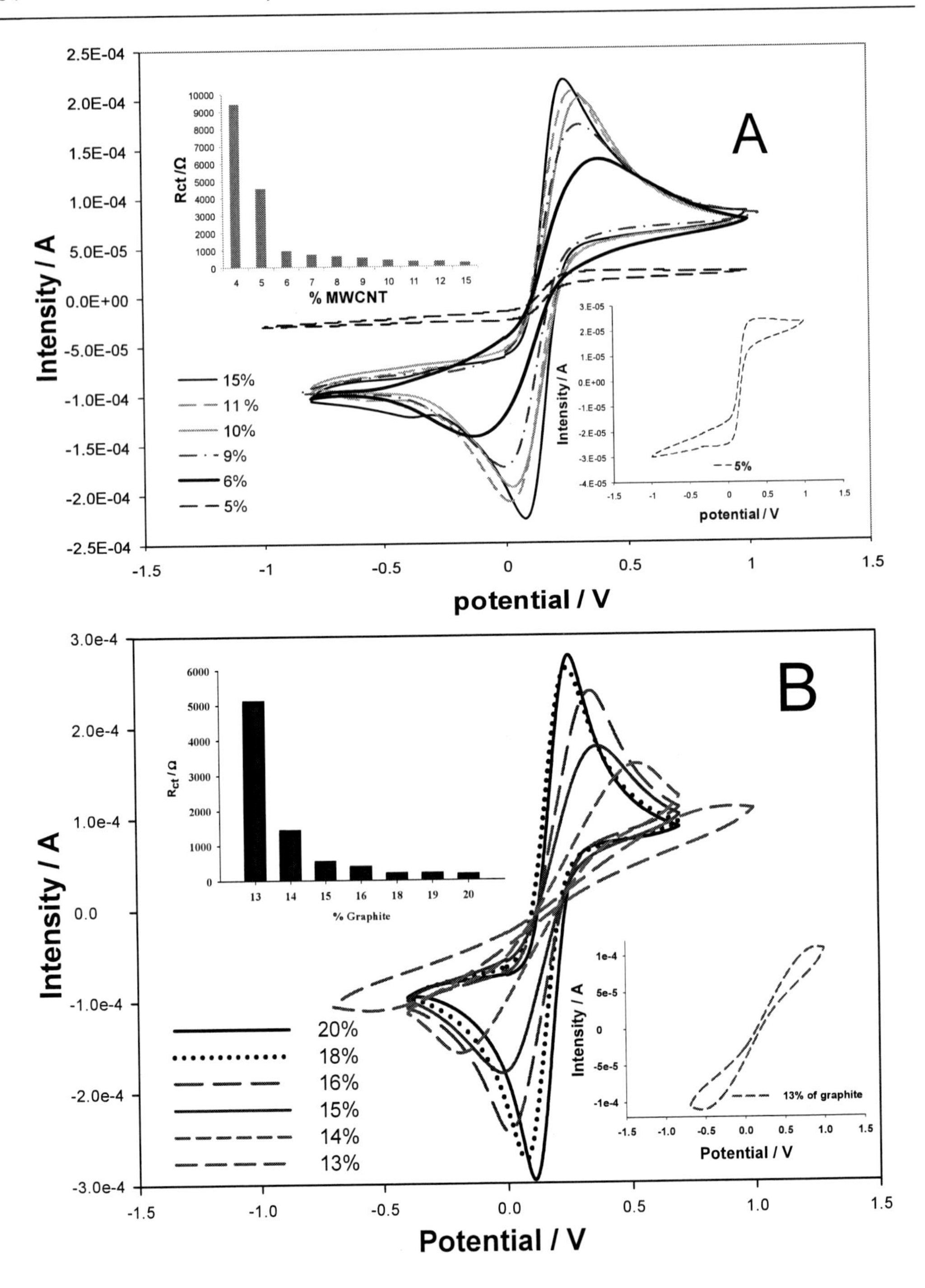

Figure 20. Cyclic voltammogram for 0.01M ferricyanide/ ferrocyanide and 0.1M KCl. Scan rate 10 mV/s. A) MWCNT composite electrodes. The upper inset shows the trend of the charge transfer resistance for different composite composition. The lower inset zooms a cyclic voltammogram obtained by a composite with 5% of MWCNT loading. B) graphite composite electrodes. The upper inset shows the trend of the charge transfer resistance for different composite composition. The lower inset zooms a cyclic voltammogram obtained by a composite with 13% of graphite loading. Adapted from ref. [67] with permission. Copyright (2010) Elsevier.

Table 2. Cyclic voltammetry parameters for the different composite compositions (MWCNT and graphite) i_o correspond to the exchange current, R_{ct} to the charge transfer resistance, i_p to peak current, A to active area and □E to the peak separation potential. $R_{ct} \cdot A$ and $R_{ct}^{EIS} \cdot A$ correspond to the R_{ct} obtained by voltammetric and EIS measurements, respectively, and normalized with respect to the active area Adapted from ref. [67] with permission. Copyright (2010) Elsevier

MWCNT	i_o (A)	R_{ct} (Ω)	i_p (A)	A (cm^2)	ΔE (V)	$R_{ct} \cdot A$ (Ω cm^2)	$R_{ct}^{EIS} \cdot A$ (Ω cm^2)	Graphite	i_o (A)	R_{ct} (Ω)	i_p (A)	A (cm^2)	ΔE (V)	$R_{ct} \cdot A$ (Ω cm^2)	$R_{ct}^{EIS} \cdot A$ (Ω cm^2)
4%	$2.68 \cdot 10^{-6}$	9418													
5%	$5.52 \cdot 10^{-6}$	4572													
6%	$2.70 \cdot 10^{-5}$	936	$1.61 \cdot 10^{-4}$	0.30	0.476	280.8	641.4								
7%	$3.36 \cdot 10^{-5}$	751	$1.96 \cdot 10^{-4}$	0.37	0.427	277.9	468.4								
8%	$4.04 \cdot 10^{-5}$	625	$2.07 \cdot 10^{-4}$	0.39	0.335	243.8	250.6								
9%	$4.63 \cdot 10^{-5}$	545	$2.33 \cdot 10^{-4}$	0.43	0.299	234.3	257.6								
10%	$6.09 \cdot 10^{-5}$	415	$2.40 \cdot 10^{-4}$	0.45	0.283	186.8	186.3								
11%	$7.24 \cdot 10^{-5}$	348	$2.53 \cdot 10^{-4}$	0.47	0.261	163.6	178.1	**13%**	$4.93 \cdot 10^{-6}$	5119	$1.11 \cdot 10^{-4}$	0.21	1.460	1058	407
12%	$7.45 \cdot 10^{-5}$	339	$2.71 \cdot 10^{-4}$	0.50	0.261	169.5	175.3	**14%**	$1.76 \cdot 10^{-5}$	1438	$1.71 \cdot 10^{-4}$	0.32	0.666	460	360
15%	$1.04 \cdot 10^{-4}$	243	$2.72 \cdot 10^{-4}$	0.51	0.169	123.9	117.9	**15%**	$4.10 \cdot 10^{-5}$	549	$2.09 \cdot 10^{-4}$	0.39	0.367	214	222
								16%	$6.26 \cdot 10^{-5}$	402	$2.87 \cdot 10^{-4}$	0.54	0.327	216	259
								18%	$1.18 \cdot 10^{-4}$	213	$3.27 \cdot 10^{-4}$	0.61	0.177	131	159
								19%	$1.19 \cdot 10^{-4}$	212	$3.28 \cdot 10^{-4}$	0.62	0.171	132	150
								20%	$1.41 \cdot 10^{-4}$	179	$3.60 \cdot 10^{-4}$	0.67	0.150	121	123

In this equation α represents the transfer coefficient which was considered to be approximately 0.5, D_{red}= 6.32 x 10^{-6} $cm^2 \cdot s^{-1}$corresponds to the diffusion coefficient of the reduced species, υ =0.01 $V \cdot s^{-1}$ represents the scan rate, A is the electroactive area and C^*_{red} = 0.01M is the bulk concentration of the electroactive species. We also evaluated the exchange current (i_o) from Tafel plots (log current vs. potential), a parameter which provides information about the reversibility of the process. From the value of the exchange current we can also evaluate the charge transfer resistance through the relation ($i_o=RT/nFR_{ct}$). The trend observed for this parameter resembles very closely the results obtained from the EIS measurements. From the comparison of the R_{ct} values extracted from voltammetry with the ones obtained by impedance measurements (table 2), it can be observed that their values agree quite well for higher carbon loads (greater than 7% of MWCNT and greater than 15% of graphite, both included). By normalizing R_{ct} with respect to the electroactive area, one can observe that this parameter is also decreasing with the increase of the active area, similar to what was observed in the impedance section. That again indicates the influence of the electrochemical anisotropy of carbon compounds which can be more noticeable as the carbon loading is increased.

7.3. Electroanalytical Characterization

The main goal of this section is to illustrate the electroanalytical advantages of optimized composites. For this purpose, we will follow with the graphite/epoxy and MWCNT/epoxy composite systems. Ascorbic acid was used as an analyte for evaluating their electroanalytical characteristics. The carbon composite response to changes in concentration of ascorbic acid was evaluated by chronoamperometric measurements. Such measurements were carried out at a set potential of 600 mV and 450 mV for the MWCNT composite and graphite composite, respectively. The analytical parameters such as the detection limit, sensitivity and the linear range were evaluated for 10% of MWCNT and 15 % of graphite composites and compared to those obtained with 20% ones, respectively (a composition used in the vast majority of previous studies reported by our research group [111, 69]. For each composition three electrodes were evaluated. Table 3 shows the calibration plot parameters for each composite electrodes using ascorbic acid as analyte. The experimental results show that when the carbon loading decreases from 20% up to 10%, of MWCNT and from 20% up to 15% of graphite, the sensitivity slightly decreases, but the linear range remarkably increases and the limit of detection (LOD) becomes one order of magnitude lower.

Comparing CNT and graphite composites, better sensitivity, slightly better LOD and linear range were obtained for composite based on MWCNT. From an electroanalytical point of view, these results suggest that MWCNT loading of 10% seems to be an optimal proportion, since it can achieve a quite good sensitivity but what is more importantly very low limits of detection and wide range of linearity.

Table 3. The calibration parameters for 20% and 10% of MWCNT and graphite composite electrode obtained using chronoamperometric technique with ascorbic acid as analyte and 0.01M KNO_3/HNO_3 as background electrolyte. Adapted from ref. [67] with permission. Copyright (2010) Elsevier

Electrodes	Sensitivity $\mu A\ L\ mg^{-1}$ (%$RSD^{95\%}_{n=3}$)	LOD mg L^{-1}	Lineal range mg L^{-1}
20% of MWCNT	0.29 (9%)	0.40	0.40 – 330
10% of MWCNT	0.20 (7%)	0.04	0.04 – 700
20% of Graphite	0.08 (7%)	0.50	0.50-300
15% of Graphite	0.030 (3%)	0.06	0.06- 400

8. Analytical Applications: Flow Injection Techniques

As already mentioned in the introduction, research fields in chemical production, pharmaceutical screening, drug investigation, medical diagnostics and environmental analysis have increased, over the past decade, their interest in rapid and on-line determinations of low analyte concentrations [1, 2] in complex samples. Recently, the emergence of new nano-materials has allowed the development of analytical sensors and detection systems capable of addressing these needs. The benefits of using improved amperometric sensors based on near-percolation composites, in terms of analytical performance, can be demonstrated for analysis of analytes at extremely low concentration. Moreover, the high malleability of the developed composite allows easy incorporation as detectors in automatic analyzers, for example for flow cell measurements.

Flow Injection Analysis (FIA) was first described in Denmark by Ruzicka and Hansen in 1975. Since then the technique has grown into a discipline covered by six monographs and more than 15,000 research papers [112, 113]. At years 1990s other technologies as Sequential Injection Analysis (SIA) was proposed [114] but it did not achieve the same popularity. FIA, the first generation of flow methodologies, is also probably the most widely utilized especially for automated routine analytical methods and to monitor and control industrial processes. FIA has achieved great popularity due to its instrumental simplicity and easy operation. In its simplest form, the sample zone is injected into a flowing carrier stream of reagent. As the injected zone moves downstream, the sample solution disperses into the carrier (Figure 21 A). The scope of the method grew from a serial assay of samples to a tool for enhancement of performance of spectroscopic and electrochemical instruments. A flow through detector placed downstream records the desired physical parameter into several units (i.e, colorimetric, absorbance, amperometric or potentiometric) depending on the detector used in each system or application. The transient signal produced when the sample reaches the detector can be related to the analyte concentration after a previous calibration system step (Figure 21 B).

The typical FIA flow rate is one milliliter per minute, typical sample volume consumption is 100 microliters per sample, and typical sampling frequency is two samples per minute. FIA assays usually result in sample concentration accuracies of a few percent.

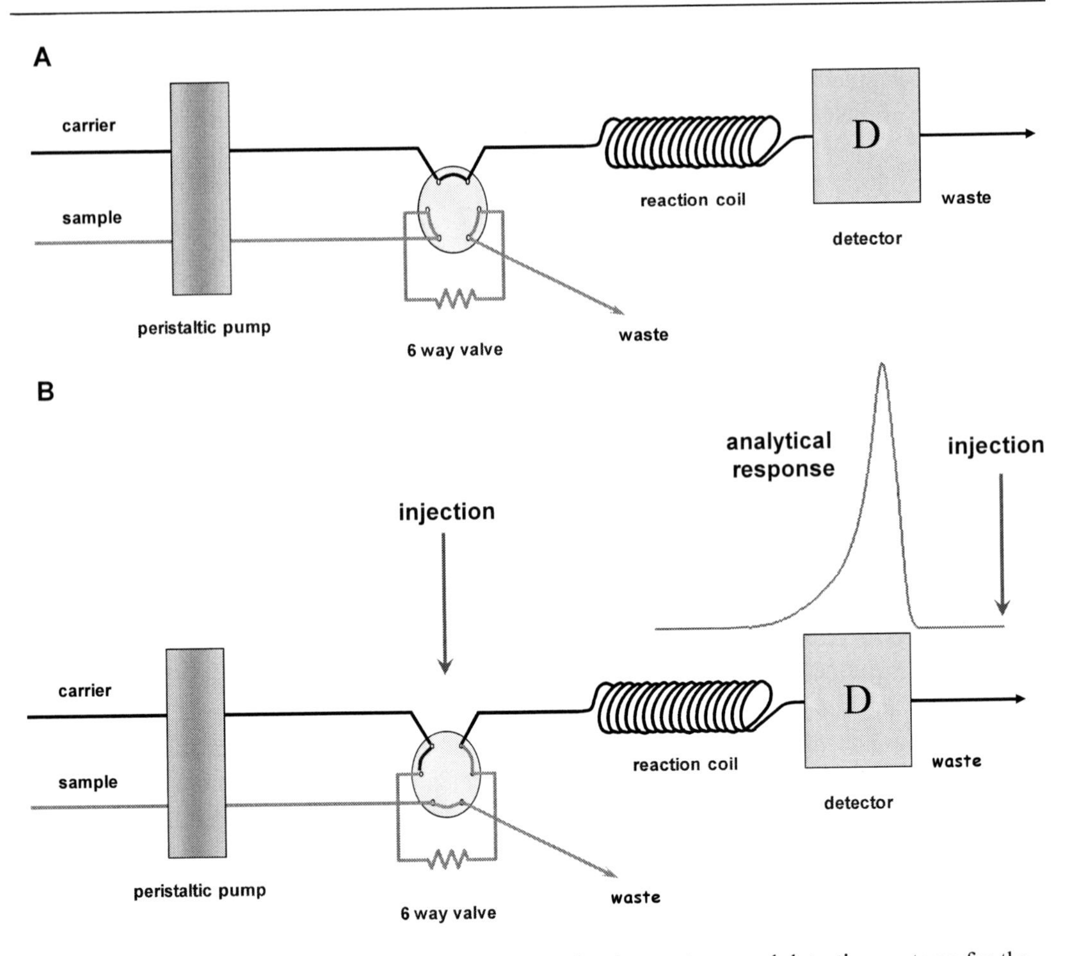

Figure 21. General diagram of a simple FIA system used to integrate several detection systems for the development of automated analytical systems. A) Valve in load position. B) Valve in injection position, then a transient signal is obtained when sample passes through the detector.

Improved electrochemical sensors can be integrated as detectors into such automatic systems based on conventional flow systems or into miniaturized analyzers. This last approximation is very interesting because the advantages of the two approaches are added. The integration of electrochemical transducers based on nanostructures using new generation materials with high-tech microfluidic systems allows the production of analyzers or chemical microanalyzers with a high analytical potential in a reduced space. In addition to reduced reagent consumption, shorter analysis time and reduced maintenance and fabrication costs, these kinds of systems also provide portability and are able to produce on-line information with temporal and spatial resolution.

In the last 20 years, great research efforts aimed at the development of microfluidic devices, key component of the Micro Total Analytical Systems (μTAS), have been made [115]. These ones include: the development of new microfabrication techniques (or the adaptation of the existing ones) for the securing of those structures [116, 117], and the study and application of new construction materials [118, 119].

In the final part of the chapter, we will show three approximations for integratinged improved electrochemical sensors based on composite material (graphite and CNTs) into automatic analyzers. The first one is a conventional FIA analyzer which incorporates an optimized graphite-epoxy composite as a detector. The second one is an improved FIA system with higher automatization grade which is enabled for on-line standards and samples preparation. This automatic system uses an optimized CNT-epoxy composite as a detector. Finally, the last one is a miniaturized analyzer constructed on green tape ceramic with a CTN-epoxy integrated as a detector. In all cases we have analyzed chlorine which is present in water of public swimming pool and tap water in low concentration range. This analyte was used as a model in order to validate the improved sensors as flow detectors into the automatic analyzers. In the last automated system, we have combined the excellent physical and electrochemical properties of CNT composite materials (high mechanical resistance and electrochemical sensitivity, as well as high signal to noise ratio, fast response, simple fabrication and low cost) with the advantages provided by the miniaturization concept, to develop an autonomous microanalyzer for real environmental applications. As a proof of concept, the microanalyzer was characterized and automated, by means of multicommutation techniques [120-122], to determine free chlorine in water samples from a public swimming pool.

Chlorine is a typical disinfectant agent used in quality water control which is added in the form of sodium hypochlorate (pKa = 7.54, at 25 °C). The pH of the swimming pool water is around 7.3, and hence at this pH, the chlorine analyte is present under two species (hypochlorate and hypochloric acid).

Optimized experimental conditions were used to detect such analyte with the developed sensors. Before chorine analysis it is necessary an online sample pre-treatment. Swimming pool water is a complex matrix which contains not only chlorine, but also compounds like chloramines, uric acid and organic matter. Additionally, due to the highly volatility of chlorine, sample preservation was a challenge in order to avoid losses. This pre-treatment consisted in diluting the sample with 0.2M potassium dihydrogen phosphate in 0.2M potassium chloride at pH 5.5. After that, the amperometric response to chlorine can be evaluated with each analytical system using as a carrier stream a solution of 0.1M potassium dihydrogen phosphate in 0.1M potassium chloride at pH 5.5. Consequently when its pH is set to 5.5 the predominant species is the acid form (HClO) [123,124].

Then, we detect HClO when sample arrives to the detector. Reactions taking place on the composite (graphite or CNTs) surface will be therefore (see reaction 1) [125]:

$$HClO + H_3O^+ + 2e^- \rightarrow Cl^- + 2H_2O \quad (1)$$

$$O_{2\,(g)} + 2H^+ + 2e^- \rightarrow H_2O_2 \quad (2)$$

Depending on the working potential, oxygen can be a possible interference. In order to detect interferences because of dissolved oxygen reduction, different chlorine voltammograms using phosphate buffer dissolution at pH 5.5, with and without oxygen, were recorded. Additions of free chlorine produced final concentrations between 1.5 mg L^{-1} and 10 mg L^{-1}. The results showed that dissolved oxygen generated a minimum positive interference (see reaction 2) so it was considered as a constant interference in all samples.

8.1. Flow Injection Analysis with Improved Amperometric Graphite-Epoxy Composite

We have developed a homemade FIA system for on-line measurements and continuous detection of free chlorine at low concentrations. This analyzer showed several advantages, such as fast response, accuracy, minimal reagent consumption in front of a standard method (such as the colorimetric detection which is a discontinuous method) [126]. In addition, the benefits of using improved graphite-epoxy composites in terms of analytical performance were demonstrated by means of the successful determination of chlorine in swimming-pool water samples where chlorine concentration is extremely low.

The FIA system used for the determination of chlorine is described in figure 20. The experimental manifold includes a three-channel flow system, one for the samples and calibration solutions, one for the conditioning solution used for the analysis of swimming-pool water sample (which adjusts the ionic strength as well as its pH), and one for the carrier solution. It includes a four-channel peristaltic pump equipped with silicon pump tubing. It also has a six-port injection valve and 0.7mm internal diameter Teflon tubing. The flow cell (see figure 7) used in this flow system is described in the section 3.1 "*Fabrication of the carbon-based composite and devices*". A double junction reference electrode Ag/AgCl ORION 900200 was integrated downstream the flow cell by means of a methacrylate assembly.

In such a flow system, the analytical response was evaluated using different standard solutions, in a concentration range between 0 mg L^{-1} and 4 mg L^{-1} prepared by dilution of a stock solution of 2000 mg L^{-1} ClO^{-}, with 0.1 M potassium dihydrogen phosphate and 0.1M potassium chloride at pH 5.5 as background dissolution.

The analytical response in the FIA systems, with optimized graphite-epoxy composite and the maximum conducting loading composite, was evaluated in a range between 0 mg L^{1} to 4 mg L^{-1} of free chlorine. The hydrodynamic experimental parameters of the FIA system were set at 2 ml min^{-1} flow rate, 600 µL injection volume and 85 cm long reaction coil. Measurements were carried out at -250 mV fixed potential. Each analysis took 2 min (time from injection to steady-state signal). The mean sensitivity of the system was -0.20 µA L mg^{-1} with a RDS of 3% using the optimized graphite-epoxy composite electrode and four calibrations at a 95% confidence level. The detection limit for this composite composition was 150 µg L^{-1} (LOD, estimated (using the *(S/N)= 3* criteria [127], n= 20 analysis of lowest ClO^{-} concentration *1/RSD= S/N* ≤3). These results were compared with those obtained with the maximum conducting loading composite (20%). In this case, the mean sensitivity was higher, -0.28 µA L mg^{-1}, with a RSD of 3% for four calibrations and 95% confidence. A LOD of 700 µg L^{-1} (estimated using the *(S/N) =3* criteria, n= 20) was achieved using low resistivity composite electrode.

The results showed that lower LOD was achieved with optimized graphite-epoxy composite electrode. This behavior may be explained by an increase in the S/N ratio. In order to evaluate this parameter, both electrodes were tested using 0.5 mg L^{-1} and 1.5 mg L^{-1} of free chlorine concentrations. For each concentration, 20 replicates were performed consecutively. The signal to noise ratio parameter was measured by means of the equation: (S/N)= 1/RSD [128]. The obtained signal to noise ratio values were 18 and 48, respectively testing the optimized graphite-epoxy composite electrode with a 0.5 mg·L^{-1} and a 1.5 mg·L^{-1} chlorine solution. For the maximum conducting loading composite electrode the signal to noise ratios were 0.6 and 12 for 0.5 mg·L^{-1} and 1.5 mg·L^{-1}, respectively. Results showed a clear higher signal to noise ratio for the optimized graphite-epoxy composite. Moreover in both composites a higher ratio was achieved in higher chlorine concentration. So the decrease in the graphite loading not only allows achieving lower detection limits but also to increase the stability and repeatability of the analytical signal. Therefore, optimized graphite-epoxy composite electrodes were chosen and applied in the real samples analysis.

8.1.1. Real Samples

We have applied the proposed system to monitor real samples from swimming-pool water analysis. To overcome the matrix effect of the swimming pool water, a new channel, upstream the sample injection was implemented in the FIA system (see Figure 22, conditioning dissolution channel). The flow rate was the same in the three flow channels (2 ml min^{-1}). This additional channel induced the dilution of the sample by a factor of 2. The system response was characterized by means of successive calibration plots. The sensitivity was -0.08 μA L mg^{-1}, with a RSD of 3% for four calibration plots and 95% confidence level. This sensitivity value decreased with respect to the one obtained without sample conditioning (-0.20 μA L mg^{-1}). In a similar way, the LOD was also affected by the dilution factor, being now 0.26 mg L^{-1} for free chlorine.

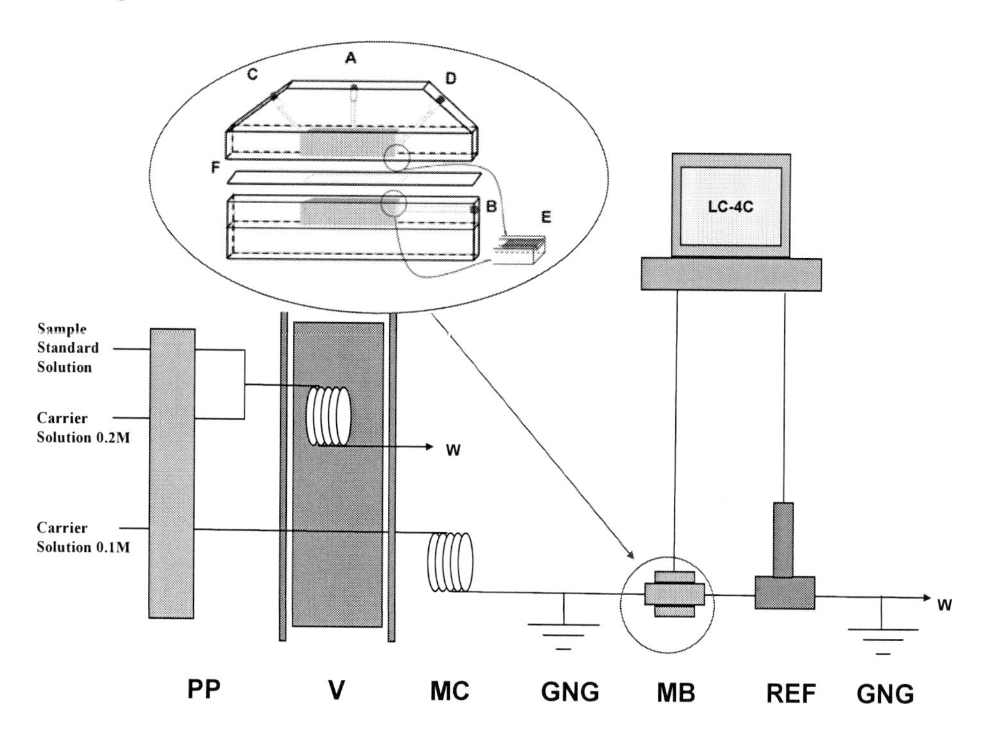

Figure 22. Experimental manifold used for the chlorine determination in a continuous flow analysis, PP, peristaltic pump; V, six-port distribution valve; MC, mixing coil; GNG, grounding electrode; MB, methacrylate block with the working and the auxiliary electrodes; REF, reference electrode; W, waste. Inset scheme of the methacrylate cell for the integration of the sensor in a flow system. A: working electrode based on optimized graphite-epoxy composite; B: stainless steel counter electrode; C/D: Inlet/outlet; E: amplification of the flow channel; F: meeting point of the perforated silicone. Adapted from ref. [71] with permission. Copyright (2009) Wiley-VCH Verlag GmbH & Co. KGaA.

Swimming pool water analysis is a practical example where it is necessary to determine low concentrations of free chlorine. According to the current legislation [129], the minimum chlorine concentration allowed in swimming-pool water is 0.5 $mg \cdot l^{-1}$.

Fifteen samples were analyzed during different days. Ten of them were directly taken from the swimming pool; meanwhile, the last five were synthetic. Synthetic samples were used as control elements. Results were compared to those obtained by the standard method. The correlation coefficients obtained with the linear regression test ($y = a + bx$) were:a= -0.04 (± 0.05) and b= 0.94 (± 0.08); r^2= 0.990. When comparing the performance of two methods with the linear regression test one would expect for the case in which they exhibit the same performance that the slope of the curve tends to 1 whereas the ordinate to 0. As can be observed the slope is close to one and the ordinate close to 0, therefore we can conclude that no significant difference was observed regarding the standard method.

8.2. Improved-FIA System with Sensitivity Amperometric CNTs-Epoxy Composite

In this part of the chapter we report the benefits of using an optimized composite electrode, based on a multiwall carbon nanotubes and epoxy resin, as working electrode in a higher automated flow system. The optimal composite electrode composition was optimized as explained before and it consists of a 10% carbon nanotubes and 90% epoxy resin. This composition provides lower limits of detection and increases the stability and reproducibility of the analytical signal compared to the 20% conventional composition electrodes. In this study free chlorine determination in water was performed following this approach in order to demonstrate the advantages of the optimized composite when compared to conventional composite electrodes [69, 111]. The optimal amperometric composite electrode was integrated in a continuous flow injection analysis (FIA) system to take advantage of all the benefits provided by this technique. The composite material was easily integrated in the flow system due to its high malleability before hardening, which provides high versatility to produce electrodes in any desired shape to match any experimental set-up. The analytical response of 10% and 20% composite electrodes was evaluated and compared. In this sense, the successful analysis of chlorine in complex matrix provided from a public swimming-pool and a tap water supply were carried out at extremely low concentrations with excellent quantification results.

Previously to the analytical application for free chlorine analysis the operational conditions of FIA system must be optimized, i.e: polarization potential and hydrodynamic variables. Figure 23 shows the hydrodynamic curve and cyclic voltammetric plots of the experiments carried out in order to determine the appropriated polarization potential. The voltammogram plots are depicted in figure 23. It can be observed that at a potential around -100 mV (inset figure 23), the free chlorine reduction over the CNT composite electrode took place without the interference of the dissolved oxygen. The hydrodynamic curve showed a plateau at which the reduction of the free chlorine took place. This plateau expanded from -50 to -125 mV and from -75 to -200 mV for both, 20% and 10% CNT, composite electrodes, respectively. Therefore, a potential of -100 mV was chosen for subsequent measurements.

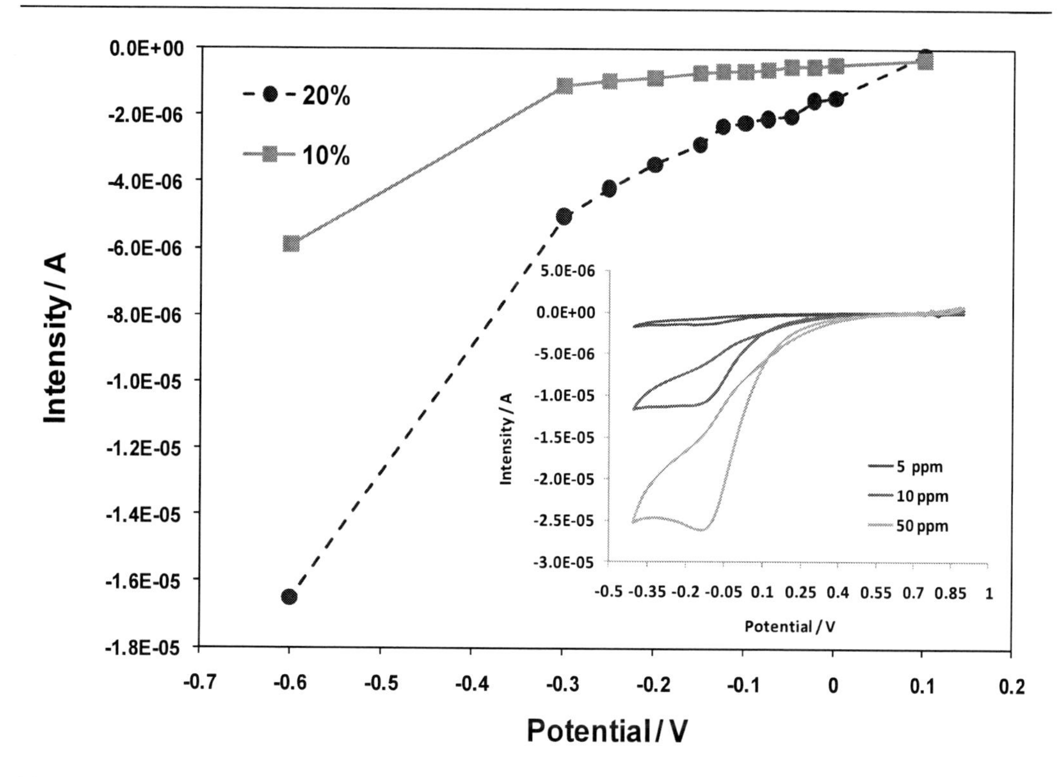

Figure 23. Free chlorine hydrodynamic curve recorded with both 10% and 20% composite electrodes. The used free chlorine concentration was 2 mg L-1. The inset shows voltammogram plots for different amounts of free chlorine in PBS at pH 5.5. Scan rate: 100 mV s-1. Adpated from ref. [130] with permission. Copyright (2010) Elsevier.

In order to find the optimal hydrodynamic experimental parameter, face centered cube experimental design was used. The design included seventeen experiments. After evaluation, these parameters were considered optimal at 600 μL, 2 mL min^{-1}, and 53 cm, respectively. These experimental conditions improved the system sensitivity, maximized its analytical response and minimized reagents consumption. The analysis time for sample was found to be one minute (time from injection to steady-state signal).

In the improved-FIA system (figure 24), the standard solutions were on-line prepared from one stock solution (5 mg L^{-1}) by means of an automatic multicommutation [120-122, 131] technique. In general, this technique is applied in continued flow systems by means of the use of discrete commutation devices (three-way solenoid valve). The definition of different frequencies of valve commutation (on/off), combined with a constant flow rate, allow us to create sequences of standard stock/deionized water (binary sampling) of n aliquots which interpenetrate themselves, producing the desired dilution for each different standard concentration. The injection volume is controlled by measuring the time that the standard stock/deionized water or sample flows through the system. In fact, here the different standard solutions were prepared by means of different duty cycle applied to the digital signal that controls the commutation system. The overall commutation binary time was 18s and the mixing ratios to obtain the desired concentrations were 0.8/0.2 s, 0.6/0.4 s, 0.4/0.6 s, 0.2/0.8 s, etc. Using these ratios the stock solution was diluted 20, 40, 60, 80%, etc. The minimum commutation time that provides a reproducible signal was 0.1 s.

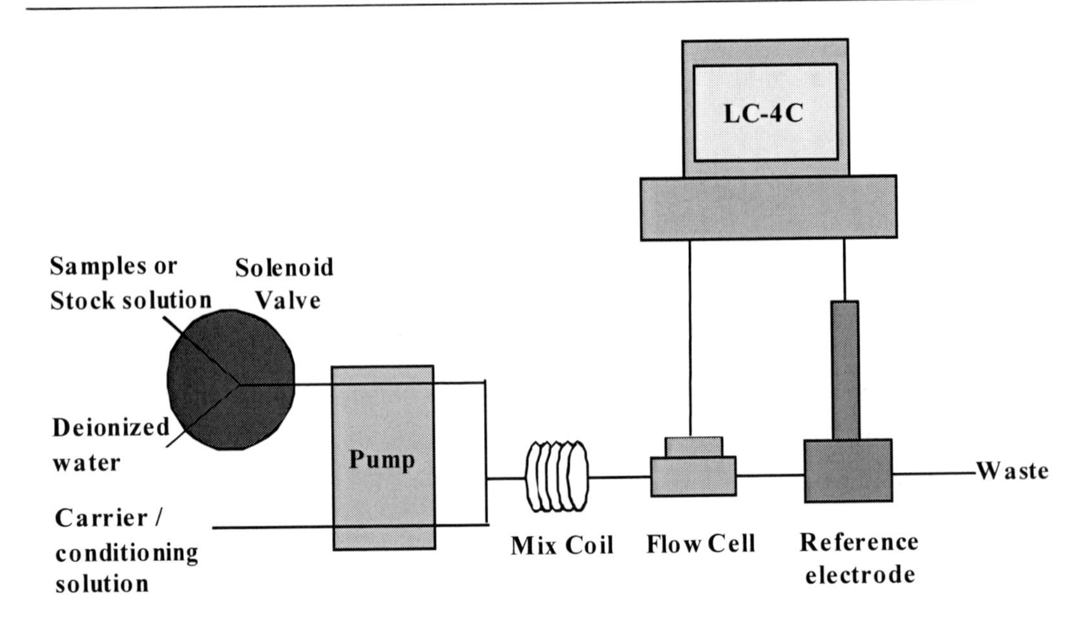

Figure 24. Experimental manifold used for chlorine determination in an improved-FIA system. The manifold of the improved-FIA system includes a three-way solenoid valve, NResearch 161T031 (NResearch Incorporated, West Caldwell, NJ, USA) controlled by a virtual instrument especially developed for this application. CNTs composite working electrode was inserted in a homemade flow cell (50 mm3 internal volume) based on two methacrylate blocks described elsewhere (see figures 8 and 20). A 0.2 M PBS and 0.2 M KCl at pH=5.5 solution was used as carrier/conditioning solution. The standard solutions/samples were also diluted (1:1) by the carrier/conditioning solution, before being analyzed. Adpated from ref. [130] with permission. Copyright (2010) Elsevier.

The main enhancement that provides the improved-FIA system was when analyses at low concentrations were performed. Figure 25 shows the amperometric recording for different injections of low concentration of free chlorine and blank solution (deionized water), for both, conventional-FIA and improved-FIA systems, using the optimized 10% composite electrode. Using the conventional-FIA system, it was not possible to detect concentrations under 150 μg L^{-1}. The small variations observed in the base line for lower concentrations could be associated to the valve commutation, which generates a transitory signal every time the sample is injected. This valve-effect masks the signal of the sensor, and avoids the system to respond to low concentrations. This drawback was overcome by replacing the injection-valve by a three-way solenoid valve. The new system was capable of analyzing free chlorine concentration of even 20 μg L-1 after manifold improvements were performed.

The CNT composite response to different free chlorine concentrations was evaluated with the amperometric technique, using the improved-FIA system. Different analytical parameters such as base line stability, repeatability of the signal, reproducibility and linear response of the system were evaluated for two electrode compositions, 20% and 10% of CNT.

It is well known that the analytical signal in FIA is determined by the difference between the maximum and the baseline. Consequently, a stable baseline is needed to make an accurate determination. To evaluate the baseline stability for both, 20% and 10% CNT composite electrodes, the background current was registered during 2 hours using the optimal experimental conditions. A stable baseline was observed for the 10% composite electrode, with a drift baseline of about 0.01 nA min^{-1}. On the other hand, the 20% composite electrode presented a baseline more unstable, with a drift baseline of 1 nA min^{-1}, 100 times higher than that obtained with 10% composite electrodes and with higher background noise.

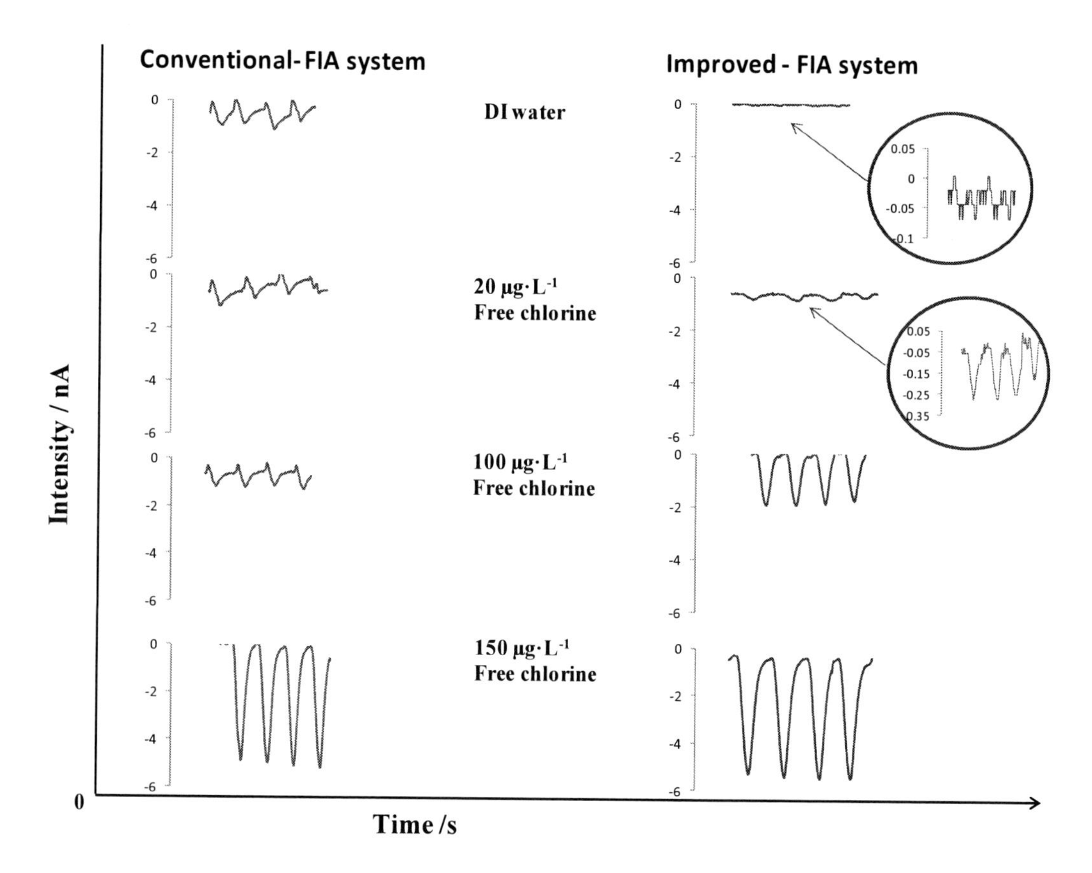

Figure 25. Different analytical signals recorded for different low concentration of free chlorine and water (blank solution), using the 10% composite electrode, for both conventional-FIA (same system shown at figure 24 but with a 10% CNTs-epoxy composite as a sensor composition) and improved-FIA system. Adpated from ref. [130] with permission. Copyright (2010) Elsevier.

The repeatability of the signal for both composite compositions was determined by successive analysis of 2 mg L^{-1} free chlorine standard solution. For the 10% composite electrode, a RSD of 2% was obtained for 30 replicates performed consecutively (95% confidence). The mean peak current value was found to be -0.300 (±0.002) µA. On the other hand, the 20% composite electrode presented a RSD of 3% for 30 replicates (95% of confidence). The mean peak current value obtained in this case was slightly higher, -0.38 (±0.01) µA, as expected.

The response of both electrodes was evaluated, using the improved-FIA system, in a range between 0 and 4 mg L^{-1}. Figure 26 shows the analytical response in this range. This concentration range was chosen because it matches the minimum and maximum legal chlorine concentration permitted by the current Spanish legislation in swimming-pool water [129] and tap water [132]. In this approach we obtained sensitivities of -0.15 (±0.01) µA L mg^{-1} (n=3) and -0.19 (±0.02) µA L mg^{-1} (n=3) for 10% and 20%, respectively. It is important to highlight that even with a lower carbon load in the 10% composite electrodes it is possible to achieve a sensitivity value similar to that obtained with 20% composite electrodes.

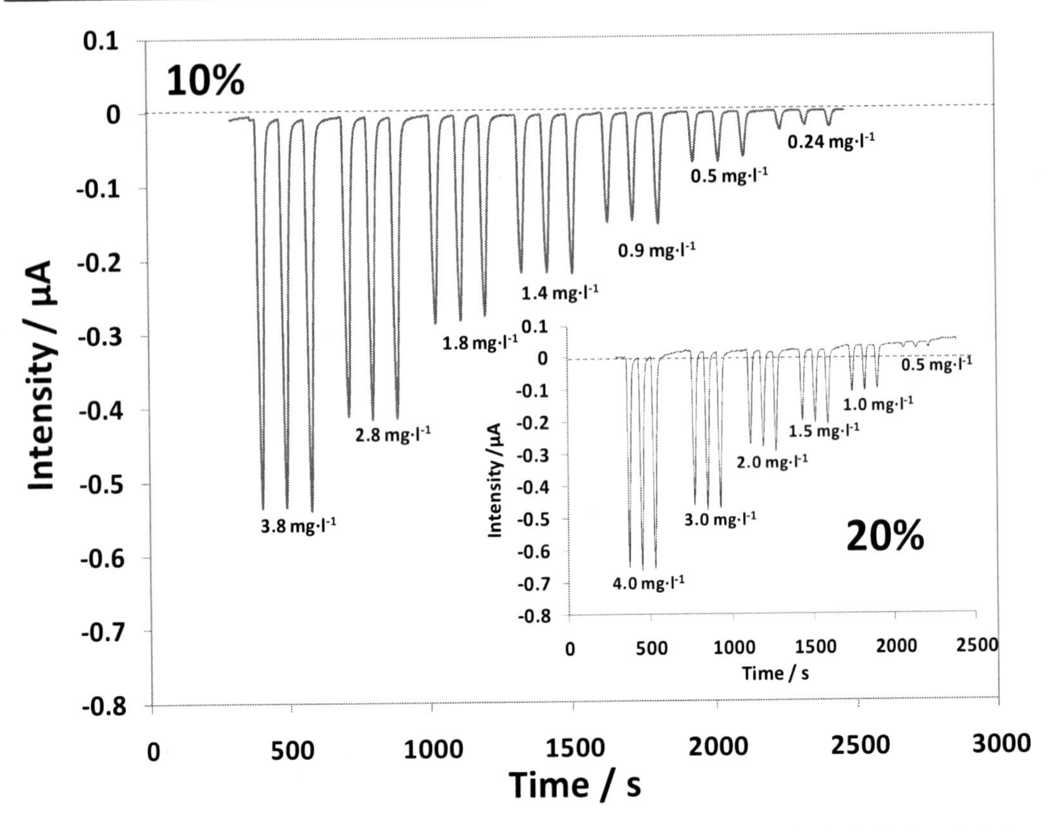

Figure 26. Records of the amperometric signal with the improved-FIA system for both 10% and 20% (inset figure) composite electrodes. Adpated from ref. [130] with permission. Copyright (2010) Elsevier.

The LOD was estimated by the *S/N =3 (n=50)* criterion [127]. Under this condition, the LOD for the 10% composite electrode is 8.5 times lower than that obtained for 20% ones. Even though the LOD for both composite electrodes were suitable to perform chlorine analysis in both swimming-pool and tap water samples, remarkable improvements in the analytical response were observed for the 10% composite electrodes compared to the 20% ones in terms of signal stability, system reproducibility and LOD,. Moreover, the LOD value achieved with the 10% composite electrode is lower or comparable to previous works using more expensive amperometric transducer materials, such as gold [133, 134] or platinum [135].

Taking into account these improvements, 10% composite electrodes were selected and subjected to further studies. An estimation of the sensor lifetime was carried out, for 10% composite electrodes, by means of different calibration experiments performed during 30 consecutive days. The mean value of the calibration plots carried out on the first working day (-0.146 (±0.008) µA L mg^{-1}, n=3, 95% of confidence) was used to calculate the control/nominal sensitive value. The upper and lower control limits were set as two times the standard deviation (*2σ*) of this value. The results showed that after one month of use the sensitivity was still within the control limits.

8.2.1. Validation of the Analytical System Using Real Samples

To validate the system performance, thirty six samples were analyzed during several days. Sixteen of them were directly collected from the swimming pool, and fourteen were doped after collected with the chlorine stock solution. Two samples were synthetically prepared and used as control. In addition, four other samples were collected from the tap water supply system. Statistical data obtained to compare results from the standard method and from the improved-FIA the recovery data show that both methods are equivalent. In order to establish the agreement between the results obtained with the proposed system and the standard method two statistical tests were applied. Statistical paired t-test and the least-square linear regression were used throughout. No significant differences were observed at 95% confidence level. In paired t-test: statistical t parameter t_{cal} (1.533) was minor that t_{tab} (2.030), while in the least square linear regression a slope of 0.99 ± 0.01 and intercept of 0.01 ± 0.02 was obtained. Correlation coefficient was r^2 = 0.990. The results of recovery data are enclosed into the range 90-108; these values show minimum difference between analytical methods. Therefore, we can show an excellent agreement between the results from the proposed system and the ones obtained from the standard method for free chlorine determination, demonstrating the high potential of the optimized composite electrode integrated in the improved-FIA system.

8.3. Miniaturized Analytical Systems with Amperometric Sensor Based on Improved CNT-Epoxy Composite

Miniaturized analytical systems can be constructed with different materials depending on the final application and detection step. Glass, silicon [120, 136], polymers [137-139] and green tape ceramics [140-143] are the materials of choice for miniaturization purposes. Amongst all of them, the green tape technology presents some advantages when compared to the rest. Unlike glass or silicon, its fabrication process is simpler, cheaper and faster, since it does not require tedious and expensive conditions such as clean room facilities or specialized staff. Unlike the majority of polymers, green tape ceramics are highly hermetic and can withstand a wide variety of chemicals reagents [144]. Furthermore, its multilayer production approach easily allows the production of complex three-dimensional structures and the integration of additional materials compatible with the fabrication conditions, such as resins and some metals (i.e. silver, platinum). In addition, its compatibility with the thick-film technology permits the integration of screen-printed conductors to define electrodes or electronic circuits [145,146]. Additional elements that are not compatible with the fabrication process can be easily integrated once the device is bourn-out. Therefore, predefined holes to place these elements should be included in the design whilst taking into account the ceramics shrinkage which is totally predictable. The set of advantages associated to the green tape technology enables the development of highly integrated analyzers that can incorporate several stages of the classical analytical procedure [147-150].

Independently of the miniaturization technology applied to develop chemical microanalyzers, the detection system constitutes a main challenge, since highly sensitive techniques are needed as a consequence of the scaling-down of the flow system and the reduction of sample volumes. Optical and electrochemical detection systems are the most commonly used [151, 152]. Despite the high sensitivity presented by optical systems, their miniaturization involves the reduction of the optical pathway and, consequently, a decrease of their sensitivity. On the other hand, the high selectivity, sensitivity and the simple implementation of electrochemical methods, mainly voltamperometric ones, along with how well these devices maintain their analytical features during the scaling-down process compared with the optical methods, have contributed to their application in analytical microsystems. Therefore, miniaturized devices that integrate voltamperometric detectors can be frequently found in the literature [153, 154].

The performance of the electrochemical detection system is highly influenced by the nature of the material used as working electrode. This electrode should provide not only a favorable signal to noise ratio but also a reproducible response. Different kinds of materials, such as gold, platinum and different allotropic forms of carbon, have been found to be useful for electrochemical determinations in microfluidic devices [152-157]. Despite the electrochemical advantages provided by carbon nanotubes as sensors, papers that deal with their integration into microfluidic devices are scarce in the literature. The first microchip integrating CNTs as sensor was reported by Wang et al. [158]. This microchip was applied to determine different standard analytes in a non-continuous flow mode using the amperometric technique. The device consisted of an exchangeable screen printed carbon electrode (SPE) modified with CNT/Nafion and integrated in a commercial glass CE (capillary electrophoresis) microchip by means of two adjustable pieces of methacrylate. Many other works reported in the literature are based on the same configuration [159-161]. Even though all these devices provided favorable results, it is important to notice that their configurations are based on the same commercially available microfluidic chip, which is modular integrated with the detection system. To obtain more robust analyzers, the monolithic integration of all platforms is desired.

Karuwan et al. [162] reported a microfluidic system based on PDMS and glass. The microfluidic channels were fabricated on the PDMS substrate using standard photolithography techniques. The detection system was fabricated on a glass substrate, later integrated to the fluidic platform, where Pt and Ag electrodes were deposited by sputtering. The working electrode consisted in CNTs grown by chemical vapor deposition (CVD). Although the system was satisfactorily applied to determine salbutamol, the fabrication of the chip involved a complex process, increasing fabrication costs. Moreover, once the electrode began folding, the chip became useless. To increase the versatility and reduce costs, an exchangeable detector is required.

We present here a microanalyzer based on Low-Temperature Co-fired Ceramics (LTCC) technology as an alternative for the fabrication of micrototal analytical systems (μTAS). Using a multilayer approach, the fast proptotyping of complex 3-D structures without liquid leakage can be easily attained using a relatively simple infrastructure. Moreover, since this material has been commonly used as an electronic substrate, it is perfectly compatible with screen-printing techniques, allowing the integration of electronic circuits, along with surface-mounted devices. Hence, by using a single technology, fluidics and electronics can be integrated to achieve a total miniaturized system [150].

8.3.1. Micro-FIA System Based on LTCC Technology with Sensitive Amperometric CNTs-Epoxy Composite Sensor

The ceramic tape in a green state mainly comprises 45% filler (mainly Al_2O_3), 40% glass and 15% organic components (solvent, plasticizer and binder). They are known as ''green'' tapes because they are mechanized in the ''green stage'', when they are still soft and malleable, as opposed to hard and cured ones.

If we compare LTCC to conventional microfabrication techniques (e.g., using glass, silicon or polymers), LTCC technology shows some additional advantages, it can be a rapid prototyping, which allows quick modifications; it has low costs of fabrication; it does not need for special and expensive fabrication facilities such as clean rooms; and sealing elements, such as epoxies, are not needed, since the ceramic layers fuse solidly after the sintering process.

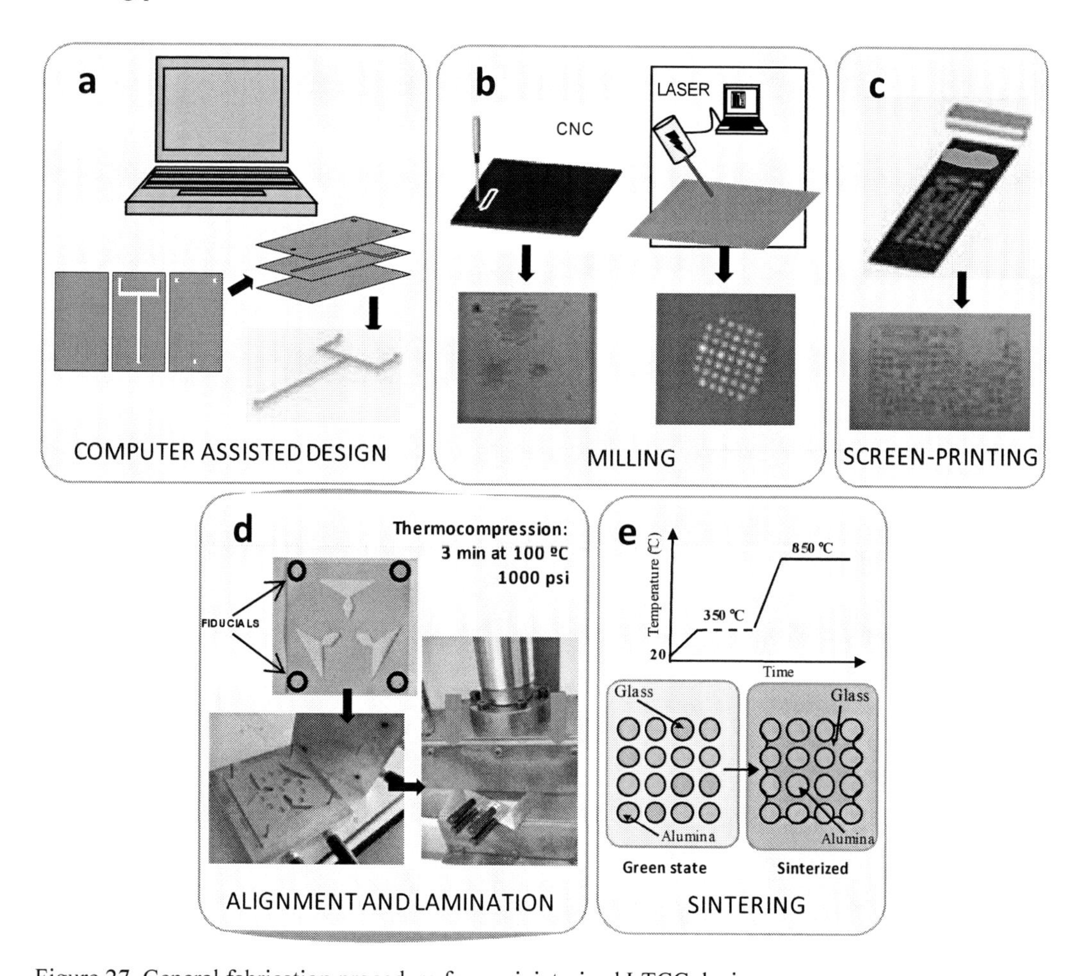

Figure 27. General fabrication procedure for a miniaturized LTCC device.

Fig. 27 shows the general fabrication procedure of an LTCC device. Since this technology is based on a multilayer approach, the desired design must be decomposed in separate layers. Each of these layers contains a certain geometrical pattern. Because the entire layers overlap, a three-dimensional structure is achieved. The layers are designed using computer-assisted design (CAD) software. After that, they can be mechanized using different methodologies (e.g., drilling, laser, jet-vapor etching or computer numerically controlled (CNC) machining). The screen-printing stage (Fig. 27 (c)), used to integrate the conductive tracks and planar passive electronic components normally needed in detection and signal-processing systems, is usually done in more than one layer. Conducting tracks on each layer are connected by means of vials filled with conductive screen-printing paste. Once all the layers have been mechanized and screenprinted, and before they undergo the sintering process in a programmable box furnace, a lamination stage is required (temperature, pressure and time).The final stage is sintering the laminated ceramics. This step can be performed in air or in a nitrogen atmosphere. The temperature profile depends on the specifications of each manufacturer. It normally involves two temperature plateaus. In the first one (200–400 °C) all the organic components burn out. The second one (600–900 °C) is the temperature at which most glasses have their vitreous transition temperature. At this point, the alumina particles can interpenetrate the original ceramic layers so that, at the end of the sintering process, no difference between those layers can be observed [140].

The process followed for the construction of our microanalyzer was the same depicted above. The device layout, designed using computer assisted design (CAD) software, consisted on eight different layers with a mixer length of 108 mm. A special layer was design to integrate platinum disk acting as counter electrode as part of the amperometric detection system embedded in the device. Circular cavities included in three different layers allow the later integration (after sinterization) of an exchangeable CNTs-epoxy composite based working electrode. A total amount of 13 layers were used for complete microdevice construction. The counter electrode consisted of a 4.8 x 4.6 x 0.125 mm platinum disk embedded in the ceramic device and sealed with sinterable glass paste. The reference electrode consisted in an Ag-based path printed under an auxiliary channel (see figure 28A-k) through which a 0.2 M KCl solution flows constantly to maintain stable the applied potential [148]. Electrical contact between the embedded electrodes and an external electronic set up was established by means of screen printed patterns and an external connector soldered to the ceramic body after sinterization (see figure 28 B).

Once burned-out, according to the predictable ceramic shrinkage, the final devices (see figure 28 B) presented an inner total volume of 250 μL. Final dimensions of inner-channel were 1.9 mm width and 400 μm height. The working electrode was separately prepared and integrated once the device was sinterized. The sensitive active area of the composite electrode (4 mm^2) was limited by the dimension of the hole located above the main flow channel in the working electrode cavity (see figure 28 A-i). The final dimensions of the cavity to be filled by the CNTs-epoxy composite were 2mm in depth and 4mm in diameter. The working electrode was integrated to the microanalyzer, in a cavity specifically designed for this purpose. The electrode was constructed separately and it was placed in the green tape ceramic device after it was burnt-out.

The microanalyzer integrates a complete amperometric detection system based on a highly sensitive CNTs-epoxy composite. The idea behind this system is to achieve a highly integrated and sensitive microanalyzer capable to operate during a considerable period of time, providing long-term stability signals, and under unattended conditions. To automate the calibration and analysis processes, it was coupled to a multicommutation system as described above. This approach allowed us to overcome some drawbacks usually presented by its counterpart in the macro scale (lack of portability, reagents consumption, maintenance and fabrication costs, etc.).

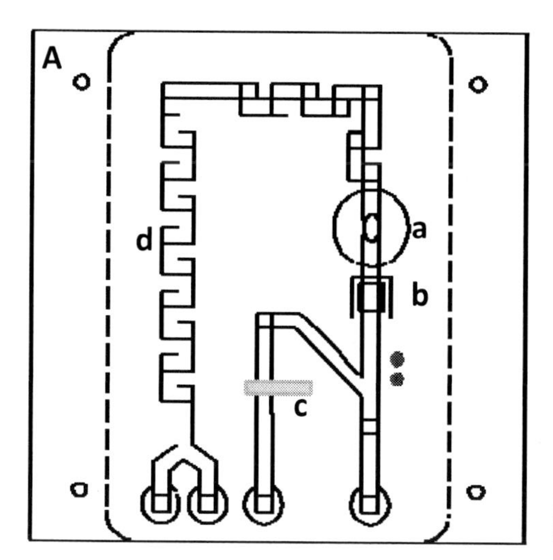

Figura 28. The CAD design with all overlapping layers and final LTCC device. A) Fluidic distribution: a) working electrode cavity, b) Pt counter electrode, c) reference electrode, d) mixer and B) Final device: e) CNT composite-based working electrode; f) Amperometric microanalyzer integrating CNT composite electrode. External connection of: 1) reference electrode, 2) counter electrode and 3) working electrode Reproduced from ref [163], with permission. Copyright (2010) Elsevier.

Experimental hydrodynamic conditions were optimized in order to analyze real samples with a range concentration of chlorine in swimming-pool waters. For this miniaturized analytical system the internal length of the mixing is fixed by the internal geometry of the microdevice (see figure 28 B), then flow rate and injection volume were optimized. The optimum hydrodynamic parameters were set at 200 μL for injection volume and 0.8 mL min^{-1} for flow rate.

The optimum working potential, -100 mV, was selected on the basis of previous results obtained using CNT composite electrodes for chlorine determination in improved-FIA system. For miniaturized systems the standard solutions prepared on-line and the samples, were pre-treated and diluted (1:1) in a 0.2 M phosphate buffer solution and a 0.2 M KCl at pH 5.5 into the microanalyzer prior to the detection cell (Figure 29). This pre-treatment step allows adjusting the ionic force and pH of the standard solution/sample, before being analyzed. In fact, here the different standard solutions were prepared by means of different duty cycles applied to the digital signal that controls the commutation system.

In these conditions, the analysis time for each sample was found to be 70 s (time from injection to steady-state signal) and the range expands from 0.2 to 4 mg L^{-1}, being the sensitivity of -0.046 (±0.001) μA L mg^{-1}. The LOD was estimated using the signal to noise (S/N = 3) criterion. Under these conditions, a LOD of 50 mg L^{-1} was achieved. This concentration range is suitable for monitoring free chlorine in swimming pool water given that this range contains the minimum and maximum legal chlorine concentration allowed by the current national legislation in swimming-pool waters.

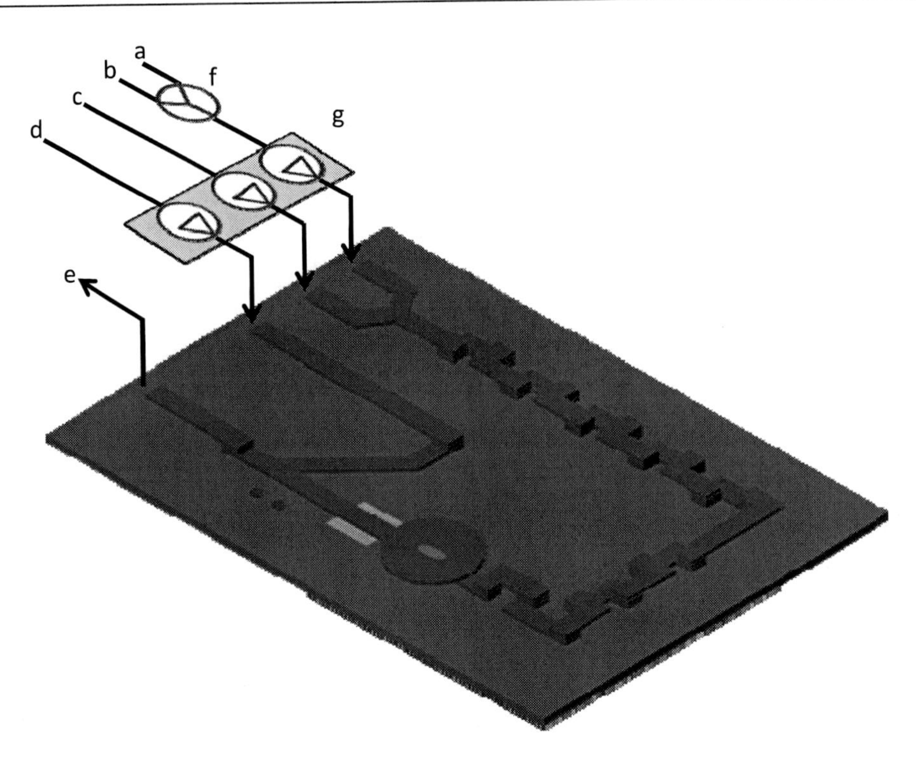

Figure 29 Experimental manifold used for chlorine determination in the ceramic microanalyzer based on CNT composite amperometric sensor. (a) Sample or stock solution; (b) deionized water; (c) carrier solution; (d) 0.2M KCl; (e) waste; (f) solenoid valve and (g) pump. Reproduced from ref [163], with permission. Copyright (2010) Elsevier.

In order to evaluate the analytical response of the microanalyzer, in terms of repeatability of the signal, a 3.7 mg L^{-1} and 1 mg L^{-1} free chlorine standard solutions were used. A RSD of 4% and 5% were obtained for 50 replicates performed consecutively, obtaining an average peak current value of -0.145(±0.002) μA and-0.0253 (±0.0004) μA, respectively. As expected, a slightly lower value of repeatability for the lowest concentration was obtained. Nevertheless, a good repeatability for both concentrations was demonstrated. The reproducibility and the robustness of microdevice were demonstrated with an estimation of the loss of analyzer sensitivity. Multiple calibration experiments were carried out with the ceramic microanalyzer during several days. A loss of sensitivity of a 30% compared to its initial value was observed after 300 analyses. During this period, the sensor was kept in the microanalyzer without renewing its surface. Regarding these results, a simple surface pretreatment step, without disassembling the sensor, was applied to increase the sensor sensitivity and to extend its lifetime. This pretreatment consisted of applying a fixed potential of 1.2 V during 30s before each calibration. The results showed that the sensitivity was recovered up to a 97% of its initial value. This pretreatment allows us to increase the sensor lifetime for at least ten days without having to disassemble it from the microsystem to renew its surface.

The analytical characteristics achieved by the integration of an exchangeable highly sensitive CNT electrode into an automated ceramic microanalyzer, in terms of linear range and reproducibility, allowed its application to the analysis of real samples, which has not been presented by similar devices previously reported [150] (see table table 4).

Table 4. Principal features of developed analytical systems using the improved composites based on different conducting carbon forms. In all cases carrier solution used was 0.2 M phosphate buffer solution in a 0.2M KCl at pH 5.5 as background electrolyte and for sampling pretreatment

Flow analyzer	Sensitivity (μA L mg^{-1})	RSD slope (%)[c]	Linear range (mg L^{-1})	LOD[a] (μg L^{-1})	Potential working (mV)	Time analysis[b] (s)	Sample consume (μL)	Carrier consume (mL min^{-1})
FIA-graphite (15%)	-0.08±0.01 (n=4)	3	0.15-4.0	150	-250	120	600	2.0
FIA-graphite (20%)	-0.11±0.02 (n=4)	3	0.7-4.0	700	-250	120	600	2.0
I-FIA-CNTs (10%)	-0.15±0.01 (n=3)	3	0.02-4.0	20	-100	60	600	2.0
I-FIA-CNTs (20%)	-0.19±0.02 (n=3)	3	-	170	-100	58	600	2.0
LTCC-CNTs (10%)	-0.046±0.001	2	0.05-4.0	50	-100	70	200	0.8
LTCC-Pt [153]	-56.6±0.2	0.3	18.0-365.4	-	350	100	500	1.5
FIA-Au [137]	-0.10±0.2	12	0.06-5.0	60	350	120	600	2.0

(a) LOD=S/N criterion [127] for minimum n=20
(b) time from injection to steady-state signal
(c) n=3, 95% confidence level

8.3.2. Validation of the Microanalyzer Using Real Samples

Regarding the real sample analysis, a total of twenty samples were analyzed. Five doped samples were prepared by dilution of the stock solution in real samples. Two synthetic samples, used as control, were prepared by dilution of the stock solution in deionized water. Samples were simultaneously analyzed using the microanalyzers proposed and the standard N,N-diethyl-p-phenylenediamine (DPD) colorimetric method [126], which consists of a Kit-commercial colorimeter (HACH, Düsseldorf, Germany) that provides measurements directly in mg L^{-1} of free chlorine. Statistical paired t-test and the least-square linear regression were used to make the statistical evaluation of the method. No significant differences were observed at 95% confidence level. In paired t-test: $t_{cal} = 1.782 < t_{tab} = 2.093$, while in the least square linear regression the slope and intercept were 1.06±0.06 and -0.04±0.05, respectively (r^2 =0.99; n= 20; 95% confidence level). The results of recovery data were enclosed into the range 91-108; these values show minimum difference between analytical methods. Excellent agreement between the obtained results with the proposed system and the standard method for free chlorine determination was obtained, demonstrating the high potential of the optimized CNT-epoxy composite electrode integrated in the miniaturized LTCC system.

8.4. Comparative Analytical Features of Analytical Proposes

In this part of the chapter we will compare and illustrate the electroanalytical features of the developed systems combined with the optimized composite loading of conducting material (graphite and CNTs). Chlorine in water was used as a target for evaluating the electroanalytical characteristics of developed analyzers. Proposed systems will be compared with similar analytical systems which use conventional detectors based on metals (Pt, Au). Table 4 shows the results obtained with the optimized composite sensors for the different automated analyzers. The experimental results show that when the optimized carbon composite electrode (15% loading of graphite and 10% loading of MWCNT) are used, the linear range remarkably increases, the LOD becomes one order of magnitude lower and the sensitivity slightly decreases compared to the maximum conducting loading composite electrode (20% loading of carbon), independently of the automated systems [71, 130, 163]. That agrees very well with the results obtained by impedance spectroscopy which showed that the range between 13-15% for graphite and 9-11% for MWCNTs exhibits a good behavior of the impedance parameters for the electroanalytical purposes with low charge transfer resistance, ohmic resistance and double layer capacitance. Moreover we have shown the versatility of this carbon-based composite to be integrated in different flow systems keep in at the same time the electroanalytical performance. It is worthy to note that the electroanalytical characteristics of such carbon-based composites are comparable or even improved with respect to other conventional and more expensive transducer materials (Au, Pt) [134].

In general, the results show that the optimization of the conducting-phase composition, according to the electrical and electrochemical features of the composite, improve the electroanalytical features of different analytical systems. Indeed, the good results with chlorine as a target could be extended to other more complex electrochemical systems. Moreover, their easy malleability makes them able to be integrated in different flow systems, even those with a high miniaturization level. Following this approach, they could be quite promising to be applied in μTAS systems or even with bio-sensing capabilities.

Conclusion

We have shown in this chapter that conducting-epoxy electrodes, based on graphite or CNTs, can be fabricated, characterized and hence optimized by means of EIS and voltammetry techniques, as well as microscopy (AFM) and chronoamperometry. We have demonstrated that EIS specially provides a versatile tool to optimize the composition of composite materials. On one side, it is possible to extract the ohmic resistance which can be related to the percolation resistivity. On the other side, other very useful parameters can be extracted which affect directly to the electroanalytical response. One of them is the charge transfer resistance, which is related with the heterogeneous electron transfer rate and which depends on the surface electrochemical reactivity. Normally, one would expect almost constant values of Rct/A for conventional metal electrodes. However this parameter changes and becomes very important for carbon based materials where electrochemical anisotropy is always present (the terminal edges provides higher electron transfer rates than the walls).

Moreover, EIS also allows the extraction of the electrode capacitance which can be correlated to the background current, an important parameter to minimize in order to enhance the signal/noise ratio. Consequently all these parameters are relevant for the composite response and by a proper evaluation of them one can choose the proportion which fulfills the electroanalytical requirements of high sensitivity, fast response and low limits of detection.

The optimized compositions range between 13-15% for graphite composites and 9-11% for CNT composites. We have shown on one hand that the electroanalytical detection of ascorbic acid was easily improved using the optimal 15%-graphite and 10%-MWCNT loadings electrodes. Although the sensitivity value was slightly inferior in the two cases, we could obtain better LOD and wider linear range than using the 20% of graphite and MWCNT loadings electrodes. Such values of optimal conductor loading allow us to fabricate attractive and robust composite electrodes with very interesting application as amperometric sensors at low analyte concentration. This strategy of characterization/optimization could be extended to the study of amperometric (bio) sensors based on composites. Thus, tailoring the graphite and CNT amounts of the composites together with a proper (bio)functionalization (either in bulk or at the surface) of (bio)receptors, redox mediators and catalysts one can get efficient (bio)chemical sensors.

Moreover, these improved sensors can be used as sensors integrated in automatic analyzers; the benefits of this approach have been shown with different developed automated systems. For instance we have demonstrated that the integration of the near-percolation composite in a flow analytical systems (FIA system, I-FIA or miniaturized system) allowed the determination of lower concentration of free chlorine than those established by law, in swimming pool water samples. This performance could not be realized with the low resistivity composite electrodes because of their higher LOD. An excellent agreement between the results obtained with the proposed systems and standard method for free chlorine determination was achieved in all studied cases.

Additionally, all developed system apart from conferring good sensitivity, they also provide robustness and automation, which make them suitable for on-line monitoring of low concentrations of others electroactive species.

The excellent performance of the flow ceramic microanalyzer opens promising perspectives for the integration of the carbon-based composites in micro total analysis system (μTAS). For instance one step further could be the achievement of μTAS with hybrid integration of external microactuators such as micropumps and microvalves, to increase the system reliability. The future work following this approach is currently under development.

References

[1] Schult, K.; Katerkamp, A.; Trau, D.; Grawe, F.; Cammann, K.; Meusel M. *Anal.Chem.* 1999, 71, 5430-5435.

[2] Wang, J. ; Rivas, G.; Cai, X.; Palecek, E.; Nielsen, P.; Shiraishi, H.; Dontha, N.; Luo, D.; Parrado, C.; Chicharro, M.; Farias, P.A.M.; Valera, F.S.; Grant, G.H. ; Ozsoz, M.; Flair, M.N. *Anal. Chimi. Acta.* 1997, 347, 1-8.

[3] Ruschan, G.R.; Newnham, R.E.; Runt, J.; Smith, E. *Sensor Actuat. B-Chem.* 1989, 20, 269-275.

[4] Navratil, T.; Kopanica, M. *C. Rev. Anal. Chem.* 2002, 32, 153-166.
[5] Navratil, T.; Kopanica, M.; Krista, J. *Chem. Anal.-Warsaw.* 2003, 48, 265-272.
[6] Navratil, T.; Senholdova, Z.; Shanmugam, K.; Barek, J. *Electroanal.* 2006, 18, 201-206.
[7] Sebkova, S.; Navratil, T.; Kopanica, M. *Anal. Lett.* 2005, 38, 1747-1758.
[8] Navratil, T.; Kopanica, M. *Chem. Listy.* 2002, 96, 111-116.
[9] Agüí, L.; Yáñez-Sedeño, P.; Pingarrón, J.M. *Anal. Chim. Acta.* 2008, 622, 11-47.
[10] Claudill, W.L.; Howell, J.O.; Wightman, R.M. *Anal. Chem.* 1982, 54, 2532-2535.
[11] Adams, R.N. *Anal. Chem.* 1958, 30, 1576.
[12] Céspedes, F.; Martínez-Fábregas, E.; Alegret, S. *Trend Anal. Chem.* 1996, 15, 296-304.
[13] Hammersley, J.M. *Ann. Math. Statist.* 1957, 28, 790-795.
[14] Scher, H.; Zallen R. *J. Chem. Phys.* 1970, 53, 3759-3761
[15] Hammersley, J.M. In *Origin of percolation theory.*, in *Percolation structures and processes*; Deutscher, G.; Zallen, R.; Adler, J.; Ed.; Adam Hilger: Bristol; The Israel Physical Society: Jerusalem; The American Institute of Physics: New York, NY, 1983.
[16] Stauffer, D.; Aharony, A. *Introduction to Percolation Theory;* . 2nd edition;.Burgess Science Press: London, 1992.
[17] Allaoui, A.; Bai, S.; Cheng, H. M.; Bai, J. B. *Compos. Sci. Technol.* 2002, 62, 1993-1998.
[18] Barrau, S.; Demont, P.; Maraval, C.; Bernes, A.; Lacabanne, C. *Macromol. Rapid Comm.* 2005, 26, 390-394.
[19] Kovacs, J.Z.; Velagala, B. S.; Schulte, K.; Bauhofer, W. *Compos. Sci. Technol.* 2007, 67, 922-928.
[20] Navarro-Laboulais, J.; Trijueque, J.; GarciaJareno, J. J.;Vicente, F. *J. Electroanal. Chem.* 1995, 399, 115-120.
[21] Li, W.; Liu, Z.Y.; Yang, M.B. *J. App. Polym. Sci.* 2010. 115, 2629-2634.
[22] Godovski, D.Y.; Koltypin, E.A.; Volka, A.V.; Moskvina, M.A. *Analyst*, 1993, 118, 997-999.
[23] Stará, V.; Kopanika, M. *Electroanal.* 1989, 1, 251-256.
[24] Kalcher, K.; Kauffmann, J.M.; Wang, J.; Svancara, I.; Vytras, K., Neuhold, C.; Yang, Z. *Electroanal.* 1995, 7, 5-22.
[25] Li, C.; Thostenson, E.T.; Chou, T-W. *Compos. Sci. Technol.* 2008, 68, 1227-1249.
[26] McCreery, R.L. *In Voltammetric Methods in Brain Systems*; Boulton, A.; Baker, G.; Adams R.N.; Ed.; Chapter 1; Human Press: Clifton, NJ, 1995.
[27] Kalcher, K. *Electroanal.* 1990, 2, 419-433.
[28] Lvovich, V.; Scheeline, A. *Anal. Chem.* 1997,69, 454-462.
[29] Khurana, M.K.; Winlove, C.P.; O'Hare, D. *Electroanal.* 2003, 15, 1023-1030.
[30] Rubianes, M.D.; Rivas, G. A. *Electrochem. Comm.* 2003, 5, 689-694.
[31] Li, J.; Chia, L.S.; Goh, N.K. *J. Electroanal. Chem.* 1999, 460, 234-241.
[32] Sampath, S.; Lev, O. *Anal. Chem.* 1996, 68, 2015-2021.
[33] Leung, E.; Cragg, P.J.; O'Hare, D.; O'Shea, M. *Chem. Comm.* 1996, 1996, 23-24.
[34] Wang, J.; Fang, L.; Lopez, D.; Tobias, H. *Anal. Lett.* 1993, 26, 1819-1830.
[35] Céspedes, F.; Martínez-Fabregas, E.; Alegret, S. *Anal. Chimi. Acta.* 1993, 284, 21-26.
[36] Weisshaar, D.E.; Tallman, D.E.; Anderson, J.L. *Anal. Chem.*, 1981, 53, 1809-1813.
[37] Mahan, B., *Química curso universitario.* Addison-Wesley Iberoamericana: Massachusett, 1988

[38] Delhaès, P.; *Graphite and precursors, the world of carbon*; Gordon and Breach Publishers: Amsterdam, 2001.
[39] Iijima, S. *Nature* 1991. 354, 56-58.
[40] Iijima, S.; Ichihashi, T. *Nature* 1993, 363, 603-605.
[41] Banks, C.E.; Compton, R.G. *Analyst* 2006, 131, 15-21.
[42] Gooding, J.J. *Electrochim. Acta.* 2005, 50, 3049-3060.
[43] Merkoçi, A., Pumera, M.; Llopis, X.; Perez, B.; del Valle, M.; Alegret, S. *Trac-Trend Anal. Chem.* 2005. 24, 826-838.
[44] Wang, J. *Electroanal.* 2005, 17, 7-14.
[45] Wildgoose, G. G.; Banks, C. E.; Leventis, H. C.;Compton, R. G. *Microchim. Acta.* 2006. 152, 187-214.
[46] Choi W.B.; Chung D.S.; Kang J.H.; Kim H.Y.; Jin Y.W.; Han I.T.; Lee Y.H.; Jung J.E.; Lee N.S.; Park G.S.; Kim J.M. *App. Phys. Lett.* 1999, 75, 3129-3131
[47] An K.H.; Kim W.S.; Park Y.S.; Choi Y.C.; Lee S.M.; Chung D.C.; Bae D.J.; Lim S.C,; Lee Y.H. *Adv. Mater.* 2001, 13, 497-500.
[48] Yao, Z.; Kane, C.L.; Dekker, C. *Phys. Rev. Lett.* 2000, 84, 2941-2944.
[49] Dai, H.J.; Hafner J.H.; Rinzler A.G.; Colbert D.T.; Smalley, R.E. *Nature* 1996, 384, 147-150.
[50] Baughman, R.H.; Cui C.X.; Zakhidov A.A.; Iqbal Z.; Barisci J.N.; Spinks G.M. ;Wallace G.G.; Mazzoldi A.; De Rossi D.; Rinzler A.G.; Jaschinski O.; Roth S.; Kertesz M. *Science* 1999, 284, 1340-1344.
[51] Appenzeller, J.; Knoch J.; Derycke V.; Martel R.; Wind S.; Avouris P. *Phys. Rev. Lett.* 2002, 89 (12), 1-4.
[52] Kymakis, E.; Alexandrou I.; Amaratunga G.A.J. *J. App. Phys.* 2003, 93, 1764-1768.
[53] Dresselhaus, M.; Dresselhaus, G.; Eklund, P. *Science of fullerenes and carbon nanotubes*. Academic Press: New York, 1996.
[54] Nalwa, H.S., *Nanostructured Materials and Nanotechnology*; Concise Edition, Academic Press: San Diego, 2002.
[55] Hiura, H.; Ebbesen, T.W.; Tanigaki, K. *Adv. Mater.* 1995 7, 275-276.
[56] Tohji, K., Goto, T.; Takahashi ,H.; Shinoda, Y.; Shimizu,N.; Jeyadevan, B.; Matsuoka, I.; Saito, Y.; Kasuya, A.; Ohsuna, T.; Hiraga, H.; Nishina, Y. *Nature* 1996, 383, 679-679.
[57] Park, Y.S.; Choi, Y.C.; Kim, K.S.; Chung, D.C.; Bae,D.J.; An, K.H.; Lim, S.C.; Zhu, X.Y.; Lee, Y.H. *Carbon* 2001, 39, 655-661.
[58] Mizoguti, E.; Nihey, F.; Yudasaka, M.; Iijima, S.; Ichihashi, T.; Nakamura K. *Chem. Phys. Lett.* 2000, 321, 297-301.
[59] Dillon, A.C.; Gennett, T.; Jones, K.M.; Alleman, J.L.; Parilla, P.A.; Heben, M.J. *Adv. Mater.* 1999, 11, 1354-1358.
[60] Martinez, M.T.; Callejas, M.A.; Benito, A.M.; Maser, W.K.; Cochet, M.; Andres, J.M.; Schreiber, J.; Chauvet, O.; Fierro, J.L.G. *Chem. Comm.* 2002, 9, 1000-1001.
[61] Duesberg, G.S.; Burghard, M.; Muster, J.; Philipp, G.; Roth, S. *Chem.Comm*. 1998, 3, 435-436.
[62] Niyogi, S.; Hu, H.; Hamon, M.A.; Bhowmik, P.; Zhao, B.; Rozenzhak, S.M.; Chen, J.; Itkis, M.E.; Meier, M.S.; Haddon, R.C. *JACS*. 2001, 123, 733-734.
[63] Dai, H.J. *Acc. Chem. Res*. 2002, 35, 1035-1044.
[64] Musameh, M.; Wang, J.; Merkoçi, A.; Lin, Y.H. *Electrochem.Comm*. 2002, 4, 743-746.

[65] Rivas, G.A.; Rubianes, M.D.; Rodríguez, M.C.; Ferreyra, N.E.; Luque, G.L.; Pedano, M.L.; Miscória, S.A.; Parrado, C. *Talanta*. 2007, 74, 291-307.
[66] Rubianes, M.D.; Rivas, G.A. *Electrochem. Comm.* 2007, 9, 480-484.
[67] Olivé-Monllau, R., Esplandiu, M.J.; Bartrolí, J.; Baeza, M.; Céspedes,F. *Sensor Actuat. B-Chem.* 2010, 146, 353-360.
[68] Morales, A.; Céspedes, F.; Muñoz, J.; Martínez-Fàbregas, E.; Alegret, S. *Anal. Chimi. Acta.* 1996, 332, 131-138.
[69] Pumera, M.; Merkoçi, A.; Alegret, S. *Sensor Actuat. B- Chem.* 2006, 113, 617-622.
[70] Céspedes, F.; Alegret, S. *Trend. Anal. Chem.* 2000, 19, 276-285.
[71] Olivé-Monllau, R.; Baeza, M.; Bartrolí, J.; Céspedes, F. *Electroanal.* 2009, 21, 931-938
[72] Oldham, K.B. *J. Electroanal. Chem.* 1981, 122, 1-17
[73] Weber, S.G. Anal.Chem. 1989, 61, 295-302.
[74] Moniruzzaman, M.; Winey, K.I. *Macromol.* 2006, 39, 5194-5205.
[75] Rosca, I.D.; Hoa, S.V. *Carbon*, 2009, 47, 1958-1968.
[76] Thompson, B.C.; Moulton, S.E.; Gilmore, K.J.; Higgins, M.J.; Whitten, P.G.; Wallace, G.G. *Carbon*, 2009, 47, 1282-1291.
[77] Xie, X.F.; Gao, L. *Carbon*, 2007, 45, 2365-2373.
[78] Navratil, T.; Barek, J. *C. Rev. Anal. Chem.* 2009, 39, 131–147.
[79] Esplandiu, M.J. *Electrochemistry on Carbon Nanotube modified surfaces in Chemically Modified Electrodes*; In Chemical modified electrodes; Alkire, R.C.; Kolb, D.M.; Lopkowski, J.; Ross, P.N.; Ed.; Wiley-VCH Verlag: Wienheim, Germany. 2009, vol.11, 57-209.
[80] Merkoçi, A. *Microchim. Acta.* 2006, 152, 157-174.
[81] Pumera, M.; Sanchez, S.; Ichinose, I.; Tang, J. *Sensor Actuat. B-Chem.* 2007, 123, 1195-1205.
[82] Wang, J.; Varughese, K. *Anal. Chem.* 1990, 62, 318-320.
[83] Wang, J.; Naser, N.; Angnes, L.; Wu, H.; Chen, L.; *Anal. Chem.* 1992, 64, 1285-1288.
[84] White, S.F; Turner, A.P.F.; Schmid, R.D.; Bilitewski, U.; Bradley, J. *Electroanal.* 1994, 6, 625-631.
[85] Yang, X.; Johansson, G.; Gortom, L. *Microchim. Acta.* 1989, 1, 9-16.
[86] Sanchez, S.; Fabregas, E.; Iwai, H.; Pumera, M. *Small*, 2009, 5, 795-799.
[87] Ye, J.S.; Wen, Y.; Zhang, W.D.; Cui, H.F.; Xu, G.Q.; Shen, F.S. *Nanotechnology.* 2006, 17, 3994-4001.
[88] Pividori, M.I.; Alegret, S. *Microchim. Acta.* 2010, 170, 227-242.
[89] Wang, Q., Tang, H.; Xie, Q.; Tan, L.; Zhang, Y.; Li, B.; Yao, S. *Electrochim. Acta.* 2007, 52, 6630-6637
[90] Du, P.; Liu, S.; Wu, P.; Cai, C. *Electrochim. Acta*, 2007, 52, 6534-6547.
[91] Creasy, K.E.; Shaw, B.R. *Anal. Chem.* 1989, 61, 1460-1465.
[92] Wang, J.; Golden, T.; Varughese, K.; El-Rayes, I. *Anal. Chem.* 1989, 61, 508-512.
[93] Skladal, P.; Mascini, M. *Biosensor. Bioelectron.* 1992, 7, 335-342.
[94] Wang, J.; Liu, J. *Anal. Chimi. Acta.* 1993, 284, 385-391.
[95] Esplandiu, M.J.; Pacios, M.; Cyganek, L.; Bartroli, J.; Del Valle, M. *Nanotechnology,* 2009, 20, 3555(02-10).
[96] Zhao, H.; O'Hare, D. *J. Phys.Chem.C.* 2008, 112, 9351-9357.
[97] O'Hare, D.; Macpherson, J.V.; Willows, A. *Electrochem. Comm.* 2002, 4, 245-250.
[98] Davies, T.J.; Banks, C.E.; Compton, R.G. *J Solid State Electr.* 2005, 9, 797-808.

[99] Davies, T.J.; Brookes, B.A.; Compton, R.G.. *J. Electroanal. Chem*. 2004, 566, 193-216.
[100] Davies, T.J.; Brookes, B.A.; Fisher, A.C.; Yunus, K.; Wilkins, S.J.; Greene, P.R.; Wadhawan, J.D.; Compton R.G. *J. Phys. Chem. B*. 2003, 107, 6431-6444.
[101] Davies, T.J.; Compton, R.G. *J. Electroanal. Chem*. 2005, 585, 63-82.
[102] Aoki, K.; Osteryoung,J. *J. Electroanal. Chem*. 1981, 125, 315-320.
[103] Caudill, W.L.;. Howell, J.O.; Wightman, R.M. *Anal. Chem.* 1982, 54, 2532-2535.
[104] Simm, A.O.; Ward-Jones, S.; Banks, C.E.; Compton, R.G. *Anal. Sci.* 2005, 21, 667-671.
[105] Welch, C.W.; Compton, R.G. *Rev. Anal. Bioanal. Chem*. 2006, 384, 601-619.
[106] Xiao, L.; Streeter, I.; Wildgoose, G.G.; Compton, R.G. Sensor Actuat. B-Chem. 2008, 133, 118-127.
[107] Navarro-Laboulais, J.; Trijueque, J.; Garcia-Jareno, J.J. *J. Electroanal. Chem.*, 1998, 444, 173-186.
[108] Navarro-Laboulais, J.; Trijueque, J.; Garcia-Jareno, J.J. *J. Electroanal. Chem.* 1998, 443, 41-48.
[109] Beaunier, L.; Keddam, M.; Garcia-Jareno, J.J. *J. Electroanal. Chem.* 2004, 566, 159-167.
[110] Faiella, G.; Piscitelli, F.; Lavorgna, M.; Antonucci, V.; Giordano, M. *App. Phys. Lett.* 2009, 95, 153106 (1-3)
[111] Pacios, M.; Del Valle, M.; Bartrolí, J.; Esplandiu, M.J. *J. Electroanal.Chem.* 2008, 619, 117-124.
[112] Ruzicka, J.; Hansen, E.H. *Anal. Chim. Acta.* 1975, 78, 145-157.
[113] FIAlab®, Leaders in Flow Injection Technology, September 2010, *http://www.flowinjection.com/method2.aspx*
[114] Ruzicka, J.; Marshall, J.D. *Anal. Chimi. Acta.* 1990, 237, 329-343.
[115] Manz, A.; Graber, N.; Widmer, H.M. *Sensor Actuat. B-Chem.* 1990, 1, 244-248.
[116] Reyes, D.R.; Iossifidis, D.; Auroux, P.A ; Manz, A. *Anal. Chem.* 2002, 74, 2623-2636.
[117] Sequeira, M.; Bowden, M.; Minogue, E.; Diamond, D. *Talanta*, 2002, 56, 355-363.
[118] H. Andersson, A. van den Berg, *Sensor. Actuat. B- Chem.* 2003, 92, 315-325.
[119] Li, J.; Ananthasuresh, G.K. *J. Micromech. Microeng.* 2002, 12, 198-203.
[120] Baeza, M.; Ibañez-Garcia, N.; Baucells, J.; Bartrolí, J.; Alonso, J. *Analyst*, 2006, 131, 1109-1115.
[121] Feres, M.A.; Fortes, P.R.; Zagatto, E.A.G. ; Santos, J.L.M.; Lima, J.L.F.C. *Anal. Chimi. Acta*. 2008, 618, 1-17.
[122] Llorent-Martinez, E.J.; Barrales, P.O.; Fernandez-de Cordova, M.L.; Ruiz-Medina, *A. Curr. Pharm. Anal.*, 2010, 6, 53-65.
[123] Del Campo, F. J.; Ordeig, O.; Muñoz, F.J. *Anal. Chimi. Acta.* 2005, 554, 98-104.
[124] Sournia-Saquet, A.; Lafage, B.; Savall, A. C. *R. Acad. Sci. II* C., 1999, 2, 497-505.
[125] Clifford White, G. *Handbook of chlorination and alternative disinfectants*, Jonh Wiley & Sons: New York, 1999.
[126] 4500-Cl G, Standard Methods for the Examination of Water and Wastewater, 21st ed. APHA, AWWA, WEF: Washington, 2005.
[127] Jin, J.Y.; Suzuki, Y.; Ishikawa, N.; Takeuchi, T. *Anal. Sci.* 2004, 20, 205-207.
[128] Skoog, D.A.; Holler, F.J., Nieman, T.A. *Principles of Instrumental Analysis*, Mc Graw Hill: Orlando, USA, 1998.

[129] Diari Oficial de la Generalitat de Catalunya, DOGC, Decret 95, 22 de Febrer 2000, *Generalitat de Catalunya*, 2000, 2338-2341.
[130] Olivé-Monllau, R.; Pereira, A.; Bartrolí, J., Baeza, M.; Céspedes, F. *Talanta*, 2010, 81, 1593-1598.
[131] Borges, S.D.; Reis, B.F. *Anal. Chimi. Acta*, 2007, 593, 39-45.
[132] Boletín Oficial del Estado, BOE num.45, Real Decreto 140/2003, 7 de Febrero 2003, 7228-7245.
[133] Ordeig, O.; Mas, R.; Gonzalo, J.; Del Campo, F.J.; Muñoz, F.J.; de Haro, C. *Electroanal.* 2005, 17, 1641-1648.
[134] Olivé-Monllau, R.; Orozco, J.; Fernández-Sánchez, C.;Baeza, M.; Bartroli, J.; Jiménez-Jonquera, C.;Cespedes, F. *Talanta.* 2009, 77, 1739-1744
[135] Okumura, A.; Hirabayashi, A.; Sasaki, Y.; Miyake, R. *Anal. Sci.* 2001, 17, 1113-1115.
[136] Wroblewski, W.; Dybko, A.; Malinowska, E.; Brzozka, Z. *Talanta*, 2004, 63, 33-39.
[137] McDonald, J.C.; Whitesides, G.M. *Acc. Chem. Res.* 2002, 35, 491-499.
[138] Romanato, F.; Tormen, M.; Businaro, L.; Vaccari, L.; Stomeo, T.; Passaseo A, A.; Di Fabrizio, E. *Microelectron. Eng.*, 2004, 73-4, 870-875.
[139] Wu, H.K.; Odom, T.W.; Chiu, D.T.; Whitesides, G.M. *J. Am. Chem. Soc.* 2003, 125, 554-559.
[140] Ibañez-Garcia, N.; Martínez-Cisneros, C.S.; Valdes, F.; Alonso, J. *TrAC-Trends Anal. Chem.* 2008, 27, 24-33.
[141] Ibañez-Garcia, N.; Puyol, M.; Azevedo, C.M.; Martínez-Cisneros, C.S.; Villuendas, F.; Gongora-Rubio, M.R.; Seabra, A.C.; Alonso, J. *Anal. Chem.* 2008, 80, 5320-5324.
[142] Achmann, S.; Hammerle, M.; Kita, J.; Moos, R. *Sens. Actuat. B-Chem.* 2008, 135, 89-95.
[143] Martínez-Mañez, R.; Soto, J.; García-Breijo, E.; Gil, L.; Ibañez, J.; Gadea, E. *Sens. Actuat A-Phys*, 2005, 120, 589-595.
[144] Wang, J.; Pumera, M. *Anal. Chem.* 2002, 74, 5919-5923.
[145] Gongora-Rubio, M.R.; Sola-Laguna, L.M.; Moffett, P.J.; Santiago-Aviles, J.J. *Sens. Actuat. A: Phys.* 1999, 73, 215-221.
[146] Gongora-Rubio, M.R.; Espinoza-Vallejos, P.; Sola-Laguna, L.; Santiago-Aviles, J.J. *Sens. Actuat A: Phys.* 2001, 89, 222-241.
[147] Llopis, X.; Ibañez-Garcia, N.; Alegret, S.; Alonso, J. *Anal. Chem.*, 2007, 79, 3662-3666.
[148] Ibañez-Garcia, N.; Mercader, M.B.; da Rocha, Z.M.; Seabra, C.A.; Gongora-Rubio, M.R.; Alonso, J. *Anal. Chem.*, 2006, 78, 2985-2992.
[149] Baeza, M.; López, C.; Alonso, J.; López-Santín, J.; Álvaro, G. *Anal. Chem.* 2010, 82, 1006-1111.
[150] Martínez-Cisneros, C.S.; da Rocha, Z.M.; Ferreira, M.; Valdés, F.; Seabra, A.; Gongora-Rubio, M.; Alonso-Chamarro, J. *Anal. Chem.*, 2009, 81, 7448-7453.
[151] Chudy, M. ; Grabowska, I.; Ciosek, P.; Filipowicz-Szymanska, A.; Stadnik, D.; Wyzkiewicz, I.; Jedrych, E.; Juchniewicz, M.; Skolimowski, M.; Ziolkowska, K.; Kwapiszewski, R. *Anal. Bioanal. Chem.*, 2009, 395, 647–668.
[152] Tanret, I.; Mangelings, D.; Vander Heyden, Y. *Current Pharma. Anal.*, 2009, 5, 101-111.
[153] Vandaveer, W.R.; Pasas, S.A.; Martin, R.S.; Lunte, S.M. *Electrophoresis.* 2002, 23, 3667-3677.

[154] Wang, J. *Talanta*. 2002, 56, 223-231.
[155] Wang, J.; Chatrathi, M.P.; Tian, B.M. *Anal. Chim. Acta*. 2000, 416, 9-14.
[156] Martin, R.S.; Ratzlaff, K.L.; Huynh, B.H.; Lunte, S.M. *Anal. Chem.* 2002, 74, 1136-1443.
[157] Nyholm, L. *Analyst*. 2005, 130, 599-605.
[158] Wang, J.; Chen, G.; Chatrathi, M.P.; Musameh, M. *Anal. Chem.* 2004, 76, 298-302.
[159] Yao, X.; Wu, H.X.; Wang, J.; Qu, S.; Chen, G. *Chem. Eur. J.* 2007, 13, 846-853.
[160] Crevillén, A.G.; Pumera, M.; González, M.C.; Escarpa, A. *Lab on Chip*. 2009, 9, 346-353.
[161] Crevillén, A.G.; Ávila, M.; Pumera, M.; González, M.C.; Escarpa, A. *Anal. Chem.* 2007, 79, 7408-7415.
[162] Karuwan, C.; Wisitsoraat, A.; Maturos, T.; Phokharatkul, D.; Sappat, A.; Jaruwongrungsee, K.; Lomas, T.; Tuantranont, A. *Talanta*. 2009, 79, 995-1000.
[163] Olivé-Monllau, R.; Martínez-Cisneros, C.S.; Bartrolí, J.; Baeza, M.; Céspedes, F. *Sensor.Actuat. B-Chem.*2010 in Press. DOI: 2010. 10.1016/j.snb.2010.10.017

In: Resin Composites: Properties, Production and Applications ISBN: 978-1-61209-129-7
Editor: Deborah B. Song

Chapter 6

FABRICATION OF EPOXY RESIN COMPOSITES WITH METAL NANOPARTICLES

*Andrey L. Stepanov**
Kazan Physical-Technical Institute, Russian Academy of Sciences,
420029 Kazan, Russian Federation
Kazan Federal University, 420008 Kazan, Russian Federation

ABSTRACT

A review of recent results on a fabrication of epoxy resin composites with metal nanoparticles using viscous properties of polymer is reported. Preparation of metal nanoparticles realized during thermal vacuum evaporation of silver onto the surface of epoxy resin at a viscosity from 20 to 120 Pa·s) having room tempetature, which is well below the glass transition temperature of the polymer. Additionally, for synthesis of metal nanoparticles the ion irradiation of viscous polymer matrix is used. The viscous epoxy resin is implanted by silver ions with diferent doses. As a result, epoxy resin layers containing silver nanoparticles in their volume are fabricated. Various types of disperse structures formed by metallic nanoparticles in the polymer are detected. The morphology of the composite material is found to be controlled by the polymer viscosity and the metal deposition time. The use of the viscous state of epoxy resin increases the diffusion coefficient of silver impurity, which stimulates the nucleation and growth of nanoparticles and allows a high filling factor of metal in the polymer to be achieved. Mechanisms of metal nanoparticle growth in viscous epoxy resin are discussed.

INTRODUCTION

Composite nanomaterials based on polymers containing metal nanoparticles (MNPs) are now of interest from a practical standpoint [1-3]. Such photonic materials are promising for application in nonlinear optics, magnetooptics, and optoelectronics in order to create effective reflectors, biosensors, catalytic systems, and so on. It is often necessary that the surface or

* E-mail: aanstep@gmail.com

near-surface layer of a polymer serve as a carrier of the required physicochemical properties and its volume retains the initial properties; as a result, e.g., flexible ultrathin displays and information carriers can be produced. Modern nanotechnologies can be used to improve and design new methods for the formation of polymer–MNP structures.

One of a methods of polymer–MNP composite material production, which is effective and rather widely applied, is based on the deposition of a thermally evaporated metal (tin, selenium, indium, gold, copper, silver) onto the surface of a molten polymer, i.e., an organic material having a temperature above glass transition temperature T_g [4-11]. The morphology of the metal–polymer transition layer (interface) was found to undergo substantial changes depending on the type of polymer, its temperature, the deposition rate, the type of metal, and the subsequent annealing of the composite material. This method can be used for various polymers, such as styrynhexomethacrylate (T_g = 60°C, T_{exp} = 60–120°C) [5-7], polymethylmethacrylate (T_g = 106°C, T_{exp} = 150°C) [8], polystyrene (T_g = 93°C, T_{exp} = 200°C) [9], and polycarbonate (T_g = 235°C, T_{exp} = 250°C) [10], where is T_{exp} temperature of experimental deposition of metal. This method uses the saturation of the surface layer of a polymer by metallic atoms to form a supersaturated solid solution of impurity atoms for nucleation and growth of MNPs. The degree of filling of the polymer volume with a metal is rather high.

As noted above [4-11], used thermal evaporation technique for MNP fabrication is characterized by heating of a polymer to a temperature $T > T_g$ in order to transform organic substrutes into a liquid or softened state. This procedure requires rather high temperatures, often more than 200°C. Unforunately, the rate of metal diffusion into the bulk of a polymer at such high temperatures is very high; as a result, it is almost impossible to control the size of MNPs in a polymer. Therefore, the reproducibility of the composite material parameters is quite low.

A new method, which is also based on the deposition of a thermally evaporated metal but onto a polymer (epoxy resin) chemically diluted by a solvent at room temperature rather than onto a very heated organic matrix is proposed and described here. As a main result, it is possible to achieve the relaxation viscous fluid state (molecular mobility) of a polymer matrix at lower temperatures (below Tg = 120°C for epoxy resin). By changing of the epoxy hardener concentration everyone can control the degree of dynamic viscosity of epoxy resin. After the finishing of metal deposition, epoxy resin transforms into a solid state in a certain time due to polymerization. Thus, the metal–polymer composite layers can be formed by thermal evaporation of silver onto the surface of viscous fluid epoxy resin and such composites with MNPs are also interesting for an application in optoelectronics for a control of light signal trasnfering along a polymer waveguide.

FORMATION OF SILVER NANOPARTICLES DURING DEPOSITION ON VISCOUS EPOXY RESIN

As was mentioned, silver nanoparticles can be synthesized in a viscous fluid epoxy matrix by thermal evaporation of the metal in vacuum onto softened (below the glass transition temperature) high molecular organic substrates at room temperature [12, 13]. For experimental demonstration of such possibility for MNPs fabrication in polymers, the amount of silver deposited onto the surface of epoxy polymer was chosen to be $4.8 \cdot 10^{-6}$ g/cm^2, the deposition rate was $4.4 \cdot 10^{14}$ atoms/cm^2, and the deposition time was 60 s. At the chosen silver deposition rate, the partial vapor pressure in a vacuum chamber was so low that metal particles could form when silver atoms drift from an evaporator to a substrate. As a viscous fluid polymer matrix, liquid epoxy resin in a mixture with dibutylphthalate and polyethylenepolyamine was used. The chosen chemical formula of polymer provided the optimum conditions of forming an epoxyamine polymer from the standpoint of the cure rate. The lifetime of the viscous fluid polymer matrix was about 2 h, and, then, its physico-mechanical properties were affected by the formation of a solid spatial network polymer structure.The dynamic viscosity of the polymer matrix during the deposition of silver was changed from 20 to 120 Pa·s. In an additional specific experiment, samples were prepared at the deposition times varying from 5 to 80 s for processing onto epoxy resin with a viscosity of 20 Pa·s.

The dependence of the dynamic viscosity μ of the epoxy polymer on the cure time t (Figure 1) was measured with a capillary viscosimeter according to a standard procedure [14, 15]. The presented curve $\mu(t)$ is typical for the various epoxies and is in consist of the epoxy curing kinetics [16]. For the time range $t < 110$ min, the low viscosity of the polymer is explained by the prevailing interdiffusion processes of the epoxy molecules, and their concentration is averaged in macrovolume. The subsequent composite curing followed by the polymer viscosity increases due to the chemical cross-linking between the molecules and the rigid polymer network formation. It is supposed [16] that the polymer network nucleation centers are "cybotaxes" – the associates comprising some epoxy molecules with a "kinetically gainful" orientation. Before the molecule are the cross-linking, the cybotaxes have no distinct shape and interface. The situation, however, changes when the polymer network is formed. Many solid microblocks appear in the liquid composite, thus providing a high degree of volume non-uniformity of physical properties in the polymer material. Upon growth of the spatial network, the viscous flow polymer transforms into a glassy one.

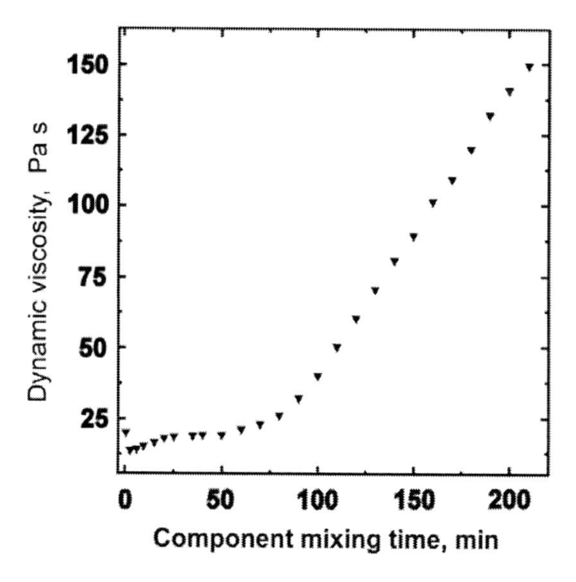

Figure 1. Epoxy dynamic viscosity dependence on component mixing time [15].

Epoxy resin layers of ~10 μm thickness were deposited by centrifuging onto thin glass substrates. Silver was deposited in a VUP-5 vacuum device. The residual vacuum during silver deposition was 10^{-3} Pa. After silver deposition, the samples were held in a vacuum for 3 h until complete curing.

When silver is deposited by thermal vacuum evaporation onto epoxy resin substrates having different viscosities, metal atoms can diffuse into the bulk of the substrate to various depths depending on its viscosity. Such silver deposition causes the formation of silver nanoparticles in the near surface volume of the polymer. The formation of MNPs in all samples is directly supported by the appearance of a characteristic selective absorption band induced by the surface plasmon resonance in silver particles in optical spectral range [1, 17, 18].

The micrographs in Figure 2 show examples of disperse MNP-containing structures studied by transmission electron microscopy (TEM) normal to the sample surface (left) and the cross sections of these structures (right) [12]. It can conventionally distinguish three types of metallic structures in the near surface volume of the polymer depending on its initial viscosity. For example, in the polymer viscosity range from 20 to 30 Pa·s, a near surface layer 110 – 120 nm thick (Figures 2a, 2b) that contains ultrafine (1.5 – 2 nm) silver particles are detected. However, such small particles cannot be resolved by TEM because of an insufficient resolution of electron microscope. The MNP sizes in this sample correspond to the minimum critical size (1.5 – 2 nm) at which plasmon optical absorption can manifest itself in MNPs [17].

As the viscosity of epoxy resin increases from 30 to 90 Pa·s, the following bilayer structure forms (Figures 2c, 2d): a surface monolayer of spherical silver nanoparticles with an average diameter of 6 – 8 nm and a layer 35 – 60 nm thick in the bulk of the polymer next to the first layer, which contains ultrafine (1.5 – 2 nm) MNPs. Similar sample characterized by a layer with ultrafine MNPs was described by the modeled optical spectra of a composite layered nanomaterial with MNPs that was an evidence of plasmon absorption of MNP in such composites [18]. Note that a bilayer structure is also fabricated when a thermally evaporated metal is deposited onto the surface of a molten (heated) polymer [5 – 8].

As was observed that the average MNP size and the distribution of the MNP sizes in the near surface layer of epoxy increase with the polymer viscosity. As an example, Figure 3 presents the histograms of the size distributions of silver nanoparticles for samples with an epoxy resin viscosity of 36 (Figure 3a) and 48 Pa·s (Figure 3b) [18]. For the same composites, Figure 4 shows optical absorption spectra with the typical selective bands of MNP plasmon absorption. The maximum absorption of a sample with a viscosity of 48 Pa·s is located at a longer wavelength compared to the more viscous sample. This fact also indicates the formation of bigger MNPs as the epoxy resin viscosity increases [1, 19].

When silver is deposited onto epoxy resin with a viscosity of 60 – 100 Pa·s, a disperse metallic thin film consisting of spherical silver nanoparticles forms on the polymer surface (Figures 2e, 2f). The formation of MNPs in the bulk of the polymer as the viscosity exceeds 100 Pa·s is not detected. At a higher viscosity of epoxy resin (> 120 Pa·s), a thin island silver film forms on the polymer surface (Figures 2g, 2h), as in the case of traditional deposition of a metal onto the surface of solid materials [20]. Thus, three types of thin films, which have different structures, can form on the surface of a polymer depending on the ratio of the volume diffusion and surface diffusion of deposited silver atoms.

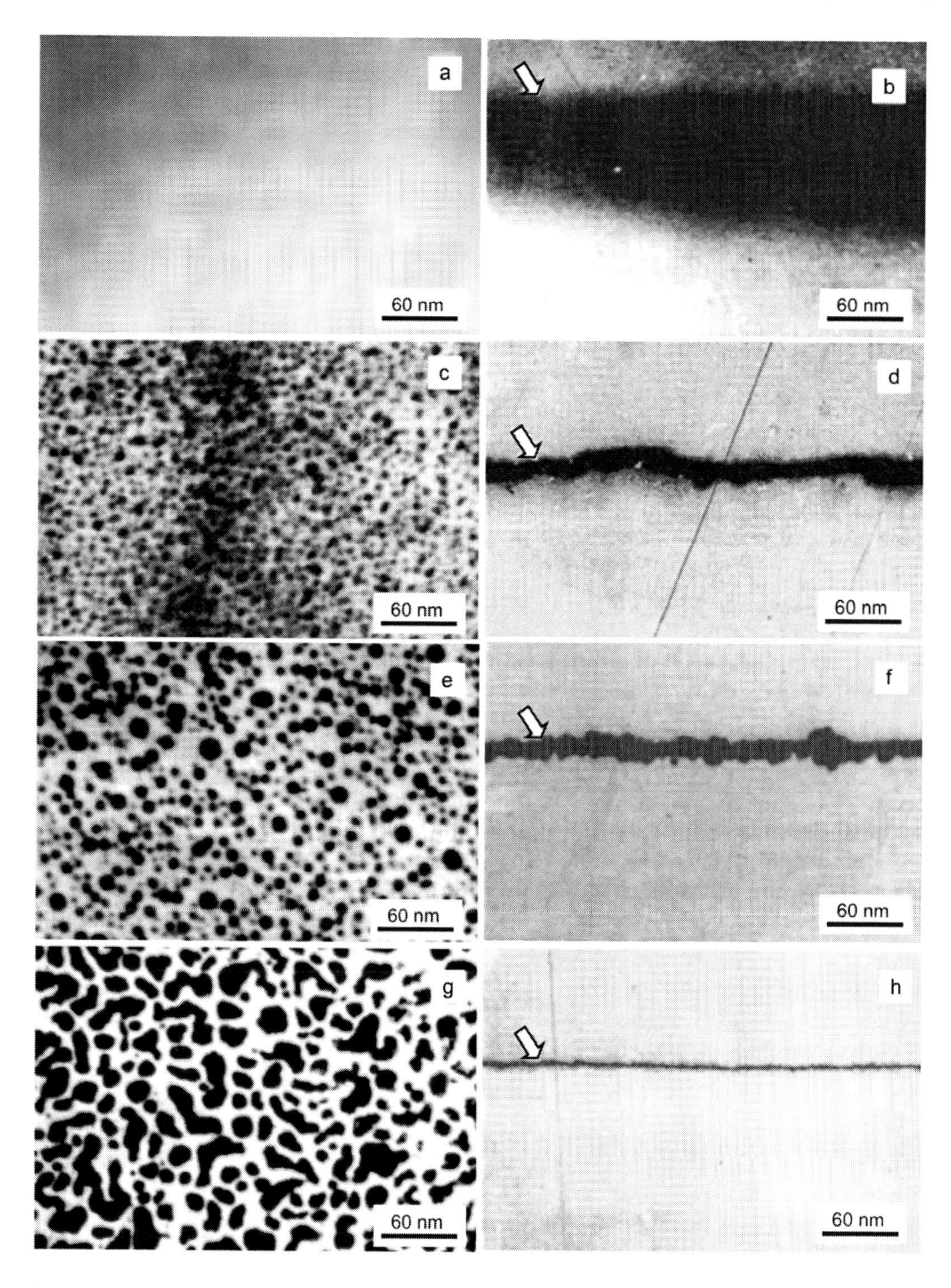

Figure 2. Micrographs of the silver disperse films synthesized by vacuum deposition onto liquid epoxy resin as a function of its viscosity. (a, c, e, g) Observation is normal to the sample surface and (b, d, f, h) cross sections of samples. The epoxy resin viscosity during deposition was (a, b) 20, (c, d) 36, (e, f) 86, and (h) 120 Pa s. Arrows indicate the position of the polymer surface [12].

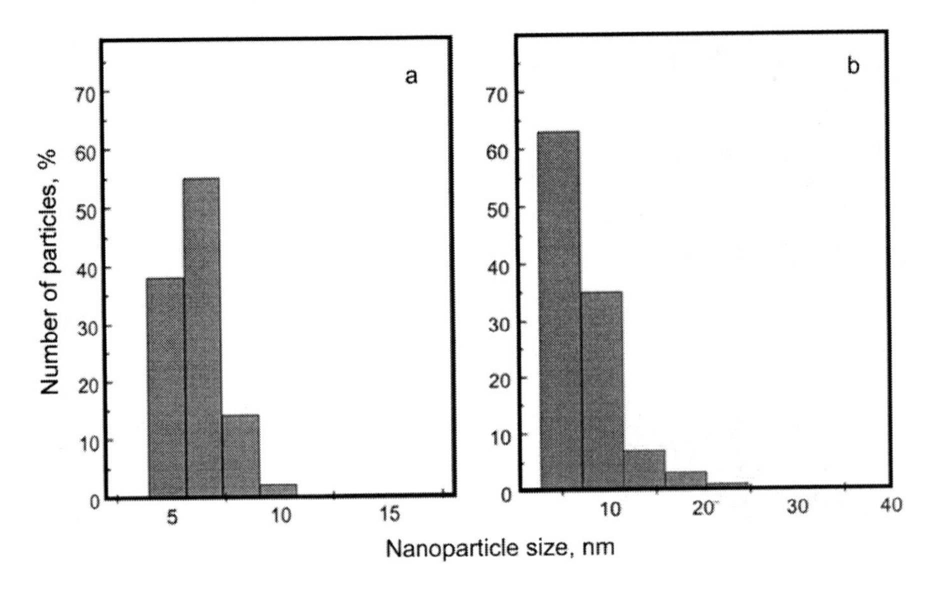

Figure 3. Histogram of the size distribution of silver nanoparticles for deposition onto epoxy resin with a different viscosity of (a) 36 and (b) 48 Pa·s [18].

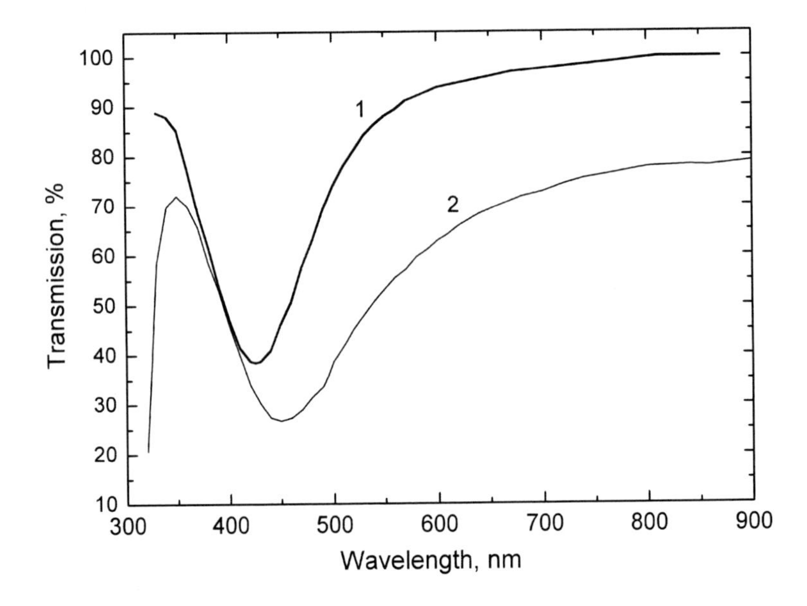

Figure 4. Optical transmission spectra for the samples prepared when silver was deposited onto epoxy resin with a different viscosity of (*1*) 36 and (*2*) 48 Pa·s. [18].

The formation of a composite layer with ultrafine MNPs when silver is deposited onto the surface of a polymer with a viscosity of 20 – 30 Pa·s at room temperature is the most interesting because the condition of polymer is close to be almost liquid. Therefore, this viscosity range of the polymer substrate was analyzed in more detail. To study the MNP growth kinetics in a viscous fluid polymer, a series of experiments on the deposition of silver onto liquid epoxy resin (20 Pa·s) at a different deposition time from 5 to 80 s were performed. In these samples, the silver particle size was estimated only optically because low TEM resolution. Since the MNP size (~ 1.5 – 2 nm) is significantly smaller than the electron free path length in a bulk silver, the average nanoparticle size in these samples can be estimated from optical plasmon absorption band width $\Delta W_{1/2}$ using the equation [17]:

$$d = 2\, V_F \,/\, \Delta W_{1/2}\,, \tag{1}$$

where d is the nanoparticle diameter and $V_F = 1.4 \cdot 10^{15}$ nm/s is the electron velocity at the Fermi surface.

Using the values of d, the optical absorption spectra of the composite materials in terms of the Bruggeman effective medium were simulated as described in the work [18, 21]. Based on the results of model simulations, the silver nanoparticle with size d density per area unit N in each sample was estimated. Figure 5 presents the dependence of N on time t of silver deposition onto liquid epoxy resin with a viscosity of 20 Pa·s.

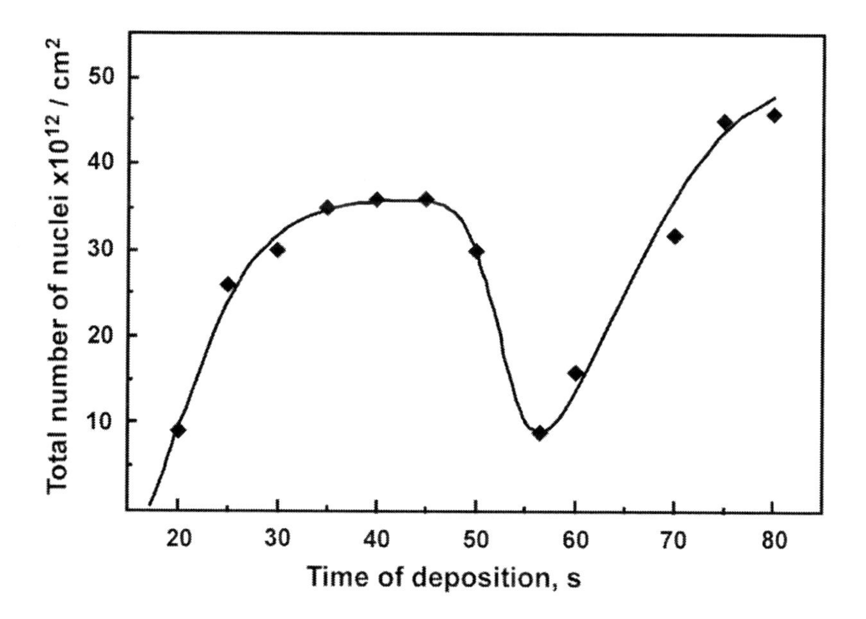

Figure 5. The number of ultrafine silver nanoparticles per unit area N when the metal is deposited onto epoxy resin with a viscosity of 20 Pa·s as a function of the deposition time (points). Simulation data is the solid line [12].

As follows from this dependence, MNPs near the polymer surface has nucleated in a time of ~ 17 s after the beginning of deposition. Also the concentration of silver atoms near the polymer surface at which optically detectable MNPs form can be estimated. During the diffusion of silver atoms, their concentration C_{Ag} at distance h from the polymer surface can be described as

$$C_{Ag} = \frac{2G}{D}\left[\left(\frac{Dt}{\pi}\right)^{1/2} e^{\frac{-2h}{4Dt}} - \frac{h}{2}\operatorname{erfc}\left(\frac{h}{\sqrt{Dt}}\right)\right], \tag{2}$$

where G is the silver atomic flux onto the polymer surface, and D is the diffusion coefficient [22].

The diffusion coefficient of silver atoms in epoxy resin can be estimated using the Stokes–Einstein formula [23]

$$D = kT/3\pi\phi\mu \tag{3}$$

where μ is the dynamic viscosity of the polymer, T is the temperature of polymer, φ is the silver atom diameter, and k is the Boltzmann constant. According to this estimation, the volume diffusion coefficient of silver atoms is $D = 8\cdot10^{-10}$ cm^2/s. Then, the value of C_{Ag} after 17-s deposition in the near surface region with a thickness on the order of the silver atom diameter is ~$7\cdot10^{20}$ atoms/cm^3, which indicates a critical supersaturation of the near surface layer in the polymer. Within this time, silver nanoparticles with an average size d = 1.6 nm fabricated, as follows from the optical data. In the next 25 s of deposition, the depth of MNPs in the polymer reaches 50 – 55 nm due to silver diffusion, the number of particles increases, and their average diameter increases monotonically by 0.1 – 0.2 nm due to the entrapment of silver atoms by the nanoparticles formed in the previous stage (Figure 5). After the next 10 – 15 s, the concentration of MNPs decreases quickly and their average size increases until 2.5 nm.

The Ostwald ripening can explain the decrease in the number of particles that accompanies their growing, when the smallest particles dissolve and the metal atoms that become free join bigger MNPs [24]. In this case, the average interparticle distance increases noticeably, and MNPs can renucleate and grow in the space that is free of nanoparticles in the substrate if deposition continues [20]. The observed increase in the number of silver particles in viscous fluid epoxy resin after 57 s from the beginning of deposition corresponds to MNP renucleation. Since the number of particles increases in a 50 nm thick layer, the average MNP size at this time decreases to 1.4 – 1.6 nm. However, longer deposition leads to an increase in both the MNP concentration and average size.

As follows from TEM study, there is a certain depth profile of the size distribution of nanoparticles. The biggest silver nanoparticles (d ~ 4 – 5 nm) are located near the surface layer of the polymer, and the smaller the MNP size are placed dipper from the surface. This behavior is clear and related to the depth profile of the impurity concentration. At the higher concentration there is the higher nanoparticle growth rate. Obviously, an increase in the epoxy viscosity at the same silver deposition rate should decrease the time it takes for a layer consisting of large MNPs to form. The silver diffusion into the bulk of the polymer decreases; the thickness of the layer with ultrafine particles decreases; and, at a certain viscosity, only a disperse surface metallic film will be form.

Discussions on the formation mechanism of a composite layer with MNPs in the near surface region of a liquid polymer ($T > Tg$) are related to the immersion of MNPs in this layer [5–11]. According to [7], two forces act on the MNPs located on the surface of a liquid polymer provided wetting is complete or partial. The van der Waals force tends to immerse MNPs in the polymer volume, whereas the entropy force (which is caused by the compression of polymer chains near particles) tends to push them to the polymer surface. The result of these forces forms a potential well in the dependence of the free energy of particles on the immersion depth. Due to this minimum, particles or their ensembles are stabilized under the surface of a viscous fluid polymer at a depth smaller than the MNP diameter. The model calculations [11] demonstrate that, as the particle size decreases, the potential well depth decreases rapidly: in particular, for gold nanoparticles smaller than 50 nm, the potential well is absent. This means that the immersion mechanism does not operate for sufficiently small MNPs. This conclusion is also true of our work, since the silver nanoparticle size did not exceed 5 nm under our experimental conditions.

When describing the results of our experiments on the thermal deposition of silver onto epoxy resin having temperature $T < T_g$ due to chemical dilution, it should be note another method that can decrease the glass transition temperature of the surface of a polymer as compared to its bulk T_g. As follows from the results of [11], when the surface of polystereneis subjected to ultraviolet radiation during the deposition of gold nanoparticles onto it from a solution, the nanoparticles are also immersed in a thin near surface layer of the polymer, which indicates a decrease in the glass transition temperature of the polymer surface. However, it is obvious that this technique differs substantially from the technological approach observed here.

Synthesis of Silver Nanoparticles in Viscous Epoxy Resin by Ion Implantation

Ion implantation is a promising method for the formation of MNPs in a polymer matrix, since it can perform a controlled synthesis of MNPs at various depths under the surface of an irradiated matrix at an unlimited content of an implanted impurity [1, 25]. Koon et al. [26] were the first to carry out studies in the field of ion synthesis of MNPs in polymers. The number of works dealing with the ion synthesis of MNPs in polymers is now continuously growing.

The following new approach has recently been discussed: it consists in the ion synthesis of MNPs in organic matrices that are in a viscous fluid relaxed state during implantation in contrast to the traditional irradiation of a glassy polymer [27, 30]. The relaxation state of viscous fluid silicone polymer implanted with Fe and Co ions was shown to affect the nucleation and growth of iron and cobalt nanoparticles. The size, shape, and crystal structure of the synthesized MNPs were found to be controlled by the diffusion mobility of metal atoms and small clusters. Various layered structures of ferromagnetic MNPs were grown. The authors of [31] assumed that, when magnetic (iron, cobalt, etc.) ions are implanted into polymers, the dipole–dipole interaction between MNPs affects the synthesis of MNPs, in particular, the formation of particles of various shapes (from a spherical to an acicular shape). This effect also manifests itself during the irradiation of viscous fluid polymers by ferromagnetic metal ions [30].

As shown early [32] the effect of the relaxation state in an irradiated viscous fluid polymer on the ion synthesis of MNPs can be also used with epoxy resin. To exclude the effect of the magnetic dipole interaction described above, diamagnetic silver ions into viscous fluid epoxy resin were implanted and fabricated metal nanostructures were studied [33]. For that glassy and viscous fluid epoxy resin was irradiated by 30_keV Ag+ ions at a current density of 4 $\mu A/cm^2$ with doses in the range of $2.2 \cdot 10^{16} - 7.5 \cdot 10^{16}$ ion/cm^2 at room temperature. For comparative experiments, epoxy resin was also implanted by 40_keV Ar^+ ions at a current density of 4 $\mu A/cm^2$ in the ion beam at doses of $3.2 \cdot 10^{16}$ and

$1.0 \cdot 10^{17}$ ion/cm^2. The average calculated penetration depth of silver ions into epoxy resin was 30 nm, and this depth for argon ions was ~50 nm. As a viscous fluid polymer matrix, a liquid epoxy resin mixed with dibutylphthalate and polyethylenepolyamine was used. The lifetime of the viscous fluid polymer matrix was about 2 h; then, its physico-mechanical properties were affected by the formation of a spatial network structure in the polymer. The viscosity of the epoxy composite was 30 Pa·s at the beginning of implantation process and increased by about 12 Pa·s upon irradiation. Thus, the ion implantation time did not exceed the composite lifetime.

Epoxy layers ~10 μm thick were deposited onto thin glass substrates by spin coating. With allowance for radiation heating, the epoxy substrate temperature was about 60°C, which is several tens of degrees Celsius lower than the upper boundary of the range of allowable nondestructive heating of this polymer. Once irradiation is finished, the epoxy composite transforms into a glassy state according to the cure (polymerization) kinetics of this compound (Figure 1).

Figure 6 shows electron micrographs of the silver ion implanted layers in epoxy resin that is in a glassy or viscous fluid state during irradiation. Implantation is seen to result in the formation of silver nanoparticles in a polymer matrix in both types of samples. Note that MNPs visible in TEM form in the viscous fluid sample at a lower dose ($2.2 \cdot 10^{16}$ ion/cm^2) as compared to the glassy polymer (~$5.2 \cdot 10^{16}$ ion/cm^2). This circumstance indicates more effective MNP nucleation in the viscous fluid material. Moreover, the MNPs formed in the viscous fluid medium at all ion doses are spherical, whereas the nanoparticles formed in the glassy matrix have an irregular droplike shape.

The electron diffraction patterns of all polymer samples with ion_synthesized MNPs consist of narrow polycrystalline rings corresponding to the face-centered cubic lattice of metallic silver superimposed on the weak broad diffuse rings of the polymer matrix. It should also be noted that the intensities of the polycrystalline rings in the glassy polymer are significantly weaker than those in the electron diffraction pattern taken from the irradiated viscous fluid polymer, which indicates a higher content of the metallic phase in the latter case. We did not detect any chemical compounds with silver ions in the irradiated matrices.

The estimates of the size parameters of the synthesized MNPs, such as mean particle size d_{mean}, particle density per unit area $N_{implant}$, and average cubic deviation Δimplant of the particle sizes from $d$mean, are given in the table 1. As is seen from the table 1 and Figure 6, d_{mean} and $\Delta_{implant}$ increase monotonically and $N_{implant}$ decreases as the dose of silver ions implanted into epoxy resin in various relaxation states increases. In the viscous fluid medium, the values of d_{mean} and $N_{implant}$ are higher and the scatter of the sizes $\Delta_{implant}$ of ion synthesized MNPs is lower as compared to the glassy polymer at the same doses. Therefore, it can be conclude that the number of implanted silver ions that form MNPs in the viscous_fluid matrix is larger; hence, the corresponding filling factor $f_{implant}$ should be higher.

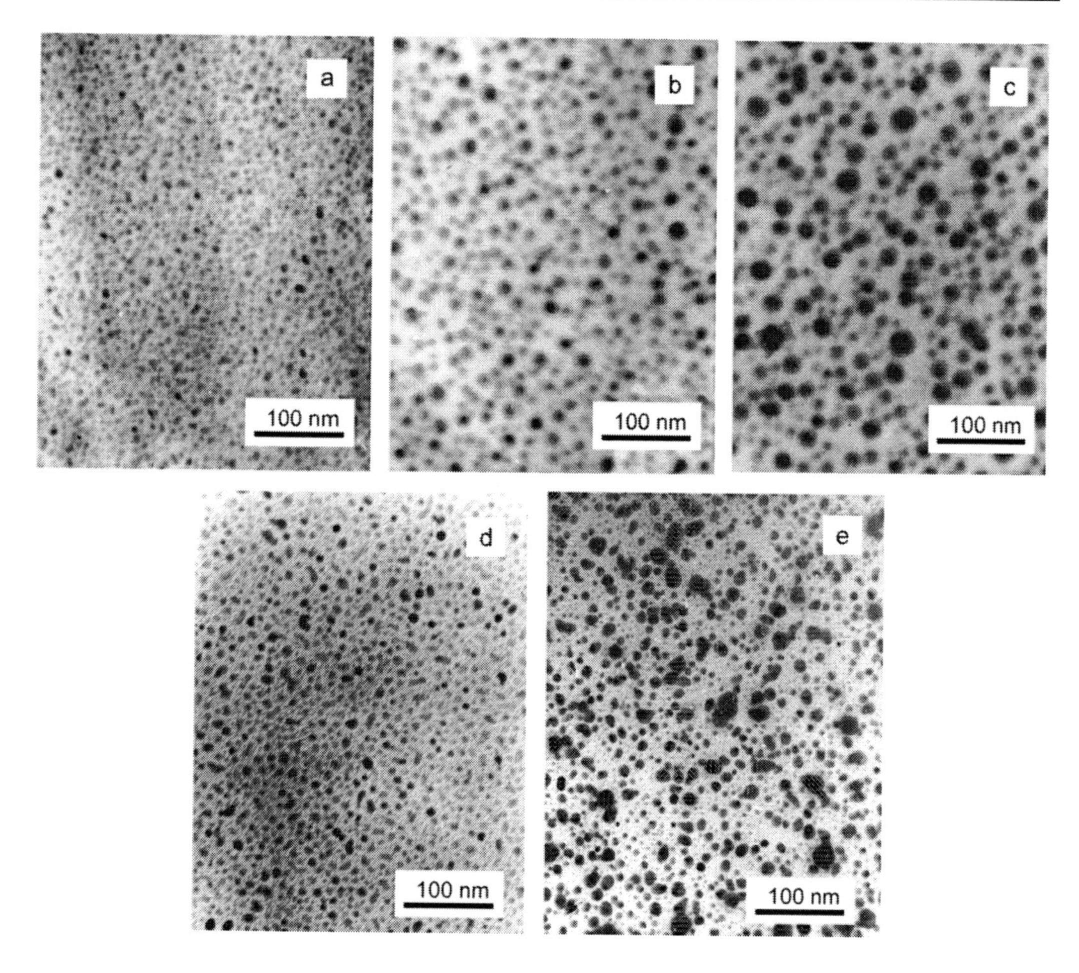

Figure 6. Micrographs of the epoxy layers with MNPs synthesized by the implantation of 30 keV silver ions at a current density of 4 μA/cm^2 in the ion beam into (a, b, c) viscous fluid and (d, e) glassy polymer matrices at a dose of (a) $2.2 \cdot 10^{16}$, (b) $5.2 \cdot 10^{16}$, (c) $7.5 \cdot 10^{16}$, (d) $5.2 \cdot 10^{16}$, and (e) $7.5 \cdot 10^{16}$ ions/cm^2 [33].

Table 1. Granulometric characteristics of the silver nanoparticles ion-synthesized in epoxy resin

Sample	State of polymer	Dose 10^{16} ion/cm^2	d_{mean} nm	$\Delta_{implant}$ nm	$N_{implant}$ $\times 10^{11}$ cm^{-2}	$f_{implant}$
1	Viscous	2.2	5.0	2.0	3.90	0.12
2	Viscous	5.2	16.3	6.0	2.10	0.35
3	Viscous	7.5	22	11.1	1.38	0.49
5	Glassy	5.2	8.3	10.2	1.20	0.25
6	Glassy	7.5	13.8	14.2	0.90	0.35

To estimate the factor of filling of an epoxy matrix with metallic silver upon ion implantation, we use the optical absorption spectra of the synthesized composite materials (Figure 7).

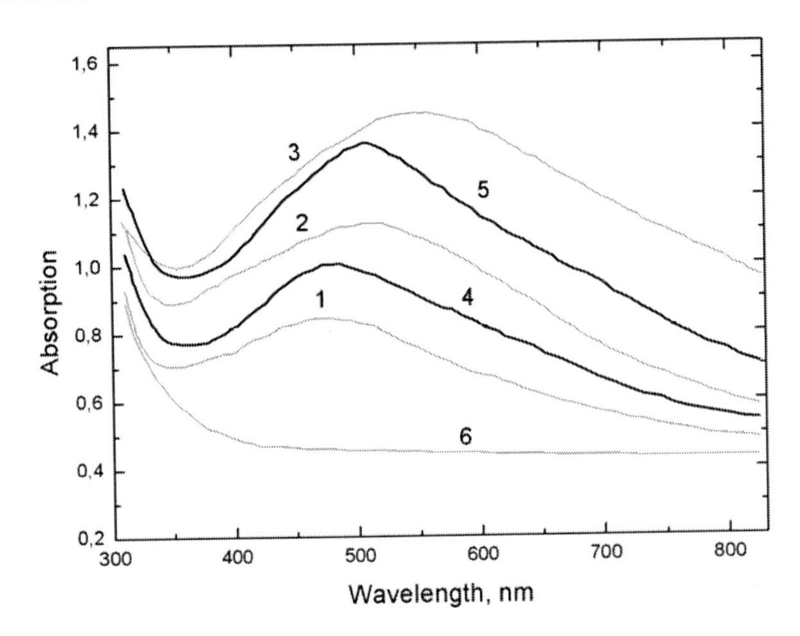

Figure 7. Optical absorption spectra (ln(1/A), where A is the transmission) of the epoxy layers with MNPs synthesized by the implantation of 30 keV silver ions at a current density of 4 μA/cm^2 in the ion beam into (*1–3*) viscous fluid and (*4, 5*) glassy polymer matrices at a dose of (*1*) $2.2 \cdot 10^{16}$, (*2*) $5.2 \cdot 10^{16}$, (*3*) $7.5 \cdot 10^{16}$, (*4*) $5.2 \cdot 10^{16}$, and (*5*) $7.5 \cdot 10^{16}$ ions/cm^2. (*6*) Absorption spectrum of the glassy epoxy resin after its irradiation by argon ions at a dose of $3.2 \cdot 10^{16}$ ions/cm^2.

As is seen from Figure 7, the optical spectra of all epoxy resin samples implanted by silver ions have a broad selective band, which is caused by the SPR absorption of MNPs [17]. This band is absent in the absorption spectrum of the epoxy resin irradiated by argon ions (Figure 7, curve *6*). All spectra exhibit intense absorption dominating in the ultraviolet region, which is caused by radiation defects and structural distortions in the polymer [34]. At a low dose ($2.2 \cdot 10^{16}$ ion/cm^2), the maximum of the SPR absorption of the viscous fluid sample is located near 490 nm (Figure 7, curve *1*). As the implantation dose increases, the SPR band absorption intensity increases monotonically and its maximum shifts toward long wavelengths, up to 570 nm (Figure 7, curves *1–3*). The changes in the optical spectra of the glassy epoxy resin with increasing dose are similar (Figure 7, curves *4*, *5*). However, the longer wavelength shift in the absorption maximum (to 520 nm) and its intensity are significantly smaller than for the viscous fluid polymer at the same implantation doses (Figure 7, curves *3* and *5*). The more intense SPR absorption bands of the ion synthesized silver nanoparticles in the viscous fluid epoxy resin also indicate more effective MNP nucleation and growth during implantation as compared to the glassy polymer. As follows from the Maxwell–Garnet theory, the position of the maximum in the absorption band of MNPs depends on the filling of the composite material with a metal [17, 35]. The position of the maximum in the SPR absorption of MNPs at a longer wavelength corresponds to a higher value of $f_{implant}$. It is assumed the absence of a strong electromagnetic interaction between MNPs (whose size is smaller than the light wavelength) in the samples under study, use the dielectric constants of silver [36] and the polymer matrix [37] at the wavelengths of the maxima in the SPR absorption of MNPs, and determine the corresponding values of $f_{implant}$ from the well known expression [17, 38]

$$\varepsilon_1^{Ag} = \frac{\left(2 + f_{implant}\right)}{\left(1 - f_{implant}\right)} n_{polymer}^2 \quad (4)$$

where ε_1^{Ag} is the real part of the dielectric constant of silver nanoparticles and *n*polymer is the refractive index of the dielectric matrix with MNPs.

The values of $f_{implant}$ calculated for the samples under study are given in the table. It is seen that, at the same implantation doses, $f_{implant}$ is significantly higher in the viscous fluid epoxy layers than in the glassy polymer.

As noted above, ion implantation in materials leads to radiation defects, such as the rupture of the covalent bonds of macromolecules, the formation of free radicals, cross bonding, oxidation, and carbonization [39]. The formation of such defects in polymers increases the absorbability in the optical region and, especially, in the ultraviolet region.

As follows from Figure 8, the ultraviolet absorption intensity and, hence, the number of radiation defects in the glassy polymer are larger than in the viscous fluid epoxy resin at the same implantation doses. The smaller number of structural defects in the implanted viscous fluid medium manifests itself in the higher optical transparency of these samples almost over the entire visible region (see Figure 8).

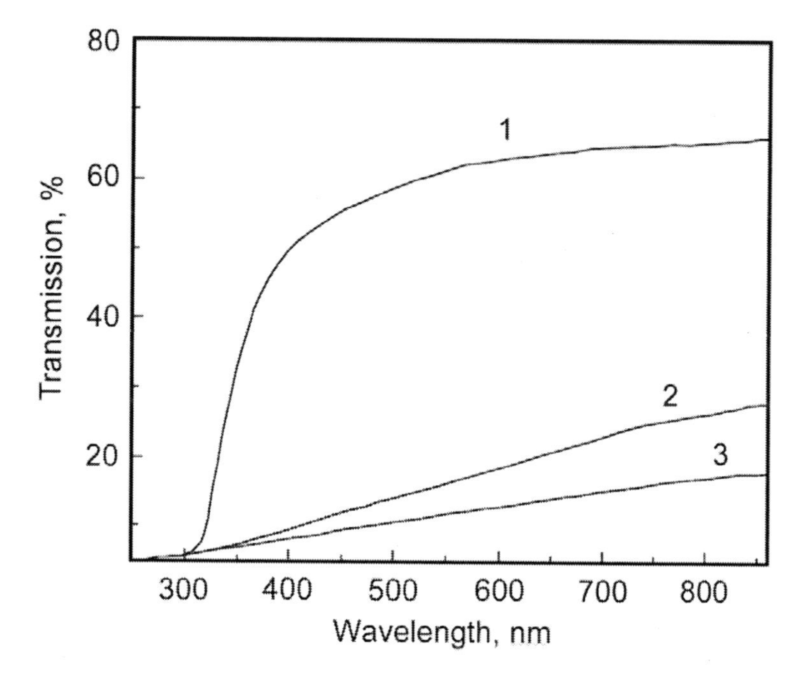

Figure 8. Transmission spectra of (*1*) unirradiated glassy epoxy resin and (*2*, *3*) the same polymer implanted by 40 keV argon ions at a current density of 4 μA/cm^2 in the ion beam at a dose of 10^{17} ions/cm^2. (*2*) Viscous fluid and (*3*) glassy states of the epoxy resin during irradiation [33].

The transmission spectra of the glassy and viscous fluid polymers irradiated by argon ions, which can only cause structural defects in the irradiated matrix, under the same conditions demonstrate higher transmission in the viscous fluid epoxy resin. Obviously, the high mobility of macromolecules in the viscous fluid matrix favors rapid restoration of broken bonds in the polymer and its molecular structure during irradiation. It was assume that the efficiency of carbonization in the viscous fluid medium during ion implantation is lower than that in the glassy polymer, since it is carbon aggregates that cause significant absorption in the longer wavelength spectral region of the irradiated polymer.

To understand the formation of MNPs in this experiment, it should be taken into account the process of a polymerization of epoxy resin. As was discussed above, during glass transition, the epoxy composite passes through several stages from the liquid state of the initial mixture to a fully solid polymer. The viscosity of epoxy resin changes in time according to these stages. At a polymer viscosity of 30 Pa·s, epoxy molecular groups begin to open and the epoxy system can be considered as a low viscosity homogeneous molecular liquid. No polymer blocks and supramolecular structure form in the polymer.

Thus, ion implantation in an epoxy polymer at a viscosity of 30 Pa·s (homogeneous liquid medium) leads to the formation of a supersaturated solution of silver atoms in the irradiated organic layer. Then, in the absence of internal defects in the matrix (which are inherent in solid materials), silver nanoparticles can homogeneously nucleate in the liquid medium similarly to the condensation of metallic drops in an inert gas atmosphere [39]. The further growth of MNPs is supported by the diffusion of impurity atoms toward their nucleation centers. Such conditions provide the formation of spherical particles and a rather uniform size distribution of MNPs (see Figure 6).

In contrast, the presence of own and additional radiation defects in glassy epoxy resin stimulates the heterogeneous nucleation of MNPs [39]. For this mechanism, the formation of MNPs and their size distribution depend on the system of defects and are also caused by the low mobility of silver atoms in the glassy polymer [40]. As a result, the MNPs acquire an asymmetrical droplike shape and the size distribution of the MNP is rather high (see table).

Droplike ion-synthesized nanoparticles were also observed earlier, e.g., in glassy polymethylmethacrylate implanted by iron ions [41]. However, it should be noted that the silver nanoparticles ion-synthesized in glassy polymers sometimes have a spherical shape [1, 25]. These polymers, which have lower softening and glass transition temperatures as compared to epoxy resin, are likely to be heated under an ion beam and to transform into a liquidlike state. As a result, impurity atoms acquire a high mobility, which leads to the formation of spherical MNPs.

During ion implantation in a viscous fluid medium at a relatively low dose ($\leq 10^{16}$ ion/cm^2), the high mobility of silver atoms can lead to the smearing of the depth profile of impurity atoms. According to the equation 3, the volume diffusion coefficient of silver atoms is five to six orders of magnitude higher than D_{Ag} in glassy polymers [42]. It is obvious that, at the early stage of implantation, a very high value of D_{Ag} can substantially decrease the local concentration of silver atoms in the near surface layer of the polymer and, thus, can decrease the efficiency of MNP nucleation. However, at a slightly lower value of D_{Ag}, the mobility of silver atoms increases the formation probability of nucleation centers, which serve as sinks for implanted impurity atoms. As is seen from the results obtained (Figure 6), MNPs nucleate in viscous fluid epoxy resin with a viscosity of 30 Pa·s at a lower implantation dose than in the glassy polymer. Thus, at the chosen viscosity of epoxy resin, the mobility of silver atoms leads to effective MNP nucleation rather than to a decrease in their local concentration in the implanted layer. It is also obvious that the growth rate of MNPs is higher in the viscous fluid polymer, which manifests itself in the formation of coarse particles (see Table 1).

The high mobility of silver atoms in the viscous fluid medium results in high efficiency of their accumulation on formed MNP nucleation centers due to an increase in the getter sphere size R_{gett}, which is proportional to D_{Ag} [43].

$$R_{gett} \propto \sqrt{D_{Ag} \cdot t_{growth}} \tag{5}$$

where t_{growth} is the MNP growth time. If t_{growth} is assumed to be the same in all irradiated samples, R_{gett} in the viscous fluid epoxy resin is three orders of magnitude higher than in the glassy polymer. According to [43], the higher the value of R_{gett}, the larger the MNP size, which was observed in our experiment (see Figure 6 and Table 1).

CONCLUSION

New method for synthesizing a disperse film that consists of silver nanoparticles on the surface and in the volume of a viscous fluid polymer (epoxy resin) kept at room temperature when the metal deposited by thermal vacuum evaporation is proposed. The structure and surface morphology of metal–polymer layers as a function of the viscosity of the polymer substrate during metal deposition is analysed. We considered The dynamics of MNP formation depending on the time of metal deposition onto the surface of a viscous fluid epoxy resin is considered and the formation mechanism of a nanostructured metal–polymer layer is discussed. Additionally it is shown that ion implantation in a epoxy resin that is in a viscous fluid state during irradiation can be used to synthesize nanoparticles of precious metals. Due to the viscous fluid state of the substrate, the diffusion coefficient of an implanted impurity without a special purpose increase in the polymer temperature can be increased. As a result, the nucleation and growth of nanoparticles are stimulated at low energy ion implantation doses and a high factor of filling of the polymer with the metal can be reached. With a viscous fluid epoxy resin, the formation of irreversible structural radiation defects in the irradiated organic material as compared to the implanted glassy polymer can be reduced and, thus, the uniformity of the size distribution of ion synthesized MNPs can be increased.

ACKNOWLEDGMENT

I wish to thank my partners and co-authors V.F. Valeev, S.N. Abdullin, V.I. Nugdin, Yu.N. Osin, R.I. Khaibullin, V.N. Bazarov, I.A. Faizrakhmanov. Also I grateful acknowledge the Alexander von Humboldt Foundation and DAAD in Germany and Austrian Scientific Foundation in the frame of Lisa Meitner Fellowship. Partly, this work was supported by the Ministry of Education and Science of the Russian Federation (FTP "Scientific and scientific-pedagogical personnel of the innovative Russia" contract No. 02.740.11.0797.

REFERENCES

[1] Stepanov, A. L. In Metal-Polymer Nanocomposites; Nicolais, L.; Carotenuto, G.; Eds., John Wiley & Sons Publ: London, 2004, pp. 241-263.

[2] Sarychev, A.; Shalaev, V. Electrodynamics of metamaterials, Wold Sci. Pub.: New York, 2007.

[3] Zhang, J. Z. Optical properties and spectroscopy of nanomaterials, Wold Sci. Pub.: London, 2009.

[4] Goffe, W. L. *Photog. Sci.* 1971, 15, 304-321.

[5] Kovacs, G. J.; Vincett, P. S. *J. Colloidal Interf. Sci.* 1982, 90, 335-351.

[6] Kovacs, G. J.; Vincett, P. S. *Thin Solid Films*, 1983, 100, 341-353.

[7] Kovacs, G. J.; Vincett, P. S. *Thin Solid Films*, 1984, 111, 65-81.

[8] Pattabi, M.; Sastry, M. S.; Sivaramkrishnan, V. *Phys. Rev. B.* 1989, 39, 9959-9964.

[9] Payne, R. S.; Swann, A.; Mills, P. J. *J. Mater. Sci.* 1990, 25, 3133-3138.

[10] Bechtolsheim, C. V.; Zaporojtchenko, V.; Faupel, F. J. *Mater. Res.* 1999, 14, 3538-3543.

[11] Rudoy, V. M.; Dement, O. V.; Yaminskii, I. V.; Sukhov, V. M.; Kartseva, M. E.; Ogarev, V. A. *Colloid. J.* 2002, 64, 746-754.

[12] Abdullin, S. N.; Stepanov, A.L.; Osin, Yu. N.; Khaibullin, I. B. *Surf. Sci. Lett.* 1998, 395, L242-L245.

[13] Stepanov, A. L.; Abdullin, S. N.; Khaibullin, R. I.; Khaibullin, I. B. RF Patent, 1997, No. 97109708.

[14] Malkin, A. Y.; Chalykh, A. E. Diffusion and viscosity of polymers, Khimiya: Moscow, 1979 [in Russian].

[15] Abdullin, S. N.; Stepanov, A. L.; Osin, Yu. N.; Khaibullin, R. I.; Khaibullin, I. B. *Surf. Coat. Technol.* 1998, 106, 214-219.

[16] Mezhikovskii, S. M. *Polym. Sci. USSR*, 1987, 29, 1571-1582.

[17] Kreibig, U.; Vollmer, M. Optical properties of metal clusters, Springer: Berlin, 1995.

[18] Stepanov, A. L. *Opt. Spektrosk.* 2001, 91, 868-872.

[19] Mie, G. *Ann. Phys.* 1908, 25, 377-425.

[20] Maissel, L. I.; Gland, R. Handbook of thin film technology, McGraw-Hill: New York, 1972.

[21] Granqvist, C. G.; Hunderi, O. *Phys. Rev. B.* 1977, 16, 3513-3534.

[22] Robertson, D.; Pundsack, A. L. *J. Appl. Phys.* 1981, 52, 455-462.

[23] Reif, F. Fundamentals of statistical and thermal physics, McGraw-Hill: New York, 1965.

[24] Zettlemoyer, A. C. Nucleation, Dekker: New York, 1969.

[25] Stepanov, A. L. *Tech. Phys.* 2004, 49, 143-154.

[26] Koon, N. C.; Weber, D.; Pehrsson, P.; Schindrel, A. I. *Mater. Res. Soc. Symp. Proc.* 1984, 27, 265-275.

[27] Khaibullin, R. I.; Osin, Y. N.; Stepanov, A. L.; Khaibullin, *Vacuum*, 1998, 51, 289-294.

[28] Khaibullin, R. I.; Osin, Y. N.; Stepanov, A. L.; Khaibullin, I. B. *Nucl. Instr. Meth. Phys. Res. B.* 1999, 148, 1023-1028.

[29] Khaibullin, R. I.; Zhikharev, V. A.; Osin, Y. N.; Zheglov, E. P.; Khaibullin, I. B.; Rameev, B. Z.; Aktas, B. *Nucl. Instr. Meth. Phys. Res. B.* 2000, 166-167, 897-902.

[30] Rameev, B. Z.; Aktas, B.; Khaibullin, R. I.; Zhikharev, V. A.; Osin, Y. N.; Khaibullin, I. B. *Vacuum*, 2000, 58, 551-560.

[31] Petukhov, V. Yu.; Ibragimova, M. I.; Khabibullina, N. R.; Shulyndin, S. V.; Osin, Yu. N.; Zheglov, E. P.; Vakhonina, T. A.; Khaibullin, I. B. *Vysokomol. Soedin.* A 2001, 42, 1973-1985.

[32] Stepanov, A. L.; Khaibullin, R. I.; Abdullin, S. N.; Osin, Yu. N.; Valeev, V. F.; Khaibullin, I. B. Proc. *Inst. Phys. Conf. Ser.* 1995, 147, 357-361.

[33] Stepanov, A. L.; Khaibullin, R. I.; Valeev, V. F.; Osin, Yu. N.; Nuzhdin, V. I.; Faiurakhmanov, I. A. Zhur. *Technich. Fiz.* 2009, 79, 77-82 [in Russian].

[34] Townsend, P. D.; Chandler, P. J.; Zhang, L. Optical effects of ion implantation, Cambridge Univ.: Cambridge, 1994.

[35] Maxwell-Garnet, J. C. *Philos. Trans. R. Soc. London*, 1904, 203, 385-411.

[36] Quinten, M. Z. *Phys. B* 1996, 1001, 211-217.

[37] Stepanov, A. L. *Opt. Spectrosc.* 2001, 91, 815-818.

[38] Arnold, G. W. *J. Appl. Phys.* 1975, 46, 4466-4473.

[39] Zettlemoyer, A. C. Nucleation, Dekker: New York, 1996.

[40] Morokhov, I. D.; Trusov, L. I.; Chizhik, S. P., Ultradispersion metallic media, Metallzrgiya: Moscow, 1977 [in Russian].

[41] Petukhov, V. Yu.; Zhikharev, V. A.; Makovskii, V. F.; Osin, Yu. N.; Mitryaikina, M. A.; Khaibullin, I. B.; Abdullin, S. N. Poverkhost 1995, 4, 27-36.

[42] Zener, C. *J. Appl. Phys.* 1949, 20, 950-954.

[43] Maurer, R. D. *J. Appl. Phys.* 1958, 29, 1-7.

In: Resin Composites: Properties, Production and Applications ISBN: 978-1-61209-129-7
Editor: Deborah B. Song

Chapter 7

EFFECT OF MONTMORILLONITE CLAY ON THERMAL AND MECHANICAL PROPERTIES OF EPOXY AND CARBON/EPOXY COMPOSITE

Yuanxin Zhou, Ying Wang and Shaik Jeelani
Tuskegee University's Center for Advanced Materials (T-CAM)
Tuskegee, AL 36088, USA

ABSTRACT

In the present investigation a novel technique have been developed to fabricate nanocomposite materials containing SC-15 epoxy resin and K-10 montmorillonite clay. A high intensity ultrasonic liquid processor was used to obtain a homogeneous molecular mixture of epoxy resin and nano clay. The clays were infused into the part A of SC-15 (Diglycidylether of Bisphenol A) through sonic cavitations and then mixed with part B of SC-15 (cycloaliphatic amine hardener) using a high speed mechanical agitator. The trapped air and reaction volatiles were removed from the mixture using high vacuum. DMA, TGA and 3-point bending tests were performed on unfilled, 1wt. %, 2wt. %, 3% and 4wt. % clay filled SC-15 epoxy to identify the loading effect on thermal and mechanical properties of the composites. The flexural results indicate that 2.0 wt% loading of clay in epoxy resin showed the highest improvement in flexural strength as compared to the neat systems. DMA studies also revealed that 2.0 wt% doped system exhibit the highest storage modulus and Tg as compared to neat and other loading percentages. However, TGA results show that thermal stability of composite is insensitive to the clay content. After that, the nanophased matrix with 2 wt.% clay was then utilized in a Vacuum Assisted Resin Transfer Molding (VARTM) set up with satin weave carbon preforms to fabricate laminated composites. The resulting composites have been evaluated by TGA, DNA, and flexural test, and 5°C increasing in glass transition temperature, 6°C increasing in decomposition temperature and 13.5% improvement in flexural strength were observed in nanocomposite. Based on the experimental result, a linear damage model has been combined with the Weibull distribution function to establish a constitutive equation for neat and nanophased carbon/epoxy.

1. INTRODUCTION

Fiber reinforced composites due to their high specific strength and specific stiffness have become attractive structural materials not only in weight sensitive aerospace industry, but also in marine, armor, automobile, railways, civil engineering structures, sport goods etc. Generally, the in-plane tensile properties of a fiber/polymer composite are defined by the fiber properties, while the compression properties and properties along the thickness dimension are defined by the characteristics of the matrix resin. Epoxy resin is the polymer matrix used most often with reinforcing fibers for advanced composites applications. The resins of this class have good stiffness, specific strength, dimensional stability, and chemical resistance, and show considerable adhesion to the embedded fiber [1]. Using an additional phase, such as inorganic fillers, to strengthen the properties of epoxy resins has become a common practice [2]. Because micro-scale fillers have successfully been synthesized with epoxy resin [3-6], nanoparticles, nanoclay, nanotubes, and nanofibers are now being tested as filler material to produce high performance composite structures with enhanced properties [7-9].

Nano-phased matrix based on organic polymers and inorganic clay minerals consisting of silicate layers such as montmorillonite (MMT) have attracted great interest because they frequently exhibit unexpected properties including reduced gas permeability, improved solvent resistance, superior mechanical and enhanced flame-retardant properties [10-15]. Different polymer/clay nanocomposites have been successfully synthesized by incorporating clay in various polymer matrixes such as, polyamides[16], polyimides [17], epoxy [18], polyurethane [19], poly(ethylene terephthalate) [20] and polypropylene [21].

Incorporation of nanoclays, in the matrix system for fiber reinforced composites also has been recently studied by several groups. Kornmann et al. [22] produced glass fiber reinforced laminates with a matrix of layered clay/epoxy system and showed that flexural strength of the composites is increased due to the presence of the nano particles in the matrix. Subramaniyan et al. [23] fabricated composites with stitched unidirectional E-glass fibers and an epoxy vinyl ester resin via vacuum assisted wet lay-up method. They observed that addition of nanoclay increased the compressive strength of glass fiber reinforced composites. Chowdhury et al. [24] investigated the effects of nanoclay particles on flexural and thermal properties of woven carbon fiber reinforced polymer matrix composites and they showed that nanoclay addition at low concentrations increased the flexural properties as well as the thermal stability of the system. The effect of fiber direction on clay distribution in the layered clay/glass fiber/ epoxy hybrid composites was studied by Lin et al. [25]. They placed the unidirectional glass fibers parallel and perpendicular to the resin flow direction and prepared laminates using vacuum assisted resin transfer molding. Their results showed that mechanical properties were deviated with the direction of resin flow and location of glass fibers, and the composite properties were improved with the clay loading. Miyagawa et al. [26] studied the influence of biobased clay/epoxy nanocomposites as a matrix for carbon fiber composites. They found that the addition of nanoclay has no effect on the flexural strength and modulus.

The primary interest of this paper was to characterize the effect of K-10 montmorillonite clay on the thermal and mechanical properties of epoxy and carbon/epoxy composite. A high intensity ultrasonic liquid processor was used to obtain a homogeneous molecular mixture of epoxy resin and nano clay. DMA, TGA and 3-point bending tests will be performed to identify the loading effect on thermal and mechanical properties of the epoxy. After optimal loading of clay in epoxy was determined, a Vacuum Assisted Resin Transfer Molding (VARTM) set up was used to fabricate nanophased carbon/epoxy composite. Flexural tests, tensile tests and fatigue test were performed to evaluate mechanical performance. Thermogravimetric Analysis (TGA) and Dynamic Mechanical Analysis (DMA) were performed to evaluate the thermal performances.

2. MATERIALS

The resin used in this study is a commercially available SC-15 epoxy obtained from *Applied Poleramic, Inc.* It is a low viscosity two phased toughened epoxy resin system consisting of part-A (resin mixture of Diglycidylether of Bisphenol-A, Aliphatic Diglycidylether epoxy toughner) and part-B (hardener mixture of, cycloaliphatic amine and polyoxylalkylamine). The inorganic clay, used in this study, was K-10 grade montorillonite obtained from Sigma-Aldrich Co. (USA) with a surface area 220-270m^2/g. 2-D plain weave carbon fiber manufactured by Fiber Materials Inc. has been used in this study. It is a 11 X 11 Plain Weave (11 X 11 indicates that in 1 sq.in of a fabric area 11 carbon fiber tows run in warp direction and 11 run in fill directions), 4060-6 type carbon fiber , 10 oz/ square yard, 6k sized fabric.

3. OPTIMIZATION OF CLAY/EPOXY RESIN

3.1. Manufacturing of Nanophased Resin

Firstly, clay was dried in oven at a temperature of 80 °C for 24 hours. Then pre-calculated amount of clay and part-A resin were carefully weighted, and mixed together in a suitable beaker. The mixing was carried out through a high intensity ultrasonic irradiation (Ti-horn, 20 kHz *Sonics Vibra Cell, Sonics Mandmaterials, Inc, USA)* for one and half hour with pulse mode (50sec. on/ 25sec. off) since part-A is insensitive to ultrasound irradiation. To avoid a temperature rise during the sonication process, external cooling was employed by submerging the beaker containing the mixture in an ice-bath. Once the irradiation was completed, part-B was added to the modified part-A then mixed using a high speed mechanical stirrer for about 10 minutes. The mix-ratio of part A and part B of SC-15 is 10:3. The rigorous mixing of part-A and part-B produced highly reactive volatile vapor bubbles at initial stages of the reaction, which could detrimentally affect the properties of the final product by creating voids. A high vacuum was accordingly applied using *Brand Tech Vacuum* system for about 20 minutes. After the bubbles were completely removed, the mixture is transferred into a Teflon-coated metal molds and kept for 24 hour at room temperature. The cured material is then de-molded and trimmed. Finally, all as-prepared panels are post-cured at 100 ◦C for 5 hours. The block diagram for manufacturing of clay/epoxy nanocomposite is shown in Figure 1.

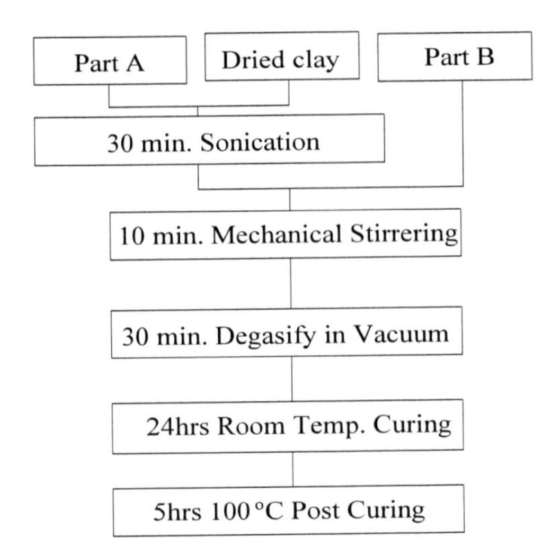

Figure 1. Processing block of clay/epoxy resin.

3.2. Thermal Properties

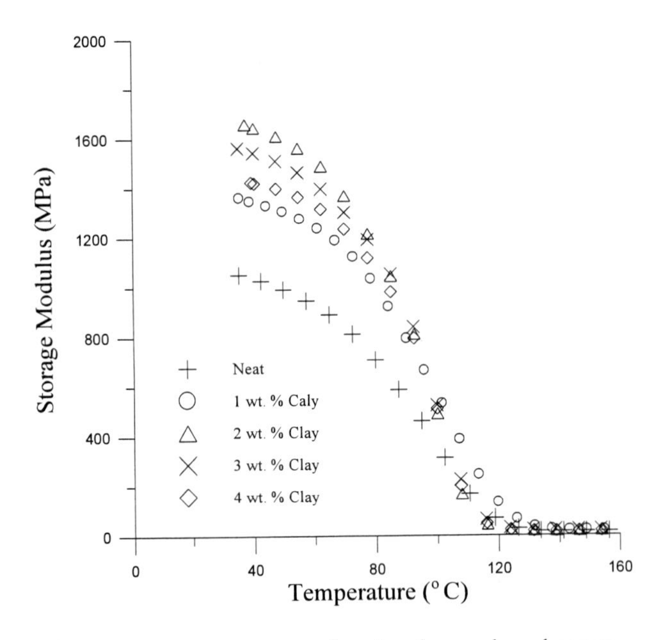

Figure 2. Storage modulus vs. temperature curves of neat and nanophased epoxy.

Dynamic Mechanic analysis (DMA) was performed on a TA Instruments 2980 operating in the three-point bending mode at an oscillation frequency of 1Hz. Data were collected from room temperature to 200°C at a scanning rate of 10°C/min. The sample specimens were cut by a diamond saw in the form of rectangular bars of a nominal $4mm \times 30mm \times 12mm$. Figure 2 illustrates the DMA plots of storage modulus versus temperature as a function of clay loading. It can be seen that the storage modulus steadily increases with an increasing clay content up to 2%. Above 2%, the storage modulus decreased with increasing clay content. The addition of 2 wt. % of clay yielded a 58% increase of the storage modulus at 30°C.

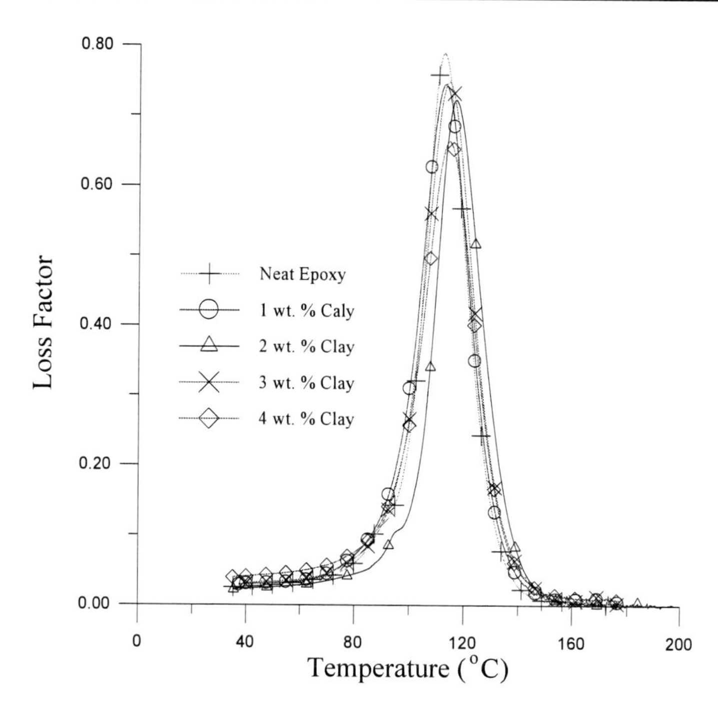

Figure 3. Loss factor versus temperature plots of clay-epoxy nanocomposite.

The loss factor, tan δ, curve of the neat epoxy and its clay/epoxy nanocomposites measured by DMA are shown in Figure 3. The maximum Tg, determined from the peak position of tan δ, was observed at 3 wt. % clay system. 3°C enhancement in Tg has been found. In Figure 2, the peak height of loss factor decreased with increasing clay content, but width of tan δ is insensitive to the clay content. The peak factor, Γ, is defined as the full width at half maximum of the tan δ peak devided by its height, can be qualitatively used to assess the homogeneity of epoxy network. The neat epoxy was observed to have a low peak factor that indicates that the crosslink density and homogeneity of the epoxy network were high. For the nanocomposites, the peak factor increased with increasing clay weight percent, as shown in Figure 4, and exhibited a broadened tanδ peak on the high temperature side of the DMA profile. The higher peak factor for the nanocomposites is indicative of lower crosslink density and greater heterogeneity, which suggests intercalation of the epoxy network into the clay layers.

Thermo gravimetric analysis (TGA) measurements were also carried out to obtain information on the thermal stability of the various nanocomposite systems. The TGA samples were cut into small pieces using ISOMET Cutter and were machined using the mechanical grinder to maintain the sample weight of about 5-20mg range. These samples were sealed in ceramic crucibles and placed inside the apparatus. The real time characteristic curves were generated by *Universal Analysis 2000-TA Instruments Inc.,* data acquisition system. Figures 5 shows the TGA of all categories of nanocomposites considered for this investigation. We define the 50% weight loss as a marker for structural decomposition of the samples. In this figure, the decomposition temperatures are almost the same, indicating the clay contents have no effect on the decomposition temperature of epoxy.

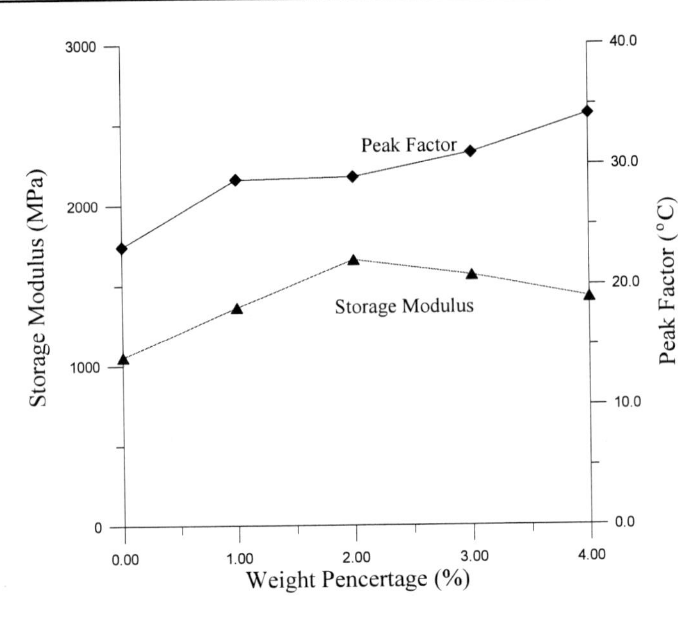

Figure 4. Effect of clay content on storage modulus and peak factor of material.

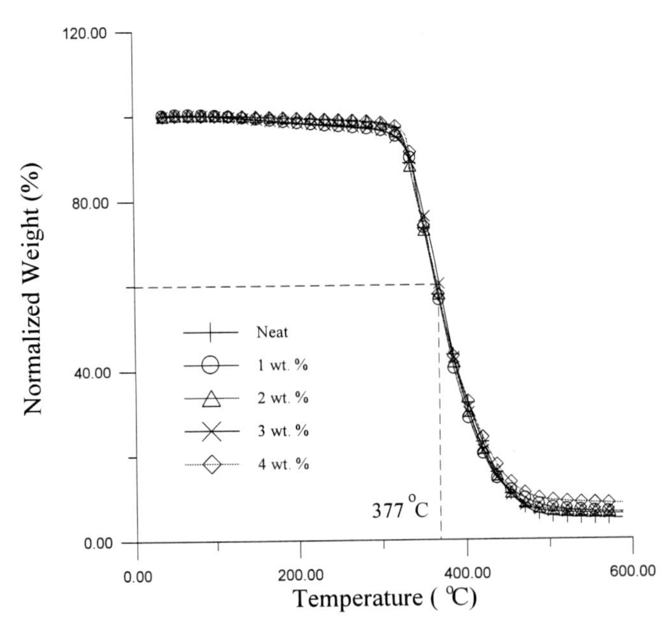

Figure 5. TGA results of clay-epoxy nanocomposite.

3.3. Flexural Response

Flexural tests under three point bend configuration were performed according to ASTM D790-86. The tests were conducted in a 10 KN servo hydraulic testing machine (MTS) equipped with Test Ware data acquisition system. The machine was run under displacement control mode at a cross head speed of 2.0 mm/min, and all the tests were performed at room temperature. Test samples were cut from the panels using a Felker saw fitted with a diamond coated steel blade. Five replicate specimens from four different materials were prepared for static flexure tests.

Typical stress strain behavior from the flexural tests is shown in Figure 6. All specimens failed immediately after the stress reached the maximum value; however, the stress-strain curves showed considerable non-linearity before reaching the maximum stress, but no obvious yield point was found in the curves. Five specimens were tested for each condition. The average properties obtained from these tests are listed in Table 1. Figure 6 shows the variation of modulus and strength with clay content. Optimal loading of clay was found at 2wt. %, and an improvement of about 31% in modulus and 27% in strength were observed with an addition of 2 wt.% of clay.

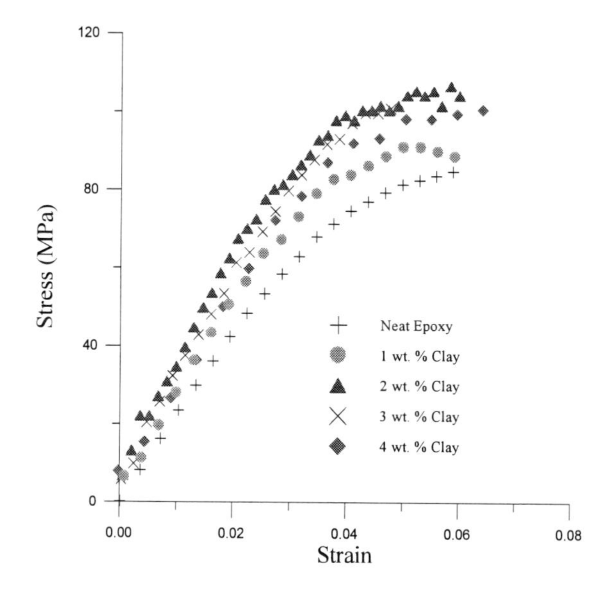

Figure 6. Stress-strain curves of materials.

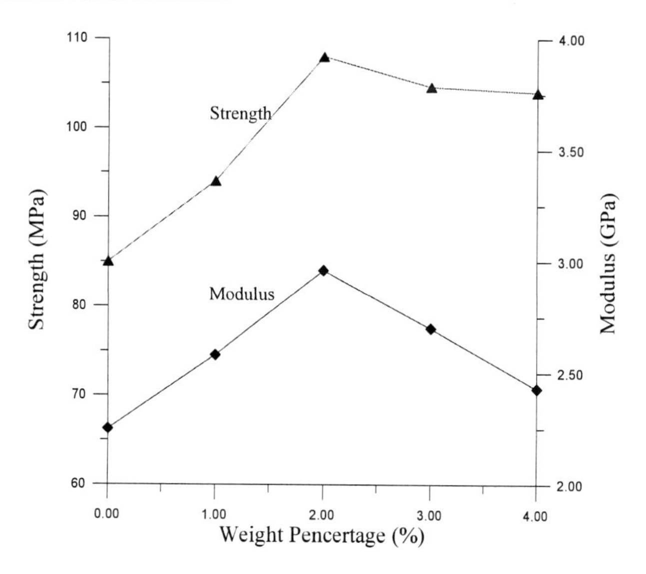

Figure 7. Effect of weight fraction of nanoclay on flexural modulus and strength of epoxy.

Table 1. Effects of clay content on the mechanical and thermal properties of epoxy and its nano composite

Material	Modulus (GPa)	Improvement in modulus	Strength (MPa)	Improvement in strength
Neat Epoxy	2.25 ± 0.11	--------	85 ± 4.3	--------
1 wt. % clay	2.58 ± 0.12	14.7%	94 ± 5.9	10.6%
2 wt. % clay	2.96 ± 0.14	31.6%	108 ± 5.6	27.1%
3 wt. % clay	2.70 ± 0.14	20.0%	105 ± 4.6	23.5%
4 wt. % clay	2.43 ± 0.11	8.0%	104 ± 4.1	23.5%

3.4. Fracture Surface

The fracture surfaces of neat epoxy and the nanocomposites were comparatively examined using SEM. It can be seen in Figure 8A that neat epoxy resin exhibits a relatively smooth fracture surface and initial crack occurred at the tension edge of specimen. River pattern in higher magnification picture in Figure 8B indicates a typical cleavage fractures, thus accounting for the low fracture toughness of the unfilled epoxy.

Compared to the case of neat epoxy, the fracture surfaces of the nanocomposites show considerably different fractographic features. As a representative example, the failure surface of the nanocomposite containing 2 wt. % clay and 4 wt. % clay are shown in Figure 9a and 9b. Generally, a much rougher fracture surface is seen upon adding clay into the epoxy matrix.

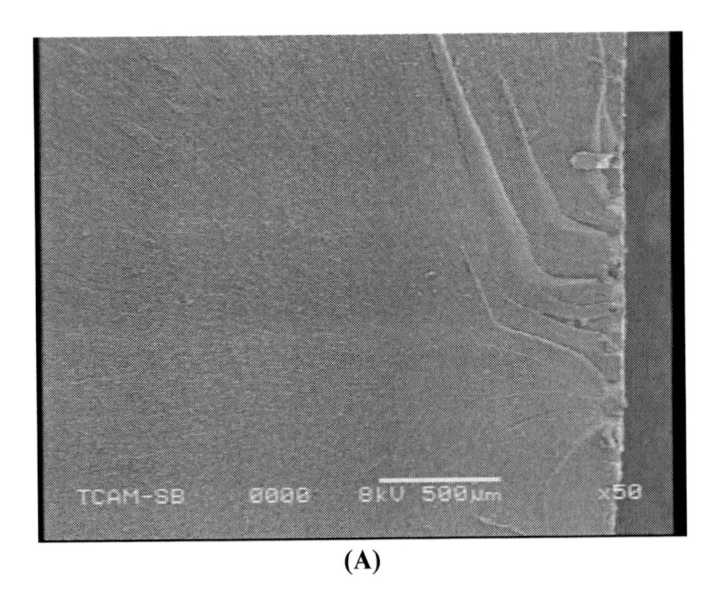

(A)

Figure 8. (Continued).

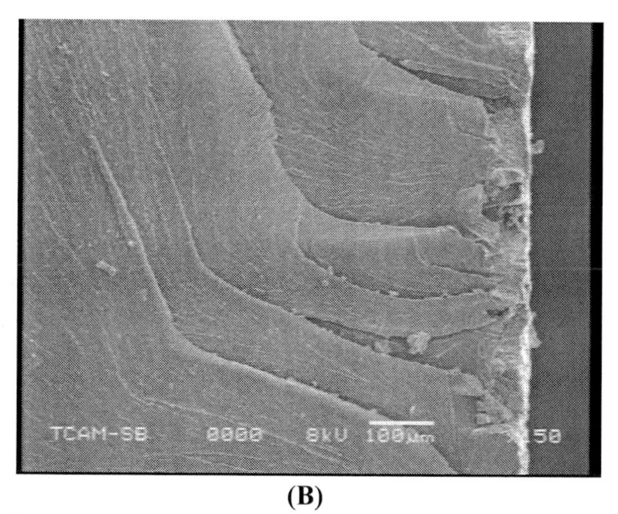

(B)

Figure 8. Fracture surface of neat epoxy.

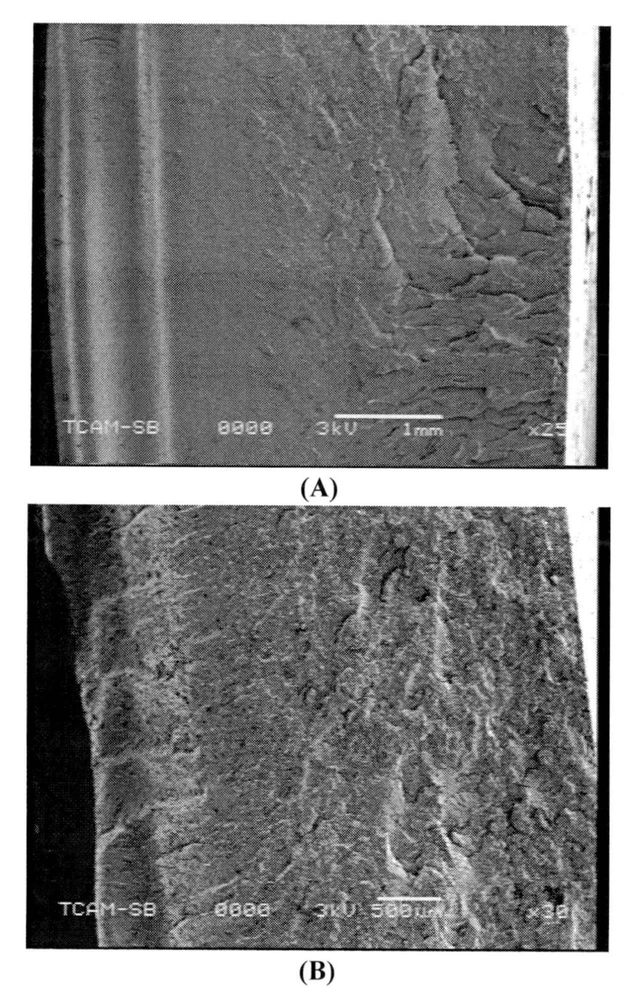

(A)

(B)

Figure 9. Initial crack occurred at the tension edge of clay/epoxy nanocomposite (A: 1 wt. % CLAY; B: 2 wt. % CLAY).

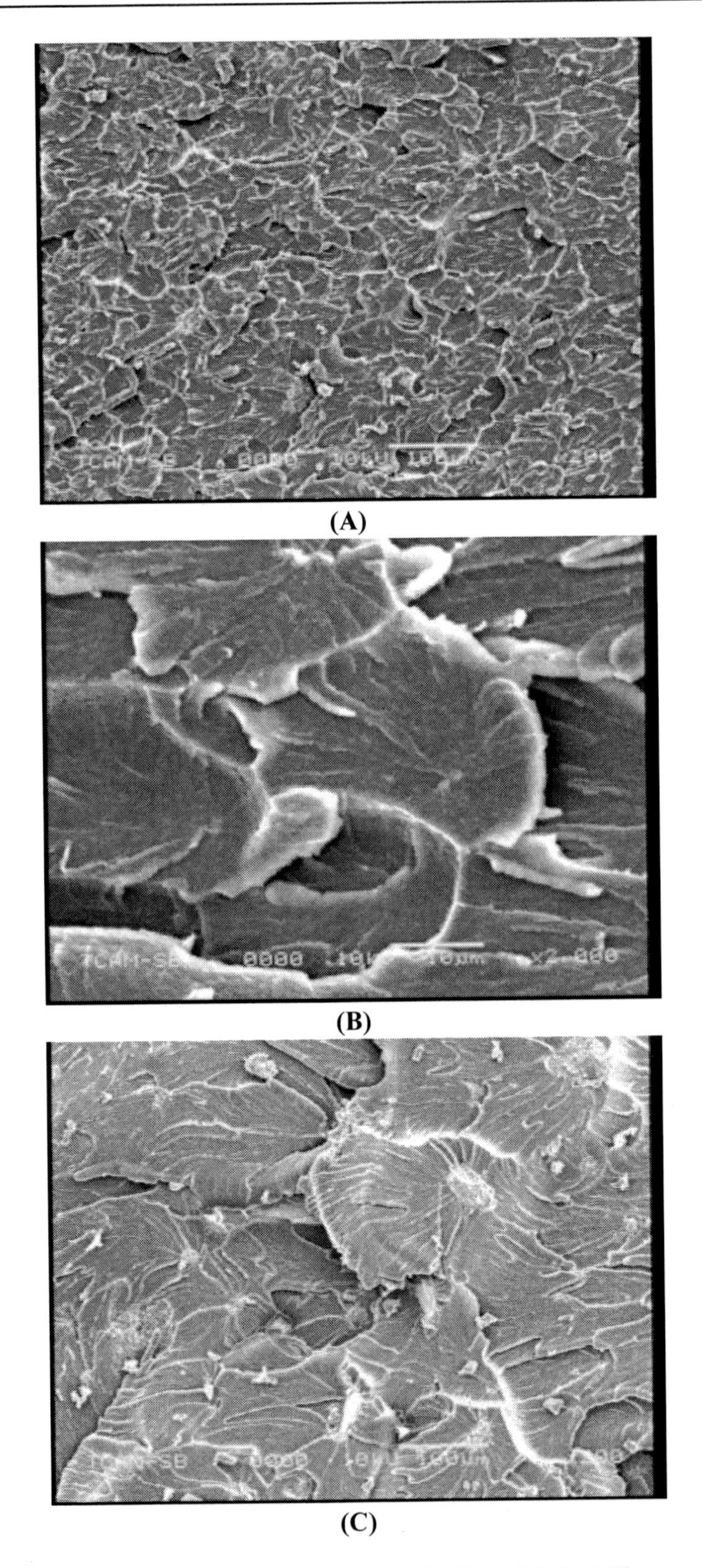

Figure 10. Dispersions of clay in the epoxy. (A: 2wt%; B: 2wt%, and C: 4 wt.%).

Figure 10a-c show the micrograph of clay-epoxy composite. The micrograph clearly indicates that, for 2 wt. % system, the clays are well dispersed in the resin. The clays are well separated and uniformly embedded in the epoxy resin. Higher magnification SEM picture in Figure 10b shows that no agglomerated particles was observed. But for 4 wt.% clay-epoxy composite, large size particles are found at the fracture surface. These large size particles are formed by the agglomeration of nano-clay. The crack initiation was caused by the stress concentration caused by the agglomerated particle.

4. Carbon/Clay/Epoxy Nanocomposite

4.1. Manufacturing of Carbon/Clay/Epoxy Nanocomposite

Based on the mechanical experimental results of clay modified Epoxy matrix, laminated composite panels were manufactured by using plain weave carbon fiber and 2wt.% clay modified Epoxy. A schematic diagram of this manufacturing process is shown in Figure 11.

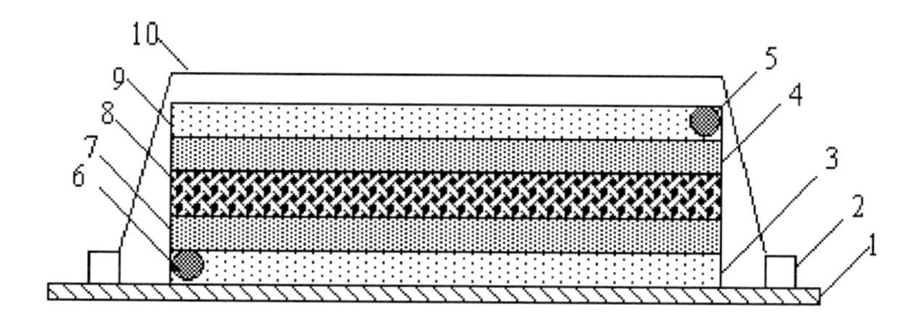

1: Aluminum plate;
2: Boundary layer;
3: Distribution mesh;
4: Porous Teflon;
5: Inlet infusion tubing;
6: Exit extracts tubing;
7: Porous Teflon;
8: Carbon finer perform;
9: Distribution mesh;
10 Vacuum bag

Figure 11. Fabrication of carbon/epoxy composite by using VARTM.

The aluminum plate was laid on a flat surface and cleaned with acetone. Mold releasing agent was added to the surface to allow easy release of the panel. Then a layer of distribution mesh and porous Teflon layer were laid on the plate respectively for uniform distribution of resin. Four layers of carbon fiber were added. Dry fabric preforms with required orientations were then laid out on the top of distribution mesh. Another porous Teflon layer and distribution media were laid on fibers. After stacking, the complete assembly was covered with a heat resistant vacuum bag, and infusion and suction lines were installed. Nanophased resin was infused from one end, and the other end was connected to vacuum pump. The vacuum bagging is the critical process in composite manufacturing and the cover should be leak proof. Final test samples were machined for mechanical characterization. All as-prepared panels were post-cured at 100°C for five hours, in a Lindberg/Blue Mechanical Convection Oven. In parallel, neat carbon/epoxy panels were fabricated by using the same method to compare with nanophased system.

4.2. Thermal Stability

Thermo gravimetric analysis (TGA) has been carried out to estimate the amount of resin present in the as-fabricated neat and nano-phased panels and their thermal stability. The weight vs. temperature curve in Figure 12 indicates that the as-fabricated panel contains 29wt. % of epoxy resin and rest is the carbon fiber and nano filler. In the present study we have considered the derivative peaks as the decomposition temperature. As seen in Figures 12, the decomposition temperatures for neat carbon/epoxy and nanphased carbon/epoxy are 351°C and 357°C, respectively. Clay has slightly effect on thermal stability of carbon/epoxy composite.

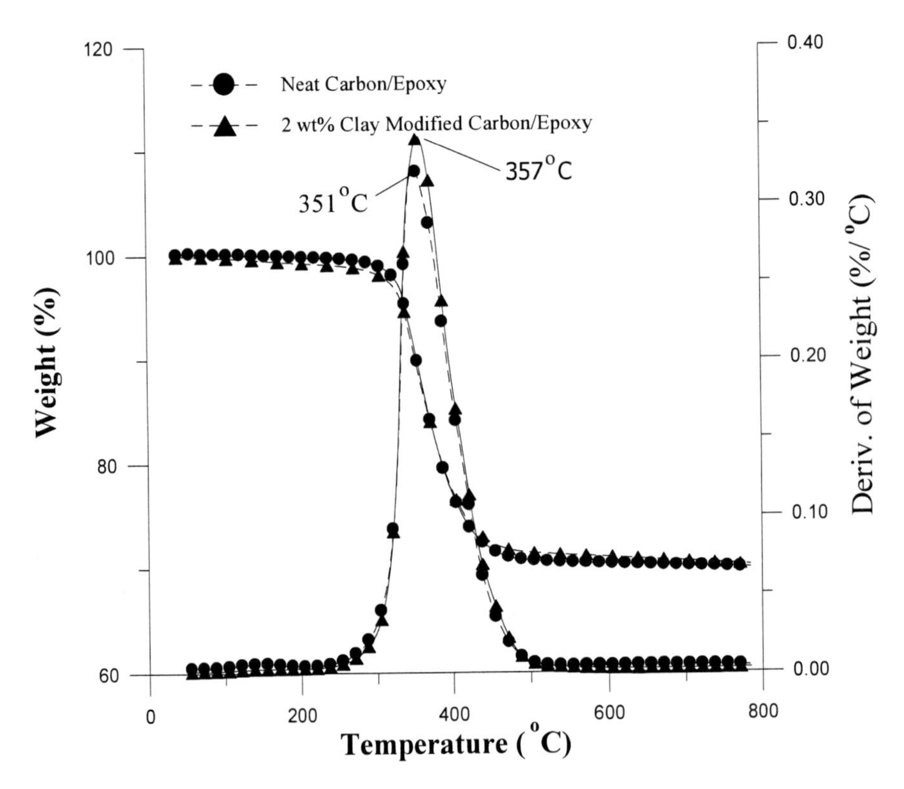

Figure 12. TGA results of neat and nanophased composite.

Figure 13 illustrates the DMA plots of storage modulus versus temperature for neat and nanophased carbon/epoxy laminate. The addition of 2 wt. % of clay yielded a 7.6% increase of the storage modulus. The loss factor, tan δ, curve of the neat and nanocomposites measured by DMA also are shown in Figure 12. Tg, determined from the peak position of tan δ, increased from 111°C to 116°C.

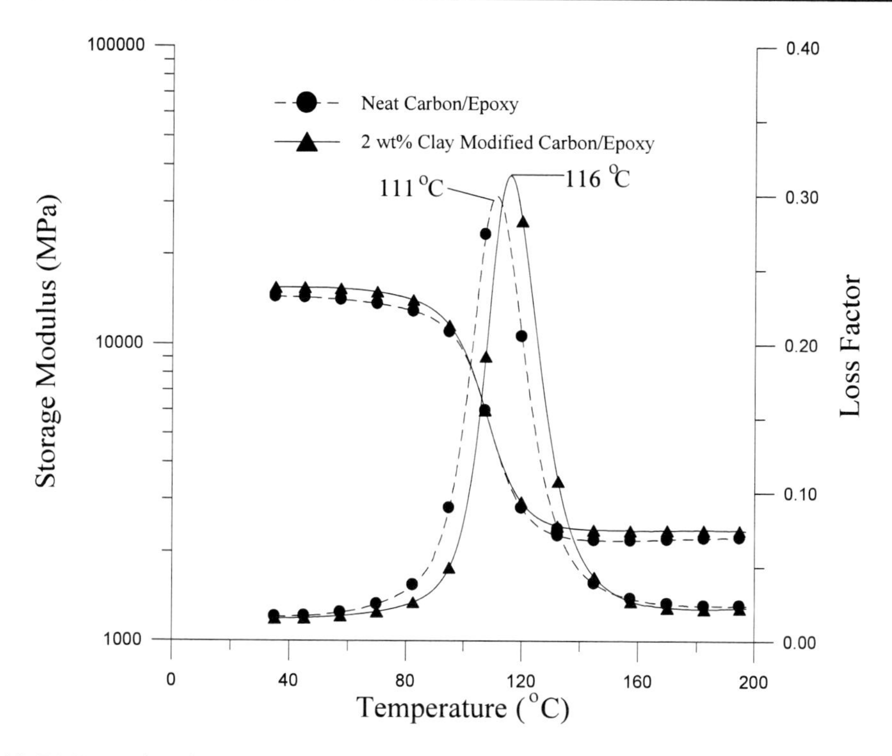

Figure 13. DMA results of neat and nanophased composite.

4.3. Mechanical Properties

Flexure, tensile and fatigue tests were performed on an MTS servohydraulic test machine equipped with a 10kN load cell. Three-point flexural tests were performed according to ASTM D790. Typical specimen dimensions were 56 mm in length, 25 mm in width, and 2 mm in thickness. The span length was 32.2mm. Dog bone-shaped specimens were used in tensile and fatigue tests according to ASTM D638. The specimen dimensions were 120 mm in overall length, 50 mm in gage length and 12.7 mm in gage width. Five specimens were tested to determine the modulus (E) and failure strength (σ_b) in flexure and tension.

Flexural stress-strain curves of neat and nanophased composites are shown in Figure 14. The curves show considerable non-linear deformation, and the irregularities in the curves were attributed to random filament breakage during loading. The specimens failed rapidly after reaching the maximum stress, and cracking noise was heard while the individual filament broke or the inter-layer delaminated. Stress-strain curves shown in Figure 14 also reveals that by infusing 2wt.% clay in the epoxy matrix, strength and failure strain were significantly improved. As shown in Table 2, the nanophased system showed approximately 13.5% increase in flexural strength, 7.9% increase in failure strain with respect to the neat system, and no changes in flexural modulus. Tensile stress-strain curves and tensile properties of neat and nanophased laminates are shown in Figures 15 and Table 3. Infusing 2wt.% CLAY in epoxy resin produced 11% enhancement in ultimate tensile strength (UTS), 4.1% enhancement in failure strain, and no changes in modulus.

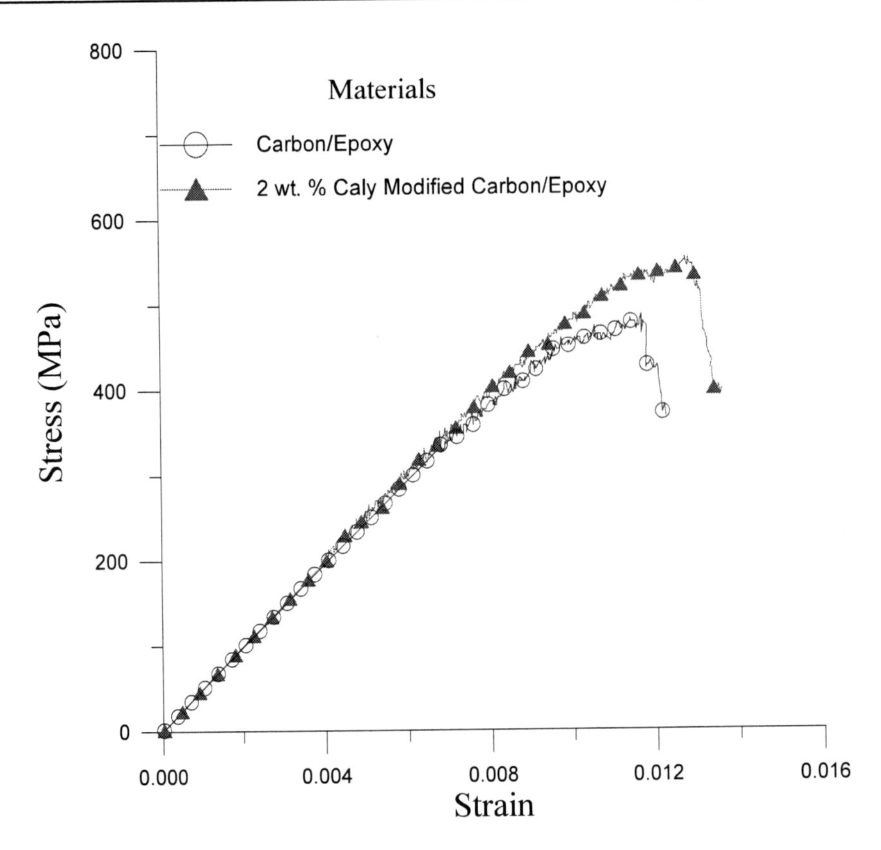

Figure 14. Flexural stress strain curve of neat and nanophased CFRP.

Table 2. Flexural Properties of Neat and Nanophased Carbon/epoxy

Properties	**Carbon Fabric /Neat Epoxy**	**Carbon Fabric / 2 wt.% Clay-Epoxy**
Modulus(GPa)	48.9(±5.1%)	49.5(±5.2%)
Gain in Modulus	--------	2.0%
Failure Strain (%)	1.17(±2.6%)	1.27(±2.4%)
Gain in Failure Strain	--------	7.9%
Strength (MPa)	488(±3.1%)	554(±3.5%)
Gain in Strength	--------	13.5%

There are two possible reasons for the observed mechanical properties of the carbon fabric reinforced laminate infused with clay. First, clay increases the strength and modulus of epoxy matrix, as was observed when clay was added to epoxy in the first part of this study. The modulus of composite laminate was not affected by clay, since modulus is dominated by the carbon fibers present in the laminate. However, the presence of clay increased the crack propagation resistance, which improves strength.

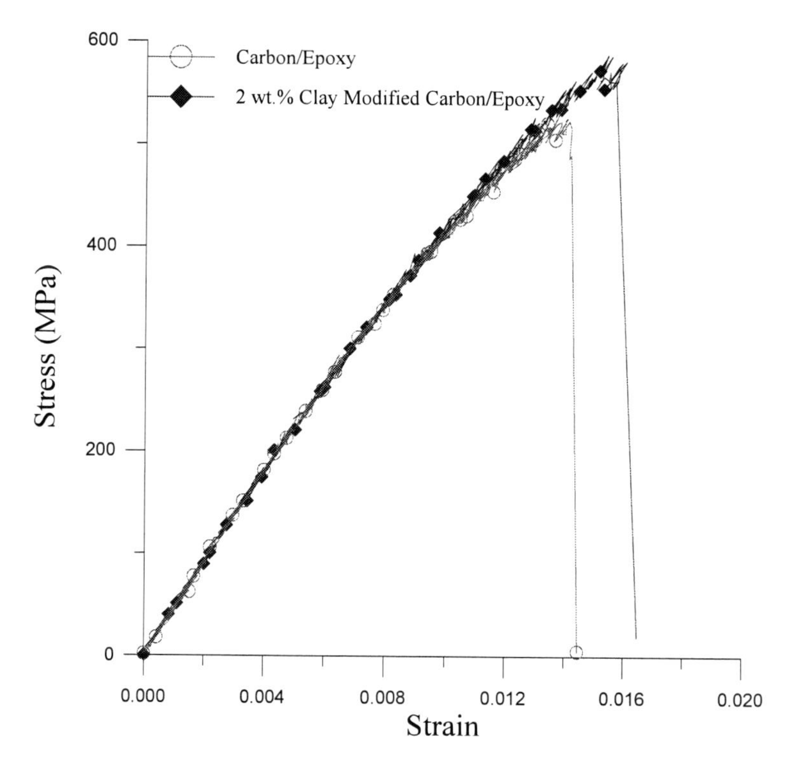

Figure 15. Tensile stress strain curve of neat and nanophased CFRP.

Table 3. Tensile Properties of Neat and Nanophased Carbon/Epoxy

Properties	**Carbon Fabric/Neat Epoxy**	**Carbon Fabric / 2 wt.% Clay-Epoxy**
Modulus(GPa)	45.2(±1.6%)	46.1(±2.5%)
Gain in Modulus	--------	2.2%
Failure Strain (%)	1.45(±4.5%)	1.54(±5.2%)
Gain in Failure Strain	--------	4.1%
Strength (MPa)	556(±2.4%)	588(±3.6%)
Gain in Strength	--------	5.7%

Stress-controlled tension-tension fatigue tests were performed at $21.5^{o}C$. The ratio of the minimum cyclic stress and the maximum cyclic stress, i.e., the R-ratio, was 0.1. A cyclic frequency of 2 Hz was used to reduce the possibility of thermal failure. Figure 16 shows the fatigue S-N curves of the neat and nanophased carbon fabric reinforced epoxy. The fatigue S-N curve of the nanocomposite was significantly higher than that of the neat carbon/epoxy. Based on the experimental data, following equations were established for the S-N curves of composites:

$$\sigma = 540(Nf)^{-0.0151} \quad \text{for neat carbon fabric/epoxy}$$

$$\sigma = 545(Nf)^{-0.0126} \quad \text{for nanophased carbon fabric/epoxy}$$

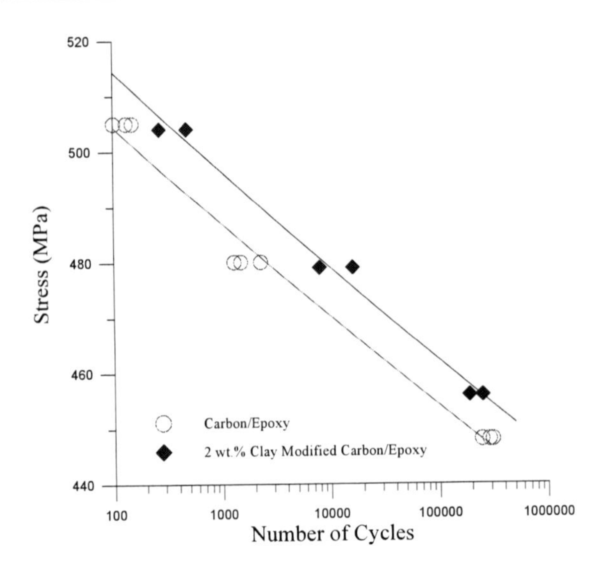

Figure 16. S-N curves of neat and nanophased CFRP.

4.4. Fracture Surface Observations

Figure 17A and 17B show the optical and SEM pictures of failed carbon fabric/nanophased epoxy composite. The neat system has the similar results. Fiber breakage, interface debonding, and delamination were observed in both neat and nanophased carbon/epoxy composites.

A higher magnification SEM photograph of the nanophased composite in Figure 18A shows that matrix material is adhering to the fibers, indicating that the interface between the carbon fibers and the epoxy matrix was strong. Fracture surface of the neat carbon fabric/epoxy composite in Figure 18b shows that the fiber surfaces did not have much matrix material adhering to them, indicating a relatively weak fiber/matrix interface. It maybe attributed to the higher strength of nanophased matrix. Figure 18 A and B show the fracture surfaces of neat and nanophased composites at higher magnifications. Neat composite exhibits a relatively smooth fracture surface. Many small dots also are shown in Figure 19a, scattered across the fracture surface. The SC-15 epoxy is similar to a second-phase rubber-toughened epoxy. These dots were caused by toughening rubber particles on the fracture surface. A crack step was observed in the nanophased composite (Figure 19B), indicating that the direction of crack propagation was changed.

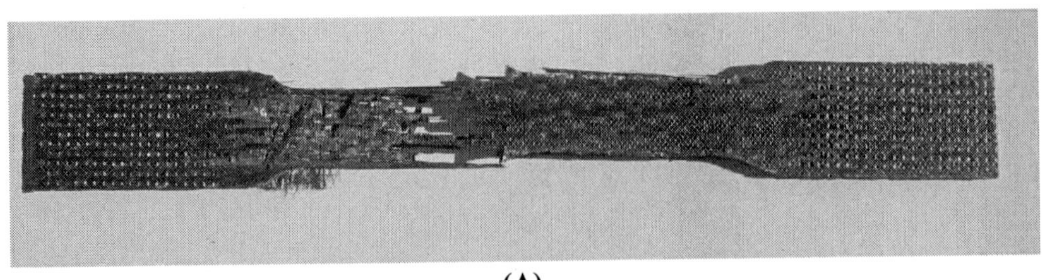

(A)

Figure 17. (Continued).

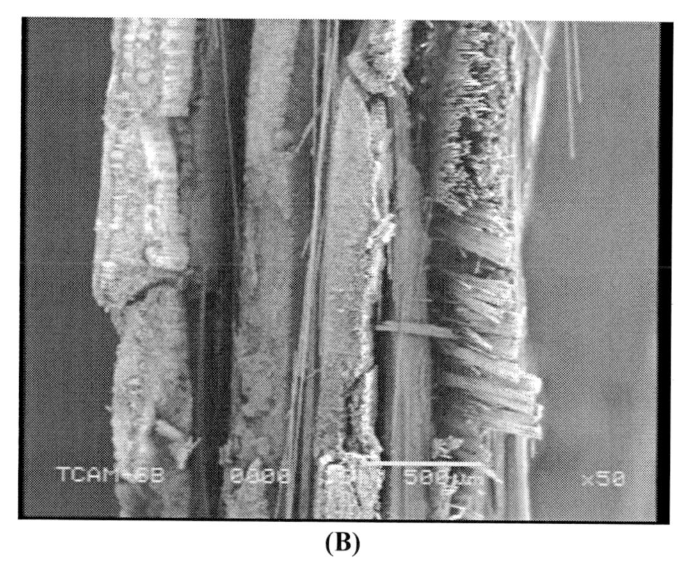

(B)

Figure 17. Optical (A) and SEM (B) picture of failed nanophased Carbon/epoxy.

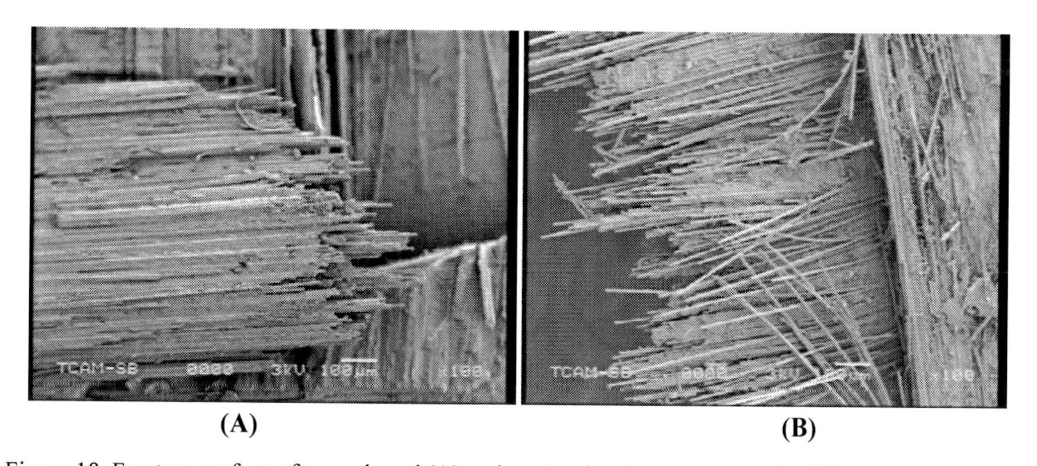

(A) **(B)**

Figure 18. Fracture surface of nanophased (A) and neat carbon/epoxy composite (B).

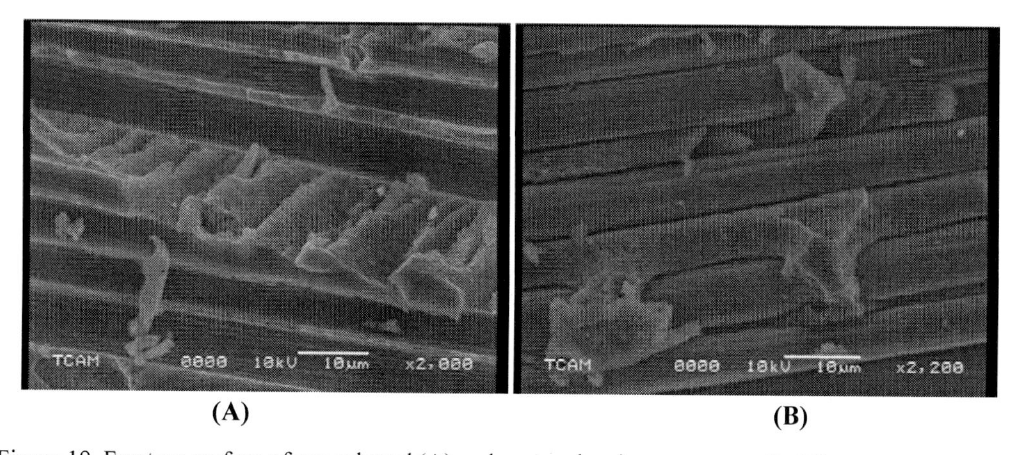

(A) **(B)**

Figure 19. Fracture surface of nanophased (A) and neat carbon/epoxy composite (B).

5. Damage Constitutive Equation for Unidirectional Composite

The failure of neat and nanophased carbon/epoxy composite involve a complicated damage accumulation process resulting from random fiber breakage, stress transfer form broken to intact fiber, interface debonding between the fiber and matrix, and laminates delamination. Based on the experimental results, following macro-damage constitutive equation was used to describe the failure process of neat and nanophased unidirectional composite.

$$\sigma = E\varepsilon(1-\varpi) \tag{1}$$

where E is elastic modulus and ϖ is a damage factor. The strength of the fiber follows the weakest link hypothesis and the strength distribution of the carbon fibers can be described by either a single Weibull function or a bimodal Weibull function. Here, we have examined the applicability of both single and bimodal Weibull strength distribution functions.

$$\varpi = G(\varepsilon) = \frac{n}{N} = 1 - \exp\left[-\left(\frac{E\varepsilon}{\sigma_0}\right)^{\beta}\right] \tag{2}$$

(single Weibull distribution)

$$\omega = G(\varepsilon) = \frac{n}{N} = 1 - \exp\left[-\left(\frac{E\varepsilon}{\sigma_{01}}\right)^{\beta_1} - \left(\frac{E\varepsilon}{\sigma_{02}}\right)^{\beta_2}\right] \tag{3}$$

(bimodal Weibull distribution)

where G is the cumulative probability of the failure, and σ_0 and β are scale parameter and shape parameter, respectively. Combining Equation (1) with Equation (2) and (3), the stress-strain curve of the coated fiber bundle can be rewritten as:

$$\sigma = E\varepsilon \exp\left[-\left(\frac{E\varepsilon}{\sigma_0}\right)^{\beta}\right] \tag{4}$$

(single Weibull distribution)

$$\sigma = E\varepsilon \exp\left[-\left(\frac{E\varepsilon}{\sigma_{01}}\right)^{\beta_1} - \left(\frac{E\varepsilon}{\sigma_{02}}\right)^{\beta_2}\right] \tag{5}$$

(bimodal Weibull distribution)

For the single Weibull distribution, we take double logarithms on both sides of Equation (4) to obtain:

$$\ln\left[-\ln\left(\frac{\sigma}{E\varepsilon}\right)\right] = \beta\ln(E\varepsilon) - \beta\ln(\sigma_0) \tag{6}$$

Equation (6) represents a linear plot of $\ln\left[-\ln\left(\frac{\sigma}{E\varepsilon}\right)\right]$ vs. $\ln(E\varepsilon)$. σ_0 and β can be calculated form the intercept and slope of this linear plot.

By taking double logarithms on both sides of Equation (5), one can obtain:

$$\ln\left[-\ln\left(\frac{\sigma}{E\varepsilon}\right)\right] = \ln\left[\left(\frac{E\varepsilon}{\sigma_{01}}\right)^{\beta_1} + \left(\frac{E\varepsilon}{\sigma_{02}}\right)^{\beta_2}\right] \tag{7}$$

The parameters σ_{01}, σ_{02}, β_1 and β_2 can be determined by the regreession analysis method.

Table 4. Weibull Parameter of neat and nanophased CFRP

Material	σ_{01} **(MPa)**	σ_{02} **(Mpa)**	β_1	β_2
Neat CFRP(F)	784.9	615.4	5.42	47.2
Nanophased CFRP(F)	813.2	682.4	7.45	52.9
Neat CFRP (T)	1940.3	682.2	1.46	21.2
Nanophased CFRP (t)	1330.5	740.7	2.45	51.0

Figure 20 shows the Weibull plots of neat and nanophased carbon/epoxy obtained from flexural tests and tensile tests. Nonlinear plots indicate that the damage process of composites obey the bimodal Weibull distribution. The values of bimodal Weibull parameters simulated form test results are listed in Table 4. The scale parameters σ_{01} and σ_{02}, which are is a measure of the nominal strength of the material, have been increased by infusion of clay. However, Weibull shape parameters β_1 and β_2, which are is a measure of the strength variability, have not been effec too mucht. By substituting the Weibull parameters into Equation (5), one can obtain the simulated stress-strain curves. The simulated curves and experimental points are shown in Figure 21 and they match well.

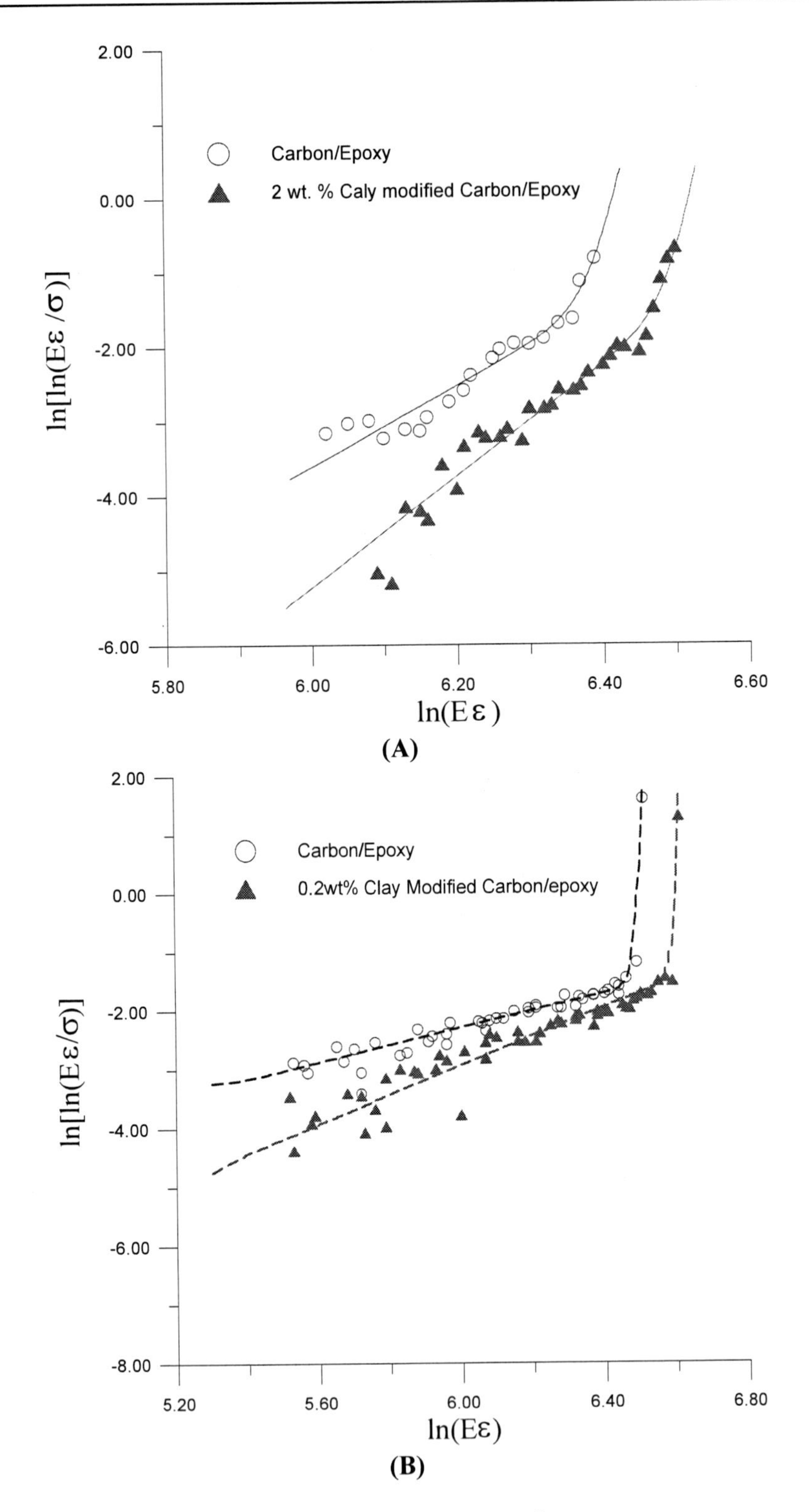

Figure 20. Weibull plots of flexural (A) and tensile (B) stress strain curves.

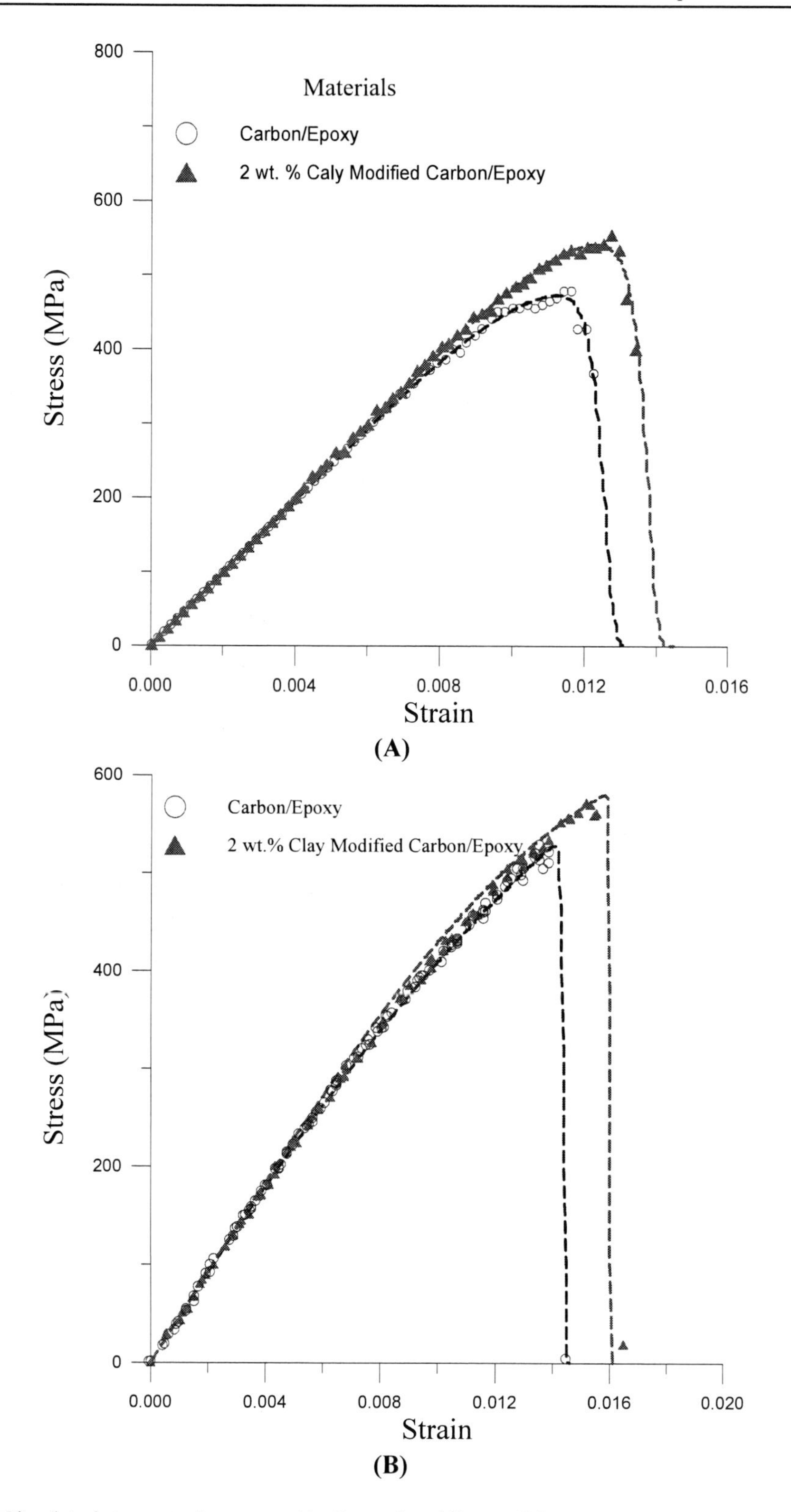

Figure 21. Simulated stress strain curves; (A: flexural and B: tensile).

Conclusion

Thermal and mechanical tests were conducted on clay modified epoxy and carbon/epoxy composite. Based on the experimental results, the following conclusions are reached:

1) Ultrasonic cavitation method is an effective to infuse clay into epoxy resin. 2.0 wt% loading of MMT clays in SC-15 epoxy resin showed the highest improvement in the strength as compared to the neat and other nanophased systems.
2) The composite fabricated with 2.0 wt.% CLAY filled epoxy produced 22.3% improvement in flexural strength and 11% improvement in tensile strength. The addition of CLAY in the epoxy matrix also improved the fatigue performance of the composite.
3) A one-dimensional damage constitutive equation has been established to describe stress-strain relationship of neat and nanophased composite. The results show that strength of both neat and nanophased carbon/epoxy obey a bimodal Weibull distribution.

Acknowledgment

The authors would like to gratefully acknowledge the support of National Science Foundation.

References

[1] J. B. Donnet, *Composites Science and Technology*, 63, (2003) pp1085-1088
[2] Day R. J., Lovell P. A., Wazzan A. A., *Composites Science and Technology*, 61 (2001) pp 41-56.
[3] Bagheri R., Pearson R. A., *Polymer*, 41 (2001) pp 269-276.
[4] Kawaguchi T., Pearson R. A, *Polymer*, 44 (2003) pp 4239-4247.
[5] Mahfuz, A. Adnan, V. K. Rangari, S. Jeelani, and B. Z. Jang, *Composites: Part A: Applied science and manufacturing*, 35 (2004) pp 519-527.
[6] Evora M.F., Shukla A., *Materials Science & Engineering*, A361: (2003) pp 358-366.
[7] Rodgers, R., Mahfuz, H., Rangari, V., Chisholm, N., and Jeelani, S., *Macromolecular Materials & Engineering*, 290(5) (2005), 423-429
[8] Farhana, P., Zhou, Y.X., Rangari, V., and Jeelani, S., *Materials Science and Engineering A,* 405(1-2) (2005), 246-253.
[9] Gojny F.H., Wichmann M.H.G, Fiedler B., Bauhofer W. and Schulte K., *Composites Part A*: Volume 36, Issue 11 , 2005, pp1525-1535
[10] K. Yano, A. Uzuki, A. Okada, T. Kurauchi and O. Kamigaito, *Journal Polym Sci Part A: Polym Chem* 31 (1993), p. 2493.
[11] R.A. Via, H. Ishii and E.P. Giannelis, *Chem Mater* 5 (1993), p. 1694.
[12] P.B. Messersmith and E.P. Giannelis, *J Polym Sci Part A: Polym Chem* 33 (1995), p. 1047.

[13] R. Krishnamoorti, R.A. Vaia and E.P. Giannelis, *Chem Mater* 8 (1996), p. 1728.
[14] M. Kawasumi, N. Hasegawa, M. Kato, A. Usuki and A. Okada, *Macromolecules* 30 (1997), p. 6333.
[15] D. F. Wu, C. X. Zhou. X. Fan, D. L. Mao and Z. Bian, *Polymer Degradation and Stability*, 87(2005), P.511-519
[16] Z.G. Wu, C.X. Zhou, R.R. Qi and H.B. Zhang, *J Appl Polym Sci* 83 (2002), p. 2403.
[17] H.L. Tyan, Y.C. Liu and K.H. Wei, *Polymer* 40 (1999) (17), p. 4877.
[18] T. Lan, P.D. Kaviratna and T.J. Pinnavaia, *J Phys Chem Solids* 57 (1996), p. 6.
[19] J. Ma, S. Zhang and Z.N. Qi, *J Appl Polym Sci* 2 (2001) (6), p. 1444.
[20] Y.C. Ke, C.F. Long and Z.N. Qi, *J Appl Polym Sci* 71 (1999), p. 1139.
[21] 21. J. B. Donnet, *Composites Science and Technology*, 63, (2003) pp1085-1088.
[22] Kornmann X, Rees M, Thomann Y, Necola A, Barbezat, Thoman R. *Compos Sci Technol.* 2005;65:2259–68.
[23] Subramaniyan AK, Sun CT. *Compos A.* 2006;12:1454–62.
[24] Chowdury FH, Hosur MV, Jeelani S. *Mater Sci Eng A.* 2006;421:298–306.
[25] Lin L, Lee J, Hong C, Yoo G, Advani SG. *Compos Sci Technol.* 2006;66:2116–25.
[26] Miyagawa H, Jurek RJ, Mohanty AK, Misra M, Drzal LT. *Compos A.* 2006;37:54–62.

In: Resin Composites: Properties, Production and Applications ISBN: 978-1-61209-129-7
Editor: Deborah B. Song

Chapter 8

THERMOOXIDATIVE AND THERMOHYDROLYTIC AGING OF COMPOSITE ORGANIC MATRICES

Xavier Colin and Jacques Verdu
ARTS ET METIERS ParisTech, PIMM, Paris, France

ABSTRACT

This chapter deals with the main causes of chemical ageing in organic matrix composites: hydrolysis, essentially in polyesters and polyamides, and oxidation in all kinds of polymer matrices. The first section is devoted to common aspects of chemical degradation of organic matrices. It is shown that chain scission and, at a lesser extent, crosslinking are especially important because they induce embrittlement at low conversions. Quantitative relationships between structural parameters and mechanical properties are briefly examined. The second section deals with diffusion–reaction coupling. In the cases of hydrolysis and oxidation, kinetics can be limited by respectively water and oxygen diffusion. Then, degradation is confined in a superficial layer, that can carry important consequences on use properties. The third section is devoted to hydrolysis. The kinetic equations are presented in both cases of non equilibrated and equilibrated hydrolysis. Structure–stability relationships are briefly examined. Osmotic cracking process, very important in the case of glass fiber/unsaturated polyester composites, is described. The last section is devoted to thermal oxidation. The simplest kinetic models are presented. The main gravimetric behaviours are explained. A mechanism is proposed for the "spontaneous cracking" in the superficial layer of oxidized samples.

Keywords: Degradation; Embrittlement; Chain scission; Hydrolysis; Water sorption; Oxidation; Diffusion–reaction coupling.

INTRODUCTION

There are many domains, for instance aeronautics, automotive or electrical engineering, where organic matrix composites are used for long times at temperatures higher than the ambient temperature, but lower than the matrix heat deflection temperature (HDT). This latter is close to the glass transition temperature T_g for amorphous polymers:

$$HDT \approx T_g - \Delta T \qquad \text{where } \Delta T \approx 5\text{–}10 \text{ K}$$

It is intermediary between T_g and the melting point T_f for semi-crystalline polymers, typically:

$$HDT \approx T_f - \Delta' T \qquad \text{where } \Delta' T \approx 40\text{–}100 \text{ K}$$

In this temperature range, common industrial fibers (glass, carbon, Kevlar, etc ...) are stable, so that composites perish always by matrix degradation. Organic matrices can undergo structural changes belonging to two categories and having eventually opposite effects on use properties. The first category corresponds to processes of which the cause is an initial thermal instability. Post-cure in thermosets, post-crystallization in semi-crystalline thermoplastics or structural relaxation (physical aging) in amorphous polymers belong to this category. Their main common characteristic is that their kinetics is atmosphere independent and only depends on thermodynamic variables (temperature and stress state). Another common characteristic of these processes is that the amplitude of their effects sharply depends on processing conditions, especially on their thermal history at the end of processing operations.

These processes, which are mainly ascribable to polymer synthesis chemistry (post-cure) or polymer physics (crystallisation, structural relaxation) will not be considered here. As a matter of fact, they can affect the polymer behaviour, especially in the early part of its lifetime, but they are generally not the main cause of its failure.

The second category of structural changes corresponds to the cases of chemical interaction between the polymer and environmental reagents, especially oxygen and water. These processes have two important common characteristics: first, they induce random chain scission, which is the cause of deep embrittlement at low conversions. Second, they are diffusion controlled, i.e. they affect only a more or less thick superficial layer and thus, induce thickness degradation gradients, a characteristic which cannot be ignored in a durability analysis.

This chapter will be divided into four main sections: The first one devoted to common aspects of chemical degradation processes, the second one devoted to reaction–diffusion coupling, the third one focused on hydrolysis processes and the fourth one on thermo-oxidative aging. Durability problems will be considered essentially from the point of view of materials science rather than chemical mechanisms. Emphasis will be put on consequences of aging on mechanical properties.

Common Aspects of Chemical Degradation Processes

Structural changes induced by chemical aging can be ranged into four categories depending on the affected structural scale (Table 1).

Table 1. The four scales of structure and the corresponding tools of investigation

Structural scale	Entity	Main analytical tools	Theoretical tools
Molecular	Group of atoms Monomer unit	IR and NMR spectrometry	Organic chemistry
Macromolecular	Chain Network strand Crosslink	Viscosimetry SEC Sol-gel analysis	Macromolecular physico-chemistry
Morphological	Crystalline lamellae Spherulite	SAXS, WAXS, DSC SEM, TEM, AFM	Materials science
Macroscopic	Skin−core structure	Visible microscopy Modulus profiling Nano- and macro-indentation	Materials science

Case A: Hydrolysis of an acrylic polymer:

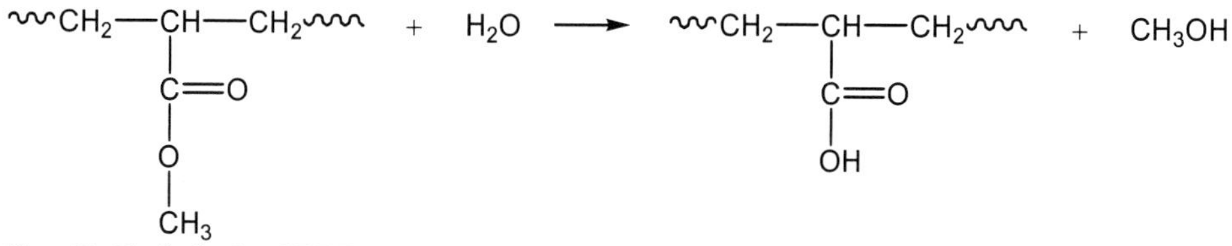

$$\sim CH_2-CH(-C(=O)-O-CH_3)-CH_2\sim + H_2O \longrightarrow \sim CH_2-CH(-C(=O)-OH)-CH_2\sim + CH_3OH$$

Case B: Hydrolysis of PET:

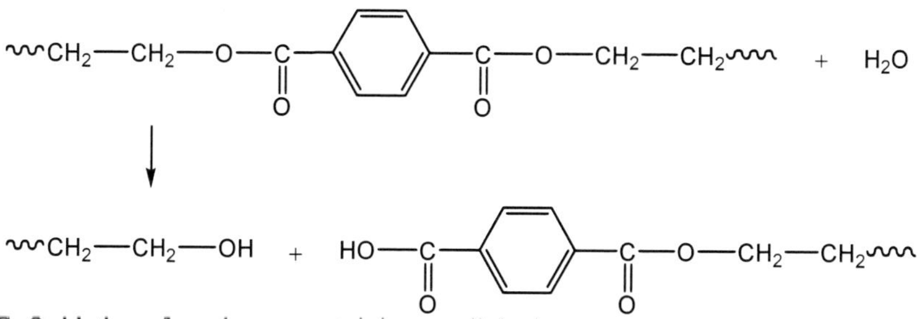

$$\sim CH_2-CH_2-O-C(=O)-C_6H_4-C(=O)-O-CH_2-CH_2\sim + H_2O$$

$$\downarrow$$

$$\sim CH_2-CH_2-OH + HO-C(=O)-C_6H_4-C(=O)-O-CH_2-CH_2\sim$$

Case C: Oxidation of a polymer containing an aliphatic segment:

$$\sim CH_2\sim + O_2 \xrightarrow{\text{several steps}} \sim C(=O)\sim \text{ or } \sim CH(OH)\sim \text{ or } \sim CH(OOH)\sim$$

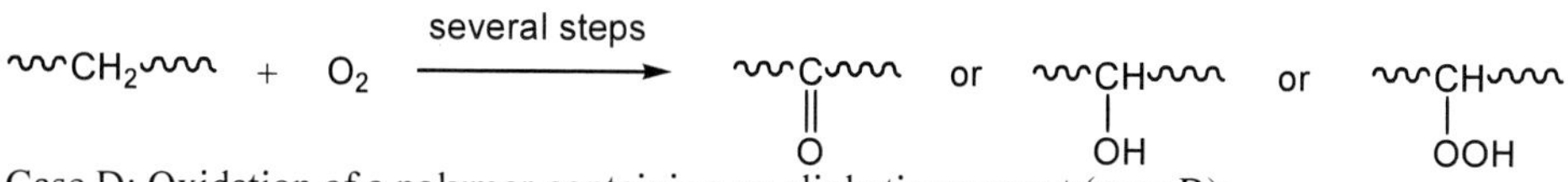

Case D: Oxidation of a polymer containing an aliphatic segment (case B):

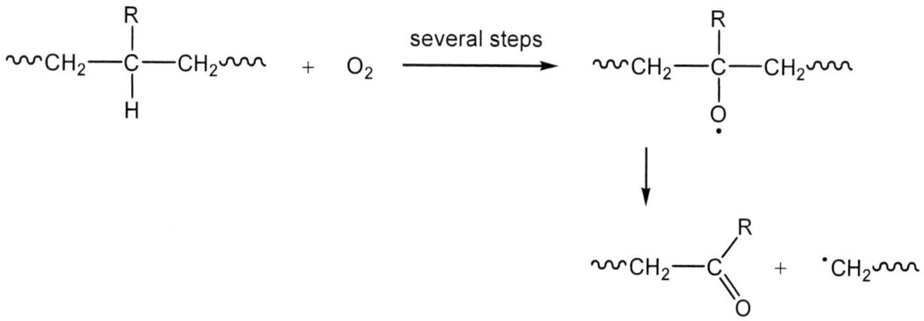

$$\sim CH_2-C(R)(H)-CH_2\sim + O_2 \xrightarrow{\text{several steps}} \sim CH_2-C(R)(O^\bullet)-CH_2\sim$$

$$\downarrow$$

$$\sim CH_2-C(R)=O + {}^\bullet CH_2\sim$$

Figure 1. Examples of hydrolysis (A, B) or oxidation (C, D) processes leading (B, D) or not (A, C) to chain scission.

The “target” of water or oxygen attack is always the molecular scale, i.e. a region of sub-nanometric dimension. Some examples of chemical transformations at this scale are presented in Fig. 1. One can distinguish two categories of chemical events: those which do not affect the structure at the macromolecular scale and those (chain scissions, crosslinking) which affect the structure at the macromolecular scale. This distinction is based on a very simple rule: Only the structural changes at the macromolecular scale can have consequences on the polymer mechanical behavior at reasonably low conversions.

Hydrolysis without chain scission occurs only in acrylic and vinylic polymers with ester side groups. These polymers families are not very frequently used in composite matrices. Oxidation leads to predominating chain scission in the great majority of cases, to predominating crosslinking in some cases such as polybutadiene (Coquillat et al. 2007). An important quantity is the yield y of chain scission (y_S) or crosslinking (y_X) expressed as the number of broken chains or crosslinks formed per oxygen molecule absorbed. There is, to our knowledge, no case of industrial polymer for which this yield would be null.

1. Changes of Side-Groups

As previously seen, a change of side-groups, for instance the replacement of an ester by an acid (case A in Fig. 1), or the replacement of a methylene by a ketone or an alcohol (case C in Fig. 1), has no effect on mechanical properties. It can however affect other physical properties, for instance:

- Color if the new group is a chromophore: Hydrolysis has generally no consequences on color. In contrast, oxidation induces yellowing in practically all the aromatic polymers because it can transform some aromatic nuclei into quinonic structures absorbing in the violet–blue part of the visible spectrum and thus appearing yellow.
- Replacement of a non polar group by a polar group, for instance an ester by an acid (case A in Fig. 1) or a methylene by an alcohol (case C in Fig. 1) is expected to have the following consequences:
  - Increase in dielectric permittivity;
  - Increase in refractive index;
  - Growth of dielectric absorption bands;
  - Increase in hydrophilicity and wettability.

These changes are rarely decisive in the case of composites.

2. Chain Scission

2.1. Random Versus Selective Chain Scission

Chain scissions can occur on peculiar sites having an especially high reactivity. In this case, they will be named “selective chain scissions”. Chain scissions can be also randomly distributed (i.e. all the repeat units are equireactive). In this case, they will be named “random chain scissions”.

The different types of scissions are schematized in Figure 2.

For composite matrices, in the context of long term hydrolytic or oxidative ageing, random chain scission predominates over all the other processes in the great majority of cases. Depolymerization occurs essentially at high temperatures, only in polymers having weak monomer–monomer bonds (e.g. poly(methyl methacrylate) and poly(oxymethylene)). Crosslinking predominates mainly in unsaturated polymers, i.e. essentially polybutadiene and its copolymers (Coquillat et al. 2007). Some exceptions are known in the domain of composite matrices, they will be examined in a short paragraph.

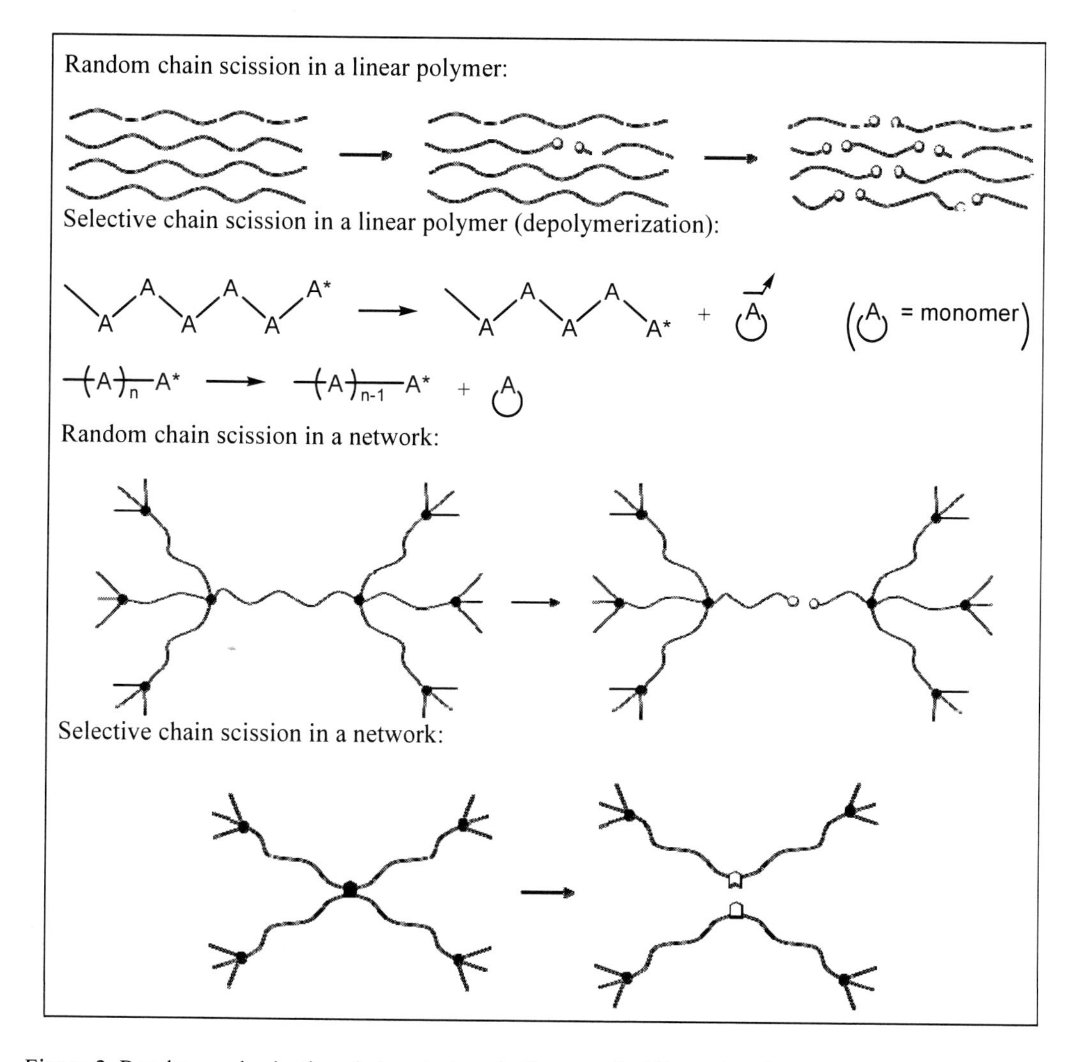

Figure 2. Random and selective chain scissions in linear and tridimensional polymers.

2.2. Random Chain Scission in Linear Polymers

The random character results from the fact that all the reactive groups of the macromolecules have an equal probability to react. This means that the probability for a chain to react is an increasing function of its length. The number (M_n) and weight (M_W) average molar masses are linked to the number S of chain scissions per mass unit by the following relationships (Saito 1958a and 1958b):

$$\frac{1}{M_n} - \frac{1}{M_{n0}} = S \tag{1}$$

$$\frac{1}{M_W} - \frac{1}{M_{W0}} = \frac{S}{2} \tag{2}$$

The polydispersity index IP varies as follows:

$$IP = IP_0 \frac{1 + SM_{n0}}{1 + \frac{IP_0}{2} SM_{n0}} \tag{3}$$

One sees that IP increases when $IP_0 < 2$, decreases when $IP_0 > 2$, but in all cases, tends towards 2 when the number of chain scissions increases. This characteristic is generally used to recognize a random chain scission.

Steric exclusion chromatography (SEC) can be used to determine the molar mass distribution (MMD) and the average values M_n and M_W. M_W can be also determined by viscosimetry using a power law:

$$\eta = K M_W^a \tag{4}$$

η is the intrinsic viscosity in the case of dilute polymer solution. In this case $a \approx 0.7$ and K depends on temperature and solvent nature.

η is the Newtonian viscosity in the case of molten polymer. In this case $a = 3.4$ and K depends on temperature and polymer chemical structure.

2.3. Consequences of Random Chain Scission on Thermodynamical Properties

2.3.1. Glass Transition Temperature T_g

According to Fox and Flory (1954):

$$T_g = T_{g\infty} - \frac{K_{FF}}{M_n} \tag{5}$$

where K_{FF} and $T_{g\infty}$ are increasing functions of the chain stiffness.

$$K_{FF} \propto T_{g\infty}^2 \text{ (Richaud et al. 2010)} \tag{6}$$

From Equs (1) and (5), one obtains:

$$\left(T_{g0} - T_g\right) = K_{FF}\, S \text{ (Richaud et al. 2010)} \tag{7}$$

T_g decreases linearly with the number of chain scissions. The effect of these latter is an increasing function of the chain stiffness. Practically, T_g changes due to degradation of flexible polymers, having typically T_g values lower than 100°C, are negligible. In contrast, they can be measured in the stiff polymers, especially those having aromatic groups in the chain.

2.3.2. Melting Point

In semi-crystalline polymers, degradation is accompanied of morphological changes (see below) and it is not easy to separate the (small) effects of molecular mass decrease from those of morphological changes. To resume, degradation induces only small variations of the melting point and the sense of this variation can vary from a polymer to another. Melting point measurement is thus not an adequate method to monitor random chain scission.

2.3.3. Elastic Properties

At the conversions of practical interest, random chain scissions have no significant effect on elastic properties of linear polymers.

2.3.4. Fracture Properties

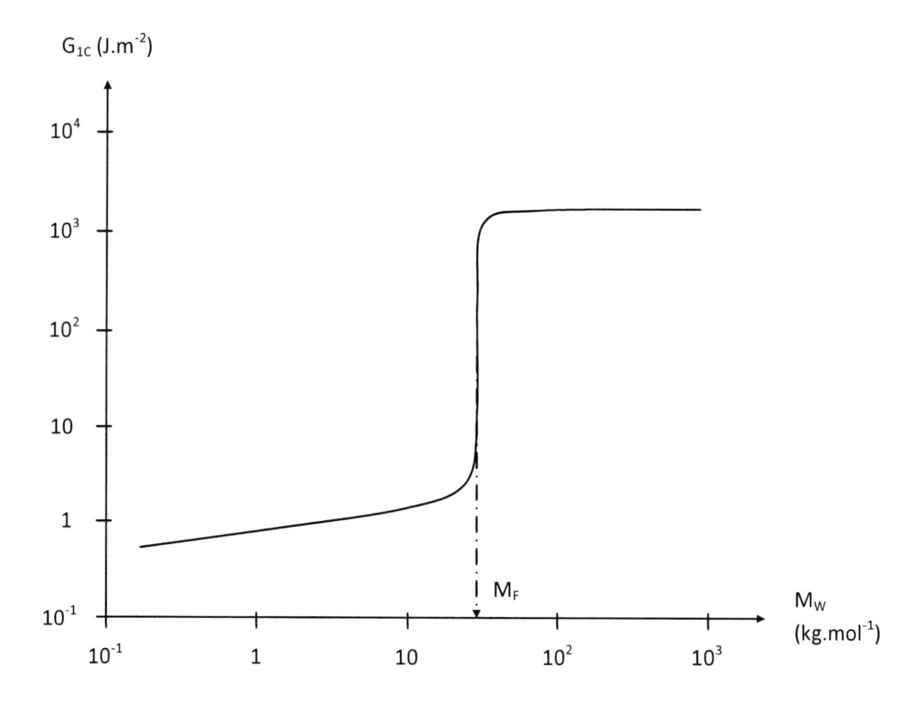

Figure 3. Shape of the variation of the crack propagation energy (critical rate of elastic energy release) with the weight average molar mass (for instance Greco and Ragosta 1987).

The effect of molar mass on toughness of linear polymers is schematized in Figure 3. For all the polymers, one can distinguish two regimes separated by a relatively sharp transition at a critical molar mass M_F. At molar masses lower than M_F, the polymers are extremely brittle, their toughness is only due to Van der Waals interactions. At molar masses higher than M_F, the polymers have a toughness often in the 10^3–10^4 J.m^{-2} range, almost independent of molar mass. In amorphous polymers, M_F is mainly linked to the entanglement density. As a matter of fact, plastic deformations responsible for the high toughness, are linked to chain drawing and this latter is possible only if the chains participate to a network, here the entanglement (topological) network.

In semi-crystalline polymers, the critical quantity is the interlamellar spacing l_a. For instance polyethylene (Kennedy et al. 1994) or polyoxymethylene (Fayolle et al. 2009) are brittle when $l_a \leq 6$ nm. Since l_a is sharply linked to molar mass, it can be considered, for these polymers also, that there is a critical molar mass M_F separating brittle and ductile domains (Fayolle et al. 2008).

It can be interesting to establish a relationship between M_F and the entanglement molar mass M_E, this latter being sharply linked to the chemical structure (Fetters et al. 1999). It appears that:

- For amorphous polymers (Kausch 2001) and semi-crystalline polymers of low crystallinity:

$$\frac{M_F}{M_E} \approx 2 \text{ to } 10 \tag{8}$$

- For relatively highly crystalline polymers (Fayolle et al. 2008):

$$\frac{M_F}{M_E} \approx 50 \tag{9}$$

According to the shape of Figure 3, the effect of random chain scission must display three characteristics:

1) If the initial molar mass is high enough, chain scission is expected to have no effect on fracture properties in the initial period of exposure, before molar mass reaches the critical value.
2) Toughness must decay abruptly of one to three decades when molar mass reaches the critical value.
3) Beyond the ductile–brittle transition, the toughness decreases continuously but very slowly. The change of toughness with the number of chain scissions S is expected to be the mirror image of Figure 3 (see Figure 4).

The critical number of chain scission S_F for embrittlement is given by Equ. 2:

$$S_F = 2\left(\frac{1}{M_F} - \frac{1}{M_{W0}}\right) \tag{10}$$

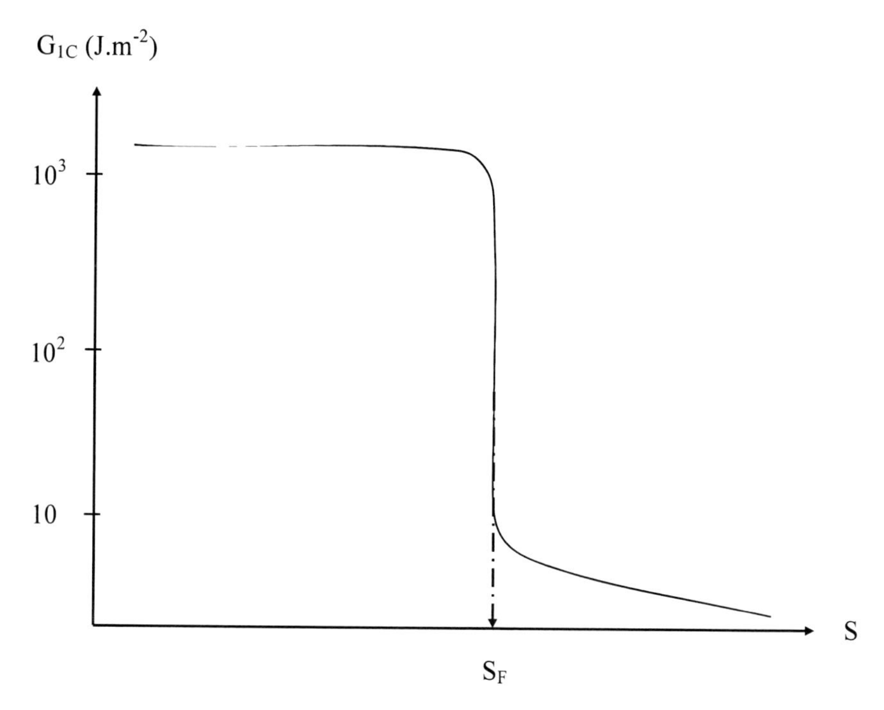

Figure 4. Expected shape of the variation of toughness with the number of chain scissions per mass unit.

In any case: $M_F > 10$ kg.mol^{-1}, so that:

$$S_F \leq \frac{2}{M_F} \approx 0.2 \text{ mol.kg}^{-1} \tag{11}$$

In common industrial thermoplastics, the monomer concentration [m] is such as:

$$[m] > 2 \text{ mol.kg}^{-1} \tag{12}$$

It appears that embrittlement occurs always at a small conversion degree of the chain scission process. In certain cases, for instance polypropylene (Fayolle et al. 2002), embrittlement occurs while no structural change is observable by IR.

2.3. Random Chain Scission in Networks

Let us consider an ideal network in which every chain is connected to network nodes at both extremities. Such chains are called: "elastically active chains" (EAC). Their concentration ν_0 is linked to the concentration x_0 of nodes by:

$$\nu_0 = \frac{f}{2} x_0 \tag{13}$$

where f is the node functionality, i.e. the number of chains connected to a node.

If the network undergoes S random chain scissions with S << ν_0, each chain scission occurs in an EAC, so that new crosslink density is given by (Pascault et al. 2002):

$$\nu = \nu_0 - jS \tag{14}$$

with j = 3 for f = 3, and j = 1 for f > 3 (see Figure 5).

Each chain scission creates two dangling chains.

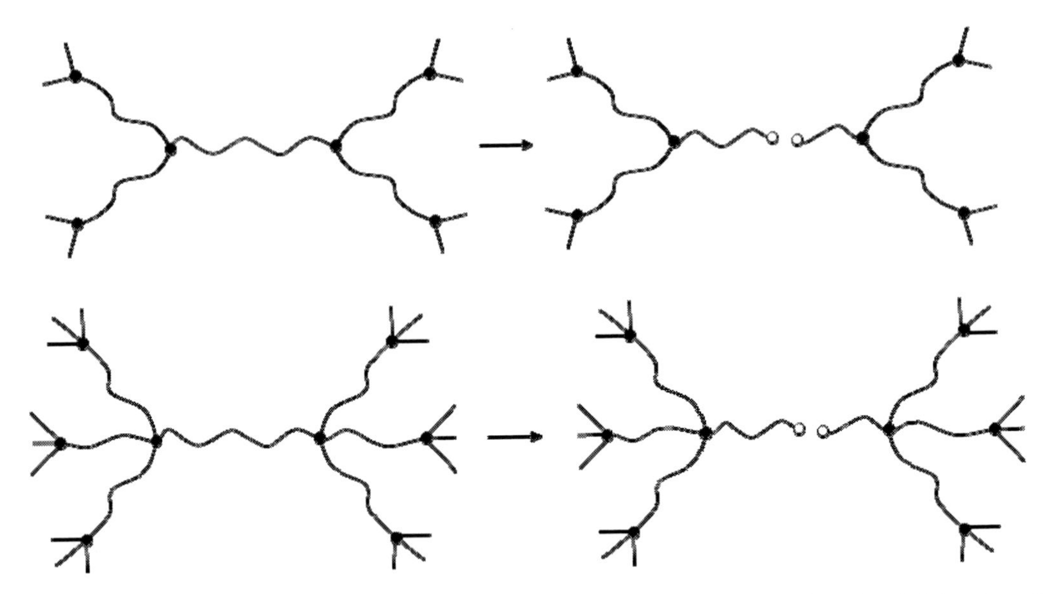

Figure 5. Schematization of a random chain scission in a network with trifunctional (above) and tetrafuntional (below) nodes.

Indeed, chain scission transforms an ideal network in a non-ideal one and the probability to have a chain scission in a dangling chain increases with the number of chain scissions (see Figure 6).

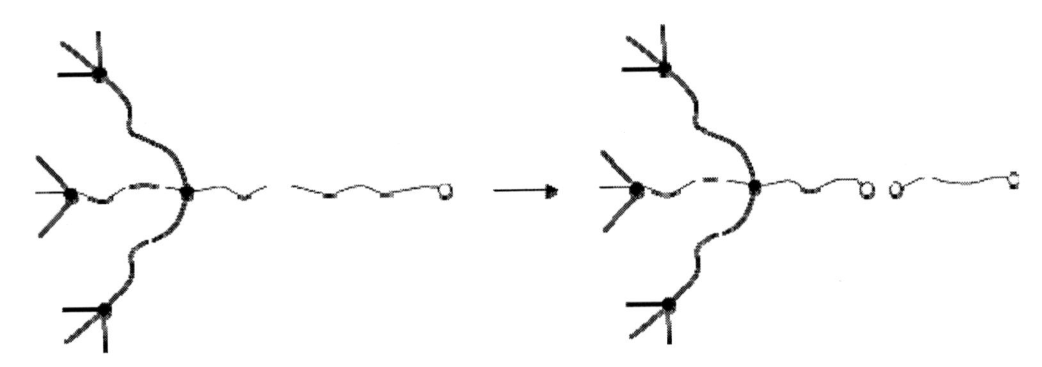

Figure 6. Schematization of a chain scission in a dangling chain.

At a given state of degradation, the mass fraction w_e of EAC is:

$$w_e = \nu M_e \tag{15}$$

Let us consider a chain scission process at a constant rate: $r = dS/dt$, for instance in a network having nodes of functionality $f > 3$. The probability to break an EAC is expected to be proportional to the EAC mass fraction so that:

$$\frac{d\nu}{dt} = -r\,w_e = -r\,M_e\,\nu \tag{16}$$

So that:

$$\nu = \nu_0 \exp - r\,M_e\,t \tag{17}$$

Although the chain scission process determined by the chemical mechanism is an apparent zero order process, the crosslink density is expected to decrease in an apparent first order process. In practice, however, the mechanical behavior is strongly altered at relatively low conversions, before the probability to have a scission in a dangling chain has reached a significant value. It is noteworthy that chain scissions on dangling chains create free chains. The amount of these latter corresponds to the extractable fraction in solvents.

Analytical methods for the determination of the number S of chain scissions per mass unit are scarce. When elastic properties in rubbery state are measurable, one can use the theory of rubber elasticity according which (Flory 1953):

$$\frac{dG}{dS} = \frac{dG}{d\nu}\,\frac{d\nu}{dS} = -jR\,T\rho \tag{18}$$

where G is the shear modulus at $T > T_g$ and ρ is the specific weight ($kg.m^{-3}$) of the polymer.

The glass transition temperature T_g is also dependent on crosslink density. According to Di Marzio (1964):

$$T_g = \frac{T_{gl}}{1 - K_{DM}\,F\nu} \tag{19}$$

where T_{gl} and F are parameters depending on the chain stiffness and K_{DM} is an universal constant ($K_{DM} \approx 2$).

$$\frac{dT_g}{dS} = -j\,\frac{T_g}{d\nu} = \frac{jK_{DM}\,F\,T_{gl}}{\left(1 - K_{DM}\,F\,\nu\right)^2} = jK_{DM}\,F\,\frac{T_g^2}{T_{gl}} \tag{20}$$

The effect of chain scissions is thus an increasing function of T_g.

Let us consider, for instance, an epoxy network based on the triglycidyl derivative of p-aminophenol (TGAP) crosslinked by diaminodiphenylmethane (DDM) in stoichiometric proportion. The characteristics (Pascault et al. 2002) are:

$T_g = 494$ K, $F = 23\ g.mol^{-1}$, $T_{gl} = 293$ K, $j = 3$ (trifunctional crosslinks).

Then:

$$\frac{dT_g}{dS} = 172 \text{ K.kg.mol}^{-1} \tag{21}$$

2.4. Consequences of Chain Scission on Mechanical Properties

The effect of chain scissions on elastic properties depends on the amplitude of the dissipation peak linked to the β relaxation. For polymers having a weak β transition, for instance styrene crosslinked polyesters or vinylesters, chain scissions have practically no effect on elastic properties in glassy state.

For polymers having a strong β transition, for instance diamine crosslinked epoxies, chain scissions lead to a modulus increase in the modulus plateau between the β transition and the glass transition (Figure 7). The phenomenon has been called "internal antiplasticization" (Rasoldier et al. 2008). It can be evidenced through nano- or micro-indentation profiles to characterize degradation gradients in the sample thickness (Olivier et al. 2009).

Chain scissions induce a decrease of fracture toughness. From this point of view, degraded networks differ from ideal networks in which fracture properties are generally a decreasing function of crosslink density (Crawford and Lesser 1999, Pascault et al. 2002). Little is known on the quantitative relationships between chain scission and embrittlement in networks.

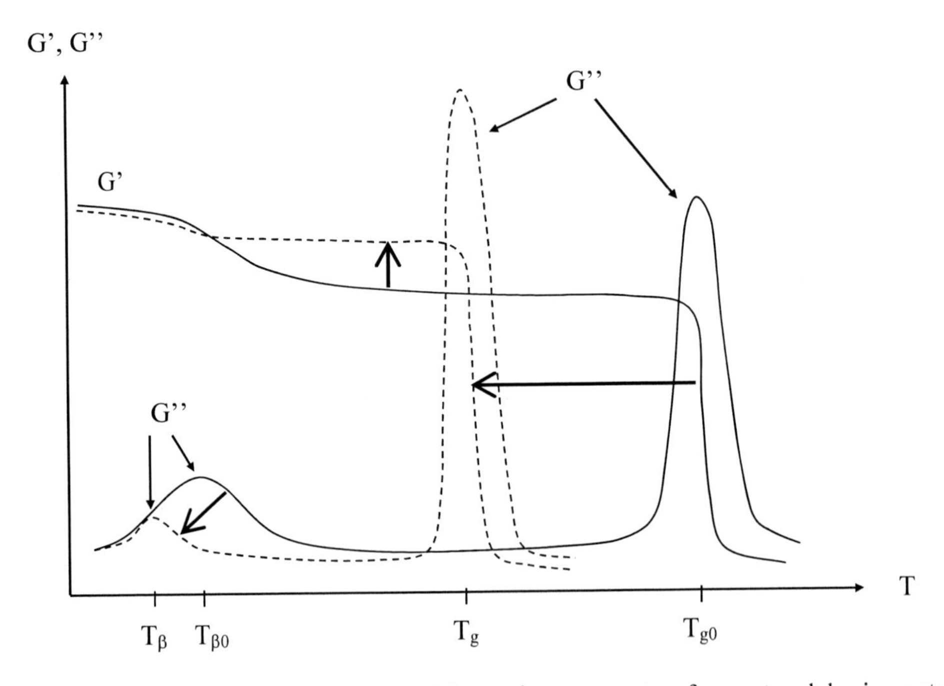

Figure 7. Storage (G') and dissipation (G'') modulus against temperature for a network having a strong β transition before (full line) and after (dashed line) chain scission.

3. Simultaneous Chain Scission and Crosslinking

3.1. In Linear Polymers

Saito's equations become:

$$\frac{1}{M_n} - \frac{1}{M_{n0}} = S - X \tag{22}$$

$$\frac{1}{M_W} - \frac{1}{M_{W0}} = \frac{S}{2} - 2X \tag{23}$$

where X is the number of crosslinks and S the number of chain scissions.

There is an "equilibrium" corresponding to the constancy of M_W, i.e. to:

$$S = 4X \tag{24}$$

For X > S / 4, crosslinking predominates over chain scissions. Polymer gelation occurs when $M_W \rightarrow 0$, i.e. when:

$$\frac{S}{2} - 2X = -\frac{1}{M_{W0}} \tag{25}$$

i.e. in the absence of chain scission for:

$$X_g = -\frac{1}{2M_{W0}} \tag{26}$$

Beyond the gel point, an insoluble fraction appears. According to Charlesby and Pinner (1959), the soluble fraction w_S is linked to the number of chain scissions and crosslinks by:

$$w_S + w_S^{1/2} = \frac{S}{2X} + \frac{1}{M_{W0}X} \tag{27}$$

3.2. Consequences of Crosslinking on Properties

In linear polymers, crosslinking affects mainly rheological properties in molten state. As a matter of fact, long branching is responsible for the disappearance of the Newtonian plateau (Figure 8).

Crosslinking induces an increase in the glass transition temperature. In the case of simultaneous chain scission and crosslinking, it can be written, in a first approach:

$$T_g = T_{g0} + k_S S + k_X X \tag{28}$$

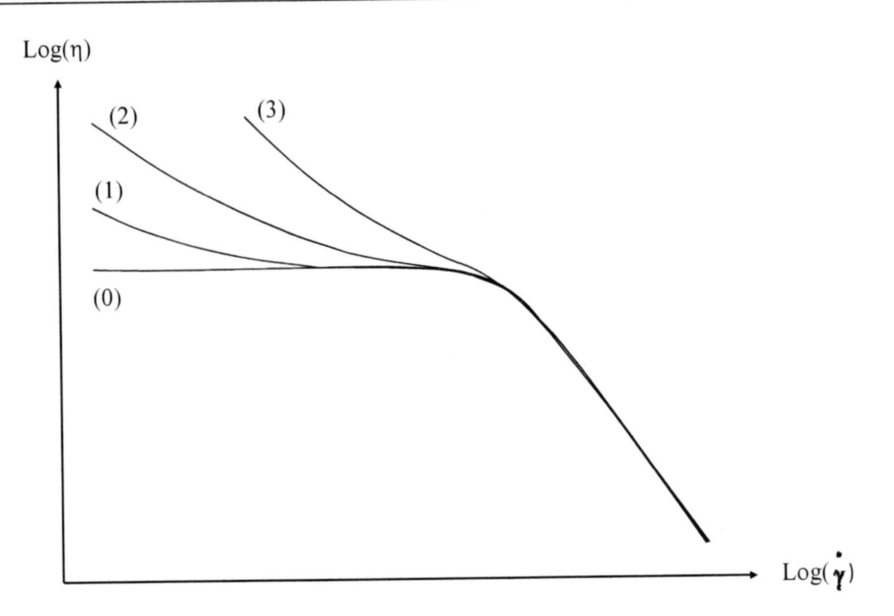

Figure 8. Shape of the curve Log(viscosity) versus Log(shear rate) for a linear polymer before (0) and after ageing leading to an increase of the degree of branching (0 < 1 < 2 < 3).

k_S is significantly higher than k_X. As an example, in bisphenol A polysulphone, $k_S/k_X \approx 2.1$ (Richaud et al. 2010). In other words, crosslinks have less influence than chain scission on T_g.

The effects of crosslinking on fracture properties are not well known. In most cases, crosslinking is expected to induce embrittlement according to the following causal chains:

1) Crosslinking → Increase in T_g → Increase in yield stress → Ductile (plastic) deformation less and less competitive with brittle deformation
2) Crosslinking → Shortening of EAC → Decrease in drawability of EAC → Reduction of the plastic zone at crack tip → Decrease in toughness

3.3. Post-Cure

In most cases of industrial thermosets, cure is not complete, reactive groups remain trapped in glassy state at the end of processing operations. In ageing conditions, they can recover mobility enough to react, because they are heated at temperatures not very far from T_g or the polymer is plasticized by water. Then, cure reactions are reactivated, the crosslink density increases in an autoretardated way and stops when all the available reactive groups have been consumed. Except for scarce cases, for instance epoxides crosslinked by unsaturated anhydrides (Le Huy et al. 1993), oxidative ageing is dominated by chain scission, so that, for thermosets, crosslink density variations during thermal ageing have the shape of Figure 9.

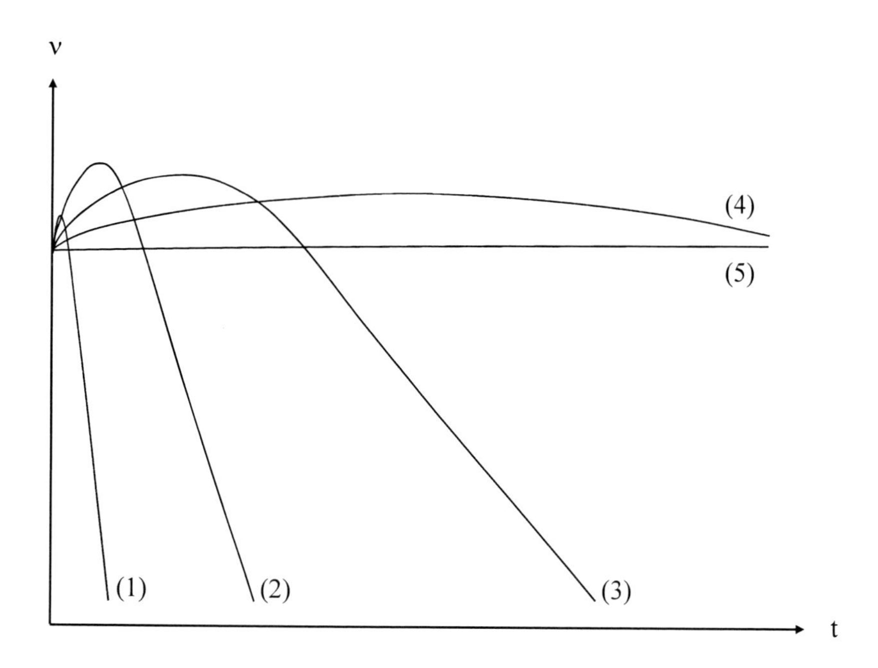

Figure 9. Shape of kinetic curves of crosslink density variations during the thermal ageing (in air) of a thermoset at various temperatures: $T_1 > T_2 > T_3 > T_4 > T_5$.

These curves can be decomposed into two components, i.e. post-cure and degradation (see Figure 10).

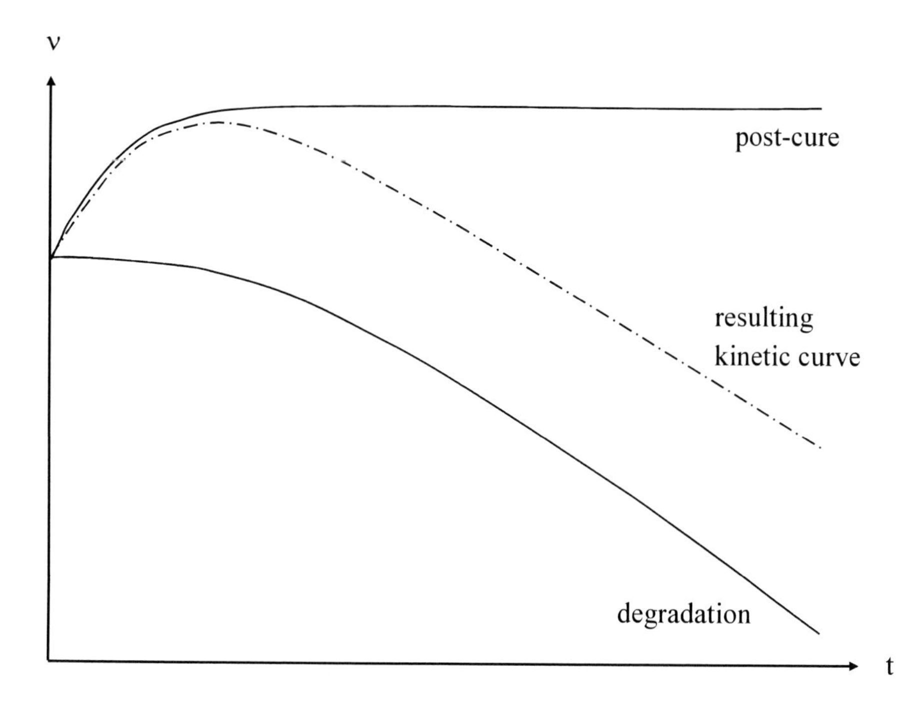

Figure 10. Schematization of combined effects of post-cure and degradation.

In the simplest cases, there is no interaction between both processes, so that their effects on crosslink density are additive. In other cases, however, oxygen or water can inhibit post-cure and a more complex behaviour can be expected.

REACTION–DIFFUSION COUPLING

In both cases of oxidation and hydrolysis, the polymer matrix reacts with a small molecule M (oxygen or water) coming from the environment. In a thin elementary layer at a distance z of the sample surface, the reactant concentration balance can be ascribed:

Reactant concentration change = Rate of reactant supply by diffusion – Rate of reactant consumption by reaction

The balance equation can be thus written, in the case of unidirectional diffusion, i.e. far from the sample edges:

$$\left(\frac{\partial C}{\partial t}\right)_z = D\frac{\partial^2 C}{\partial z^2} - r(C) \tag{29}$$

where C is the reactant concentration, D is the coefficient of reactant diffusion in the polymer and r(C) is the rate of reactant consumption expressed as a function of the reactant concentration.

The resolution of Equ. 29 needs the knowledge of two physical quantities: the equilibrium concentration C_S of reactant and its coefficient of diffusion D in the polymer, and one chemical data: the concentration dependence r(C) of the reactant chemical consumption.

1. Reactant Transport Properties

Concerning first transport properties in matrices and their relationships with polymer structure, there are various book chapters and monographs (Crank and Park 1968, Hopfenberg 1974, Van Krevelen 1976, Bicerano 2002). The main differences between oxygen and water properties can be summarized as follows: The oxygen solubility in polymers is always low, typically $\leq 10^{-3}$ mol.l^{-1}, practically insensitive to small structural changes. The coefficient of oxygen diffusion is of the order of $10^{-11 \pm 2}$ m^2.s^{-1} at ambient temperature and its apparent activation energy in the [30–60 kJ.mol^{-1}] interval. Oxygen transport properties are practically always determined from permeability measurements. It can be reasonably assumed that, during an ageing experiment at constant temperature, D is independent of C and of reaction conversion at reasonably low conversions, for instance before and just after embrittlement.

Some relationships between the polymer structure and water transport characteristics are illustrated by Table 2.

The main trends of structure–property relationships can be briefly summarized as follows:

a) Three main types of groups can be distinguished:

Table 2. Molar mass of the constitutive repeat unit, water mass fraction at equilibrium at 50°C and 50% RH, coefficient of diffusion in the same conditions, and number of moles of water per constitutive repeat unit. Compiled data from Bellenger et al. (1994), Tcharkhtchi et al. (2000) and Gaudichet et al. (2008)

Polymer	code	M ($g.mol^{-1}$)	m_{equ} (%)	$D \times 10^{12}$ ($m^2.s^{-1}$)	n ($mol.mol^{-1}$)
Poly(methyl methacrylate)	PMMA	100	1.28	0.36	0.071
Poly(ethylene terephthalate)	PET	192	0.55	0.54	0.059
Polycarbonate	PC	254	0.25	5.4	0.035
Polyamide 11	PA11	183	1.5	0.13	0.153
Poly(bisphenol A) sulphone	PSU	442	0.52	8.97	0.128
Polyethersulphone	PES	232	1.8	2.79	0.232
Polyetherimide	PEI	592	1.4	0.97	0.460
Polypyromellitimide	PPI	382	5.0	0.1	1.061
Polyimide	IP960	486	4.2	0.83	1.134
Epoxy	DGEBA−Etha	858	2.0	1.03	0.953
Epoxy	DGEBD−Etha	578	6.8	0.11	2.183
Unsaturated polyester	UP	334	0.83	0.88	0.154
Vinylester	VE(D)	980	1.7	0.6	0.926
Vinylester	VE(C)	550	0.5	0.5	0.153

G1: Hydrocarbon and halogenated groups of which the contribution to hydrophilicity is negligible. Polymers containing only these groups (polyethylene, polypropylene, polystyrene, elastomers, etc ...) absorb less than 0.5 wt% water.

G2: Groups of relatively low polarity (ethers, ketones, esters, etc ...). Polymers containing only these groups (with hydrocarbon ones) absorb generally less than 2 wt% water. Physical effects of water absorption (plasticization, swelling) are generally negligible. Polymers containing the ester group (polyalkylene terephthalates, unsaturated polyesters, anhydride crosslinked epoxies, etc ...) are however susceptible to hydrolysis (see below). Polymers containing methacrylic esters (polymethyl methacrylate, vinylesters) are generally resistant to hydrolysis.

G3: Highly polar groups able to establish strong hydrogen bonds with water (sulphones, alcohols, amides, acids, etc ...). These polymers can absorb up to 5 wt% water, that can induce considerable physical changes, for instance T_g decreases of about 10 K per percent water absorbed, swelling and damage by swelling stresses occur during the sorption or desorption transients.

b) In each polymer family containing one type of hydrophilic group, for instance polyamides, polysulphones, polyimides, amine crosslinked epoxies, etc... the equilibrium water concentration increases non linearly with the concentration of polar groups. A theory based on the hypothesis that water is doubly bonded has been proposed to explain this behavior (Tcharkhtchi et al. 2000, Gaudichet-Maurin et al. 2008).

c) The relationships between diffusion coefficient and polymer structure are not fully understood, but what is clear is that in a given family, D is a decreasing function of the water equilibrium concentration (Thominette et al. 2006). This dependence indicates that water–polymer hydrogen bonds slow down diffusion as well in polyethylenes (McCall et al. 1984) as in epoxies (Tcharkhtchi et al. 2000).
d) Diffusion is thermally activated, apparent activation energies are generally in the 20–60 $kJ.mol^{-1}$ interval. Equilibrium water concentrations depend only slightly on temperature, that can be explained by considerations of heat of solubility (Merdas et al. 2002).

2. Chemical Reactant Consumption

2.1. Hydrolysis

Let's consider now the term representing the chemical reactant consumption in the reaction–diffusion equation (Equ. 29). In the simplest case of hydrolysis, r(C) appears as a simple first order equation:

$$\frac{\partial C}{\partial t} = D\frac{\partial^2 C}{\partial z^2} - k[A]_0 C \tag{30}$$

where k is the second order rate constant of the water–polymer reaction, $[A]_0$ is the concentration of hydrolysable groups, considered constant at reasonably low conversions.

A more complex equation is needed in the case where hydrolysis is equilibrated (see below). When Equ. 30 is an acceptable approximation, the integration for a symmetric sheet of thickness L gives:

$$C = C_S \frac{\cosh J\left(z - \frac{L}{2}\right)}{\cosh \frac{JL}{2}} \tag{31}$$

where $J = \frac{k[A]_0}{D}$

where the origin of z is taken at a sample edge.

The water concentration and thus, also, the hydrolysis rate, decreases in a pseudo-exponential way from the edges, where $C = C_S$, to the middle of the sample where $C = C_m$ (minimum).

When $C = 6\ J^{-1}$, $C_m / C_S \approx 0.1$. Thus, for $L >> 6\ J^{-1}$, the sample behaves as a sandwich made of an undegraded core with two degraded superficial layers.

In the case of equilibrated hydrolysis, for instance for PA11 (Jacques et al. 2002), a degradation gradient appears at the beginning of exposure, but the sample tends to homogenize as the hydrolysis rate slows down.

2.2. Oxidation

The case of oxidation is more complex because the mechanism is a branched radical chain of which the kinetic modeling was considered out of reach for a long time. The first attempts were made at the beginning of 80's by Seguchi et al. (1981 and 1982) and Cunliffe and Davis (1982) in the simplest case of constant initiation rate, steady-state for radical concentration, long kinetic chain and low conversion. All these simplifying assumptions are more or less valid in some cases of radiochemical oxidation, but they are questionable in the case of thermal oxidation.

Assuming their validity, the rate of oxygen consumption can be expressed by an hyperbolic function of oxygen concentration:

$$r(C) = \frac{a\,C}{1 + b\,C} \tag{32}$$

where a and b can be expressed in terms of rate constants of the elementary reactions participating to oxidation.

The reaction–diffusion equation becomes then:

$$\frac{\partial C}{\partial t} = D\frac{\partial^2 C}{\partial z^2} - \frac{a\,C}{1 + b\,C} \tag{33}$$

There is no analytical solution for this equation, but approximations can be obtained for extreme cases, for instance:

$$C >> b^{-1} \Rightarrow r(C) = \frac{a}{b} = r_S \tag{34}$$

$$\text{And } C << b^{-1} \Rightarrow r(C) = a\,C \tag{35}$$

In the second case, the same solution as for hydrolysis (Equ. 30) will be obtained. In the first case, however, the integration leads to a parabolic shape of the oxygen concentration profile:

$$C = C_S + \frac{r_S}{2D}(z - L)z \tag{36}$$

The concentration in the middle of the sample is:

$$C = C_S - \frac{r_S\,L^2}{8D} \tag{37}$$

3. Case of Composites

In the case of composites, new problems linked to the anisotropy of diffusion paths, the eventual role of interfacial diffusion and the role of preexisting or swelling induced damage appeared in the middle of 70's. The interest was mainly focused on the effect of humidity on carbon fiber/amine crosslinked epoxy composites of aeronautical interest. For the pioneers of this research (Shen and Springer 1976), the determination of diffusion kinetic laws appeared as the key objective. Various studies revealed that, in certain cases, diffusion in composites cannot be modeled by a simple Fick's law and that Langmuir's equation is more appropriate. Carter and Kibler (1978) proposed a method for the parameter identification. At the end of 70's, the kinetic analysis of water diffusion into composites became a worldwide research objective. A great quantity of experimental data was obtained. The results can be summarized as follows:

a) Concerning the effect of fiber anisotropy on diffusion, a model for unidirectional composites was proposed by Kondo and Taki (1982). This model makes full account of the fact that water diffusivity is more privileged in the fiber direction than in the transverse one:

$$D_{//} = \frac{1 - V_f}{1 - 2\sqrt{V_f / \pi}} D_{\perp} \tag{38}$$

where $D_{//}$ and $D_{\perp}$ are the respective diffusion coefficients in the longitudinal and transverse fiber directions.

Colin et al. (2005) showed that such models can also be used to predict oxygen diffusivity in composites. More recently, Roy and Singh (2010) showed that these models can be improved to take into account physical discontinuities such as highly permeable fiber/matrix interface or fiber/matrix debonding due to oxidative shrinkage and erosion.

b) Concerning Langmuir's mechanisms, it was assumed, for a long time, that water was trapped in "defects" resulting from damage or preexisting, eventually located at interface. Tcharkhtchi et al. (2000) found that unreacted epoxide groups undergo a reversible hydrolysis:

$$—CH_2—CH(—O—)CH_2 \;+\; H_2O \rightleftharpoons —CH_2—CH(OH)—CH_2—OH$$

Epoxide groups appear thus as "water traps" and are responsible for a Langmuir component in diffusion kinetic curves. Since industrial composites are rarely fully cured, it can be assumed that epoxide hydrolysis was often the cause of Langmuir's behavior in previous studies.

Recently, however, Derrien and Gilormini (2006) have found that Langmuir's behavior could be simply linked to the stress state induced by water diffusion.

c) Concerning eventual interfacial processes, they raised up an abundant literature. Various techniques were used to characterize interfaces and interphases (Shradder and Block 1971, Di Benedetto and Scola 1980, Ishida and Koenig 1980, Rosen and Goddard 1980, Ishida 1984, Di Benedetto and Lex 1989, Thomason 1990, Hoh et al. 1990, Schutle and Mc Donough 1994). Round Robin tests showed that no one analytical method is able to give unquestionable results (Pitkethly et al. 1993). Even in the cases where the interface response to humid aging has been unambiguously identified from studies of model systems (Kaelble et al. 1975 and 1976, Salmon et al. 1997), it seems difficult, at this sate of our knowledge, to build a non-empirical kinetic model of water effects on interface/interphase in composites.

HYDROLYSIS

1. Uncatalysed Hydrolysis Mechanisms and Kinetics in Pure Water

Hydrolysis is an ionic, step by step mechanism. In a first approach, the mechanism can be written:

$$\text{(H)A–B} + H_2O \rightarrow \text{AH} + \text{A–OB} \qquad (k_H)$$

$$\text{(R)AH} + \text{A–OB} \rightarrow \text{A–B} + H_2O \qquad (k_R)$$

The hydrolysis rate is thus:

$$\frac{d[A-B]}{dt} = k_R[AH][B-OH] - k_H[A-B][H_2O] \qquad (39)$$

If S is the number of broken links, it can be written:

$$\frac{dS}{dt} = -\frac{d[A-B]}{dt} = -k_R(a_0+S)(b_0+S) + k_H w(e_0-S) \qquad (40)$$

where $a_0 = [AH]_0$, $b_0 = [B-OH]_0$, $e_0 = [A-B]_0$ and $w = [H_2O]$.

In the simplest cases, at reasonably low conversions, w is constant (no diffusion control) and $a_0 = b_0$, so that:

$$\frac{dS}{dt} = -k_R(a_0+S)^2 + k_H w(e_0-S) \qquad (41)$$

The equilibrium corresponds to $S = S_e$ such as:

$$k_R(a_0+S_e)^2 - k_H w(e_0-S_e) = 0 \qquad (42)$$

Two cases can be distinguished:

Case A: Hydrolysis reaches high conversions (case of most polyesters).

In this case, the reversible reaction can be neglected in a first approach and:

$$\frac{dS}{dt} = k_H w (e_0 - S) \tag{43}$$

$$\text{And } S = e_0 (1 - \exp - k_H w t) \tag{44}$$

At low conversions ($S << e_0$):

$$S \approx e_0 k_H w t = r_H t \tag{45}$$

Hydrolysis behaves as a pseudo zero order process with a quasi-constant rate r_H.

Case B: Equilibrium occurs at a low conversion (case of polyamide 11 (Meyer et al. 2002, Jacques et al. 2002). Then, since ($S << e_0$), Equ. 41 can be simplified as:

$$\frac{dS}{dt} = -k_R (a_0 + S)^2 + k_H w e_0 \tag{46}$$

It is noteworthy that, in a linear polymer with no other chain end than AH and B–OH:

$$a_0 = \frac{1}{M_{n0}}, \ a_0 + S = \frac{1}{M_n} \text{ and } a_0 + S_e = \frac{1}{M_{ne}}$$

where M_{n0}, M_n and M_{ne} are the respective values of the number average molar mass at the beginning of exposure, after a time t of exposure and at equilibrium.

Resolution of the differential equation (46) leads to:

$$M_n = M_{n0} \frac{M_{n0}^{-1} + M_{ne}^{-1} + (M_{ne}^{-1} - M_{n0}^{-1}) \exp - K t}{M_{n0}^{-1} + M_{ne}^{-1} - (M_{ne}^{-1} - M_{n0}^{-1}) \exp - K t} \tag{47}$$

$$\text{where } M_{ne} = \left(\frac{k_R}{k_H e_0 w} \right)^{1/2} \tag{48}$$

$$\text{and } K = 2 (k_R k_H e_0 w)^{1/2} \tag{48}$$

If the temperature dependence of water solubility can be represented by an Arrhenius law:

$$w = w_0 \exp - \frac{E_w}{RT} \tag{49}$$

and the activation energies of k_R and k_H are respectively E_R and E_H, one sees that the apparent activation energy of the equilibrium molar mass is:

$$E_{Mne} = \frac{1}{2}(E_R - E_H - E_w) \tag{50}$$

The apparent activation energy of the composite rate constant K is:

$$E_K = \frac{1}{2}(E_R + E_H + E_w) \tag{51}$$

Generally, $E_R \approx E_H >> E_w$, so that $E_{Mne} << E_K$. The equilibrium value of molar mass is considerably less affected by temperature variations than the degradation rate at the onset of exposure. As an example, in PA11:

$E_K = 97$ kJ.mol^{-1} and $E_{Mne} = -6.5$ kJ.mol^{-1} (Jacques et al. 2002).

In thermosets, hydrolysis processes have been studied essentially in unsaturated polyesters crosslinked by styrene (UP). The problem raised up an abundant literature owing to the great technological importance of glass fiber–UP composites in boats, tanks, pipes, swimming pools, etc ... Before the beginning of 90's, however, kinetic modeling was considered out of reach. Mortaigne et al. (1992) used rubbery modulus to determine the crosslink density according to Equ. 18. They used networks differing by the length of polyester prepolymers, i.e. by the concentration of dangling chains, to calibrate the measurements. Comparative studies (Bellenger et al. 1995, Belan et al. 1997) allowed to establish the following hierarchy of hydrolysis rate constants:

Unreacted fumarate > Reacted fumarate > Orthophthalate > Isophthalate >> Methacrylate(52)

The diol structure (and, indeed, the styrene concentration) affects mainly the hydrophilicity but does not influence significantly the hydrolysis rate constant, except for ethylene glycol which is more reactive than other diols. Vinylesters are, at least, one order of magnitude less reactive than UP.

A comparison of apparent zero order rate constants: $r_H = k_H e_0 w$, for linear and tridimensional polymers is given in Table 3.

These comparisons show clearly that there is no influence of polymer architecture (tridimensional versus linear polymers) on hydrolysis kinetics. The observed differences are mainly due to differences in ester reactivity. Indeed, global hydrolysis rates depend also on water concentration, i.e. of water activity. In a first approach, one can consider that hydrolysis rate is proportional to the relative hygrometry.

Table 3. Comparison of apparent zero order rate constants at 100°C and apparent activation energy for two linear (PC and PET) and two tridimensional polymer families (UP and VE) of ester containing polymers

Polymer	Code	r_H at 100°C (mol.l^{-1}.s^{-1})	E_{act} (kJ.mol^{-1})	Reference
Polycarbonate	PC	6.7×10^{-9}	75	Pryde et al. 1982
Poly(ethylene terephthalate)	PET	6.0×10^{-8}	107	McMahon et al. 1959
Unsaturated polyester	UP	$(2 \text{ to } 15) \times 10^{-7}$	70 ± 10	Mortaigne et al. 1992
Vinylester	VE	$(2 \text{ to } 10) \times 10^{-9}$	–	Ganem et al. 1994

2. Catalysed Hydrolysis Mechanisms and Kinetics

Hydrolysis can be catalyzed as well by H^+ as by OH^- ions. As an example, in polyesters:

$$P{-}C({=}O){-}O{-}R + H^+ \longrightarrow P{-}C({=}O){-}\overset{+}{O}H{-}R$$

$$P{-}C({=}O){-}\overset{+}{O}H{-}R \longrightarrow P{-}C^+({=}O) + R{-}OH$$

$$P{-}C^+({=}O) \xrightarrow{H_2O} P{-}C({=}O){-}OH + H^+$$

$$P{-}C({=}O){-}O{-}R + OH^- \longrightarrow P{-}C(OH)(O^-){-}O{-}R$$

$$P{-}C(OH)(O^-){-}O{-}R \longrightarrow P{-}C({=}O){-}OH + RO^-$$

$$RO^- \xrightarrow{H_2O} ROH + OH^-$$

Indeed, the first source of H^+ ions can be the acidic chain ends which accumulate into the polymer matrix during hydrolysis and which are expected to induce an autoacceleration of process. In fact these acids are weak and their dissociation is relatively disfavored in polymer matrices of low polarity. In PA11, they have no autocatalytic effect (Jacques et al. 2002). In PET, there is a relatively low autoacceleration which could be due only to the increase of polymer hydrophilicity linked to the accumulation of acid and alcohol groups resulting from hydrolysis (Ballara et al. 1989). The existence of an eventual autocatalysis in UP's is not obvious because it can be masked by osmotic cracking (see below). In contrast, an autocatalytic behavior has been recently observed in aliphatic polyesters of the polyadipate type (Coquillat et al. 2010).

Catalytic effects of H^+ or OH^- ions are, in contrast, clearly observed when the materials are immersed in acidic or basic solutions. It is noteworthy that ions are practically insoluble in nonpolar polymer matrices, so that two types of behavior can be observed:

- In the case of bases, there is only a superficial attack similar to erosion.
- In the case of acids, only undissociated species are soluble in the polymer, so that organic acids can be more active than mineral ones because they are more soluble. Concerning strong acids such as HCl, it has been observed that their effect increases almost exponentially when the pH of the solution decreases (Serpe et al. 1997), that can be explained by the fact that the undissociated form of the acid exists only at low pH value (Ravens 1960, Merdas et al. 2003).

3. Osmotic Cracking

Blistering of boat hulls made of glass fiber/polyester laminates led to worldwide catastrophic consequences on boat industry in the 80's. Blisters result from crack growth under the gel coat, parallel to the free surface. They were first observed by Ashbee and Wyatt (1969). Several interpretations were proposed in the following years, but it was finally recognized that the crack propagation results from an osmotic process (Fedors 1980). The mechanism can be briefly described as follows: If a crack is formed in a subcutaneous region, it is rapidly filled by water coming from the bath. Water dissolves small organic (and eventually inorganic) molecules present in the polymer matrix. Then, the polymer layer separating the crack from the bath acts as a semi-permeable membrane, permeable to water but considerably less permeable to the solutes. Then, an osmotic pressure grows in the crack. According to Van't Hoff (1888):

$$\Delta p = RT\sum_{i=1}^{n} C_i \tag{53}$$

where C_i is the concentration of the ith solute in water.

Above a certain critical value depending on the polymer toughness, this pressure induces crack propagation (Walter and Ashbee 1982, Sargent and Ashbee 1984). Osmotic cracking can be put in evidence by gravimetry on relatively thin samples (Mortaigne et al. 1992) (Figure 11).

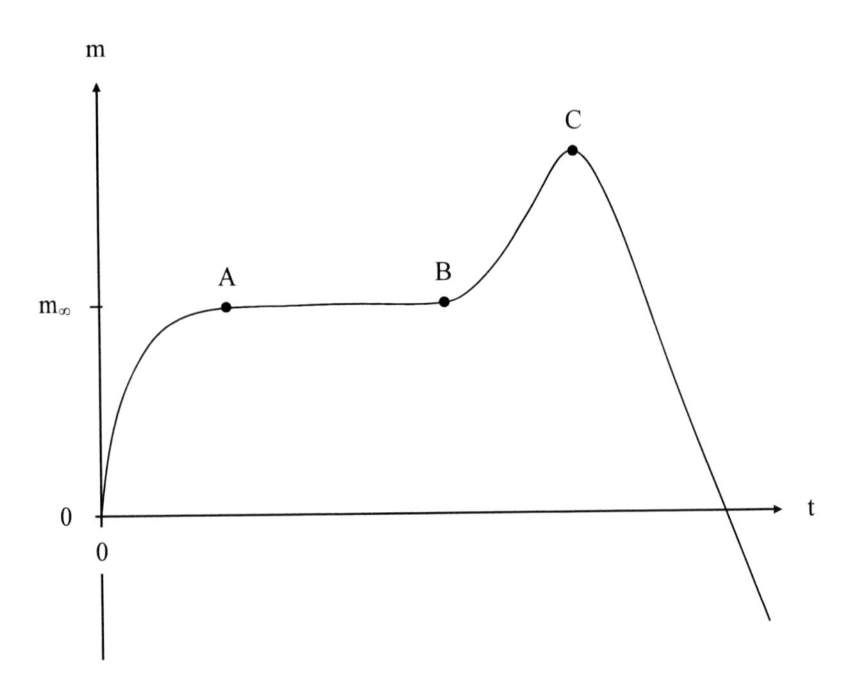

Figure 11. Shape of gravimetric curves in the case of osmotic cracking of thin samples. In A, the sorption equilibrium is reached. In B, cracking begins and increases the capacity of water absorption. In C, the cracks coalesce and the soluble organic molecules are lost in the bath.

In the 80–90's, there were several studies aimed to establish the influence of initial composition (for instance, Abeysinghe et al. 1983) and the polymer structure (for instance, Mortaigne et al. 1992), on the osmotic cracking phenomenon. It was shown that it is faster and faster as:

- The initial concentration of organic molecules (unreacted monomers, initiator fragments, etc …) increases.
- The initial concentration of dangling chains increases, i.e. the prepolymers molar mass decreases.
- The hydrolysis rate increases.

Various explanations were proposed for the crack initiation mechanism. The only one consistent with all the available experimental data was proposed by Gautier et al. (1999) and can be briefly summarized as follows: Hydrolysis on dangling chains (initially present or resulting from hydrolysis events) generates small organic molecules. These latter (acid and alcohol terminated) are more polar than the polymer, so that their solubility in the matrix is limited. When they reach a critical concentration, they demix to form highly polar micro-domains. Since these latter are more hydrophilic than the polymer, they absorb water and the osmotic cracking process begins. It is noteworthy that this process occurs because organic molecules have a smaller diffusivity than water, so that they accumulate in the timescale of hydrolysis. The resistance to osmotic cracking can be reduced by:

- Limiting the initial concentration of small molecules (initiators, catalysts, etc …).

- Reducing the concentration of dangling chains, but the initial prepolymers length must be limited to facilitate processing.
- Reducing hydrolysis rate by: First, a decrease of polymer hydrophilicity by using nonpolar diols such as, for instance, neopentylglycol, and by increasing, when it is possible, the styrene content; Second, a decrease of ester reactivity by an appropriate choice of diacids (isophthalate rather than orthophthalate, etc ...).

Unfortunately, the "weakest point" of the polymer structure is the fumarate ester of which the replacement is difficult to envisage.

THERMAL OXIDATION

1. Introduction: General Aspects of Thermal Oxidation Processes

Thermal oxidation displays three very important characteristics:

i. It results from a radical chain process. An important characteristic of such processes is that it is possible to envisage an inhibition, i.e. a polymer stabilization using, for instance, radical traps (antioxidants).
ii. Radical chains are mainly initiated by hydroperoxide decomposition. This latter reaction is mainly characterized by a low activation energy (typically 80–140 $kJ.mol^{-1}$) linked to the low dissociation energy of the O–O bond (140–160 $kJ.mol^{-1}$). The activation energy of the thermolytic decomposition of polymers in the absence of oxygen is higher than 200 $kJ.mol^{-1}$. This is the reason why, when oxygen is present, oxidation always predominates over all the other reactions, at least in the temperature domain of practical interest, i.e. typically below 200°C in most cases.
iii. Oxidation is kinetically controlled by the oxygen diffusion into the polymer. In other words, in bulk samples, oxidation is restricted to a superficial layer of thickness ℓ. As it has been shown in the third section of this chapter, ℓ can be estimated using a scaling law. Moreover, the temperature effect on oxygen diffusivity D and on oxidation pseudo first rate constant K can be approximated by an Arrhenius law. It can be thus written (Audouin et al. 1994):

$$\ell = (D/K)^{1/2} = \left(\frac{D_0 \exp - E_D/RT}{K_0 \exp - E_K/RT}\right)^{1/2} = \ell_0 \exp - \frac{E_\ell}{RT} \qquad (54)$$

where $\ell_0 = \left(\frac{D_0}{K_0}\right)^{1/2}$ and $E_\ell = \frac{1}{2}(E_D - E_K)$

Since, generally, $E_D < E_K$, the apparent activation energy of ℓ is negative, i.e. ℓ is a decreasing function of temperature, that induces two important consequences:

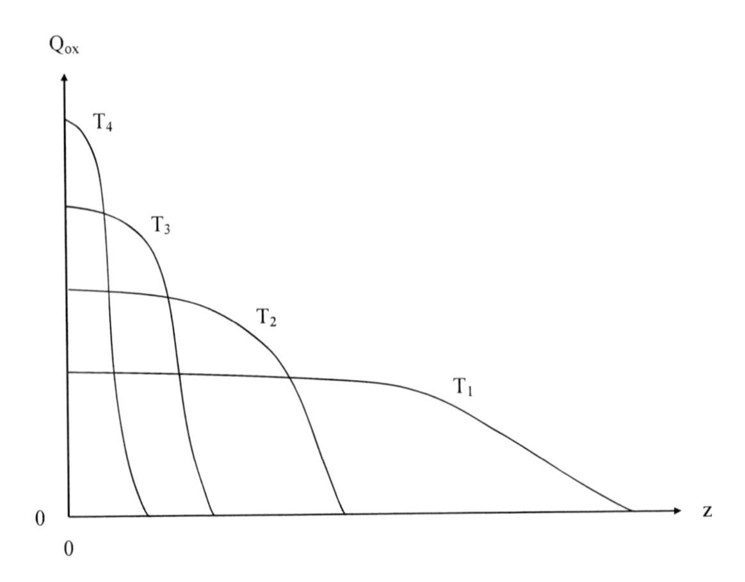

Figure 12. Schematic shape of oxidation profiles at four temperatures: $T_1 < T_2 < T_3 < T_4$.

First, the thickness of oxidation profiles is expected to vary with temperature as schematized in Figure 12.

Q_{ox} is the quantity of grafted oxygen. Z is the depth of the layer in the sample thickness.

It appears thus that if the chosen mode of accelerated ageing is simply to increase temperature, one can expect not only a quantitative change, i.e. an increase of oxidation rate, but also a qualitative change which can have important consequences on certain use properties, for instance mechanical ones.

Second, let us consider the thermolytic decomposition of polymer. It is, generally, a radical process inhibited by oxygen for the reason which can be schematized as follows:

(Th) $P° \rightarrow$ thermal decomposition reactions (k_{Th})

(II) $P° + O_2 \rightarrow PO_2°$ (k_2)

The rate constant of oxygen addition to radicals (k_2) is extremely high (10^8–10^9 l. $mol^{-1}.s^{-1}$) and its activation energy is very low, close to zero. Thus, the reaction (Th) can compete with reaction (II) only at very high temperature and low oxygen concentration. Indeed, in the sample core where oxygen has no access, thermal decomposition can occur. This is the reason why thermal ageing tests made over a large temperature interval, for instance thermogravimetry, lead to an Arrhenius plot having the shape of Figure 13.

One can clearly distinguish two domains: (I) At low temperature, predominating oxidation; (II) At high temperature, predominating thermolytic decomposition.

This characteristic, as well as the one schematized in the previous Figure, shows that ultra-accelerated ageing tests at very high temperature are not pertinent when the objective is to predict the behaviour at service temperature, generally below the glass transition temperature of polymer. In this latter temperature domain, thermolytic decomposition is generally negligible, thermal ageing is mainly due to oxidation (eventually combined with post-cure).

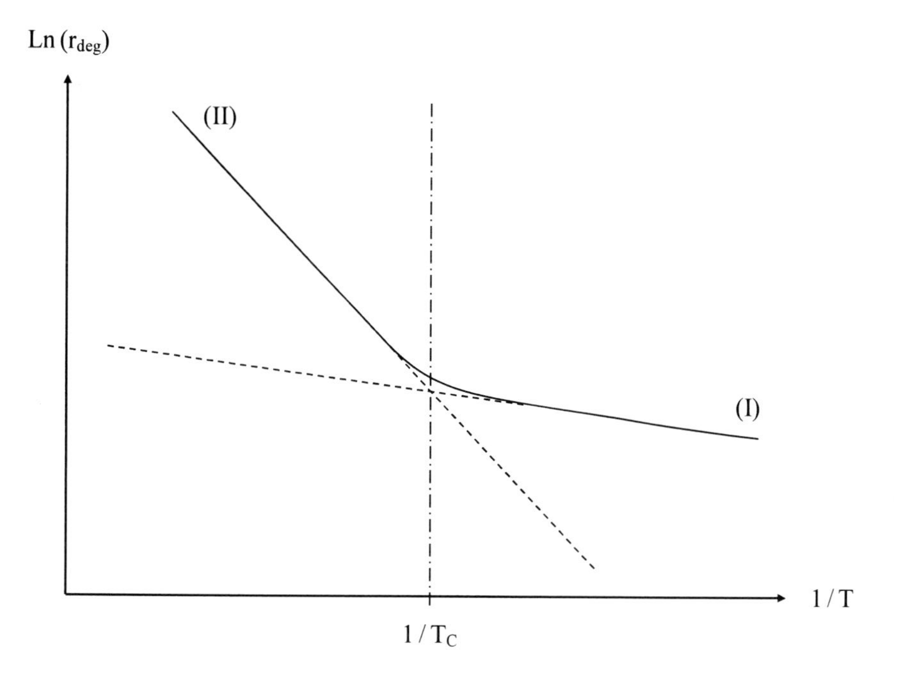

Figure 13. Shape of the Arrhenius plot of polymer degradation over a large temperature interval.

2. Oxidation Mechanisms and Kinetics

The discovery of oxidation radical chain mechanism is often attributed to the English (RAPRA) group of Bolland (1946) although the true pioneer of the discipline is rather Semenov (Nobel Award 1956) (Semenov 1935). The first schemes start from the assumption of an undefined radical source, producing radicals at a rate r_i in a medium where oxygen is in excess and hydroperoxides are stable. The mechanistic scheme is thus:

(I)Radical source→ P° (r_i)
(II)P° + O_2→ PO_2° (k_2)
(III)PO_2° + PH→ POOH + P° (k_3)
(VI)PO_2° + PO_2°→ inactive products (k_6)

The kinetic scheme derived from the mechanism is composed of two equations:

$$\frac{d[P°]}{dt} = r_i - k_2[O_2][P°] + k_3[PH][PO_2°] \quad (55)$$

$$\frac{d[PO_2°]}{dt} = k_2[O_2][P°] - k_3[PH][PO_2°] - 2k_6[PO_2°]^2 \quad (56)$$

Assuming steady-state for radical concentration:

$$\frac{d[P°]}{dt} + \frac{d[PO_2°]}{dt} = 0 \qquad (57)$$

$$\Rightarrow r_i = 2k_6[PO_2°]^2$$

$$\Rightarrow [PO_2°] = \left(\frac{r_i}{2k_6}\right)^{1/2} \qquad (58)$$

Thus:

$$\frac{d[O_2]}{dt} = -k_2[O_2][P°] = -\left(r_i + k_3[PH](r_i/2k_6)^{1/2}\right) \qquad (59)$$

The kinetic chain length Λ is the number of propagation events per initiation event:

$$\Lambda = \frac{k_3[PH](r_i/2k_6)^{1/2}}{r_i} = \frac{k_3[PH]}{(2r_i k_6)^{1/2}} \qquad (60)$$

It is often assumed that $\Lambda >> 1$, so that:

$$\frac{d[O_2]}{dt} \approx -\frac{d[POOH]}{dt} = -k_3[PH]\left(\frac{r_i}{2k_6}\right)^{1/2} \qquad (61)$$

This model was used for a long time to interprete thermal oxidation results, despite the fact that many key assumptions are not appropriate. The most important one is, no doubt, the assumption that hydroperoxides are stable. As a matter of fact, hydroperoxides are highly unstable and decompose into radicals in three main pathways:

– Unimolecular decomposition:

$$POOH \rightarrow PO° + HO° \qquad (k_{1u})$$

– Bimolecular decomposition:

$$POOH + POOH \rightarrow PO° + PO_2° + H_2O \qquad (k_{1b})$$

– Catalyzed decomposition:

$$POOH + M^{i+} \rightarrow PO° + HO^- + M^{(i+1)+} \qquad (k_{1c-})$$

$$POOH + M^{(i+1)+} \rightarrow PO_2° + H^+ + M^{i+} \qquad (k_{1c+})$$

– Balance reaction:

$POOH \rightarrow \frac{1}{2} PO° + \frac{1}{2} PO_2° + \frac{1}{2} H_2O$ (k_{1c})

HO° and PO° radicals are several orders of magnitude more reactive than $PO_2°$ ones. They abstract rapidly hydrogens to the polymer. PO° radicals can also rearrange into P° ones (see below). Thus, the above reactions can be rewritten:

Unimolecular: $POOH \rightarrow 2P°$ (k_{1u})
Bimolecular: $POOH + POOH \rightarrow P° + PO_2°$ (k_{1b})
Catalyzed: $POOH \rightarrow \frac{1}{2} P° + \frac{1}{2} PO_2°$ (k_{1c})

In a first approach, these mechanisms will be not distinguished and it will be written:

$\delta\ POOH \rightarrow \alpha\ P° + \beta\ PO_2°$ (k_1)

Since hydroperoxides are especially unstable, they are expected to play a predominant role in initiation, so that the above mechanistic scheme becomes:

$\delta\ POOH \rightarrow \alpha\ P° + \beta\ PO_2°$ (k_1)
$P° + O_2 \rightarrow PO_2°$ (k_2)
$PO_2° + PH \rightarrow POOH + P°$ (k_3)
$PO_2° + PO_2° \rightarrow$ inactive products (k_6)

The corresponding kinetic scheme must be composed of three equations describing the concentration changes of P°, $PO_2°$ and POOH:

$$\frac{d[P°]}{dt} = \alpha k_1 [POOH]^{\delta} - k_2[O_2][P°] + k_3[PH][PO_2°] \quad (62)$$

$$\frac{d[PO_2°]}{dt} = \beta k_1 [POOH]^{\delta} + k_2[O_2][P°] - k_3[PH][PO_2°] - 2k_6[PO_2°]^2 \quad (63)$$

$$\frac{d[POOH]}{dt} = -\delta k_1 [POOH]^{\delta} + k_3[PH][PO_2°] \quad (64)$$

There is a steady-state when the POOH formation by reaction (III) equilibrates their destruction by reaction (I). Then:

$$\frac{d[POOH]}{dt} = 0 \quad (65)$$

$$\Rightarrow [PO_2°]_S = \frac{\delta k_1}{k_3[PH]} [POOH]_S^{\delta} \quad (66)$$

where the subscript S indicates the steady-state.

Tobolsky and coll. (1950) used an ad hoc assumption to solve the problem: They assumed that there is a steady-state for radical concentration (Equ. 57), so that:

$$(\alpha+\beta)k_1[POOH]^{\delta} = 2k_6[PO_2°]^2$$

$$=> [PO_2°]_S = \left(\frac{\alpha+\beta}{2k_6}k_1\right)^{1/2}[POOH]^{\delta/2} \quad (67)$$

For the uncatalyzed oxidation, $\alpha + \beta = 2$ in both cases, so that:

$$[POOH]_S = \left(\frac{k_3[PH]}{\delta(k_1k_6)^{1/2}}\right)^{2/\delta} \quad (68)$$

$$\text{And } [PO_2°]_S = \frac{k_3[PH]}{\delta k_6} \quad (69)$$

The behavior during the transient depends on the molecularity (δ) of the POOH decomposition and the initial POOH concentration $[POOH]_0$. It is schematized in Figure 14.

The kinetics display an induction period which can be defined as the time to reach the steady-state. Its characteristics differ with the mode of POOH decomposition (Audouin et al. 2000). In the case of unimolecular POOH decomposition, the conversion increases pseudo-exponentially. The induction time t_i is a decreasing function of the initial POOH concentration displaying an asymptote:

$$t_i \rightarrow \frac{3}{k_{1u}} \text{ when } [POOH]_0 \rightarrow 0 \quad (70)$$

In the case of bimolecular POOH decomposition, the conversion increases abruptly at the end of the induction period and the induction time increases logarithmically when the initial POOH concentration decreases. At low $[POOH]_0$ values:

$$t_i \rightarrow \frac{3}{k_3[PH](k_{1b}/2k_6)^{1/2}} Ln\frac{[POOH]_S}{[POOH]_0} \quad (71)$$

where $[POOH]_S$ is the steady-state POOH concentration.

In fact, indeed, both POOH decomposition modes coexist and the unimolecular mode becomes always predominant at low POOH concentration. This is the reason why the t_i variation with $[POOH]_0$ displays always the shape of Figure 15.

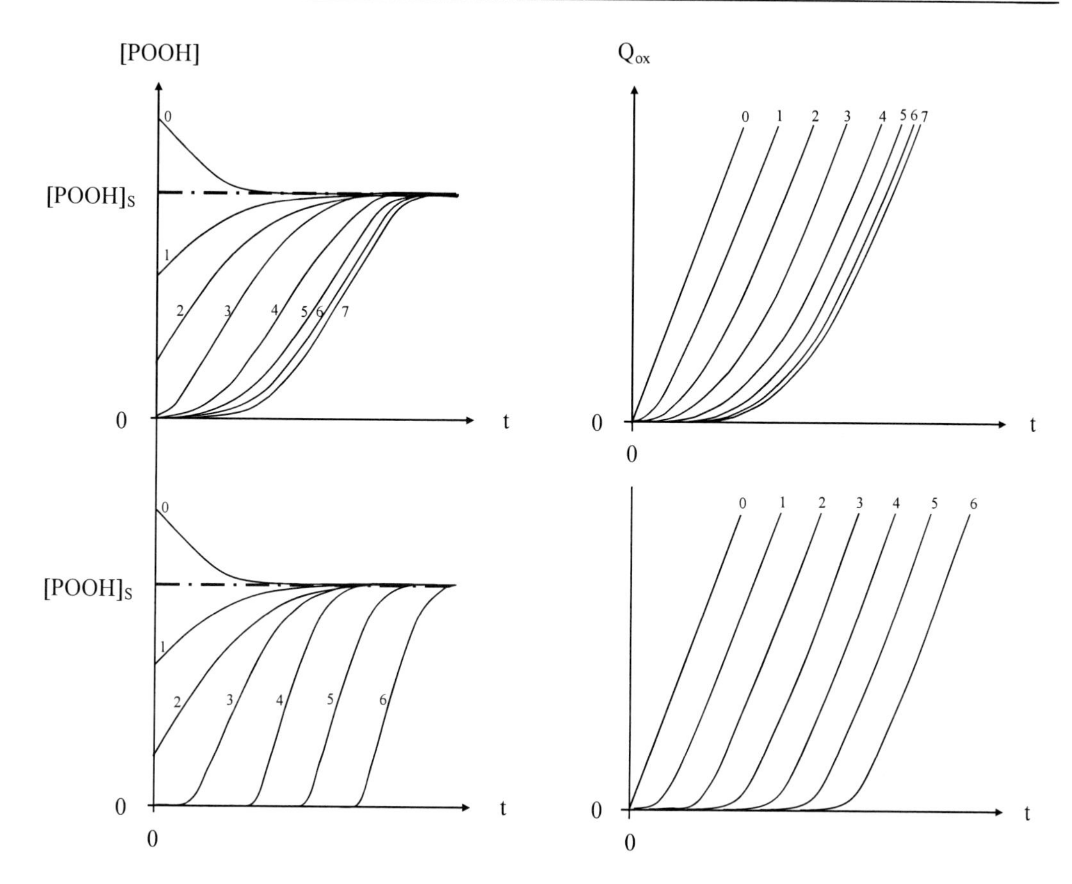

Figure 14. Shape of theoretical kinetic curves of hydroperoxide build-up (left) or stable oxidation products build-up (right) in the case of unimolecular (up) and bimolecular POOH decomposition (bottom), for various initial hydroperoxide concentrations $[POOH]_0$ decreasing from 0 to 7.

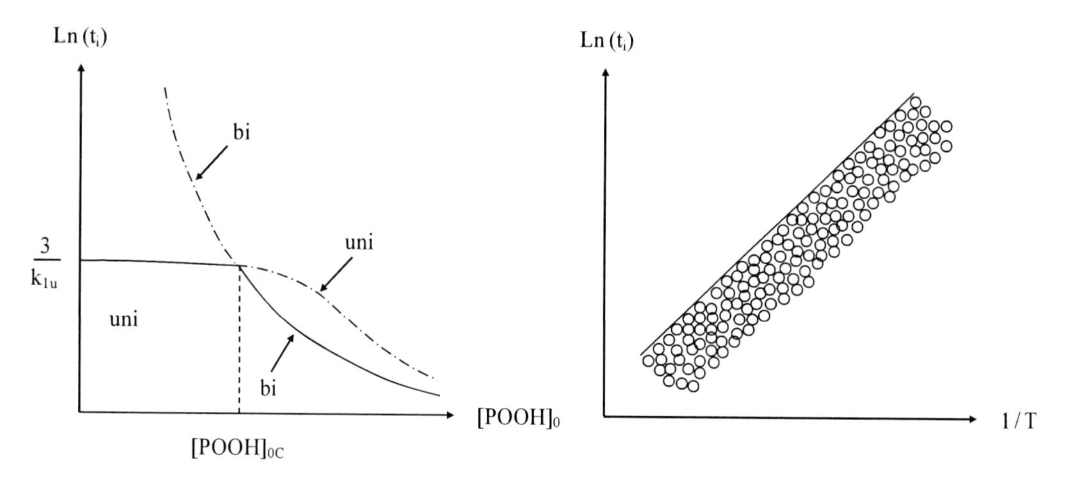

Figure 15. Effect of the initial POOH concentration on the induction period (left) in the case of pure unimolecular (uni) and pure bimolecular (bi) POOH decomposition. The full line represents the true dependence. Consequence of this dependence on the scatter of induction time values represented on an Arrhenius plot (right).

One can define a critical POOH concentration: $[POOH]_{0C}$ separating the unimolecular regime from the bimolecular one. Schematically, when $[POOH]_0 \leq [POOH]_{0C}$, the induction time is maximum and independent of $[POOH]_0$. When $[POOH]_0 > [POOH]_{0C}$, the induction time decreases logarithmically with $[POOH]_0$. This dependence explains the characteristic feature of Arrhenius plots of compiled t_i values coming from various sources (Richaud et al. 2008). The clouds of points display a relatively sharp upper boundary corresponding to the asymptotic value of t_i. They display also a relatively low scatter linked to the fact that, in bimolecular regime, t_i varies slowly with $[POOH]_0$.

It appears thus necessary to take into account both uni and bimolecular POOH decomposition processes in kinetic model, that imposes the use of numerical methods for their resolution.

3. Kinetic Modeling

The following mechanistic scheme is valid, at least in a first approach, in a wide variety of cases:

(Iu) $POOH \rightarrow 2P° + H_2O + \gamma_1 S + \upsilon V$ (k_{1u})
(Ib) $POOH + POOH \rightarrow P° + PO_2° + H_2O + \gamma_1 S + \upsilon V$ (k_{1b})
(II) $P° + O_2 \rightarrow PO_2°$ (k_2)
(III) $PO_2° + PH \rightarrow POOH + P°$ (k_3)
(IV) $P° + P° \rightarrow$ inactive products $+ \gamma_4 X$ (k_4)
(V) $P° + PO_2° \rightarrow$ inactive products (k_5)
(VI) $PO_2° + PO_2° \rightarrow$ inactive products $+ O_2$ (k_6)

The following system of differential equations can be derived:

$$\frac{d[P°]}{dt} = 2k_{1u}[POOH] + k_{1b}[POOH]^2 - k_2[O_2][P°] + k_3[PH][PO_2°] - 2k_4[P°]^2 \quad (72)$$

$$-k_5[P°][PO_2°]$$

$$\frac{d[PO_2°]}{dt} = k_{1b}[POOH]^2 + k_2[O_2][P°] - k_3[PH][PO_2°] - k_5[P°][PO_2°] - 2k_6[PO_2°]^2 \quad (73)$$

$$\frac{d[POOH]}{dt} = -k_{1u}[POOH] - 2k_{1b}[POOH]^2 + k_3[PH][PO_2°] \quad (74)$$

$$\frac{d[PH]}{dt} = -\mu_{1u}k_{1u}[POOH] - \mu_{1b}k_{1b}[POOH]^2 + k_3[PH][PO_2°] \quad (75)$$

$$\frac{d[O_2]}{dt} = D\,\frac{\partial^2[O_2]}{\partial z^2} - k_2[O_2][P°] + k_6[PO_2°]^2 \tag{76}$$

$$\frac{dS}{dt} = \gamma_1 k_{1u}[POOH] + \gamma_1 k_{1b}[POOH]^2 \tag{77}$$

$$\frac{dX}{dt} = \gamma_4 k_4[P°]^2 \tag{78}$$

γ_1 and γ_4, μ_1 and μ_2 are dimensionless yield values, generally lower than unity.

V is the "average volatile molecule" of which the molar mass M_V corresponds to the number average molar masses of true volatile molecules. υ is the number average yield of volatile molecules.

The boundary conditions are:

At t = 0, for all the sample thickness: $[P°] = 0$, $[PO_2°] = 0$, $[POOH] = [POOH]_0$ and $[O_2] = [O_2]_S$ (equilibrium oxygen concentration for the oxygen partial pressure under consideration).

At t > 0: $[O_2] = [O_2]_S$ at superficial layers: z = 0 and z = L.

These equations allow to determine the weight variation. In an elementary layer, it can be written (Colin et al. 2002):

$$\left(\frac{dm}{dt}\right)_z = \frac{32}{\rho}\frac{d[O_2]}{dt} - \frac{18}{\rho}\frac{d[H_2O]}{dt} - \frac{\upsilon M_V}{\rho}\frac{d[V]}{dt} \tag{79}$$

So that:

$$\left(\frac{dm}{dt}\right)_z = \frac{32}{\rho}k_2[O_2]_z[P°]_z - \left[\frac{32}{\rho}k_6[PO_2°]_z^2 + \left(\frac{18}{\rho} + \frac{\upsilon M_V}{\rho}\right)\left(k_{1u}[POOH]_z + k_{1b}[POOH]_z^2\right)\right] \tag{80}$$

It is noteworthy that mass gain (linked to oxygen reaction with radicals) occurs only in propagation processes, whereas mass loss occurs exclusively in initiation or termination processes. It results that the balance weight loss / weight gain must be sharply linked to the kinetic chain length Λ. An interesting property of the mechanistic scheme under consideration is that Λ decreases continuously during the induction time to reach a value close to unity in steady state. One can generally expect three main kinds of gravimetric behaviors (in isothermal conditions) as schematized in Figure 16.

When the induction period phenomenon is well marked, the initial kinetic chain length is high, mass gain predominates over mass loss in the early period of exposure. When the end of the induction period approaches, the kinetic length decreases at a point where mass loss becomes predominant. In steady-state, the mass loss rate is constant. At long term, however, the whole oxidation rate decreases as a consequence of substrate consumption, and the mass loss rate decreases (regime I).

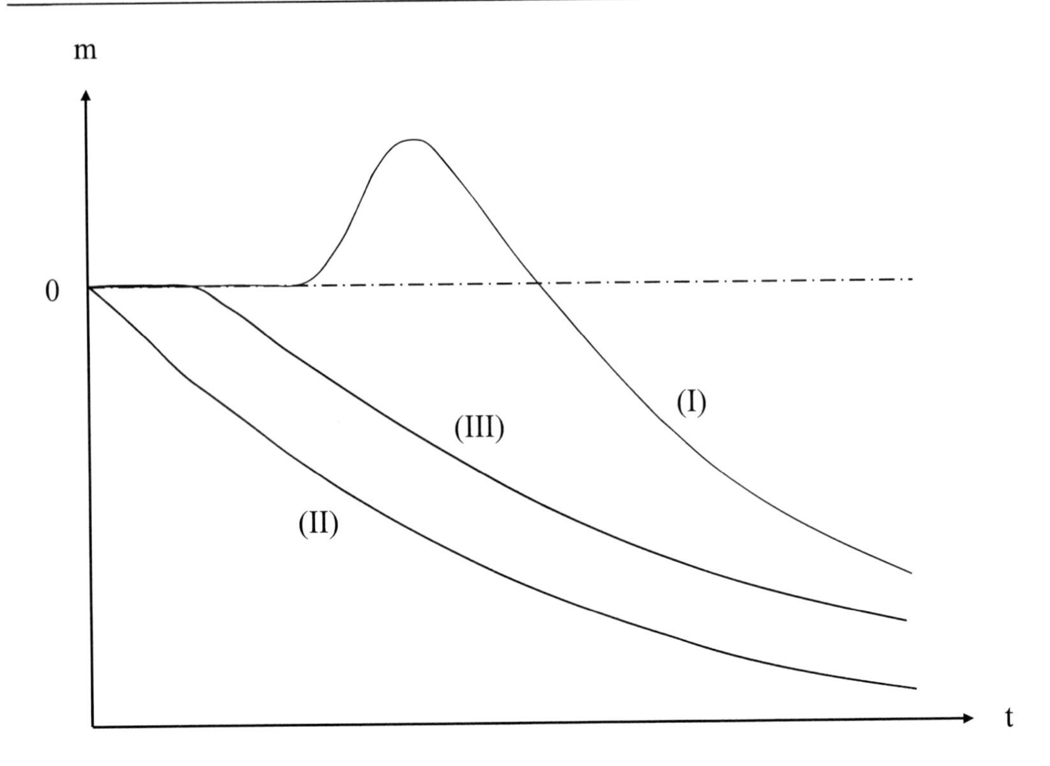

Figure 16. Shape of isothermal gravimetric curves for polymer oxidation (see text).

In the absence of significant induction period, the initial kinetic chain length is low and mass loss predominates as soon as exposure begins (regime II).

Indeed, an intermediary behaviour (regime III) can be observed. It is noteworthy that a given polymer can degrade in regime (I) at low temperature and in regime (II) at high temperature. In regime (I), the amplitude of mass gain depends on the size and yield of volatile molecules.

The ability of the above kinetic model to simulate gravimetric curves having a wide variety of shapes, including non monotonous ones, can be considered as a strong argument in favour of the model validity, especially if it predicts at the same time oxidation thickness profiles.

4. Spontaneous Cracking Induced by Oxidation

Oxidation in a superficial layer induces a weight change, as seen above, and, at least in polymers based predominantly on C and H atoms, a density (ρ) increase due to the fact that oxygen atom is heavier than C or H atoms (Pascault et al. 2002). There is thus a volume change of the oxidized layer:

$$\frac{\Delta v}{v} = \frac{\Delta m}{m} - \frac{\Delta \rho}{\rho} \tag{81}$$

Since $\Delta\rho / \rho$ is generally positive, it can be seen that the oxidized layer tends to shrink if $\Delta m / m$ is negative, i.e. when weight loss predominates over weight gain. Since shrinkage is hindered by the adhesion of the oxidized layer to the sample core, a stress state is induced. At the same time, the oxidized layer undergoes embrittlement (section 2). One can thus schematize the sample behaviour as follows: The local stress σ in the oxidized layer increases with the exposure time, whereas the ultimate stress σ_u decreases as a result of polymer degradation. Cracking must occur when σ becomes equal to σ_u. Gravimetric behaviours (I) and (II) are expected to induce two distinct cracking behaviours (Figure 17).

In the case of gravimetric behaviour of type I, stresses develop only at the end of the induction period, when mass loss begins. Cracks can thus initiate only after the end of induction period. But chain scission occurs during the induction period, so that the oxidized layer can be noticeably embrittled when cracks initiate. As a result, cracks are expected to propagate rapidly through the oxidized layer. In the case of gravimetric behaviour of type II, stresses develop as soon as exposure begins. One can thus imagine cases where cracking begins at a moment where the toughness of the oxidized layer remains relatively high. Then, cracks are expected to propagate comparatively slowly, at a rate mainly determined by the local changes of polymer toughness. Amine crosslinked epoxies are typical of this last behaviour, whereas poly(bismaleimide) are typical of the former one (Colin et al. 2005). In composites, indeed, matrix shrinkage is expected to induce also fiber debonding (Lafarie-Frenot et al. 2010).

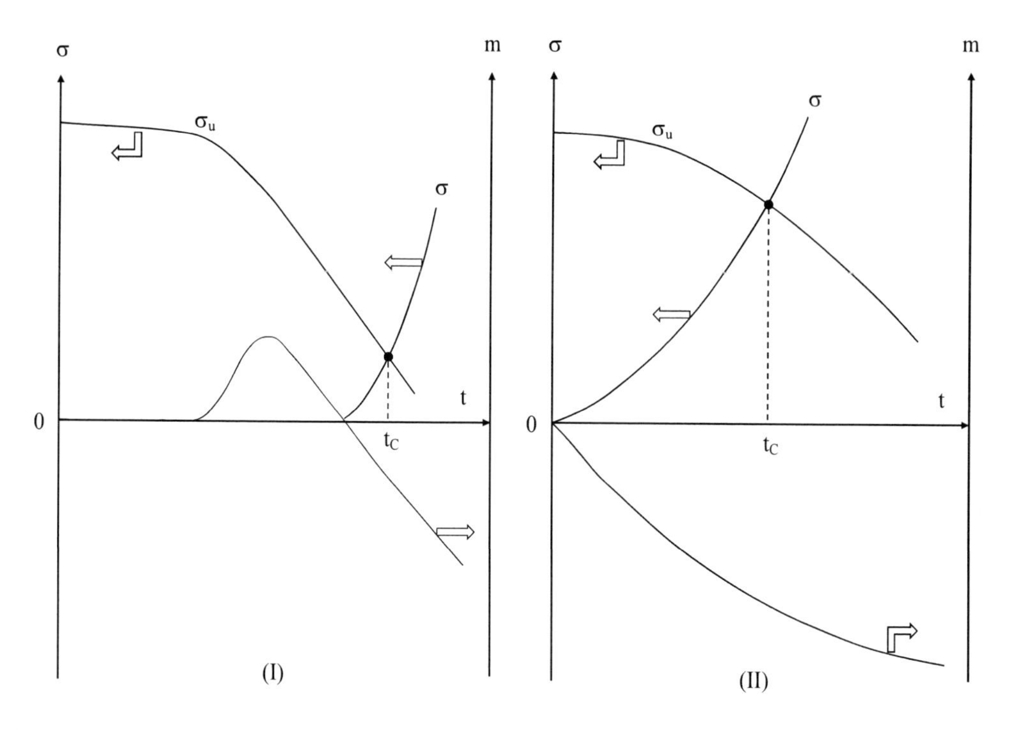

Figure 17. Spontaneous cracking conditions in the case of gravimetric behaviour of type (I) (left) and (II) (right).

CONCLUSION

The main modes of physico-chemical ageing for organic matrix composites are hydrolysis (especially important for the most common class of resin composites, i.e. glass fibers/unsaturated polyesters) and thermal oxidation (responsible for superficial cracking and embrittlement in many parts of aeronautic engines or structural parts of supersonic aircrafts). Both hydrolysis and oxidation processes have in common to induce chain scission, this latter being responsible for embrittlement at low conversions. The quantitative relationships between macromolecular structure and number of chain scissions on one side, and between macromolecular structure and fracture properties on the other side, are recalled. Embrittlement mechanisms are relatively well elucidated in the case of linear polymers, but remain an open problem in thermosets.

Hydrolysis and oxidation have also in common to have their kinetics eventually controlled by the diffusion of a small molecule (water or oxygen) coming from environment. This control leads to the confinement of degradation processes in more or less thick superficial sample layers, that carries important consequences on use properties. The characteristics of oxidized layers can be predicted from equations coupling diffusion and reaction rate processes. These equations are relatively simple in the case of hydrolysis, but can be highly complex in the case of oxidation where they need several simplifying assumptions to be analytically solved.

Hydrolysis mechanisms and kinetics were established long time ago. Complications in kinetic modelling can appear in the presence of catalytic processes, especially because the solubility of catalytic species (ions) in polymer matrices is difficult to determine. In the case of unsaturated polyesters, osmotic cracking induced by hydrolysis is a very important process, responsible for blistering of boats, swimming pools, etc … Its mechanism is now relatively well understood. Oxidation mechanisms were established one half century ago, but kinetic modelling remained a challenging objective for a long time, especially when reaction−diffusion coupling is needed.

The recent introduction of numerical methods allowed to suppress many oversimplifying assumptions and opens now the way to efficient lifetime predictions. The first decade of 21th century will probably appear, in the future, as the transition period between the empirical area of polymer ageing and the modern area where lifetime prediction will be a fully scientific approach.

REFERENCES

Abeysinghe H.J., Ghotra J.S., Pritchard G., Substances contributing to the generation of osmotic pressure in resins and laminates, *Composites* 14, 57-61 (1983).

Ashbee K.H.G., Wyatt R.C., Water damage in glass fibre/resin composites, *Proceedings of the Royal Society* A312, 553-564 (1969).

Audouin L., V. Langlois, J. Verdu and J.C.M. De Bruijn, Role of oxygen diffusion in polymer ageing: kinetic and mechanical aspects, *Journal of Materials Science* 29(3), 569-583 (1994).

Audouin L., Achimsky L., Verdu J., "Modelling of Hydrocarbon Polymer Oxidation", in *Handbook of Polymer Degradation*, 2nd edition, S. Halim Hamid ed., Marcel Dekker, New-York, Chap. 20, pp. 727-763, 2000.

Ballara A., Verdu J., Physical aspects of the hydrolysis of polyethylene terephthalate, *Polymer Degradation and Stability* 26, 361-374 (1989).

Belan F., Bellenger V., Mortaigne B., Verdu J., Relationship between the structure and hydrolysis rate of unsaturated polyester prepolymers, *Polymer Degradation and Stability* 56, 301-309 (1997).

Bellenger V., Verdu J., Ganem M., Mortaigne B., Styrene crosslinked vinylesters. II: Water sorption, water diffusion and cohesive properties, *Polymer and Polymer Composites* 2(1), 17-25 (1994).

Bellenger V., Ganem M., Mortaigne B., Verdu J., Lifetime prediction in the hydrolytic ageing of polyesters, *Polymer Degradation and Stability* 49(1), 91-97 (1995).

Bicerano J., *Prediction of Polymer Properties*, 3rd edition, Marcel Dekker, New-York, 2002.

Bolland J.L., Kinetic studies in the chemistry of rubber and related materials. I. The thermal oxidation of ethyl linoleate, *Proceedings of the Royal Society* A186, 218-236 (1946).

Carter H.G., Kibler K.G., Lagmuir-type model for anomalous moisture diffusion in in composite resins, *Journal of Composite Materials* 12, 118-131 (1978).

Charlesby A., Pinner S.H., Analysis of the solubility. Behaviour of irradiated polyethylene and other polymers, *Proceedings of the Royal Society* A249, 367-386 (1959).

Colin X., Marais C., Verdu J., Kinetic modeling and simulation of gravimetric curves: application to the oxidation of bismaleimide and epoxy resins, *Polymer Degradation and Stability* 78(3), 545-553 (2002).

Colin X., Mavel A., Marais C., Verdu J., Interaction between cracking and oxidation in organic matrix composites, *Journal of Composite Materials* 39(15), 1371-1389 (2005).

Coquillat M., Verdu J., Colin X., Audouin L., Nevière R., Thermal oxidation of polybutadiene, *Polymer Degradation and Stability* 92(7), 1326-1333, 1334-1342 and 1343-1349 (2007).

Coquillat M, Audouin L, Verdu J, to be published in 2010.

Cunliffe A.V., Davis A., Photo-oxidation of thick polymer samples. Part II: The influence of oxygen diffusion on the natural and artificial weathering of polyolefins, *Polymer Degradation and Stability* 4(1), 17-37 (1982).

Crank J., Park G.S., *Diffusion in polymers*, Academic Press, London, 1968.

Crawford C.D., Lesser A.J., Brittle to ductile fracture toughness mapping on controlled epoxy networks, *Polymer Engineering and Science* 39(2), 385-392 (1999).

Derrien K., Gilormini P., Interaction Between Stress and Diffusion in Polymers, in *Proceedings of the DSL 2006 Conference, Defect and Diffusion Forum* 258/260, pp. 447-452, 2006.

Di Benedetto A.T., Scola D.A., Characterization of S-glass/polysulfone adhesive failure using ion scattering spectroscopy and secondary ion mass spectrometry, *Journal of Colloid and Interface Science* 74(1), 150-162 (1980).

Di Benedetto A.T., Lex P.J., Evaluation of surface treatments for glass fibers in composite materials, *Polymer Engineering and Science* 29(8), 543-555 (1989).

Di Marzio E.A., On the second-order transition of rubber, *Journal of Research of the National Bureau of Standards: Section A: Physics and Chemistry* 68, 611-617 (1964).

Fayolle B., Audouin L., Verdu J., Initial steps and embrittlement in the thermal oxidation of stabilized polypropylene films, *Polymer Degradation and Stability* 75(1), 123-129 (2002).

Fayolle B., E. Richaud, X. Colin and J. Verdu, Review: Degradation-induced embrittlement in semi-crystalline polymers having their amorphous phase in rubbery state, *Journal of Materials Science* 43, p. 6999-7012 (2008).

Fayolle B., Verdu J., Piccoz D., Dahoun A., Hiver J.-M., G'Sell C., Thermooxidative ageing of polyoxymethylene. Part 2: Embrittlement mechanisms, *Journal of Applied Polymer Science* 111, 469-475 (2009).

Fedors R.F., Osmostic effects in water absorption by polymers, *Polymer* 21, 207-212 (1980).

Fetters L.J., Lohse D.J., Graessley W.W., Chain dimensions and entanglement spacings in dense macromolecular systems, *Journal Polymer Science: Part B: Polymer Physics* 37, 1023-1033 (1999).

Fox T.G., Flory P.J., The glass temperature and related properties of polystyrene. Influence of molecular weight, *Journal of Polymer Science* 14(75), 315-319 (1954).

Flory P.J., *Principles of Polymer Chemistry*, Cornell University Press, Ithaca, New York, 1953.

Ganem M., Mortaigne B., Bellenger V., Verdu J., Hydrolytic ageing of vinylester materials. Part 1: Ageing of prepolymers and model compounds, *Polymer Networks and Blends* 4(2), 87-92 (1994).

Gaudichet-Maurin E., Thominette F., Verdu J., Water sorption characteristics in moderately hydrophilic polymers. Part 1: Effect of polar groups concentration and temperature in water sorption in aromatic polysulphones, *Journal of Applied Polymer Science* 109(5), 3279-3285 (2008).

Gautier L., Mortaigne B., Bellenger V., Verdu J., Osmotic cracking nucleation in hydrothermal-aged polyester matrix, *Polymer* 41(7), 2481-2490 (1999).

Greco R., Ragosta G., Influence of molecular weight on fracture toughness and fractography of glassy polymers, *Plastics and Rubber Processing and Applications* 7(3), 163-171 (1987).

Hoh K.-P., Ishida H., Koenig J.L., Multi-nuclear NMR spectroscopic and proton NMR imaging studies on the effect of water on the silane coupling agent/matrix resin interface in glass fiber-reinforced composites, *Polymer Composites* 11(3), 192-199 (1990).

Hopfenberg H.B., *Permeability of plastic films and coatings in gases, vapours and liquids*, Plenum Press, New York, 1978.

Ishida H., Koenig J.L., A Fourier-transform infrared spectroscopic study of the hydrolytic stability of silane coupling agents on E-glass fibers, *Journal of Polymer Science: Part B: Polymer Physics* 18, 1931-1943 (1980).

Ishida H., A review of recent progress in the studies of molecular and microstructure of coupling agents and their functions in composites, coatings and adhesive joints, *Polymer Composites* 5(2), 101-123 (1984).

Jacques B., Werth M., Merdas I., Thominette F., Verdu J., Hydrolytic ageing of polyamide 11. I. Hydrolysis kinetics in water, *Polymer* 43(24), 6439-6447 (2002).

Kaelble D.H., Dynes P.J., Crane L.N., Maus L., Interfacial mechanisms of moisture degradation in graphite-epoxy composites, *The Journal Adhesion* 7(1), 25-54 (1975).

Kaelble D.H., Dynes P.J., Maus L., Hygrothermal ageing of composite materials. Part 1: Interfacial aspects, *The Journal Adhesion* 8(2), 121-144 (1976).

Kausch H.H., Heymans N., Plummer C.F., Decroly P., *Matériaux Polymères. Propriétés Mécaniques et Physiques*, Presses Polytechniques et Universitaires Romandes, Lausanne, Switzerland, 2001.

Kennedy M.A., A.J. Peacock and L. Mandelkern, Tensile properties of crystalline polymers: linear polyethylene, *Macromolecules* 27(19), 5297-5310 (1994).

Kondo K., Taki T., Moisture diffusivity of unidirectional composites, *Journal of Composite Materials* 16(2), 82-93 (1982).

Lafarie-Frenot.M.C, Grandidier.J.C, Gigliotti.M, Olivier.L, Colin.X, Verdu.J and Cinquin.J. *Polym. Degrad. Stab.* (2010), 95, 965.

Le Huy H.M., Bellenger V., Verdu J., Paris M., Thermal oxidation of anhydrode cured epoxies. III. Effect on mechanical properties, *Polymer Degradation and Stability* 41(2) 149-156 (1993).

McCall D.W., Douglass D.C., Blyler Jr. L.L., Johnson G.E, Jelinski L.W., Bair H.E., Solubility and diffusion of water in low-density polyethylene, *Macromolecules* 17(9), 1644-1649 (1984).

McMahon W., Birdsall H.A., Johnson G.A., Camilli C.J., Degradation studies of polyethylene terephthalate, *Journal of Chemical and Engineering Data* 4, 57-79 (1959).

Merdas I., Thominette F., Tcharkhtchi A., Verdu J., Factors governing water absorption by composite matrices, *Composites Science and Technology* 62, 487-492 (2002).

Merdas I., Thominette F., Verdu J., Hydrolytic ageing of polyamide 11. Effect of carbon dioxide on polyamide 11 hydrolysis, *Polymer Degradation and Stability* 79, 419-425 (2003).

Meyer A., Jones W., Lin Y. Kranbuehl D., Characterizing and modelling the hydrolysis of polyamide 11 in a pH 7 water environment, *Macromolecules* 35, 2784-2798 (2002).

Mortaigne B., Bellenger V., Verdu J., Hydrolysis of styrene crosslinked unsaturated polyesters. Effect of chain ends, *Polymer Networks and Blends* 2, 187-195 (1992).

Olivier L., C. Baudet, D. Bertheau, J.-C. Grandidier and M.-C. Lafarie-Frenot. Development of experimental, theoretical and numerical tools for studying thermo-oxidation of CFRP composites. *Composites: Part A: Applied Science and Manufacturing* 40, 1008-1016 (2009).

Pascault J.-P., Sautereau H., Verdu J., Williams R.J.J., *Thermosetting Polymers*, Marcel Dekker, New-York, Chap. 10, pp. 282-321 and Chap. 14, pp. 420-467, 2002.

Pitkethly M.J., Favre J.-P., Gaur U., Jakubowski J., Mudrich S.F., Caldwell D.L., Drzal L.T., Nardin M., Wagner H.D., Di Landro L., Hampe A., Armistead J.P., Desaeger M., Verposet I., *Composites Science and Technology* 48, 205-214 (1993).

Pryde C.A., Kelleher P.G., Hellman M.Y., Wentz R.P., The hydrolytic stability of some commercially available polycarbonates, *Polymer Engineering and Science* 22(6), 370-375 (1982).

Rasoldier N., Colin X., Verdu J., Bocquet M., Olivier L., Chocinski-Arnault L., Lafarie-Frenot M.-C., *Composites: Part A: Applied Science and Manufacturing* 39, 1522-1529 (2008).

Ravens D.A.S., The Chemical reactivity of poly(ethyelene terephthalate). Heterogeneous hydrolysis by hydrochloric acid, *Polymer* 1, 375-383 (1960).

Richaud E., Colin X., Fayolle B., Verdu J., Induction period in the low temperature thermal oxidation of saturated hydrocarbons: Example of polyethylene, *International Journal of Chemical Kinetics* 40(12), 769-777 (2008).

Richaud E., Ferreira P., Audouin L., Colin X., Verdu J., Monchy-Leroy C., Radiochemical ageing of poly(ether ether ketone), *European Polymer Journal* 46, 731-743 (2010).

Rosen M.R., Goddard E.D., FDT: a technique for direct study of water attack at the silane-fiber interface, *Polymer Engineering and Science* 20, 413-425 (1980).

Roy S., Singh S., Analytical modelling of orthotropic diffusivities in a fiber reinforced composite with discontinuities using homogenization, *Composites Science and Technology*, in press (2010).

Saito O., On the effects of high energy radiation to polymers. I. Crosslinking and degradation, *Journal of the Physical Society of Japan* 13(2), 198-206 (1958a).

Saito O., On the effects of high energy radiation to polymers. II. End-linking and gel fraction, *Journal of the Physical Society of Japan* 13(12), 1451-1464 (1958b).

Salmon L., Thominette F., Pays M.-F., Verdu J., Hydrolytic aging of polysiloxane networks modelling the glass fiber epoxy-amine interphase, *Composites Science and Technology* 57, 1119-1127 (1997).

Sargent J.P., Ashbee K.H.G., Very slow crack growth during osmosis in epoxy and in polyester resins, *Journal of Applied Polymer Science* 29(3), 809-822 (1984).

Schradder M.E., Block A., Tracer study of kinetics and mechanism of hydrolytically induced interfacial failure, *Journal of Polymer Science: Part C: Polymer Letters* 34, 281-291 (1971).

Schutte L.L., Mc Donough W., Shioya M., Mcauliffe M., Greenwood M., The use of a single-fibre fragmentation test to study environmental durability of interfaces/interphases between DGEBA/mPDA exposy and glass fibre: the effect of moisture, *Composites* 25(7), 617-624 (1994).

Seguchi T., Arakawa K., Hayakawa W., Watanabe Y., Kuryama I., Radiation induced oxidative degradation of polymers. I. Oxidation region in polymer films irradiated in oxygen under pressure, *Radiation Physics and Chemistry* 17(4), 195-201 (1981).

Seguchi T., Arakawa K., Hayakawa W., Watanabe Y., Kuryama I., Radiation induced oxidative degradation of polymers. II. Effects of radiation on swelling and gel fraction of polymers, *Radiation Physics and Chemistry* 19(4), 321-327 (1982).

Semenov N.N., *Chemical kinetics and chain reactions*, Oxford University Press, London, 1935.

Serpe G., Chaupart N., Verdu J., Ageing of polyamide 11 in acid solutions, *Polymer* 38(8), 1911-1917 (1997).

Shen C.H., Springer G.S., Moisture absorption and desorption of composite materials, *Journal of Composite Materials* 10(1), 2-20 (1976).

Tcharkhtchi A., Bronnec Y., Verdu J., *Polymer* 41(15), Water absorption characteristics of diglycidylether of butane diol-3,5-2,4-diaminotoluene networks, 5777-5785 (2000).

Thomason J.L., Investigation of composite interphase using dynamic mechanical analysis: artifacts and reality, *Polymer Composites* 11(2), 105-113 (1990).

Thominette F., Gaudichet-Maurin E., Verdu J., Effect of structure on water diffusion in moderately hydrophilic polymers, in *Proceedings of the DSL 2006 Conference, Defect and Diffusion Forum* 258/260, pp. 442-446, 2006.

Tobolsky A.V., Metz D.J., Mesrobian R.B., Low temperature autoxidation of hydrocarbons: the phenomenon of maximum rates, *Journal of American Chemical Society* 72(2), 1942-1952 (1950).

Van Krevelen D.W. and Hoftyzer P.J., *Properties of Polymers. Their Estimation and Correlation with Chemical Structure*, 3rd edition, Elsevier, Amsterdam, Chap. 18, pp. 412-413, 1976.

Van't Hoff J., The function of osmotic pressure in the analogy between solutions and gases, *Philosophical Magazine* 26(5), 81-105 (1888).

Walter E., Ashbee K.H.G., Osmosis in composites materials, *Composites* 13(4), 365-368 (1982).

In: Resin Composites: Properties, Production and Applications ISBN: 978-1-61209-129-7
Editor: Deborah B. Song

Chapter 9

POSTERIOR COMPOSITE RESIN RESTORATION: NEW TECHNOLOGY, NEW TRENDS

***Matheus Coelho Bandéca**, *Victor Rene Grover Clavijo, Luis Rafael Calixto, and José Roberto Cury Saad*[1]**
Restorative Dentistry, School of Dentistry, State University of Sao Paulo, Araraquara, Brazil.

ABSTRACT

The composites have been widely used over the last years, providing highly aesthetic restorations. The silorane composite is a new technology that replaces the conventionally used methacrylate resin matrix within conventional dental composite, thereby providing lower polymerization shrinkage, excellent marginal integrity, up to 9 minutes operatory light stability and low water sorption that substantially decreased exogenic staining. These qualities of the composite alleviate clinical problems such as marginal staining, microleakage, secondary caries, enamel micro-cracks and post-operative sensitivity. Only the specific self-etching adhesive must be applied to bond the silorane composite to the tooth enamel and dentin. The technique of silorane composite is different from that used in those methacrylate composites, which becomes important to prevent problems during its use. This paper aims to describe properties, application technique of new silorane-based composites and the difference between methacrylate- and silorane-based composites. After reading this paper, the reader will be able to apply such technique in their daily practice with maximum performance of these restorative materials.

* Address: Matheus Coelho Bandéca,
Rua Humaitá, 170,
Araraquara, Brazil, 14801-385
E-mail to: Matheus.Bandéca@utoronto.ca
www.matheusbandeca.com.br

INTRODUCTION

The challenge of more efficient restorative materials in terms of aesthetics and functionality leads to the emergence of new materials. The amalgam has been used in dentistry since 1826 and has excellent physical and mechanical properties, being highly resistant to wear and masticatory efforts and, to insoluble oral fluids. Another point is its easy handling and insertion, low cost, besides having a low sensitivity technique [2-5]. However, it is becoming increasingly less utilized as it compromises aesthetics and does not provide adhesion to tooth structure. Besides, it contains mercury in its composition, which may be absorbed by the organism [6-9].

The silicate cement supplied anti-cariogenic features, but had low solubility, loss of translucency, surface cracks and low mechanical properties that made it contraindicated for use in large restorations [9].

The acrylic resin had a great evolution along time in relation to direct restorative materials, as they showed insolubility in the oral environment and low cost, but provided high levels of leakage even associated with enamel acid etching [10].

This scenario began to change in 1963 when Bowen [11] linked silica particles treated with vinyl silane and a resin matrix of Bisphenol-Glycidyl Methacrylate (Bis-GMA), giving origin to the resins.

Due to the good performance presented in relation to mechanical, physical and optical properties, the resins are currently used in anterior and posterior teeth.

Along time, the resins were undergoing changes in order to overcome some disadvantages that the material presented. Therefore, their organic content has variations in its quantity and composition, without modifying their main array-based methacrylate.

A difference between the composites is in relation to its viscosity. When the composite has a large amount of organic matrix (41-53% volume and 56-70% by weight) in its content, this is called flowable composite resin [12]. The lowest viscosity in this material ensures a lower microleakage. Thus, fluid resins are indicated as base and padding between the tooth / restoration interface.

However, the high organic content present in this material may also have disadvantages by providing higher polymerization shrinkage and reducing its wear resistance [13].

On the other hand, when the composite has a larger amount of inorganic matrix, the material shows more resistance to wear and polymerization shrinkage of less than 2%. In this case, the resin is called compressible resin (greater than 80% by weight).

The composite has a higher viscosity due to the low amount of organic matrix in its content, which determines its handling and viscosity. The condensable composite resins have the disadvantage of presenting more microleakage, a greater difficulty in handling, lower flexibility, and its indication is restricted to posterior teeth [14].

The latest composite resins have demonstrated that not only the amount of load was altered, but the changes also occurred in relation to the shape, composition and distribution of inorganic particles.

Despite all the progress made by the composites up to this point, the material continued with its methacrylate-based organic load, allowing contraction during the composite resin curing process. This polymerization shrinkage occurs because the conversion of monomer molecules into a polymer chain is accompanied by an approximation of these molecules, which causes a decrease in volume. This decrease in volume generates a force opposing the binding force between the restorative material and tooth structure, causing contraction of the material.

In order to prevent this polymerization shrinkage from affecting the bond between the restorative material and tooth structure, the insertion of composite resin should be done incrementally and in small portions, especially in posterior teeth [15].

In order to eliminate this disadvantage of methacrylic composites, a new composite resin was created. This material is intended to be used in posterior teeth, as it provides a polymerization shrinkage of less than 1%. This was possible because this new resin contains another organic component in its composition, called Silorane by the manufacturer.

The organic matrix Silorane originates from two main chains called siloxane and oxirane. This matrix undergoes a slight expansion and then contraction during its polymerization, and this expansion-contraction process results in a contraction of less than 1%.

The use of composite resin is performed in increments of 2 mm thick and can be carried in small portions. The need for more than one increment of composite resin, when the silorane-based resin is used, will only be required when the cavity has a size larger than the maximum thickness of resin allowed for the complete polymerization of the material. Then, the cavity size determines the amount of resin portions to be used, and the amount of increment is not related to polymerization shrinkage.

Chemical Properties of Silorane

Silorane is a new class of compounds for use in dentistry. This name is derived from two constituent molecules in the resin matrix. One of them is the siloxane, known in industry for being used in the manufacturing of inks. Its main feature is the ability to be hydrophilic, which makes inks waterproof. The other one is Oxirane, which has been used in many technical areas, especially when large forces and challenging physical environments are expected, such as in the manufacturing of sports equipment like tennis rackets and skis, or in the automotive industry and aviation. Oxirane polymers are also known for its low contraction and excellent stability in relation to the various physical and chemical-physical forces and influences.

The polymerization process occurs through a cationic ring-opening reaction that results in lower polymerization shrinkage compared to methacrylate resins, which polymerize via a radical addition reaction to double bonds. The ring opening step during the composite resin polymerization reduces significantly the amount of polymerization shrinkage that occurs during the polymerization process.

During the polymerization process, the molecules need to get closer to one another to form a chemical bond. The process results in a loss of volume, that means, polymerization shrinkage. In contrast to the reactive groups of linear methacrylates, the chemistry of silorane ring opening starts with the cleavage of the rings. This process is gaining ground and compensates the volume loss that occurs in subsequent steps, when chemical bonds are formed. In general, the ring-opening polymerization process results in reduced polymerization shrinkage.

Silorane-Based Adhesive System

An exclusive adhesive system is necessary to bind the silorane-based composite resin to the dental structure. It is suggested by the manufacturer because the resin is highly hydrophobic, which requires an adhesive system also containing a highly hydrophobic adhesive to adhere to the silorane-based resin. Unlike the adhesive, the primer should be hydrophilic in order to penetrate the tooth structure and humidify it. This transition between the tooth structure (hydrophilic) and composite resin (hydrophobic) is provided through this restorative system.

The manufacturer has chosen to use a self-conditioning adhesive system for providing fewer errors during their use, simplicity of use and low post-operative sensitivity.

The self-etching primer contains acid monomers (phosphate methacrylate and Vitrebond copolymers) that are responsible for conditioning dental substrate. The primer also contains monomers and HEMA BisGMA as a solvent system containing water and ethanol to wet and penetrate the dental substrate, and a system based on camphorquinone photoinitiator for a complete and rapid polymerization. Silica particles treated with silane with primary particles of about 7 nm were added to improve mechanical strength and film-forming properties of the self-etching Primer Silorane Adhesive System. These particles are finely dispersed to prevent decantation.

The additional conditioning of the enamel is not necessary in cases where the enamel is prepared. However, if the adhesive is applied to make the enamel intact, the phosphoric acid etching of the prepared enamel is recommended (Figure 1).

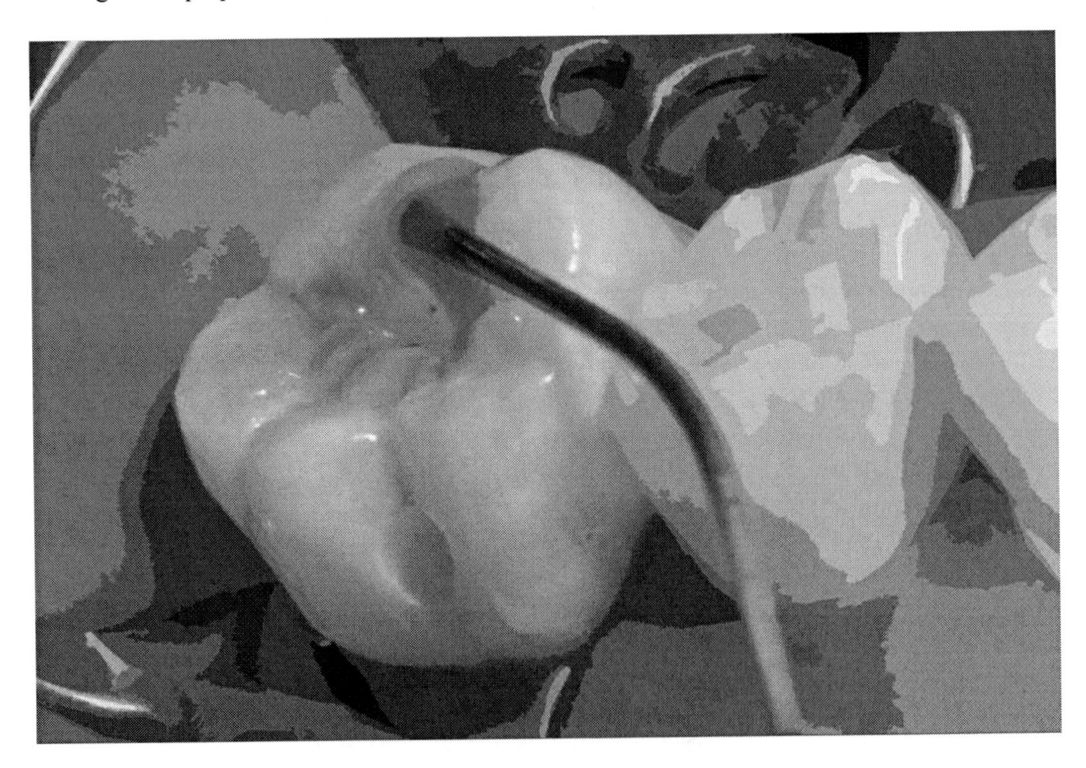

Figure 1. 35% phosphoric acid etching of the enamel is recommended to make it intact.

Advantages of the Silorane-Based Composite Resin

1) Allows a longer operative work with lighting (9 minutes);
2) Excellent marginal integrity;
3) Reduced risk of marginal gaps and microleakage;
4) Reduced risk of marginal staining;
5) Reduced risk of secondary caries;
6) Reduced cusp deflection;
7) Reduced risk of stress-induced enamel fracture;
8) Reduced risk of post-operative sensitivity.

Disadvantages of the Silorane-Based Composite Resin

1) The use of composite resin is associated with its own adhesive system;
2) Aesthetic limitation due to the low amount of colors available.

Application Technique

1) Application of the self-etching primer adhesive system for 15 seconds using a microbrush, followed by gentle air dispersion and 10 seconds of light curing;
2) Application of the adhesive bond system with a microbrush, followed by gentle air dispersion and 10 seconds of light curing.

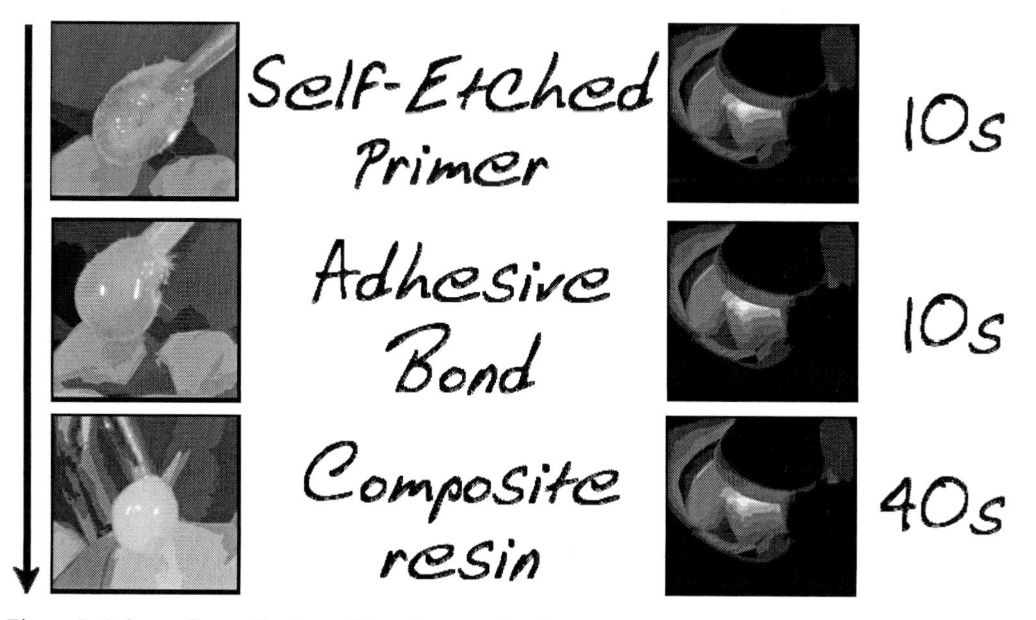

Figure 2. Scheme for application of the silorane adhesive system.

Case Reports

Silorane-Based Composite Resin – Occlusal Cavity

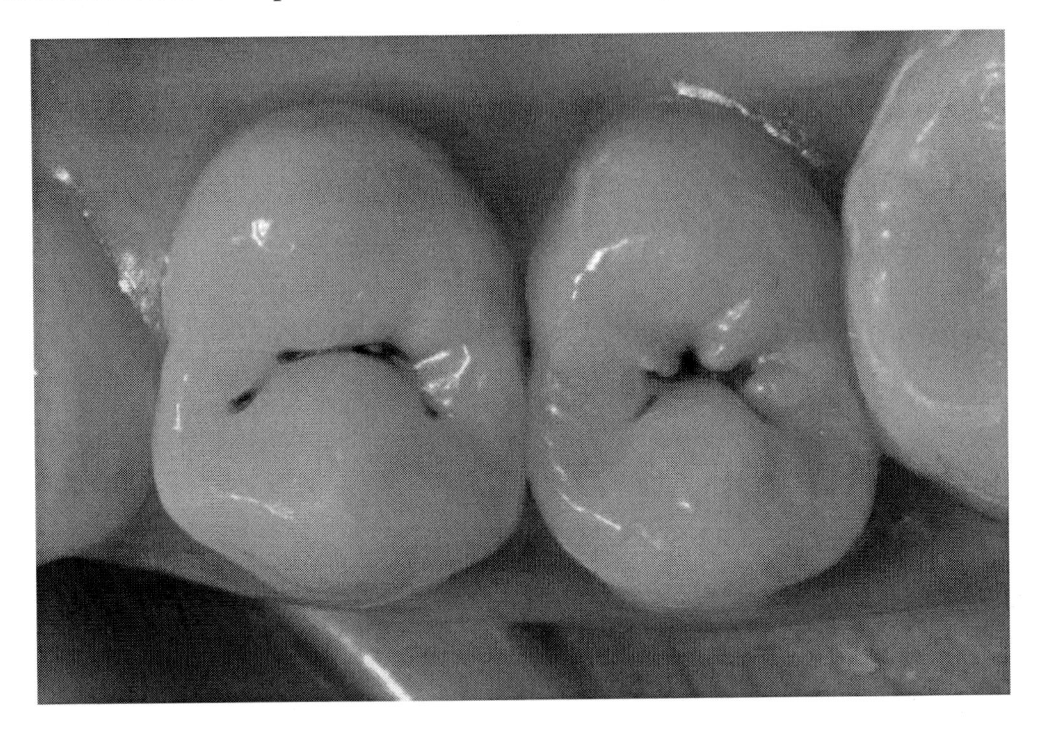

Figure 3. Initial situation.

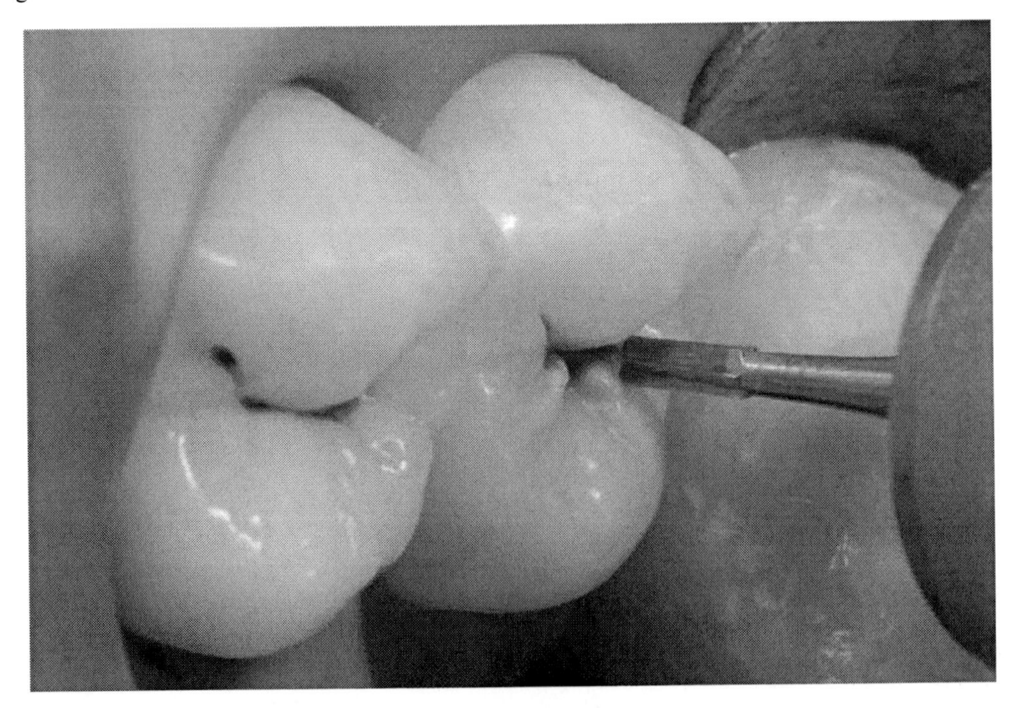

Figure 4. Caries removal was performed with a carbide bur.

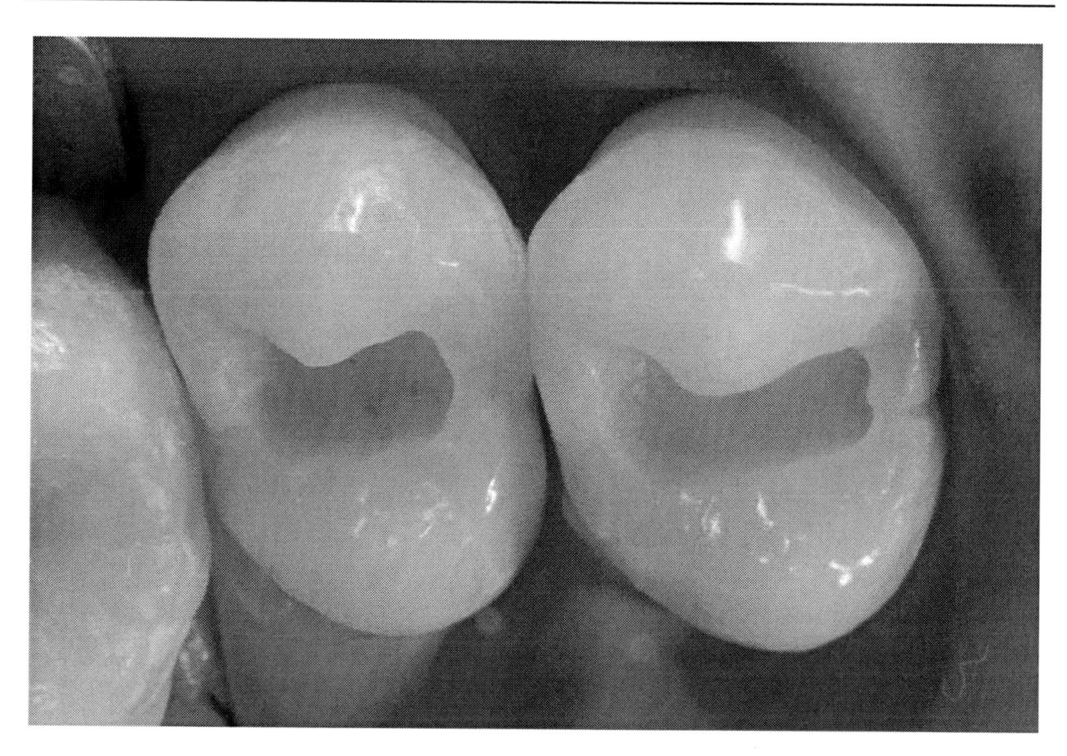

Figure 5. Occlusal view of cavity.

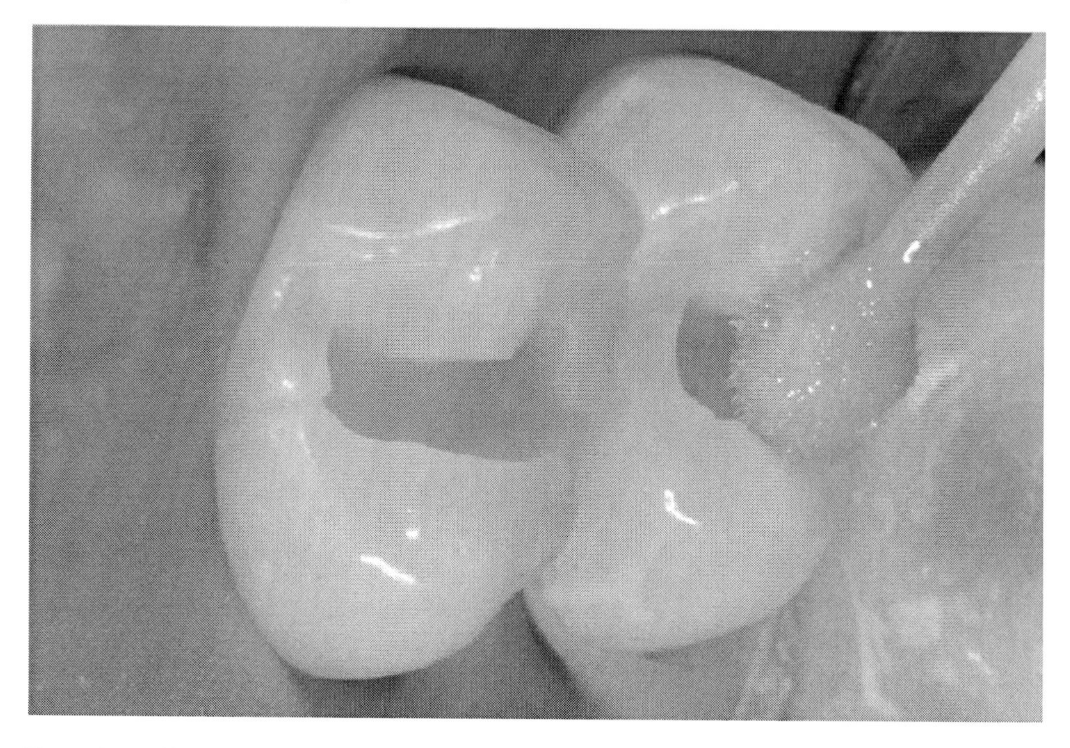

Figure 6. Application of the self-etching primer adhesive system for 15 seconds with a microbrush, followed by gentle air dispersion.

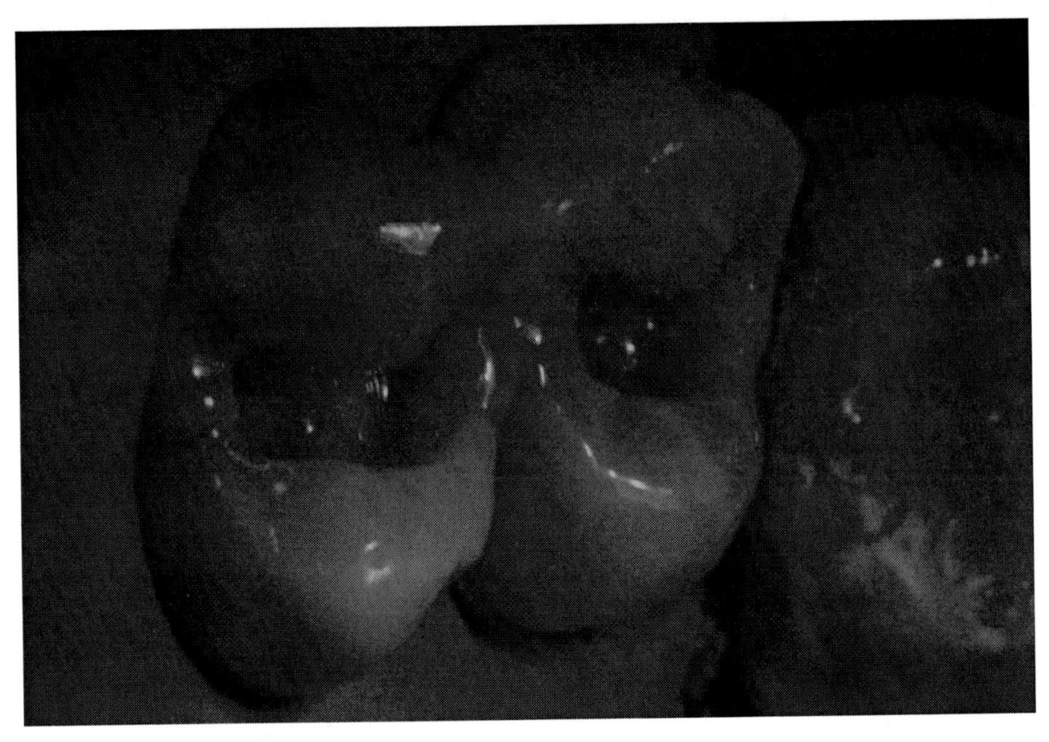

Figure 7. Polymerization of the self-etching primer adhesive system with high-power LED during 10 seconds.

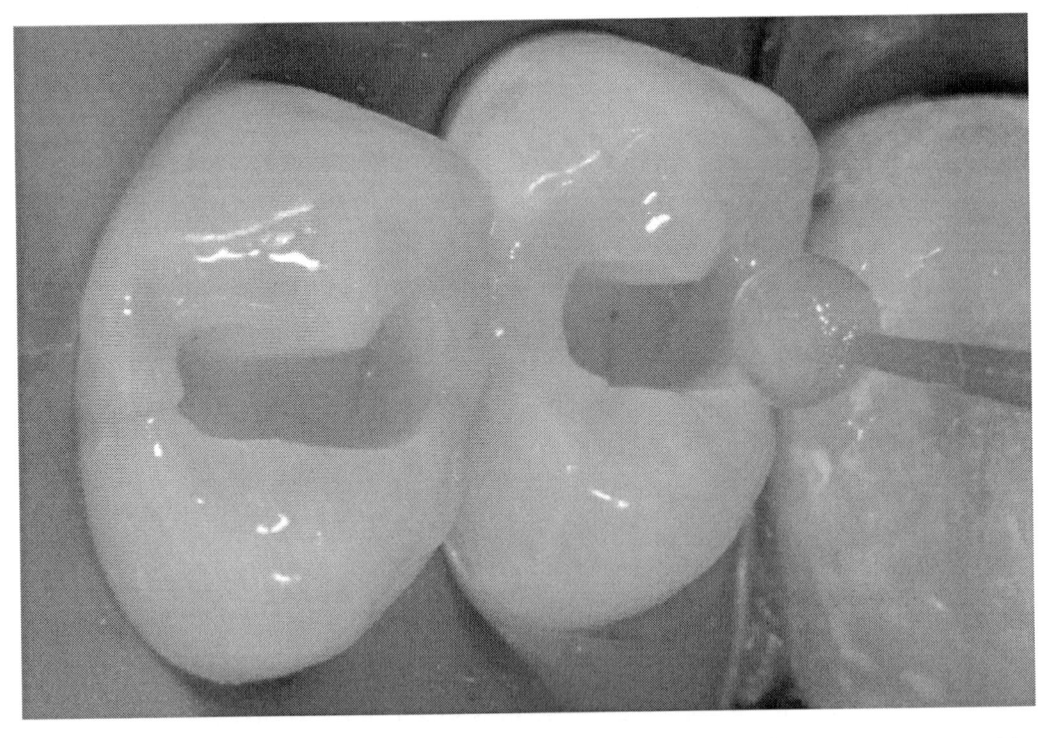

Figure 8. Application of the adhesive bond system for 15 seconds with a microbrush, followed by gentle air dispersion.

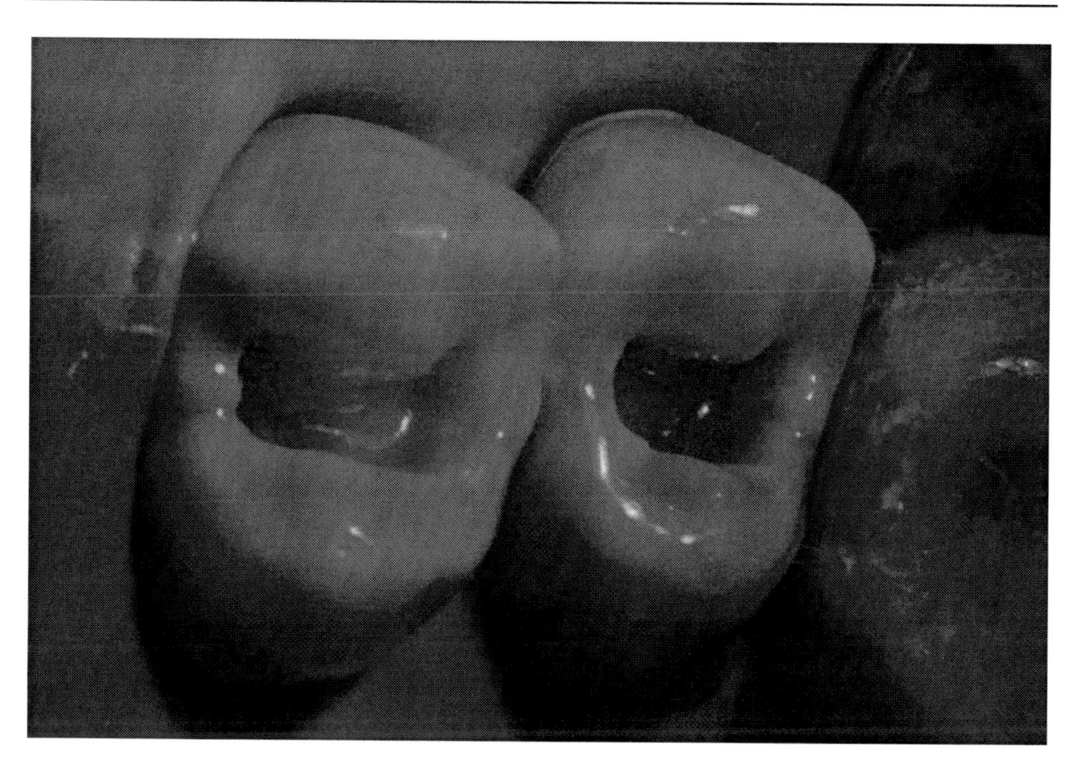

Figure 9. Polymerization of the adhesive bond system with high-power LED during 10 seconds.

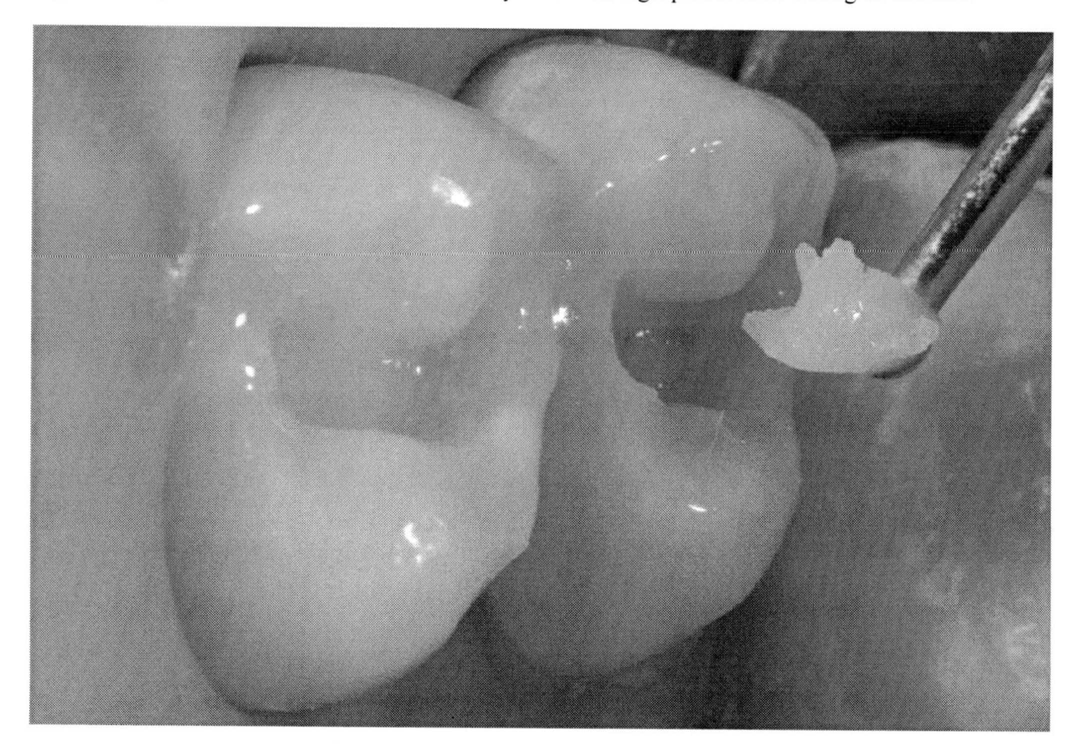

Figure 10. Application of the first layer of composite resin. The layer should be 2 mm in diameter.

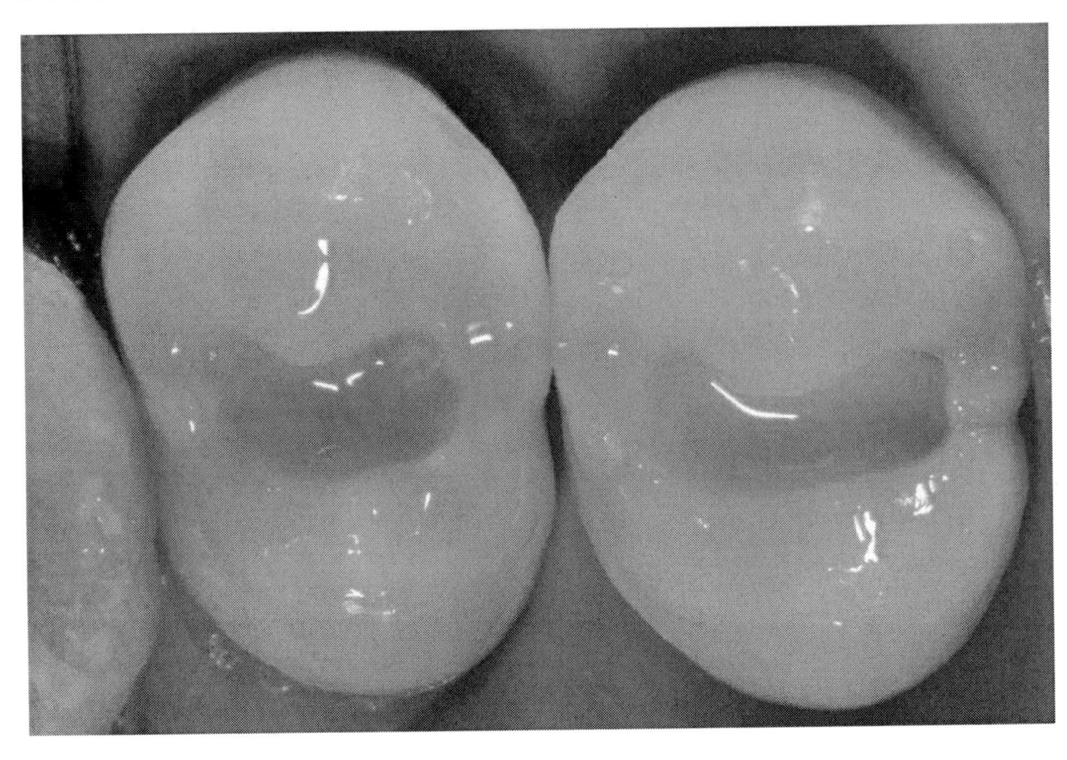

Figure 11. Occlusal view of the first layer of composite resin.

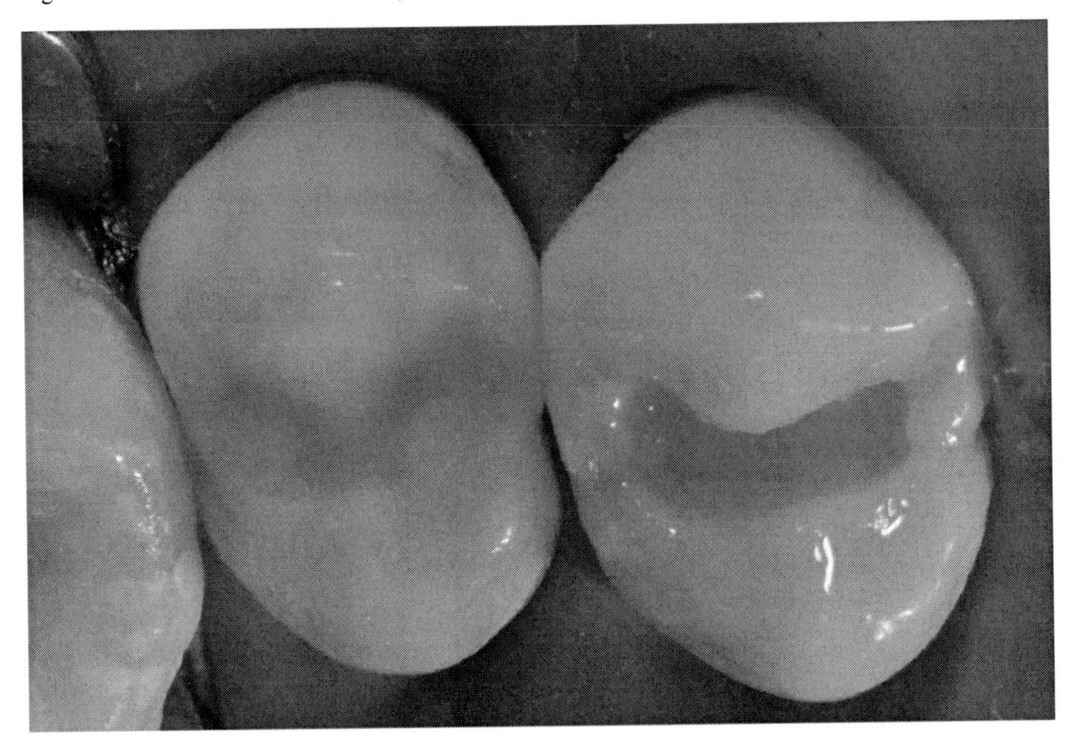

Figure 12. First restoration completed.

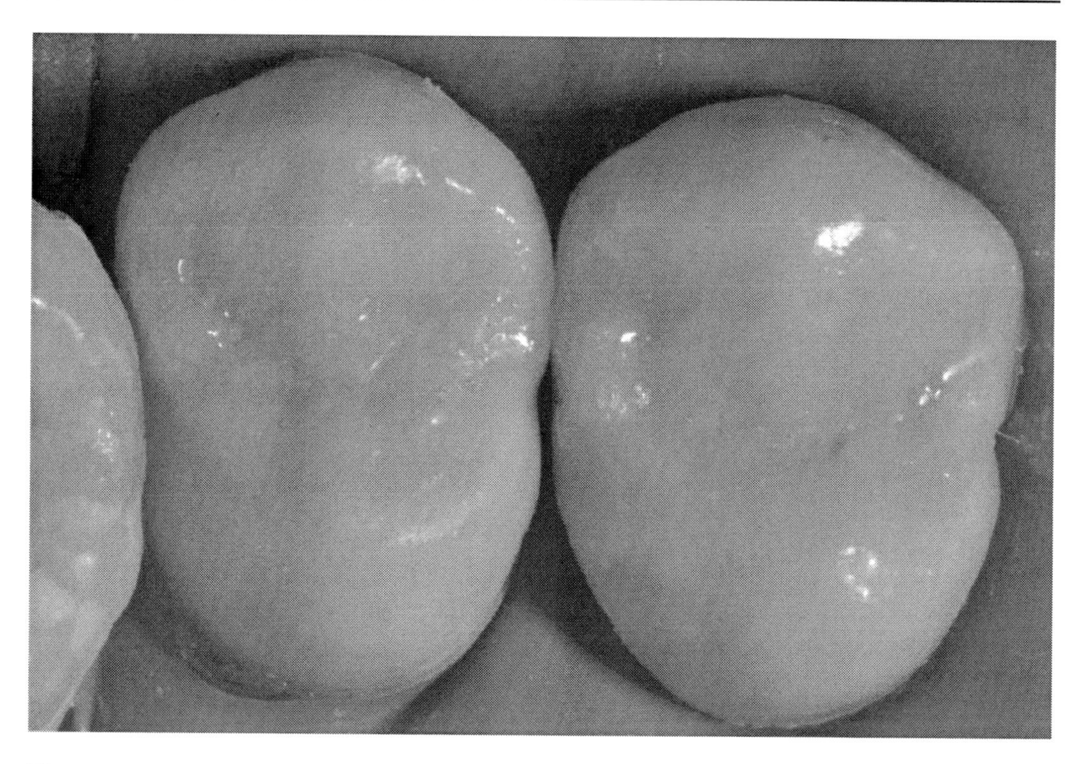

Figure 13. Restorations completed.

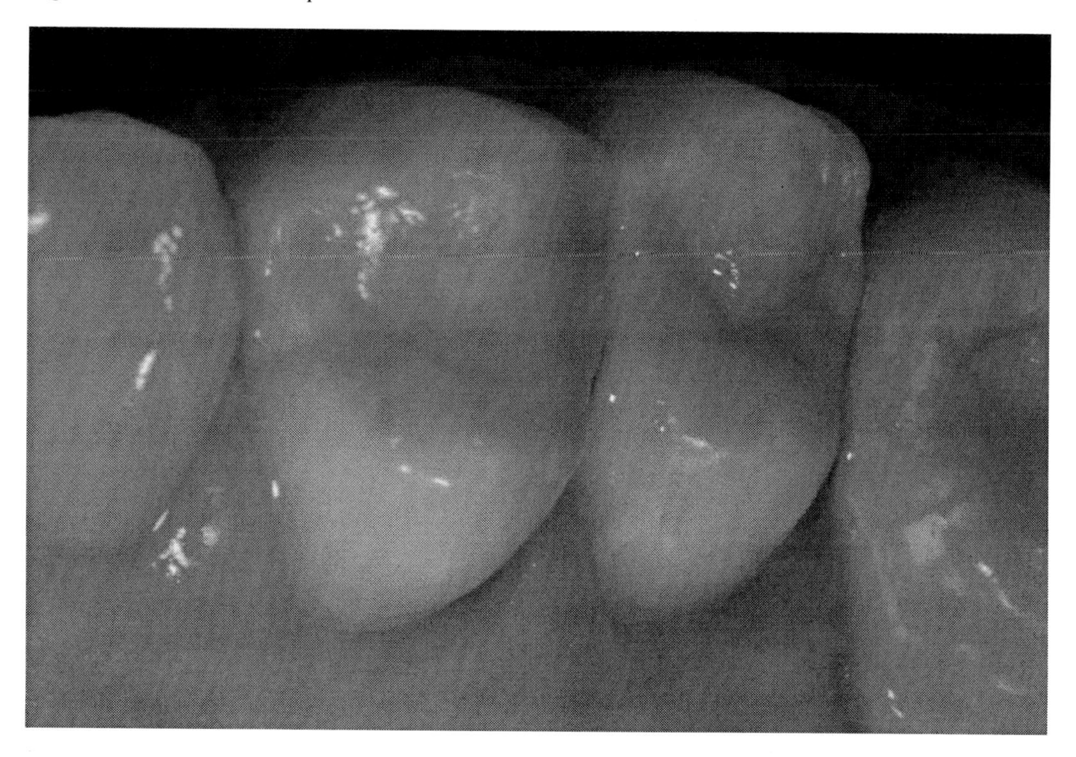

Figure 14. Restorations completed after 1 week.

Silorane-Based Composite Resin – Disto-Occlusal Cavity

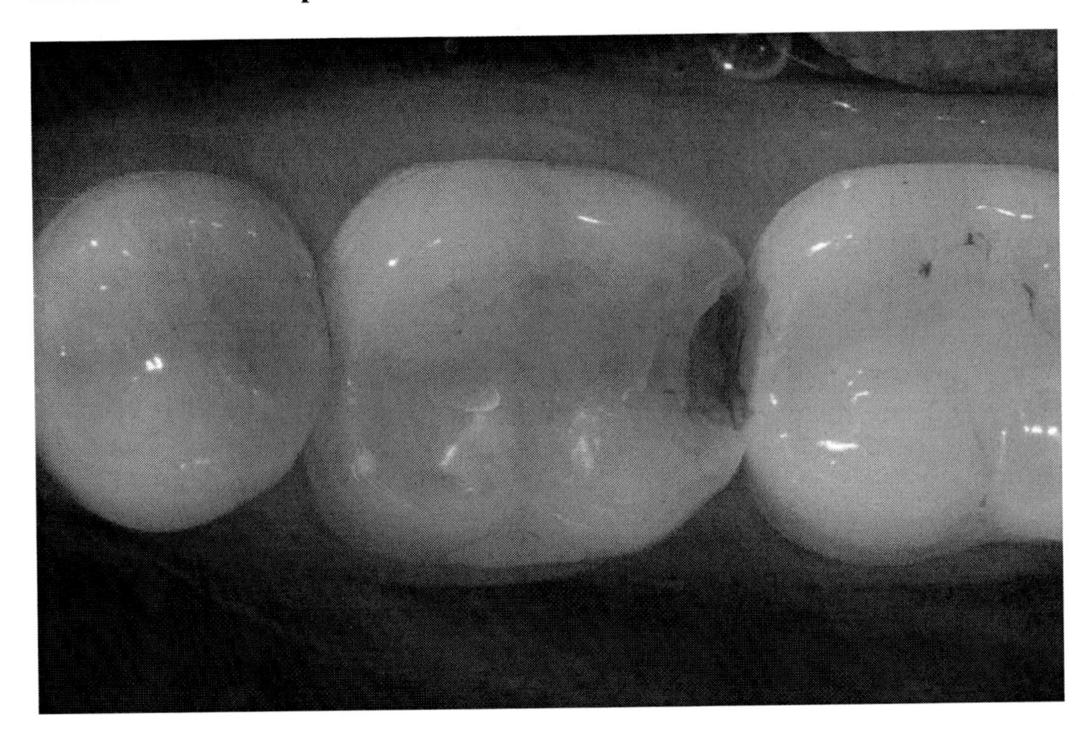

Figure 15. Initial view.

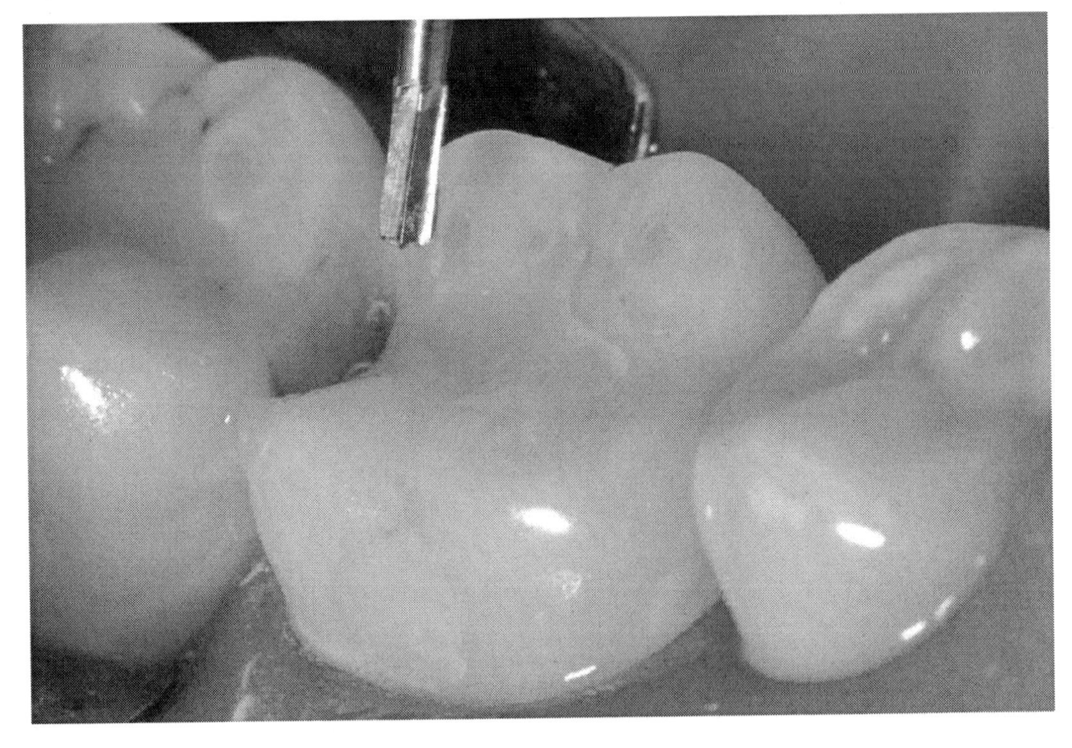

Figure 16. Removal of restoration and caries was performed with a carbide bur.

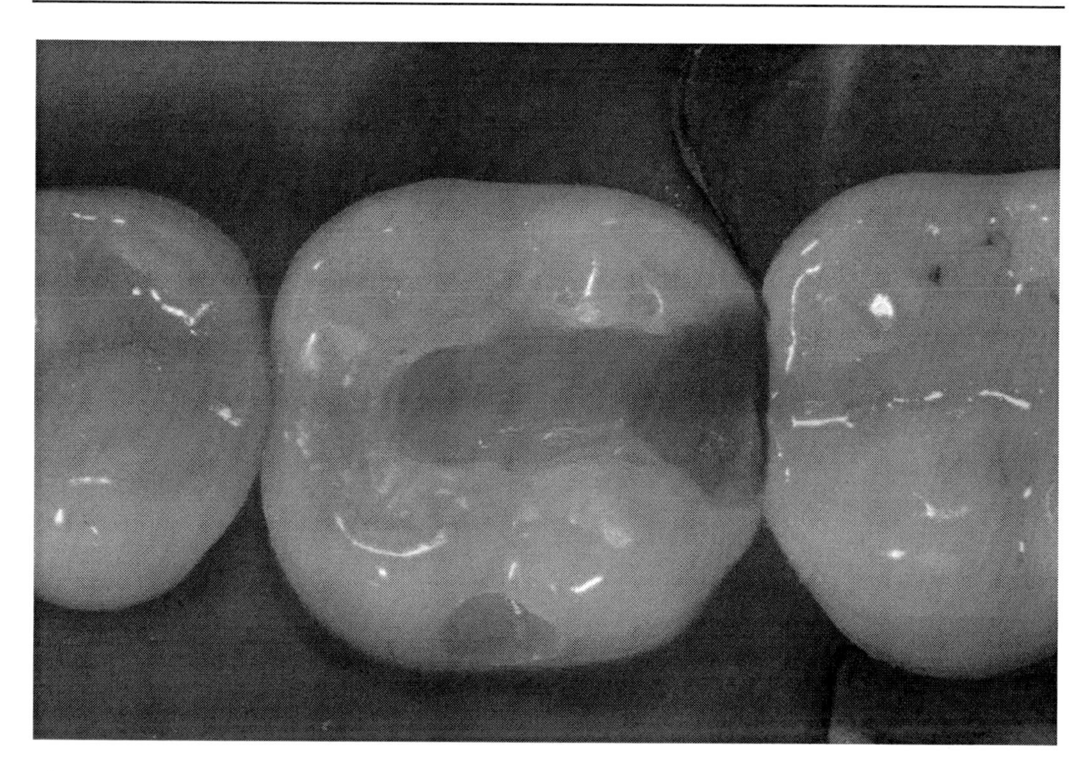

Figure 17. Occlusal view of cavity.

Figure 18. Application of the self-etching primer adhesive system for 15 seconds with a microbrush, followed by gentle air dispersion.

Figure 19. Polymerization of the self-etching primer adhesive system with high-power LED during 10 seconds.

Figure 20. Application of the adhesive bond system for 15 seconds with a microbrush, followed by gentle air dispersion.

Figure 21. Polymerization of the adhesive bond system with high-power LED during 10 seconds.

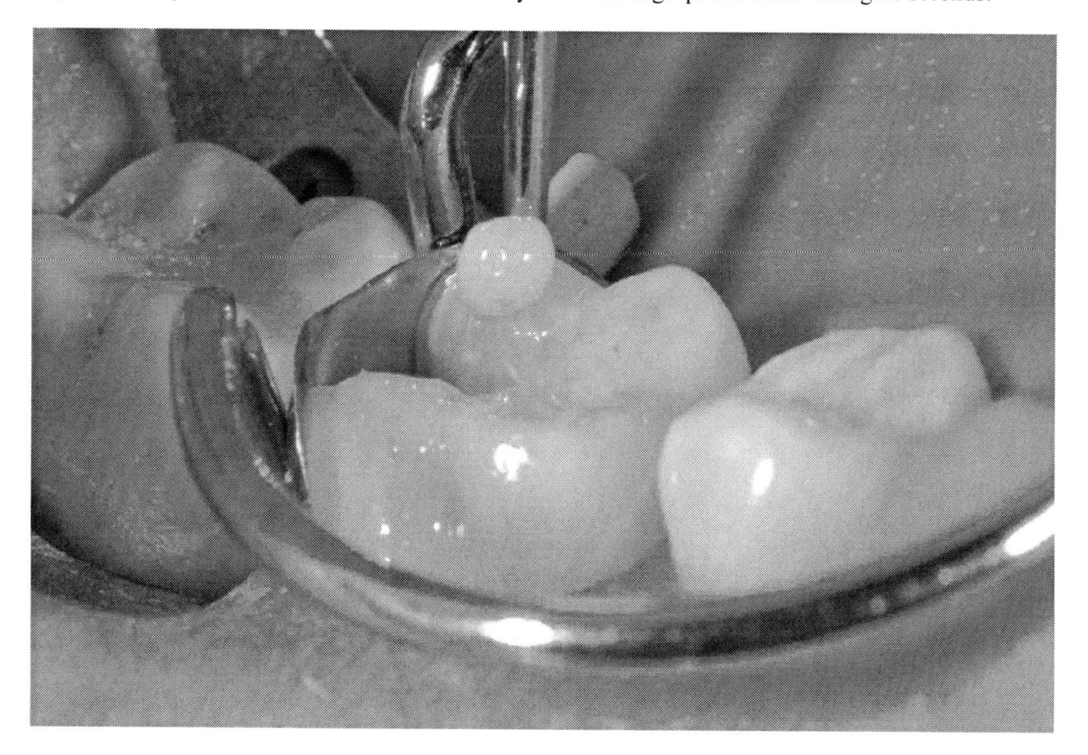

Figure 22. Application of composite resin to construct the marginal ridge and coverage of the pulp wall.

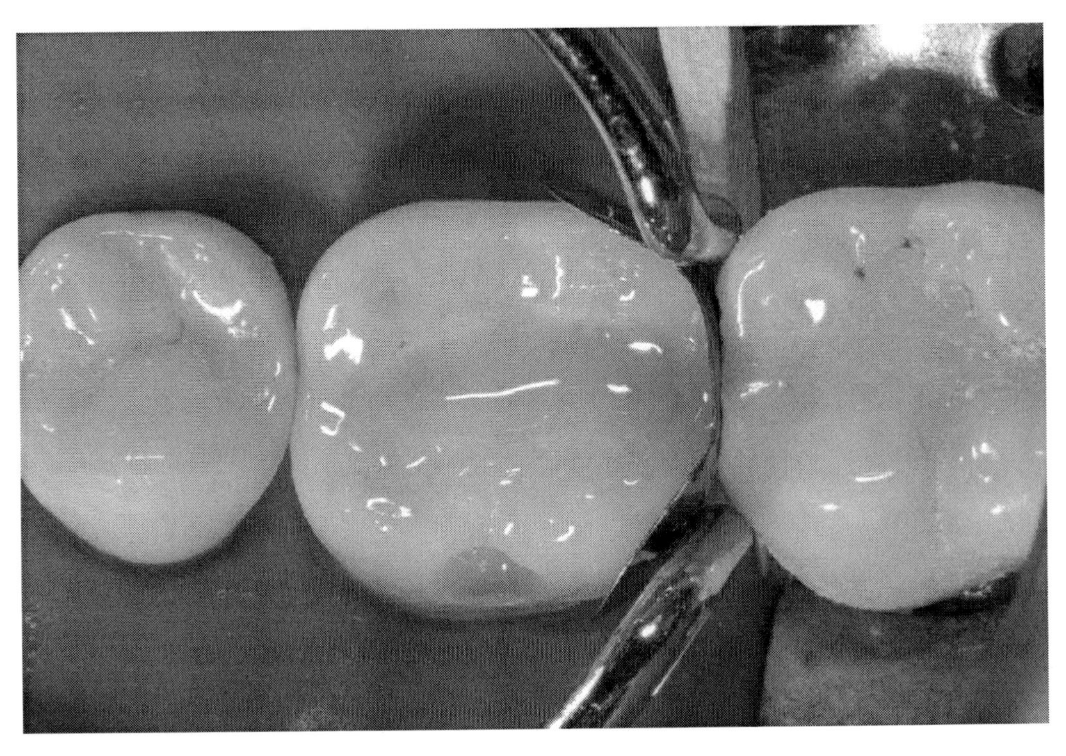

Figure 23. Occlusal view after confection of the marginal ridge and coverage of the pulp wall.

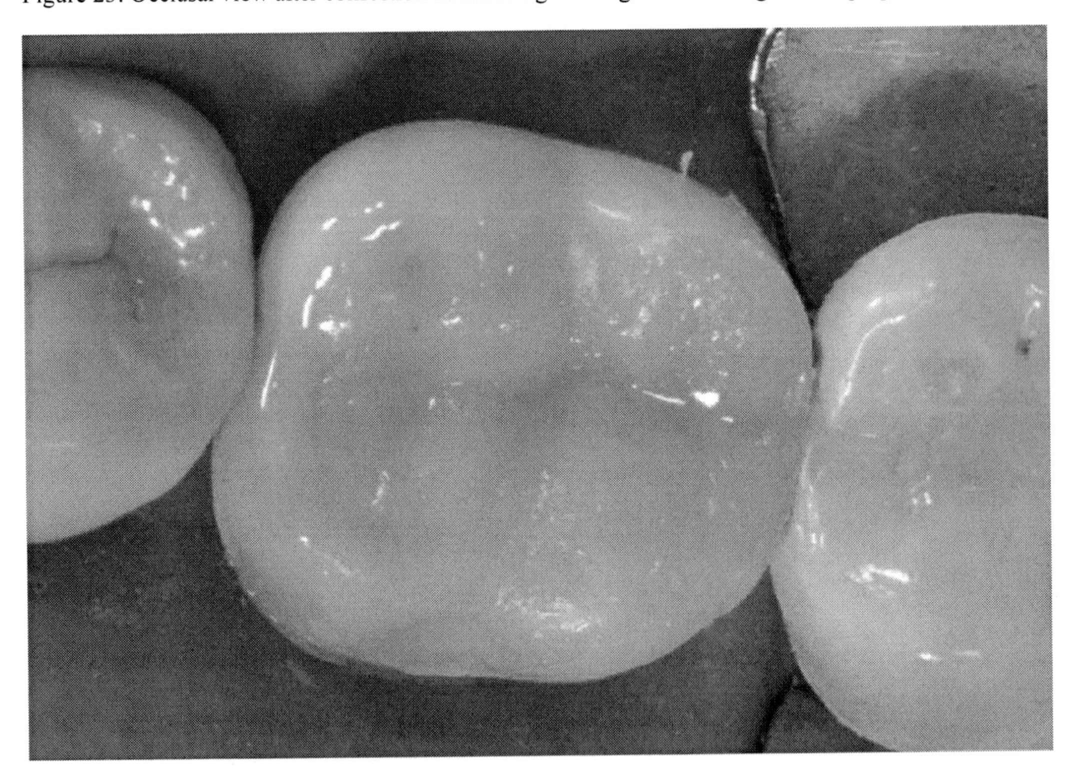

Figure 24. Restoration completed.

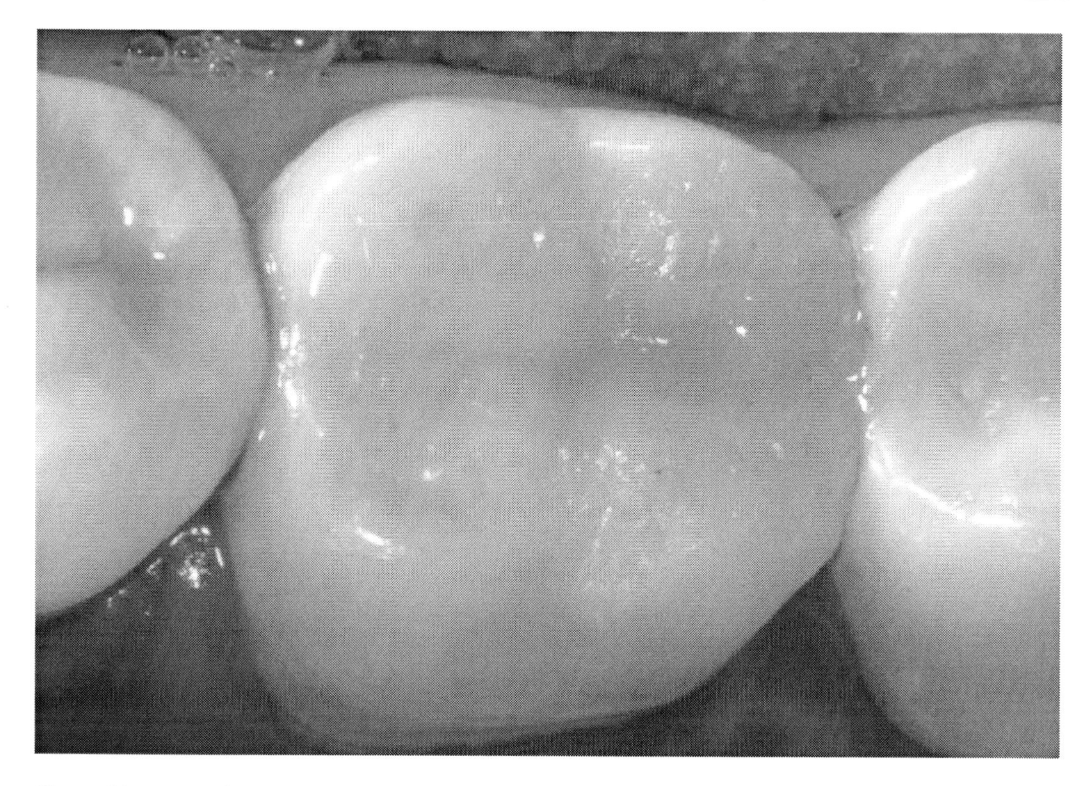

Figure 25. Restorations completed after 1 week.

REFERENCES

[1] Anusavice KJ, Phillips RW. Phillips' science of dental materials. 11th ed. St. Louis: W.B. Saunders; 2003.

[2] Charlton DG, Murchison DF, Moore BK. Incorporation of adhesive liners in amalgam: effect on compressive strength and creep. *Amer. J. Dent.* 1991; 4:184-188.

[3] Gordan VV, Mjör IA, Hucke RD, Smith GE. Effect of different liner treatments on postoperative sensitivity of amalgam restorations. *Quintessence Int.* 1999; 30:55-59.

[4] Gwinnett AJ, Baratieri L, Monteiro S, Ritter AV. Adhesive restorations with amalgam: guidelines for the clinician. *Quintessence Int.* 1994; 25:687–695.

[5] Marchiori S, Baratieri LN, de Andrada MA, Monteiro Júnior S, Ritter AV. The use of liners under amalgam restorations: an in vitro study on marginal leakage. *Quintessence Int.* 1998; 29:637-642.

[6] Mahler DB, Engle JH, Simms LE, Terkla LG. One-year clinical evaluation of bonded amalgam restorations. *J Am Dent Assoc.* 1996; 127:345-349.

[7] Mahler DB. The amalgam-tooth interface. *Oper Dent.* 1996; 21:230-236.

[8] Roulet JF. Benefits and disadvantages of tooth-coloured alternatives to amalgam. *J Dent.* 1997; 25:459-473.

[9] Wilson AD. Specification test for the solubility and disintegration of dental cements: a critical evaluation of its meaning. *J Dent Res.* 1976; 55:721-729.

[10] Craig RG. Chemistry, composition, and properties of composite resins. *Dent Clin North Am.* 1981; 25:219-239.

[11] Bowen RL. Properties of a silica-reinforced polymer for dental restorations. *J Am Dent Assoc*. 1963; 66:57-64.

[12] Bayne SC, Thompson JY, Swift EJ Jr, Stamatiades P, Wilkerson M. A characterization of first-generation flowable composites. *J Am Dent Assoc*. 1998; 129:567-577.

[13] Kugel G. Direct and indirect adhesive restorative materials: a review. *Am J Dent*. 2000; 13:35D–40D.

[14] Peris AR, Duarte S Jr, de Andrade MF. Evaluation of marginal microleakage in class II cavities: effect of microhybrid, flowable, and compactable resins. *Quintessence Int*. 2003; 34:93-98.

In: Resin Composites: Properties, Production and Applications ISBN: 978-1-61209-129-7
Editor: Deborah B. Song

Chapter 10

ADVANCED NUMERICAL SIMULATION OF COMPOSITE WOVEN FABRIC FORMING PROCESSES

A. Cherouat and H. Bourouchaki*
University of Technology of Troyes, Charles Delaunay Institute/GAMMA3-INRIA Project Team, Troyes, France

ABSTRACT

Different approaches used for the simulation of woven reinforcement forming are investigated. Especially several methods based on geometrical and finite element approximations are presented. Some are based on continuous modelling, while others called discrete or mesoscopic approaches, model the components of the dry or prepregs woven fabric. Continuum semi discrete finite element made of woven unit cells under biaxial tension and in-plane shear is detailed. In continuous approaches, the difficulty lies in the necessity to take the strong specificity of the fibre directions into account. The fibres directions must be strictly followed during the large strains of the fabric. In the case of geometrical or continuum approaches (semi-discrete) the directions of the fibres are naturally followed because the fibres are modeled. Explicitly, however, modeling each component at the mesoscopic scale can lead to high numerical cost. During mechanical simulation of composite woven fabric forming, where large displacement and relative rotation of fibres are possible, severe mesh distortions occur after a few incremental loads. Hence an automatic mesh generation with remeshing capabilities is essential to carry out the finite element analysis. Some numerical simulations of forming process are proposed and compared with the experimental results in order to demonstrate the efficiency of the proposed approaches.

Keywords: Composite woven fabric, finite element analysis, geometrical approach, composite forming, remeshing procedure

---

* University of Technology of Troyes, Charles Delaunay Institute/GAMMA3-INRIA Project Team, 12 rue Marie-Curie, BP 2060, 10010 Troyes, France, Email: abel.cherouat@utt.fr, Tel +33 3 25 71 56 74 & Fax. +33 3 25 51 59 11 20

1. Introduction

Polymer composite reinforced by woven fabric (glass, carbon or kevlar) is known to have high specific stiffness and, in combination with automatic manufacturing processes, make it possible to fabricate complex structures (aircraft, boat, automotive, etc) with high level of weight and cost efficiency. As known, the substitution of metal alloys by composite materials, in general, reduces structural mass by 20-30%. The mass increase is due also to the numerous variety of semi-products (roving, fabrics, knitted fabrics, braids pre-impregnated or not) permitting the development of new structures. Fabrication processes, also, have undergone substantial evolution in recent years. Although the traditional lay-up process will remain the process of choice for some applications, new developments in Resin Transfer Molding (RTM) or Sheet Molding Compound (SMC), low temperature curing prepregs and low pressure molding compounds have matured significant are reached, and are now being exploited in high technology areas such as aerospace industry. For example, by using such composites, the automotive industry can realize improved fuel economy through vehicle-weight reduction by replacing the currently used steel and aluminium parts with thermoplastic composites with the added benefit of a corrosion-resistant material. The choice of manufacturing process depends on the type of matrix and fabric, the temperature required to form the part and the cost effectiveness of the process. In particular, thermo-forming is a promising manufacturing process for producing high-volume low-cost composite parts using commingled fibreglass/polypropylene woven fabrics [1-3].

The simulation of the manufacturing of a textile reinforced composite part with a Liquid Composite Moulding-like process, which involves draping (or deep-drawing) and impregnation of the preform, includes several stages (see Figure 1). First, a mould is designed with CAD software, and the CAD model is meshed. Then, a draping (or deep-drawing) simulation tool is used to compute the deformations of the textile layers inside the mould. As a result, for every element of the mesh, textile parameters like the shear angle and the thickness of the layer are available. With these parameters given, the local (meso-scale) properties for every element is determined (pre-processing), and given as input for the macroscopic structure after resin polymerization simulation. The result of the macro-simulation is then post-processed to optimize the mechanical properties of composite structure. The numerical simulation of composite forming is an efficient means of evaluating factors related to manufacturing processes and an efficient help to design pre-forming sequence for the manufacturing of fabric reinforced composites. It is possible to detect main problem occurring during the shaping deformation and to obtain good quantitative information on the forming process [2-8].

Different levels of modeling intervene in the simulation of woven reinforcement forming (1) architecture design level, (2) preliminary design level, (3) mechanical level by computational software and (4) optimisation level [6-8]. Most of these levels are integrated and take into account specific constraints of manufacturing processes. The particular form of composite fabrication (Pre-preg) begins with the pre-impregnation of reinforcement materials with a resin. The combining of these two materials occurs prior to the moulding process and therefore enables a very accurate reinforcement to resin ratio to be achieved. Pre-preg materials are used extensively in the aerospace or automotive industry due to their ability to maximize strength to weight ratios. Pre-pregs are pliable and therefore able to be cut into various shapes or patterns prior to processing into the moulded products. But for the manufacturing of non-developable composite part (part that can not make flat un-stretched), a new problem intervenes in the design chain resulting from the number of parameters influencing the global behavior of composite forming process. The ability to define, in advance, the ply shapes and material orientation allowed the engineers to optimize the composite structural properties of the composite products for maximum strength, maximum material utilization and maximum lay-up efficiency [10-20].

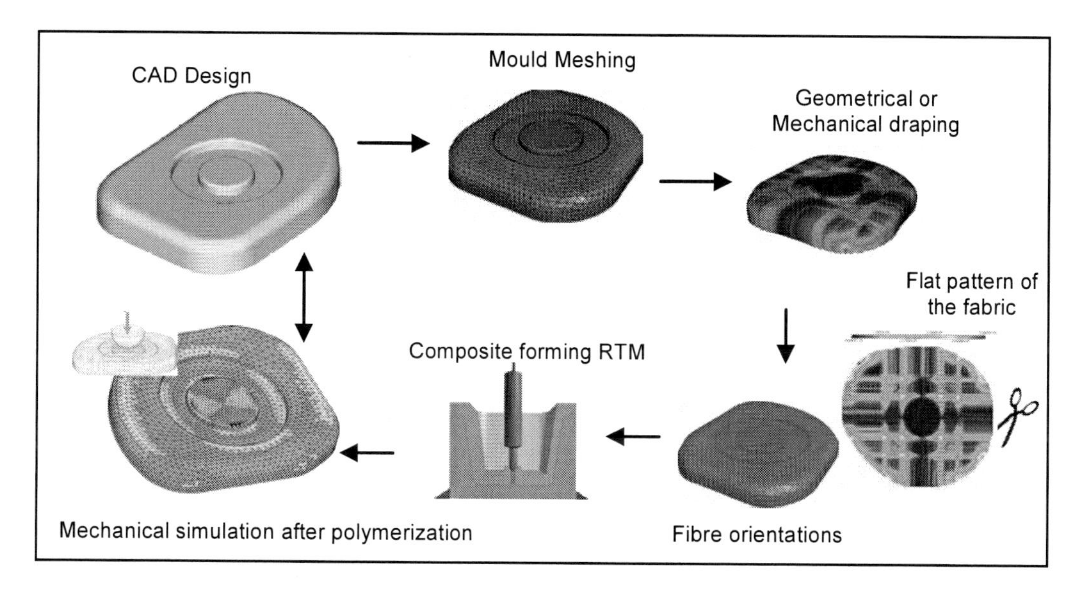

Figure 1. Illustration of the manufacturing of a textile reinforced composite part.

The composite manufacturing process involves large displacements and rotations and large shear of weft and warp fibres, which can have a significant effect on the processing and structural properties of the finished product. The formulation of new and more efficient numerical models for the simulation of the shaping composite processes must allow for reduction in the delay in manufacturing of complex parts and an optimization of costs in an integrated design approach [3, 20-25]. Several modelling approaches have been developed to account for the evolution of the orthotropic directions during high shearing, and these approaches include the geometrical and the finite element approaches. The geometrical approach so called fishnet algorithms is used to determine the deformed shape of draped fabrics. This method, where the fabric is placed progressively from an initial line, provides a close enough resemblance to handmade draping [26-30]. They are very fast and fairly efficient in many prepreg draping cases. Nevertheless, this method has major drawbacks. They account neither for the mechanical behaviour of the fabric nor for the static boundary conditions. This last point is very important in the case of forming with punch and die (such as in the pre-forming of the RTM process). The loads on the tools, especially on the blank-holder, influence the quality of the shaping operation, and therefore, need to be considered in simulations and therefore, need to be considered in simulations [31-34].

The alternative to the geometrical approach consists of a mechanical analysis of the fabric deformation under the boundary conditions prescribed by the forming process. This requires a specific model of the woven reinforcement and its mechanical behaviour. The mechanical behaviour of woven fabrics is complex due to the intricate interactions of the fibres. It is a multi-scale problem. The macroscopic behaviour is very much dependent on the interactions of fibres at the mesoscale (scale of the woven unit cell) and at the micro-scale (level of the fibres constituting yarns). Despite of a great amount of work in the field, there is no widely accepted model that accurately describes all the main aspects of fabric mechanical behaviour. The main model families come from the multi-scale nature of the textile. A first family of models is obtained by homogenizing the mechanical behaviour of the underlying meso-structure and considering the fabric as an anisotropic continuum [9, 12, 15-20]. If these models can easily be integrated in standard finite element using conventional shell or membrane elements, then the identification of homogenized material parameters is difficult, especially because these parameters change when the fabric is strained and when, consequently, the directions and the geometry (crimp, transverse sections…) of the fibres change. Some of these approaches will be described, especially a non-orthogonal constitutive model [35-40] and an anisotropic hypoelastic continuous behaviour for fibrous material based on an objective derivative using the rotation of the fibre [34-35]. Conversely, some authors present fully discrete models of fabrics [41-44]. Each yarn or each fibre is modelled and is assumed to be a straight or a curved beam or truss. Sometimes they are modeled as 3D domains [42]. Springs are often used to model warp and weft yarn interactions. In the objective of fabric forming simulations, some authors extend the discrete modelling to the whole textile structure that is represented by a network of interwoven trusses or beams with different tensional and rotational springs. Accounting for the simplicity of each component, the whole textile structure deformation can be computed.

Nevertheless, the computational effort needed is relatively significant. At present this method is restricted to simple geometry of the local yarn and relatively simple mechanical behaviour. When a fine model of the fibrous yarns is used, the analysis can only consider a small part of the textile reinforcement such as a few woven or knitted cells. The semi-discrete approach is a compromise between the above continuous and discrete approaches [20, 25, 36-39]. A finite element method is associated to a mesoscopic analysis of the woven unit cell. Specific finite elements are defined that are made of a discrete number of woven unit cells. The mechanical behaviour of these woven cells is obtained by experimental analyses or from 3D FE computations of the woven cell. The nodal interior loads are deduced from this local behaviour and the corresponding strain energy in the element deformation.

In current study, the geometrical and the finite element semi-discrete approaches are used to simulate the deformation of composite fabrics by shaping process. These approaches, while giving good results and being efficient in terms of computing time, are generally somewhat complex and sometimes very challenging to implement into commercially available FEA packages.

The geometrical approach is well adapted to preliminary design level. It is based on geometrical aspects of the warping. Our method is based on a modified "MOSAIC" algorithm, which is suitable to generate a regular quad mesh representing the lay-up of the curved surfaces (giving the exact fibre orientations). The method is implemented in the GeomDrap software [45] which is now integrated in the ESI-Pam software [23]. This software provides a fibre quality chart (showing the fibre distortions, the rate of falling and the rate of draped surface) to predict local folding due to overlapping of fibers in the shear exceeds limit value (up to 60° in some cases). It can be used to optimize the draping process (with respect to the above quality measure) by improving the lay-up directions or the marker data location. The lay-up of complex curved surfaces can be made in a few seconds [17, 27-33].

The mechanical approach is base on a meso-structural description for finite deformations and geometrical non-linearity. The not polymerized resin has a viscous behavior and the reinforced fibers are treated as either unidirectional non linear elastic behavior. The unit cell of the mesoscopic model used here for a plain-weave fabric consists of bi-component finite elements. The tensile load is carried by the 1-D elements that will capture the changes in the orthotropic directions during the shearing. The 2-D element accounts only for the shearing resistance of the fabric and hence has no tensile stiffness. The bi-component finite elements for modeling composite fabric behavior are based on 3D membrane finite elements representative of resin behavior and truss finite elements representative of warp and weft fibers behavior. The efficiency of the proposed model resides in the simplicity of its finite element discretization and the performance of its mechanical background [46-47-50].

Due to large displacement due to forming process, the bi-component finite element mesh representing the workpiece undergoes severe shear of fibres hence, necessitates remeshing or the generation of a new mesh for the deformed/evolved geometric representation of the computational domain [51-55]. It is therefore necessary to update the mesh in such a way that it conforms to the new deformed geometry and becomes dense enough in the critical region while remaining reasonably coarse in the rest of the domain. In this paper we give the necessary steps to remeshing a mechanical composite structure subjected to large displacement. An important part is constituted by geometric and physical error estimates. A three dimensional finite element analysis has been performed using the explicit finite element formulation.

2. Geometrical Approach of Composite Fabric Forming

The draping of composite fabric using a mechanical approach requires the resolution of equilibrium PDE's problems by the finite element method (see section 3). In general, in the case of complex surfaces, the boundary conditions are not well defined and the contact between the surface and the fabric is difficult to manage. Furthermore, the resolution of such a problem can be too long in CPU time and is detrimental to the optimization stage of draping regarding the initial fibre directions. All of these facts lead us to consider rather a geometrical approach which is very fast and more robust allowing simultaneously to define the stratification sequences and the flat pattern for different plies and to predict difficult impregnated areas which involve manual operation like dart insertion or, on the contrary, the shortage of fabric. Based on technical criteria (mould surface covering, fabric drape covering and fibre angular distortion), this approach can constitute the pre-dimensionning or the pre-optimization stage for the manufacturing of complex composite parts. The geometrical approach is based, in general, on the fishnet method for which a fabric mesh element is subjected only to shear deformations. The difficulty of such a method is the mapping of the fabric mesh element onto any surface. Within this context, several algorithms (see [26-27, 30, 32, 55] for a synthesis), approximating the geometry of a fabric mesh element plotted onto the surface, are proposed. In particular, the edges of the fabric mesh element are approximated by line segments representing a pure estimate in the highly curved area.

In this study, we propose a new geometrical draping simulation algorithm which takes into account the true geometry of the fabric mesh element plotted onto the surface. Such a fabric mesh element is then defined by a curved quadrilateral whose edges are geodesic lines with the same length plotted onto the surface to drape. Given three vertices of the fabric mesh element on the surface, we propose an optimization algorithm to define the fourth vertex of the fabric mesh element. This algorithm allows us to drape the surface using an advancing front approach from the data of an initial start point between the fabric and the surface and the initial fibre directions at this point. In this section, continuous and discrete formulations of geometrical forming are presented. For the second formulation, we propose an algorithm of composite fabric draping without any approximation on the geometry of surface to be draped.

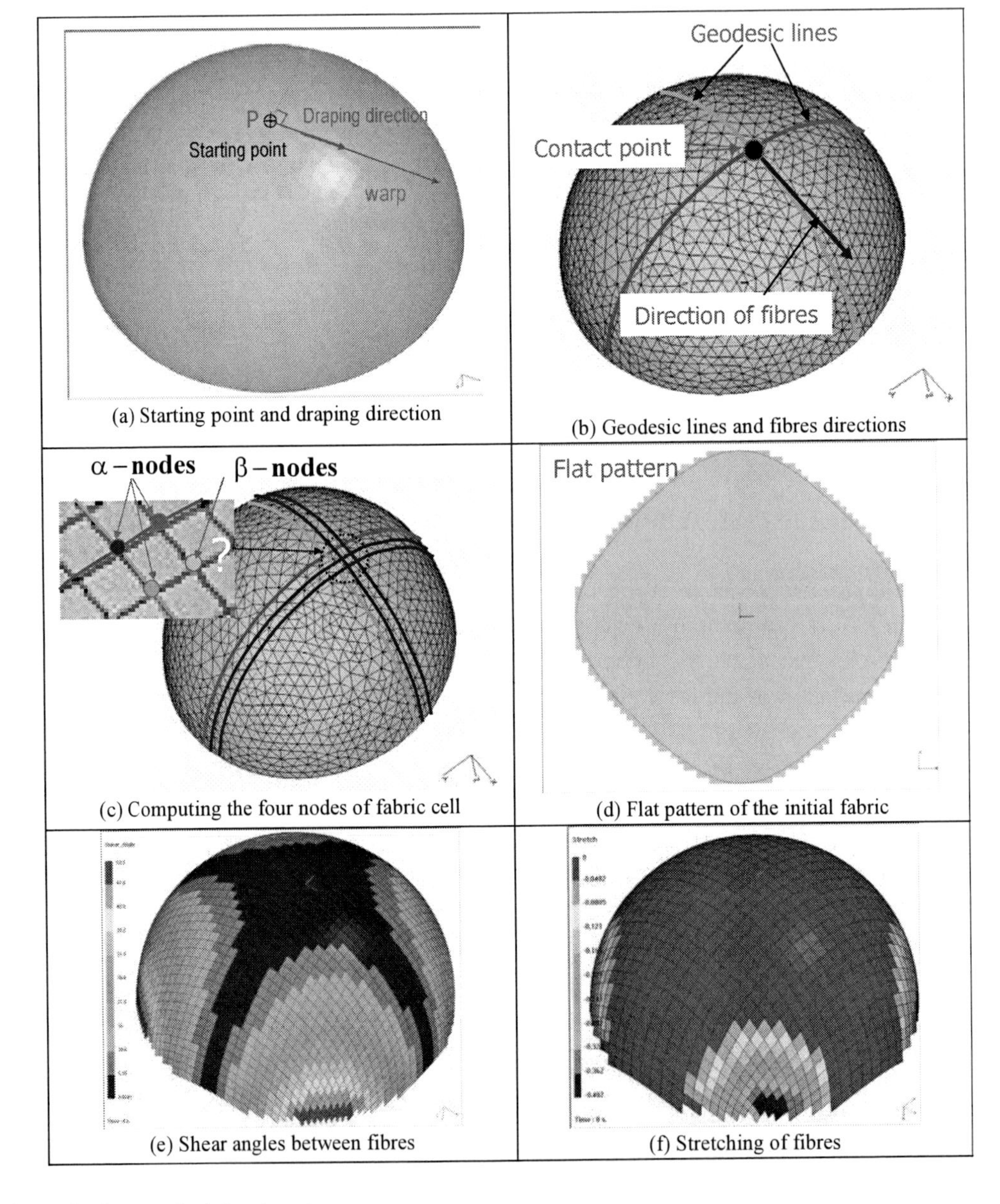

Figure 2. Geometrical draping steps.

2.1. Discrete Formulation

First, we present the mathematical formulation of the geometrical draping and then we propose an algorithm scheme to solve the draping problem. Let denote by Σ the surface of the part to drape and we assume that a geometrical mesh $\mathbf{T}_\Sigma$ of surface is known. Let $\mathcal{F}$ be the woven composite fabric modeled by two families (warp and weft) of mutually orthogonal and inextensible fibre described by the local coordinates $x = (\xi, \eta)$. These families constitute regular quadrilateral fabric mesh $\mathbf{T}_\mathbf{F}$ of the fabric $\mathcal{F}$ (Figure 2 gives example of draping steps of complex part). The problem of geometrical draping of $\mathcal{F}$ onto the surface Σ consists of calculating each node displacement of fabric mesh $\mathbf{T}_\mathbf{F}$ with a point of the surface mesh $\mathbf{T}_\Sigma$ such that the lengths of the edge of the corresponding mesh $\mathbf{T}_\mathbf{F}^\Sigma$ on the surface are preserved (no extensible). This problem presents infinity of solutions depending on:

1. Starting point associated with a node of fabric $\mathbf{T}_\mathbf{F}^\Sigma$.
2. Initial warp and weft orientation α.

Thus, to ensure a unique solution, we suppose that the points of impact on the part surface as well as the fabric orientation are given. The draping scheme is given by the following step [55]:

1. associate a starting point (corresponding to the point of impact of the machine to drape) on the surface on geometrical part mesh $\mathbf{x}_0^\Sigma = (\xi_0, \eta_0)$
2. compute step by step the warp nodes of $\mathbf{T}_\mathbf{F}^\Sigma$, classified as $\alpha-$**nodes**, from the starting point, associated with nodes (ξ, η_0) of $\mathbf{T}_\mathbf{F}$,
3. compute step by step the weft nodes of $\mathbf{T}_\mathbf{F}^\Sigma$, classified also as $\alpha-$**nodes**, from the starting point, associated with nodes (ξ_0, η) of $\mathbf{T}_\mathbf{F}$,
4. compute cell by cell all the other nodes of $\mathbf{T}_\mathbf{F}^\Sigma$, classified as $\beta-$**nodes**, from x_0 and the nodes associated with nodes (ξ, η_0) and (ξ_0, η) of $\mathbf{T}_\mathbf{F}$.

The nodes of $\mathbf{T}_\mathbf{F}^\Sigma$ associated with nodes (ξ, η_0) and (ξ_0, η) of $\mathbf{T}_\mathbf{F}$ and the $\alpha-$**nodes** are located on the surface along the geodesic lines emanating from the point of impact. Regarding the $\beta-$**nodes**, various algorithms are proposed [33-34, 42]. Most of them use an analytical expression of the surface and formulate the draping problem in terms of non-linear partial differential equations. Other algorithms are also proposed to simplify these equations by using a discrete approximation of the surface by flat triangular face (i.e. a mesh of the surface). Based on this latter approach we propose a new algorithm. The $\beta-$**nodes** are computed by solving an optimization problem corresponding to determine a vertex of an equilateral quad plotted on the surface from the data of the three other vertices. This optimization problem formulates the direction of the geodesic lines emanating from the searched vertex [45, 55, 57].

Consequently, two problems arise:

1. Problem 1: determine the geodetic exit of a given point of surface according to a given orientation.
2. Problem 2: determine the geodetic exits of these points intersecting itself mutually according to given two points of surface and lengths (these geodetic is given according to their orientations).

2.2. Numerical Examples of Geometrical Approach

Three draping simulation examples are given. These simulations are performed using the geometrical analysis computer code GeomDrap [45] and the FEA computer code Abaqus/explicit [58]. For each example, we assume that a mesh of the piece to drape is given. The first and second examples show the influence of the fibre orientations draping in the draping process. The last example shows the efficiency of the proposed method to simulate geometrically the draping of B-Pillar composite part.

The first example concerns the geometrical draping of a base plate piece. The figure 3 shows the piece as well as a mesh of this piece. The centroid of this piece is chosen as the point of impact for which two different fibre orientations (0°/90°) and $(\pm 45°)$ are specified. Figure 4 shows the resulting 3D surface lay-up for the (0°/90°) fibre orientation and the 2D corresponding flat patterns. Likewise, figure 5 shows the draping results for the $(\pm 45°)$ fibre orientation. One can notice that, in the considered cases, the surface of the piece is draped globally. However, in the second case, a smaller area of the flat fabric is used (cf. table 1). This result shows the importance of the fabric orientation in the draping process.

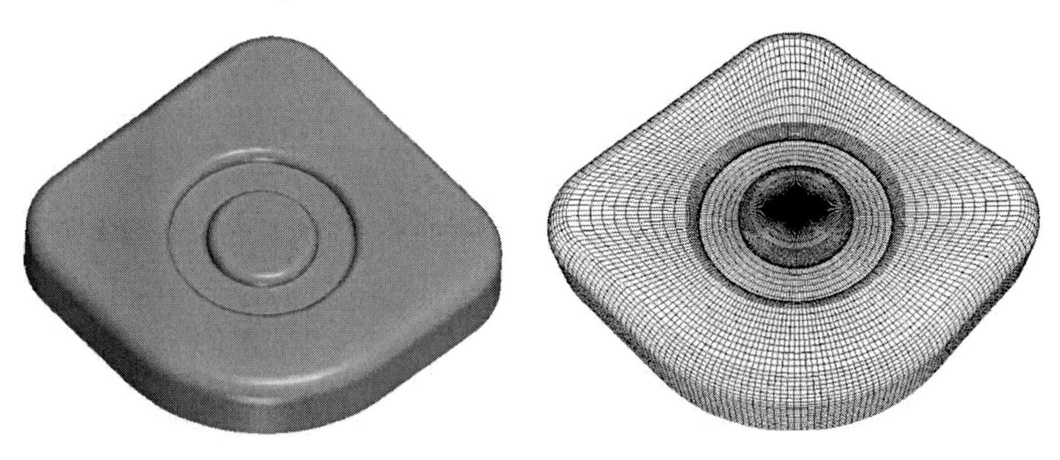

Figure 3. CAD of finite element meshing of the base plate part.

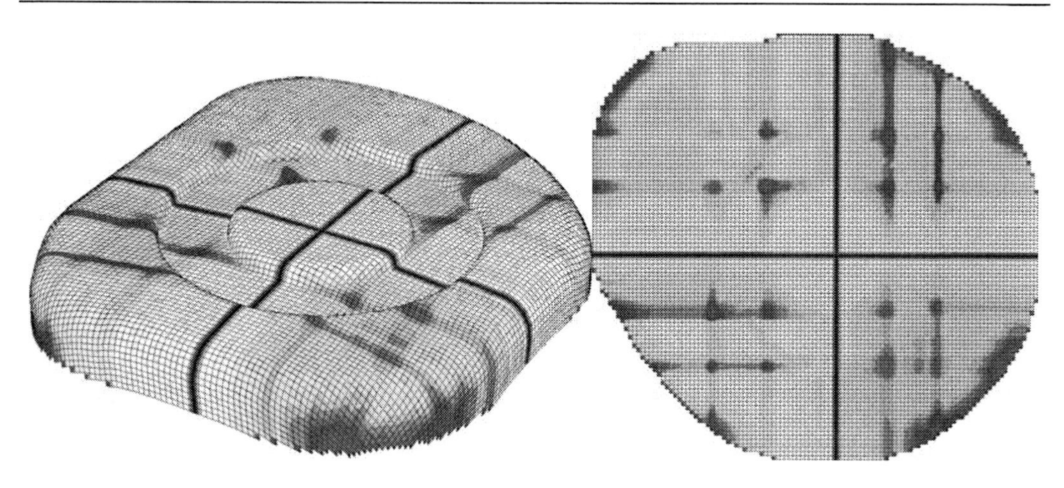

Figure 4. 3D surface lay-up (0°/90°) and 2D corresponding flat pattern.

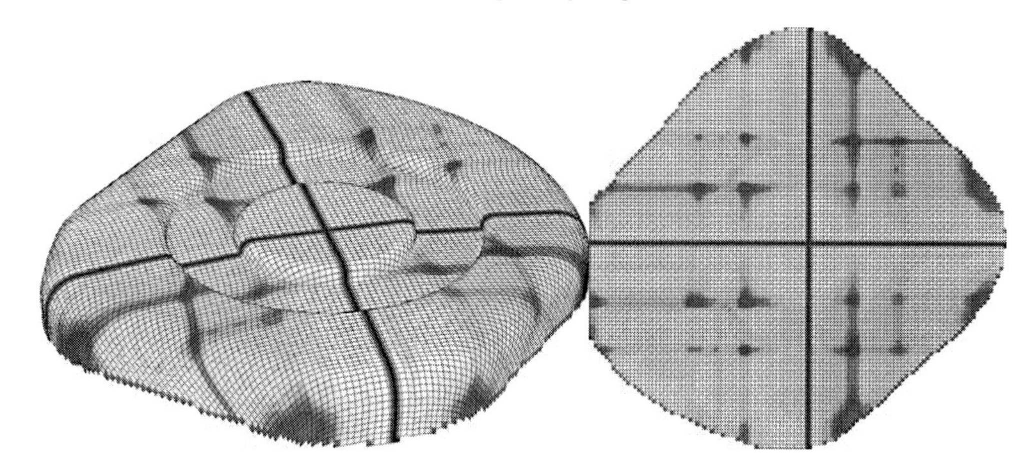

Figure 5. 3D surface lay-up (-45°/45°) and 2D corresponding flat pattern.

Table 1. Numerical results of the geometrical draping

Fabric	length of fabric	drape quality	surface covering	speed grid/s	fall rate	CPU s
(0°/90°)	120	51.0	96.7 %	27494	19.4 %	0.34
(-45°/45°)	120	58.4	95.2 %	26368	30.8 %	0.35

The second example is the draping of complex shape (car hood). The centroid of the part is chosen as the starting point from which the (0°/90°) and $(\pm 45°)$ fibre orientations are specified. Figure 6 shows the resulting 3D draping for the two orientations. We can note that all part surface is completely draped. Figure 7 presents shaded contours interpolated from the map of the fiber distortions of (0°/90°) and $(\pm 45°)$ fiber orientations. The fiber distortions for both (0°/90°) and $(\pm 45°)$ draping are very small but the maximum shear angle localization are different.

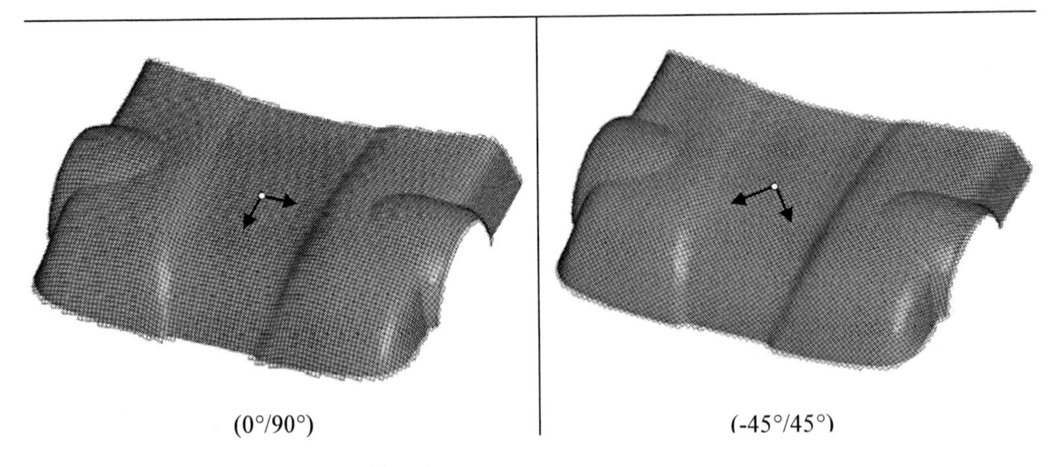

Figure 5. 3D geometrical draping of hood car part.

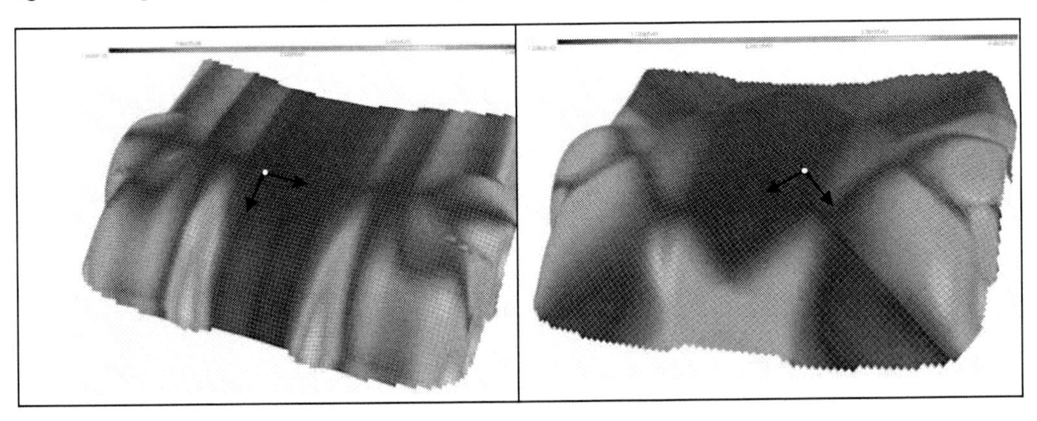

Figure 6. Iso-values of shear angles between fibres.

The last example concerns the geometrical draping of the B-Pillar composite part. The B-pillar is a major structural component for side impact car safety. The use of carbon composite as an alternative in the vehicle B-Pillar can reduce the risk of injuries on the occupant. The aim of this simulation is to demonstrate the capacity of the used model to drape completely complex shape and the effect of the chose of the starting point and the initial fibre orientation of the draping process. The centroid of the part is chosen as the starting point from which the (0°/90°) fibre orientations are specified (Figure 7a) and in order to drape completely the b-pillar shape. Figure 7b shows the resulting 3D draping corresponding to the fibre orientations. We can note that all part surface is completely draped. The fiber distortions draping are very large and the maximum shear angle localization are shown in Figure 7. The B-pillar shape is completely draped with 20689 quadrilateral elements. The choice of the starting point (impact point) has a considerable effect on draping of the part. The point optimized to drape the part completely is on the high part.

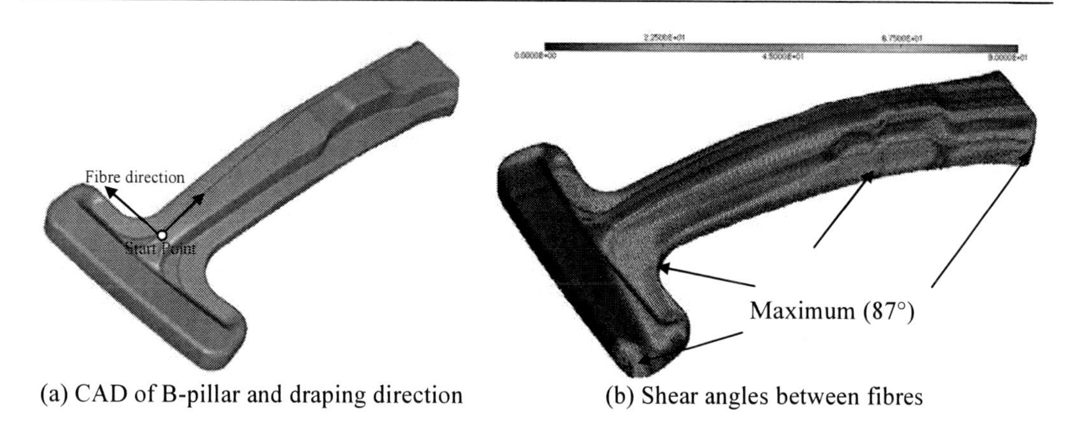

(a) CAD of B-pillar and draping direction

(b) Shear angles between fibres

Figure 7. Geometrical draping of B-pillar composite part.

3. Mechanical Approach

The mechanical behaviour of composite fabrics is complex and it is a multi-scale problem due to the interactions of the fibres or yarns. The macroscopic behaviour is very much dependent on the interactions of fibres at the meso-scale (scale of the woven unit cell) and at the micro-scale (level of the fibres constituting yarns). Despite of a great amount of work in the field, there is no widely accepted model that accurately describes all the main aspects of fabric mechanical behaviour. The main model families come from the multi-scale nature of the textile.

During the forming process of woven fabric, the two main mode of deformation at the mesoscopic scale are the stretching of the fibres due fibres undulation and the in-plane shearing of the fabric resulting in a change of the angle between the warp and the left yarns. In the deep-drawing or the draping of woven fabrics, the in-plane shear of fibres is the principal mode of deformation and is very different than the sheet metal [59-65]. Figure 8 shows the evolution of two straight lines draw alternatively on warp and weft fibre directions during the forming deformation. These lines become curved but remain continuous. The absence of inter-yarn sliding ensured by the fabric weaving, viscoelastic behavior of resin and friction fiber/fiber and fiber/resin) can be observed over the main areas of the fabric (i.e. far enough from the free edges of the fabric). Also, for the composite fabrics based on high modulus, the compressive as well as bending stiffness are negligible compared to the in-plane membrane stiffness. The assumption is that each cross connection of straight warp and weft fibre before deformation remains cross connected during the deformation. The basic assumptions for the mechanical forming are that the woven fabric is considered as a continuous 3D material. The warp and weft fibres are represented by a truss which connecting points are hinged and the membrane resin is coupled kinematically to the fabric at these connecting points.

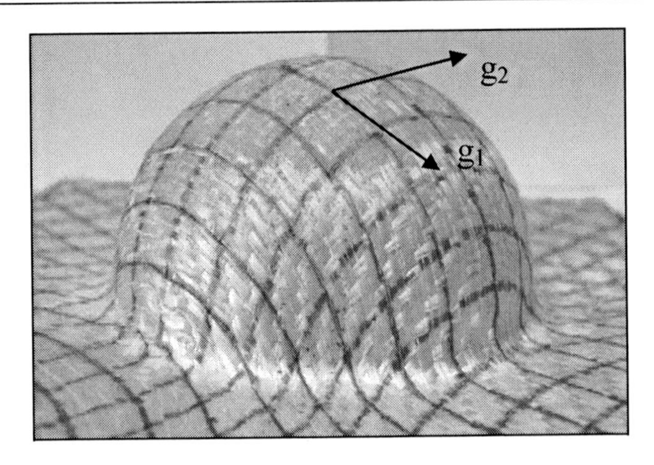

Figure 8. Woven fabric deformation mode.

3.1. Model Description

In the unit cell of the mesoscopic model used here for a plain-weave fabric each fibre and resin is modelled and is assumed to be a straight or a curved beam or truss. The tensile load is carried by the 3D truss or beam elements that will capture the changes in the orthotropic directions during the shearing. The 3D membrane or shell element accounts only for the shearing resistance of the fabric and hence has no tensile stiffness. The interaction warp-weft yarn and resin-fibres is negligible. In this approach, the stress and strain of a continuous material are related to fibrous reinforcement using the constitutive relation in a non-orthogonal frame directed by the fibre directions. In this study we consider two yarn directions and we use them to define the non-orthogonal frame. This approach uses the Green Naghdi frame. It is an orthonormal frame which is rotated by R, the rotation of the polar decomposition in which the local stress increment computations at finite strain are made [66-70].

Two reference frames have to be considered. The $\mathbf{e}_i$ unit vectors define the local orthogonal reference frame that rotates with the continuum material, and the $\mathbf{g}_i$ basis vectors form a non-orthogonal frame that follows the fibre direction. Here, $(\mathbf{g}_1, \mathbf{g}_2)$ correspond to warp and weft directions, respectively.In this case, for each connecting point $\vec{\mathbf{X}}^{\mathbf{f}}$ of warp and weft yarns is associated a material position space of a resin $\vec{\mathbf{X}}^{\mathbf{m}}$. At the connecting points we have $\vec{\mathbf{X}}^{\mathbf{f}} = \vec{\mathbf{X}}^{\mathbf{m}} = \vec{\mathbf{X}}$ before deformation. The current position of these points is obtained by:

$$\begin{cases} \mathbf{d}\vec{\mathbf{x}}^{\mathbf{f}} = \mathbf{F}^{\mathbf{f}}\left(\vec{\mathbf{X}},\mathbf{t}\right)\mathbf{d}\vec{\mathbf{X}}^{\mathbf{f}} & \textbf{fibres} \\ \mathbf{d}\vec{\mathbf{x}}^{\mathbf{m}} = \mathbf{F}^{\mathbf{m}}\left(\vec{\mathbf{X}},\mathbf{t}\right)\mathbf{d}\vec{\mathbf{X}}^{\mathbf{m}} & \textbf{resin} \end{cases} \tag{1}$$

where $\mathbf{F}^{\mathbf{f}}$ and $\mathbf{F}^{\mathbf{m}}$ are the deformation gradient tensor of fibre and resin respectively. The relationship of the no sliding inter-fibre can write at each connecting point as

$\vec{x} = \vec{x}(\vec{X}, t) / \vec{x}^f = \vec{x}^m = \vec{x}$. The gradient of transformation of the pre-impregnated woven fabric **F** and the pseudo gradient of transformation of the fibre $\mathbf{F}^f$ are defined by the function

$$\begin{cases} F_{ij}^f = \lambda_L g_i \otimes g_{0j} & \text{fibres} \\ F_{ij}^m = \dfrac{\partial \mathbf{x}^m}{\partial \mathbf{X}^m} e_{0i} \otimes e_{0j} & \text{re sin} \end{cases} \tag{2}$$

where λ_L^f is the longitudinal elongation of each fibre and g_{0i} and g_i are respectively the fibre orientations in the initial C_0 and the current C_t configurations and are the resin . The relative rotation of fibre can be associated to the rotation of the rigid body of the median line of the fibre $\mathbf{R}$. We can note this assumption by the following kinematic relation:

$$\mathbf{R} = g_i \otimes g_{0i} \tag{3}$$

Using the above assumptions, the mechanical deformation of composite fabric depends on the relative movement of fibres and the deformation of not polymerized resin.

$$\begin{cases} \lambda_i^f = \sqrt{g_{0i} F^{fT} F^f g_{0i}^f} & \text{fibres} \\ U^m = \sqrt{F^{mT} F^m} & \text{re sin} \end{cases} \tag{4}$$

The shaping problem imposes the use of incremental formulation in finite deformations. In finite deformation analysis a careful distinction has to be made between the coordinate systems that can be chosen to describe the behaviour of the body. The rate constitutive equations for finite strain use objective derivatives [65-66]. The problem of the integration of strain rate tensors is a central one in large deformations. The rate of deformation tensor of woven fabric is obtained by:

$$\begin{cases} D^f = \left(\dfrac{\dot{\lambda}_i^f}{\lambda_i^f} \right) \left(g_i \otimes g_i \right) & \text{fibres} \\ D^m = \dfrac{1}{2} \left(\dot{F} F^{-1} + F^{-T} \dot{F}^T \right) \left(e_{0i} \otimes e_{0i} \right) & \text{re sin} \end{cases} \tag{5}$$

The rate equations for finite strains use objective derivatives [36-38, 66]. The approaches traditionally developed in finite element codes for anisotropic metal at large strains are based on Jaumann corotational formulation or the Green–Naghdi approach. In these models, a rotation is used both to define an objective derivative for the hypoelastic model law and to update the orthotropic frame. The rotations used in Green–Naghdi and Jaumann derivatives are average rotations of the material (polar rotation and corotational rotations respectively). The frame associated with Green-Naghdi's derivative is defined, at the material point considered. The rotation $\mathbf{R}^f$ is used to update the initial constitutive axes of warps or weft fibres $g_{0i} \otimes g_{0i}$ to the current constitutive axes and the rotation $\mathbf{R}$ is used to update the initial constitutive axes of resin $e_{0i} \otimes e_{0i}$. The stretching tensors Eq. (5) are written in the rigid body rotation frames. The longitudinal component is $\bar{D}_L^f = \dot{\lambda}_L^f / \lambda_L^f$ and the transversal components $\left(\bar{D}_T^f, \bar{D}_3^f\right)$ are obtained by the unidirectional behaviour of fibre as $\bar{D}_T^{fR} = \bar{D}_3^{fR} = -\nu_{LT}\bar{D}_L^{fR} \Rightarrow \lambda_T^f = \lambda_3^f = \left(\lambda_L^f\right)^{-\nu_{LT}}$ and ν_{LT} is the Poison's ratio of fibre. Using Green-Naghdi's objective tensor stress, the stress rate $\bar{\sigma}_L^f$, depending on the stretching deformation $\bar{D}_L^f$, and the stress rate tensor of the membrane resin $\bar{\boldsymbol{\sigma}}^m$, depending on the tensor deformation rate $\bar{\mathbf{D}}^m$ and the elastic properties $\mathbf{C}^m$, can be written at each time as: The stress and strain relationship of the woven fabric composites are defined in the non-orthogonal material coordinate frame. Therefore, coordinate transformations of stress and strain into the orthogonal coordinate system should be considered in the non-orthogonal constitutive model as $\sigma = \bar{\sigma}^f + \bar{\sigma}^m$:

$$\sigma = \begin{pmatrix} \sum\limits_{weft} E_L^f(\lambda_1^f)\frac{\dot{\lambda}_1^f}{\lambda_1^f} - \nu_{LT}E_L^f(\lambda_2^f)\sum\limits_{weft}\frac{\dot{\lambda}_2^f}{\lambda_2^f} & 0 & 0 \\ 0 & \sum\limits_{warp} E_L^f(\lambda_2^f)\frac{\dot{\lambda}_2^f}{\lambda_2^f} - \nu_{LT}\sum\limits_{weft} E_L^f(\lambda_1^f)\frac{\dot{\lambda}_1^f}{\lambda_1^f} & 0 \\ 0 & 0 & 0 \end{pmatrix} + \begin{pmatrix} \frac{E^m}{1-\nu_m^2} & \frac{\nu_m E^m}{1-\nu_m^2} & 0 \\ \frac{\nu_m E^m}{1-\nu_m^2} & \frac{E^m}{1-\nu_m^2} & 0 \\ 0 & 0 & G^m \end{pmatrix} \begin{Bmatrix} \bar{D}_{11}^m \\ \bar{D}_{22}^m \\ \bar{D}_{12}^m \end{Bmatrix} \quad (6)$$

The constitutive law of fibres is nonlinear and is written in terms of longitudinal modulus of stretching $E_L^f\left(\lambda_L^f\right)$, the compressive stiffness $\left(\lambda_L^f \leq 0\right)$ of fiber is supposed negligible $E_L^f = 0$. Later is function of elongation of warp and weft fibre $\left(\lambda_1^f, \lambda_2^f\right)$, effective elastic modulus of fibre $\bar{E}_f$ and undulation factor ε_{sh}; $\left(E^m, \nu_m\right)$ are the membrane elastic properties. To determine the stress state in a membrane material at a given time, the deformation history must be considered. For linear viscoelastic materials, a superposition of hereditary integrals describes the time dependent response. Let $G^m(t)$ be the shear stress relaxation modulus of the not polymerized resin and $G^{m^\infty} = G^m(t = \infty)$ the limit value. The viscoelastic behaviour of not polymerized resin is formulated in the time domain by the hereditary integral and using the relaxation time τ_k and the shear modulus relaxation, which are material parameters G^{m^k}. Hereditary integrals with Prony series kernels can be applied to model the shear behavior of the not polymerized resin (Abaqus 2004). The behavior of fibre and resin can be written as:

$$\begin{cases} E_L^f\left(\lambda_L^f\right)=\bar{E}_f\left(1-\mathrm{Exp}\left(\dfrac{-\dot{\lambda}_L^f}{\lambda_L^f\varepsilon_{sh}}\right)\right) & \text{fibres} \\ G^m(t)=G^m(\infty)+\sum_1^k G^{m^k}\mathrm{Exp}\left(\dfrac{-t}{\tau_k}\right) & \text{re sin} \end{cases} \tag{7}$$

3.2. Finite Element Formulation

Each material point is moving as in a continuum, ensured by the non-sliding of fibers due to fabric weaving and resin behavior. Therefore, a nodal approximation for the displacement can be used. The deformation of composite fabric is described within the frame of membrane assumptions. The energy of deformation $\Pi(\dot{u})$ is obtained by a summation of membrane strain energy of not polymerized resin and elastic tensile strain energy of fibers:

$$\delta\Pi(\dot{u})=h_0\int_{S_0^m}\bar{\sigma}^m:\delta\bar{D}^m ds+\sum_{\text{fibres}}S_0^f\int_{L^f}\bar{\sigma}_L^f:\delta\bar{D}_L^f dl-\int_{\Gamma_\sigma}\bar{t}.\delta\dot{u}d\Gamma \tag{8}$$

where h_0 denote the initial thickness of fabric, L^f the length of fiber and S_0^f the initial effective cross section of fibre and $\bar{t}$ is the external surface load applied along Γ_σ of woven fabric. The effective cross section of the fibre S_0^f and the effective surface S_0^m of the membrane resin, that assumed no void between the fibre and the resin, was used are calculated by using the fibre volume fraction V_f of the woven fabric ($V_{fabric}=V_fS_0^fL^f+(1-V_f)h_0S_0^m$).

The global equilibrium of the fabric is obtained by minimizing the total potential energy $\Pi(\dot{u})$. The effect of spatial equilibrium of composite material on the actual configuration is established in terms of nonlinear equations: kinematic non-linearity, material non-linearity and contact with friction non-linearity. It is linearized for each load increment by an iterative Newton method. It should be emphasized that during the motion, nodes and elements are permanently attached to the material points with which they were initially associated. Consequently, the subsequent motion is fully described in terms of the current nodal positions as:

$$x=\sum_{k=1}^{m}N^k(\xi,\eta)X^k+\sum_{k=1}^{m}N^k(\xi,\eta)u^k \tag{9}$$

where u^k are the nodal displacements of each connecting point, $N^k(\xi,\eta)$ are the standard shape functions (of membrane of truss element) and (m) denotes the number of nodes.

The discretization of rate deformation tensor can be obtained by introducing Eq.5 into the definition of ($\bar{D}_L^f$ and $\bar{\mathbf{D}}^m$) and given in Eq. 8 to give:

$$\begin{cases} \bar{D}_L^f = \dfrac{\dot{\lambda}_L^f}{\lambda_L^f} = \dfrac{1}{\alpha_n^2}\left(\dfrac{dx^T}{d\xi}\dfrac{d\dot{u}}{d\xi}\right) = \left[B_{fibres}\right]\dot{u}_n & \text{fibres} \\ \bar{D}^m = \dfrac{1}{2}R^T\left(\dfrac{d\dot{u}}{d\xi}\dfrac{d\xi}{dx} + \dfrac{\partial \dot{u}^T}{d\xi}\dfrac{d\xi}{dx}\right)R = \left[B_{resin}\right]\dot{u}_n & \text{resin} \end{cases} \tag{10}$$

where $\alpha_n = \sqrt{\dfrac{d\mathbf{X}^{fT}}{d\xi}\dfrac{d\mathbf{X}^f}{d\xi} + 2\dfrac{d\mathbf{X}^{fT}}{d\xi}\dfrac{d\mathbf{u}}{d\xi} + \dfrac{d\mathbf{u}^T}{d\xi}\dfrac{d\mathbf{u}}{d\xi}}$ and $\left[B_{resin}^e\right]$ and $\left[B_{fibres}^e\right]$ are the geometric or strain-displacement matrix of membrane resin and truss fibres.

In a finite element approximation, the only independent variables in the equations of linearized virtual work are the displacements of the material points. Substituting the element coordinate and displacement interpolations into the equilibrium equations, for a given set of elements we obtain the state that forces acting on a fabric equals the mass times the acceleration of the body:

$$\sum_e \left[M^e\right]\{\ddot{u}\} + \sum_e \left(\left\{\Re_{int}^e\right\} - \left\{\Re_{ext}^e\right\}\right) = \{0\} \tag{11}$$

where $\left[\mathbf{M}^e\right]$ is the consistent composite mass matrix and $\left(\left\{\Re_{\mathbf{int}}^{\mathbf{e}}\right\} - \left\{\Re_{\mathbf{ext}}^{\mathbf{e}}\right\}\right)$ is the so called quasi-static equilibrium residual :

$$\begin{cases} \left[M^e\right] = \rho_r h_0 \displaystyle\int_{S^m} \left[N^e\right]^T \left[N^e\right] ds + \sum_{fibres} \rho_f S_0^f \int_L \left[\tilde{N}^e\right]^T \left[\tilde{N}^e\right] dl \\ \left\{\Re_{int}^e\right\} - \left\{\Re_{ext}^e\right\} = h_0 \displaystyle\int_{S^m} \left[B_{resin}^e\right]^T \left\{\bar{\sigma}^{mR}\right\} ds - \sum_{fibres} S_0^f \int_{L^f} \bar{E}_L^f \left[B_{fibres}^e\right]^T \left\{\bar{\sigma}^{fR}\right\} dl - \int_{\Gamma_f} \left[N^e\right]^T \left\{\bar{t}\right\} ds \end{cases} \tag{12}$$

where $\left[N^e\right]$ and $\left[\tilde{N}^e\right]$ are the matrix of the nodal interpolation functions both associated elements and $\left(h_0, S_0^f\right)$ are the initial thickness and surface of resin and fibres, respectively. The index e refers to the eth element.

According to the different modes of deformation occurring in the prepreg fabric during the shaping process, bi-component finite elements are developed to characterize the mechanical behavior of thin composite structures. The bi-component element is based on an association of 3D linear membrane finite elements (T3 and Q4) combined with a complementary truss linear finite elements. The global stiffness of composite fabric is obtained by the summation of elementary stiffness matrix of warp fiber, elementary stiffness matrix of weft fiber and elementary stiffness matrix of resin. These finite elements are complementary in the finite element discretization (isoparametric and use three DOF per node) and use the same mechanical formulation in finite deformations (Green- Naghdi's approach). The non linear constitutive equation of fibre behavior is implemented in the Abaqus/Explicit using VUMAT user's subroutine [17-20].

The governing equilibrium Eq. 11 is solved as a dynamic problem using explicit integration. This is achieved by using the central difference method to approximate the velocity and the acceleration in the next time step, using only information from the previous step, where all state variables are known. This approach has proven to be, in particular, suitable to highly non-linear geometric and material problems, particularly where a large amount of contact between different structural parts occurs. In the present work, the Dynamic Explicit (DE) resolution procedure is used within the general purpose FE code Abaqus/Explicit. The DE algorithm available in Abaqus/Explicit for solving the algebraic system works by using the lumped form of the mass matrix [58, 66]. The major disadvantage of the explicit scheme is that it is only stable for time steps small enough. This can cause for instance the energy balance to be changed. One simple and conservative criteria for a stable time increment:

$$t_{stable} = \frac{\rho_r L^e}{\bar{E}^f} \tag{13}$$

where L^e is the minimal length of membrane element, ρ_r the resin density and $\bar{E}^f$ Young's modulus of fibre. The interpretation of this requirement is that the time increment must be shorter than the time it takes for a propagating wave in the material to cross the shortest side of an element. Thus, the element size and the critical time increment are connected. The smaller the elements are, the shorter the time increment must be. Because of the difference in wave speed for different materials, the critical element size is larger for a stiff material (e.g. steel) than for a softer material (e.g. polymer).

3.3. Application of Mechanical Approach

3.3.1. Uniaxial Test

Due to the importance of the composite fabric behaviour on material formability, tensile test of pre-impregnated fabric is proposed in order to study the influence of fibre orientation, fibre undulations due to fabric weaving and resin behaviour. At processing, the impregnated fabric is idealised as a viscous material subject to the kinematic constraints of incompressibility and inextensibility in the fibre direction. The pre-impregnated fabric tested in this study was a satin 5 with aramid woven fabric. The fabric was impregnated with epoxy resin using a hot-melt pre-pregging process (the mechanical properties are given in Table 2). A lower loading velocity will generate lower viscous forces at the intra-ply shearing of fabric. In this experiment, a displacement in the vertical direction is imposed at the moving extremity of the rectangular composite specimen with three layers of fibres (length =150mm and width=30mm and thickness=2mm). The uniaxial tensile test (see Figure 9) is carried out for different orientations of fibres with the loading direction (0, 15, 30 and 45°). The experimental effort imposed by the tensile machine is compared to the numerical values for different fibre orientations in Figure 10. The behaviour of the not polymerized resin and the initial fibre orientation influenced largely the global response of the fabric during tensile loading. In this figure we can show the good correlation between the model and the experimental results. The agreement between predicted and experimental values is good and proves the validity of the proposed model of pre-impregnated woven fabric behaviour. The numerical model described above clearly shows the strong non linearity of this behaviour law. It takes into account the mechanical characteristics of a viscoelastic resin, the anisotropic behaviour of fabric and the geometrical non linearities due to the high deformability of fibres (straightening and relative rotation).

Table 2. Mechanical properties of the pre-impregnated composite fabric

E_f (MPa)	ε_{sh}	ρ (g/cm^3)				
130000	0.005	1,45				
time **(s)**	0.01	0.1	1	10	100	1000
Shear modulus of resin G^{mk}	0.02332	0.023332	0.083509	0.11723	0.14423	0.178

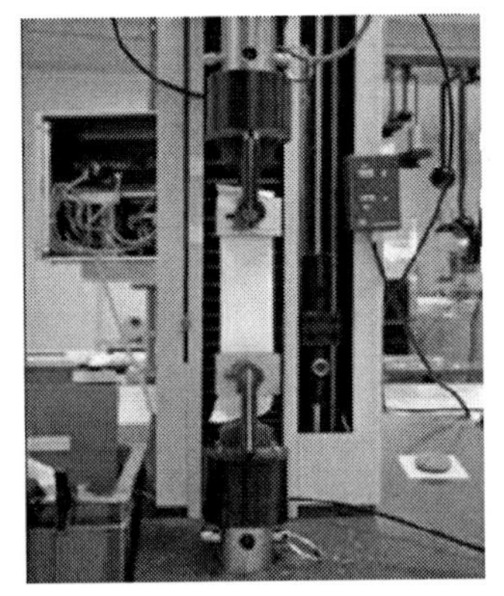
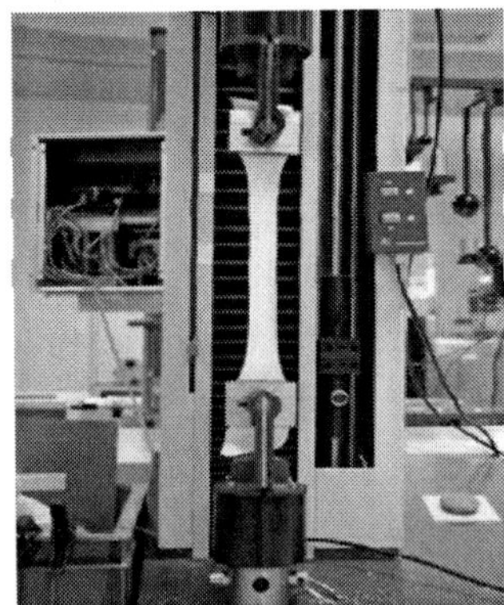
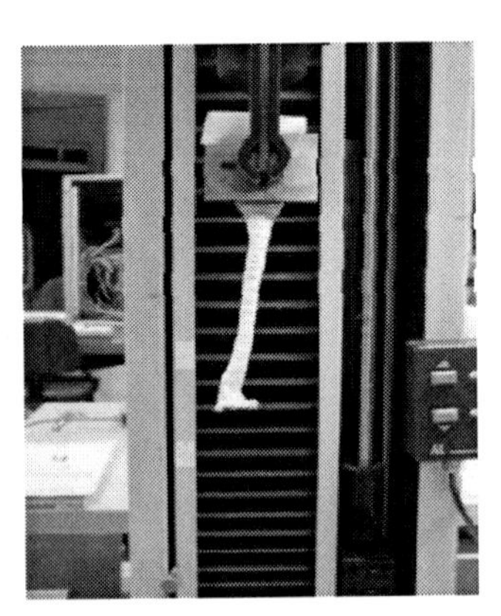

Figure 9. Tensile apparatus of pre-impregnated composite fabric specimen.

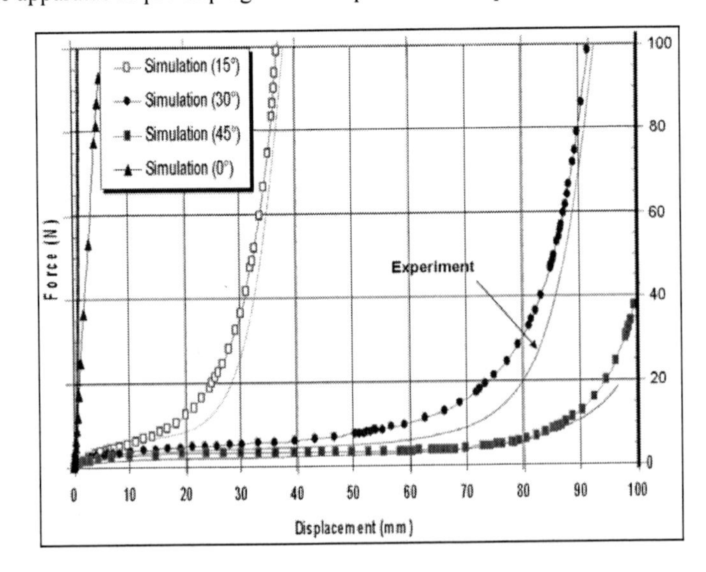

Figure 10. Load force- displacement of the fabric specimen for different fibre orientations.

3.3.2. Biaxial Test

In order to determine and analyse the undulation and interaction effects, a biaxial tensile apparatus has been built in order to test woven materials in both the warp and weft directions. The loads are measured using load cells located very close to the specimen in order to avoid the influence of friction within the device. The strains are measured using extensometers devices which permit the verification of the homogeneity of the strain field in the effective part of the specimen. The set of tests which have been performed on different fibre fabrics (carbon twill 2x2), mainly used in aeronautics, has shown features of the biaxial behaviour and the importance of the undulations due to the weaving. For a given fabric, the response is strongly influenced by the strain ratio in warp and weft directions due to large fibre undulations [46]. Especially, the non-linearity in warp and weft direction is increasing. The tendency of the fibre to straighten is more impeded when the strain in the perpendicular direction is large (see Table 3). The biaxial specimen, as shown in Figure 11, was modelled in 2D with 400 four nodes membrane elements (representative of elastic behaviour of not polymerized matrix) and 1600 two nodes truss elements (representative of elastic non linear behaviour of fibres) of Abaqus element library. The load was applied at the same time on each strip. The overall agreement of the proposed non-linear numerical model is very good in comparison with the experimental results (see Figure 12).

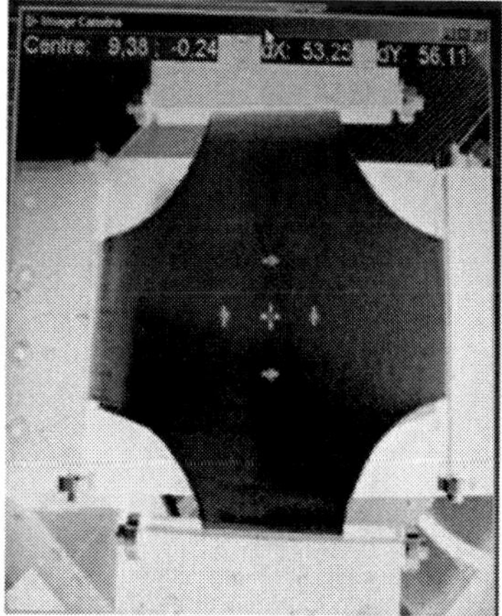

Figure 11. Device for a biaxial tensile test on cross fabric specimen.

Table 3. Mechanical parameters of the composite fabric carbon twill 2x2

	Warp direction	Weft direction
Maxi strain	132%	80%
Maxi displacement	80 mm	80 mm
Maxi force	620 N	1180 N
Maxi Stress	35.42 MPa	67.42 MPa
Longitudinal Young's modulus $\left(\overline{E}_f\right)$	15.72 MPa	27.79 MPa
Undulation factor ε_{sh}	0.015	0.20

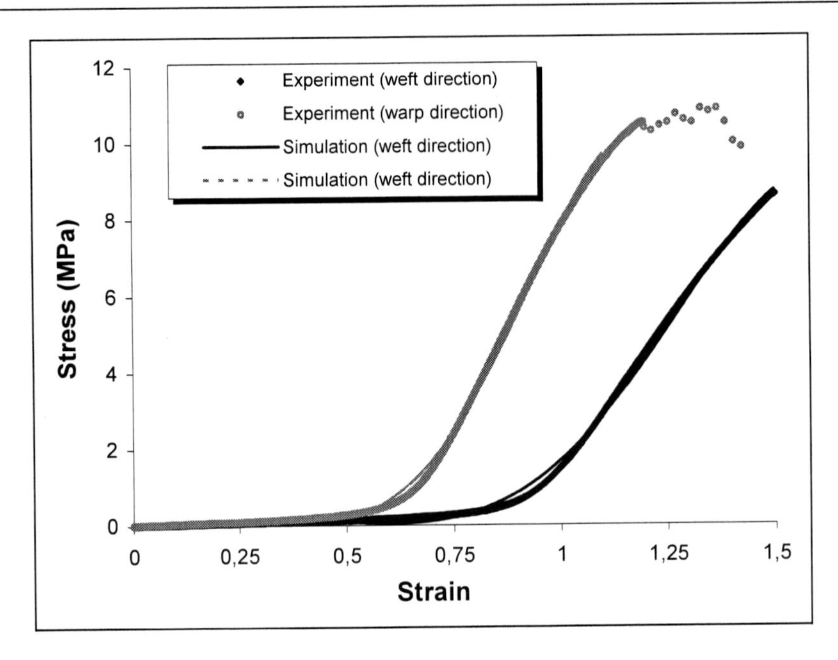

Figure 12. Effect of the fibre undulation of the force versus displacement response.

3.2.4. Deep-Drawing of Prepreg Woven Fabric

The numerical analysis of prepreg composite fabric deformation by deep-drawing process is performed by utilizing the commercial FEM-package ABAQUS/EXPLICIT. The resin is modeled by using 1600 membrane finite elements (linear triangular element M3D3) and warp and weft fibres are modeled by 3200 truss finite elements (linear element T3D2). The rigid surface is modeled by 1600 Bezier patches (three nodes R3D3 and four nodes R3D4). The behavior of the resin is assumed to be isotropic viscoelastic and the behaviour of the fibre is supposed as elastic.

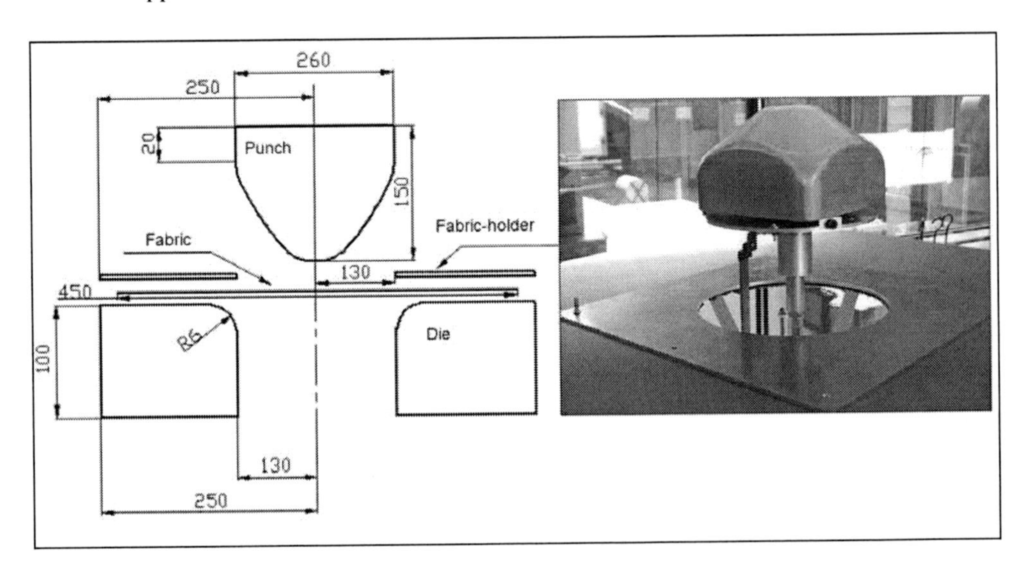

Figure 13. Geometry of deep-drawing tools.

The first forming example concerns the 3D deep-drawing of aramid pre-impregnated fabric with conical tools (see Figure 13). The mechanical properties of the used material are given in Table 2. Figure 14 reports the experimentally obtained shapes with respect to 0°/90° and ±45° fibre orientation for different punch displacements (initial, and final for 100mm of punch displacement). The corresponding final computed shapes are shown in Figure 15. The evolution of the predicted shear angle variations mechanical approach is compared on Figure 16 with the experimental values. We notice that these shear angle values are very large >38° for mechanical approach along the median line for ±45° fabric and along the diagonal line for 0°/90° fabric. But along the median lines of 0°/90° and along the diagonal lines of ±45° the angular distorsions are very small <6°.

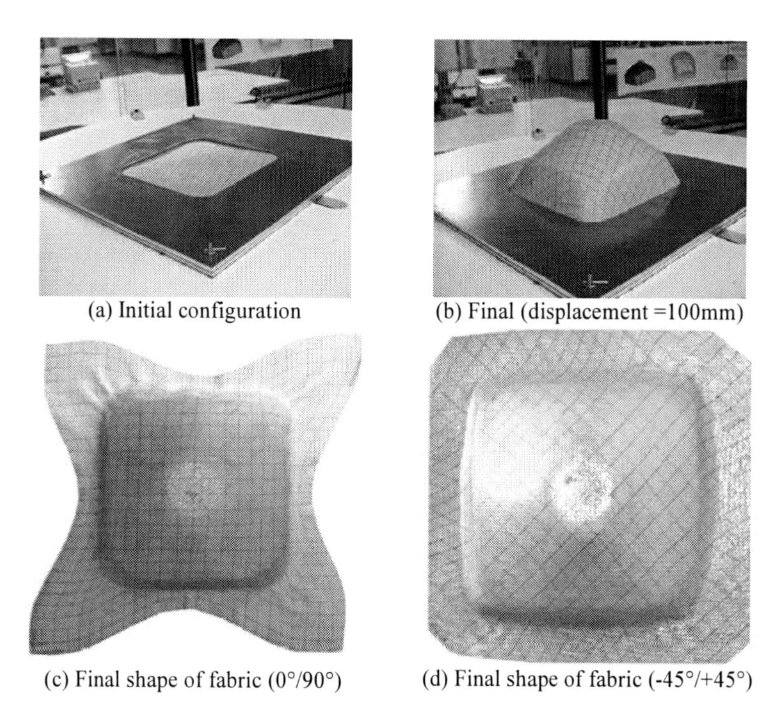

(a) Initial configuration (b) Final (displacement =100mm)

(c) Final shape of fabric (0°/90°) (d) Final shape of fabric (-45°/+45°)

Figure 14. Experimental results (a) (0°/90°) and (b) (-45°/45°).

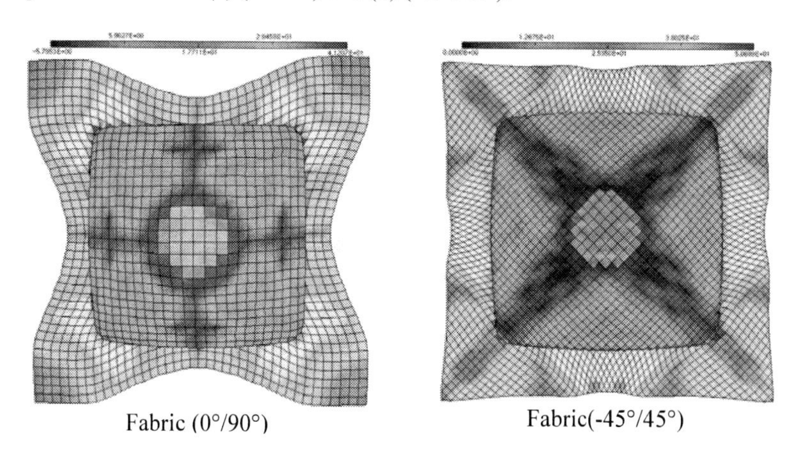

Fabric (0°/90°) Fabric(-45°/45°)

Figure 15. Predicted results (a) (0°/90°) and (b) (-45°/45°).

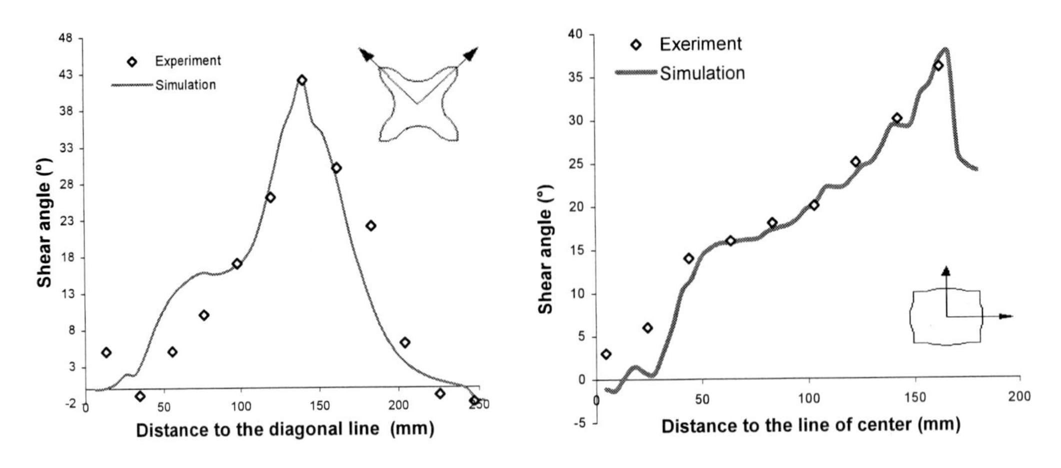

Figure 16. A comparison between experimental and predicted shear angles.

3.2.5. Deep-Drawing of Dry Woven Fabric

The second example is the 3D deep-drawing of dry glass woven fabric by hemispherical punch by experimental and mechanical approaches. The initial shape of the glass fibre fabric is a square (360x360mm). Its edges are free but a pressure equal to 2MPa is applied on the binder and the friction coefficient between the glass fabric and the steel tools is 0.27. The forming simulation has been performed in ABAQUS/Explicit using rigid tools. As mentioned before the shear resistance model is assigned to reduced membrane elements, and truss elements parallel to the membrane edges represent the high tensile stiffness in the yarn direction. Figures 17a and 18a report respectively the experimental obtained shapes with respect to (0°/90°) and $(\pm 45°)$ fibre orientations, while figures 17b and 18b show the final shape using the numerical approach. Likewise, the numerical simulation agrees with the experimental results. The angular fibre distorsion exceeds 38° along the diagonal axis for (0°/90°) fibre orientations and 50° along the diagonal axis for $(\pm 45°)$ fibre orientations. Furthermore we can notice that the final shape obtained with (0°/90°) fibre direction is very different from the (-45°/+45°). Another interesting result of the numerical calculation is the angular distortion variations between warp and weft fibres.

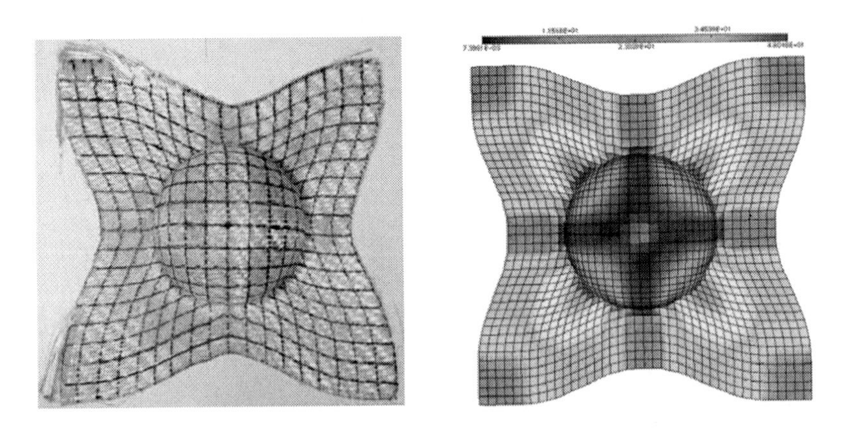

Figure 17. A comparison between (a) experimental and (b) predicted shape shear angles of (0°/90°).

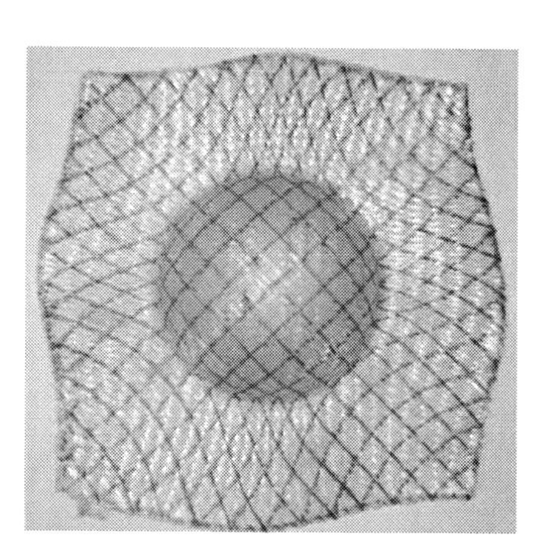

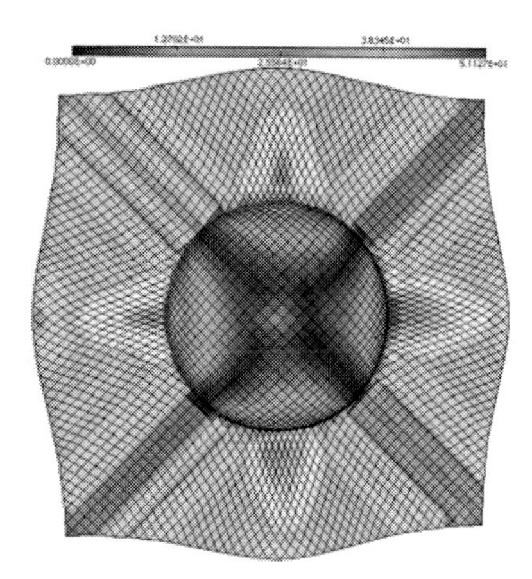

Figure 18. A comparison between (a) experimental and (b) predicted shape and shear angles of $(\pm 45°)$.

4. Remeshing Procedure of Composite Fabric

The shape of elements in the finite element analysis may be the most important of many factors which induce a discretizing error. In particular, the efficiency of adaptive refinement analysis depends on the shape of the elements, and so estimating the quality of element shape is requisite during the adaptive analysis being performed. Unfortunately, most posterior error estimates can not evaluate the shape error of an element, so that some difficulties remain in the application of an adaptive analysis. The goal of this section is to develop a fully automatic procedure to update the geometry of the deformed fabric and generate its discretization in an effort to develop an automated modeling system for forming analyses. There are several key technical issues that affect the overall accuracy of the simulation that must be considered in the automation of the process. The use of finite element methods for problems where the domain evolves during the simulation presents a number of challenges that are not present in solutions for a fixed domain. Two high level issues that must be considered when automating process simulations (i) *when* to remesh, and, (ii) *how* to remesh.

The criteria used to trigger a remesh are collectively called the *remeshing criteria*. Four sources of errors that influence the decision to remesh are:

(i) Geometric approximation errors (due the discretization of the workpiece and the tools)
(ii) Element distortion errors (error due to element shape).
(iii) Mesh discretization errors (errors in the solution due to use of specific basis functions on individual finite elements).
(iv) Mesh rezoning errors (an issue critical to the successful use of remeshing in history dependent solution procedures is the transfer of the appropriate solution variables between meshes.).

The impact of the different types of errors encountered based on metrics to measure them will key a remeshing step. The process of remeshing focuses on controlling these errors so that the simulation can continue. The objective of the geometry update procedure is to take a finite element mesh representing the current state of the workpiece, and evolve the geometric representation of the workpiece to represent its current deformed shape. This geometric representation is then discretized to generate a mesh consisting of valid, high quality elements focused on controlling the mesh discretization and distortion errors [71-76]

The simulation of the composite fabric forming is based on an iterative process. At first, a coarse initial mesh of the part is generated with bi-component finite elements (quadrilateral element representing the resin behavior and truss elements representing the fiber behavior). At each step, a finite element computation is realized in order to simulate numerically the composite forming process for a small tools displacement [52, 55, 63]. Then, the remeshing is applied after each deformation increment, if necessary, according to the following scheme:

(i) Definition of a physical size map based on the adaptation of the mesh element size with respect to one of the mechanical fields,
(ii) Definition of a geometrical size map based on the geometric curvature of the boundary,
(iii) Adaptive remeshing of the domain based on refinement and coarsening techniques with respect to the physical and geometrical sizes map.

4.1. Definition of a Physical Size Map

A physical size map is defined by calculating a physical size (h_D) for each element of the part. This physical size is defined with respect of one of the mechanical field. A critical value has been defined from which the minimal size must be reached. For the other elements, a linear size variation can be used. For a given element, if the ratio between the average size of its edges ($\overline{h}$) and its physical size (h_D) is greater than a given threshold, the element must be refined. During the step of remeshing, the refinement is repeated as long as the physical size is not reached.

4.3. Definition of a Geometrical Size Map

The geometrical size map indicates if a boundary membrane element must be refined or not. The geometric curvature is estimated at each boundary vertex of the domain. If this curvature has been modified during the deformation of the computational domain, all elements sharing this boundary vertex must be refined. At each iteration load the "curved" elements are identified using a geometrical criterion which, for a given element, represents the maximal angular gap between the normal to the element and the normals at its vertices. An element is thus considered to be "curved" if the corresponding angular gap is greater than a given threshold (for example 8 degrees). The normal vector $\vec{\nu}$ at a node P can be defined as the normalized average of the unit normal vectors $\vec{N}_i$ (i=1,..m) to elements sharing node P.

$$\vec{V} = \frac{\sum_{i=0}^{m} \vec{N}_i}{\left\| \sum_{i=0}^{m} \vec{N}_i \right\|} \tag{14}$$

The computation of normal vector to the element depends on the element shape (triangle or quadrilateral) and on the element type (ordinary or extraordinary). The normal vector $\vec{N}$ to an ordinary triangle $P_1P_2P_3$ is the unit normal vector to its supporting plan:

$$\vec{N} = \frac{\overrightarrow{P_1P_2} \wedge \overrightarrow{P_2P_3}}{\left\| \overrightarrow{P_1P_2} \wedge \overrightarrow{P_2P_3} \right\|} \tag{15}$$

The normal vector to an extraordinary triangle is the average of the normal vectors to its two subdividing triangles.

In order to avoid the creation of poor shaped triangles, the shape quality of triangles resulting from the subdivision of each extraordinary element is considered during the refinement operation. If one of the triangle qualities is smaller than a given threshold, the extraordinary element is subdivided in four ordinary elements. The quality Q of a given triangle can be defined by the following formula:

$$Q = \alpha \frac{\left\| \overrightarrow{P_1P_2} \wedge \overrightarrow{P_1P_3} \right\|}{\overline{P_1P_2}^2 + \overline{P_1P_3}^2 + \overline{P_2P_3}^2} \tag{16}$$

where $\alpha = 3.4641002$ is a normalization coefficient in order to obtain a shape quality equal to 1 for an equilateral triangle.

A quadrilateral element is geometrically distorted if its vertices are not coplanar. To measure the distortion, we can consider the angle between the two triangles constituting the quadrilateral element. In fact, there are two possibilities to define this pair of triangles according to the cutting diagonal. Actually, we consider that a quadrilateral element is distorted if one of these angles is smaller than 145°. In this case, the quadrilateral element is subdivided by its two defining triangles. Such element is curved and its size is almost minimal.

4.3. Adaptive Remeshing Based on Refinement and Coarsening Techniques

The adaptive remeshing technique consists in improving the mesh by coarsening and refinement methods in order to conform to the geometry of membrane and truss elements and the mechanical fields of the current part surface during deformation. Two consecutive steps are executed:

(i) Coarsening step during which the mesh is coarsened with respect to the physical size map,
(ii) Refinement step during which the mesh is refined according to the geometrical size map and then to the physical size map
(iii) Define the deformed truss finite elements representing the fixed warp and weft fibres discretization.
(iv) Transfer the mechanical field of resin and fibres from the old mesh to the new mesh.

The refinement technique consists in a uniform subdivision into four new elements. Each triangular or quadrilateral membrane element which needs to be refined is subdivided in four. Truss linear elements associated to the refined membrane element are subdivided in two new truss elements. This procedure allows conserving the fibres orientation during the remeshing (Figure 19). An element is refined if it is a boundary element which needs to be refined (element which belongs to the list of the geometrical size map) or if its size is greater than its physical size (physical size map). There is only one element subdivision which allows preserving the element shape quality: the uniform subdivision into four new elements. For this subdivision, a node is added is the middle of each edge of the element. Boundary elements which belong to the geometrical size map are first refined. The refinement is then applied according to the physical size map. In this case, the refinement procedure is repeated as long as the physical size map is not reached. After each refinement procedure (geometrical criterion or physical criterion), an iterative refinement to restore mesh conformity is necessary. Indeed, after applying the subdivision according to the geometrical or physical criteria, adjacent elements to subdivided elements must be modified. A procedure of subdivision has been proposed for the adjacent elements in order to stop the propagation of the homothetic subdivision (see [55] for more information).

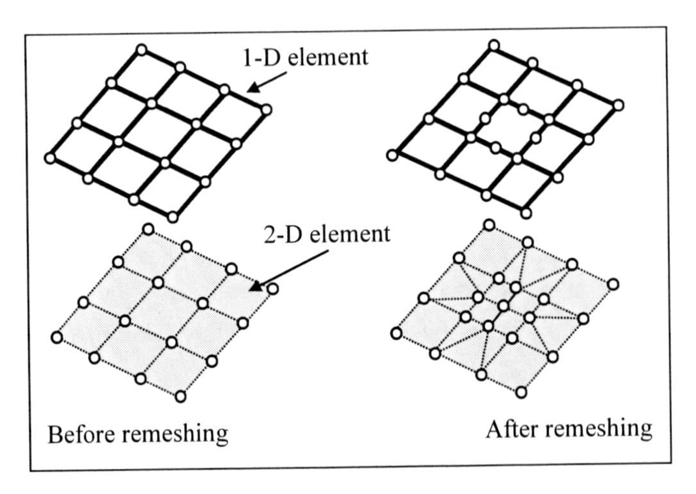

Figure 19. Remeshing of associated membrane and truss finite elements.

4.4. Numerical Application of Remeshing Procedure

The initial coarse mesh of the membrane resin is composed by 900 quadrilateral membrane elements representative of not polymerized matrix (930 truss element representative of weft fibre and 930 truss element representative of warp fibre), this mesh is refined in the curved areas resulting from the deep-drawing process. The minimal element size h_{min}=1mm and the angular geometrical threshold is 3°. The contour of the final shape after deformation is in good agreement with the experimental results. We can see that, the initial thin fabric is computed using an initial coarse mesh, the mesh is again refined uniformly and the adaptive mesh refinement procedure is activated where elements are created automatically in regions of large curvature to even more accurately represent the complex material flow (large stretching) around the die radius. Figure 10 show the final shape draping of conical part using the geometrical approach. One can notice that, in the considered cases (0°/90°) and (-45°/45°), the surface of the conical piece is draped globally (cf. Table 2).

4.4.1. Pressure Burst Tester

The burst tester is designed for measuring the bursting strength of sample materials subjected to an increasing hydrostatic pressure. This pressure is applied to a circular region of the specimen via an elastic diaphragm. The specimen is firmly held round the edge of this circular region by a pneumatic clamping device. When the pressure is applied, the specimen deforms together with the diaphragm. This method is used to determine the pressure required to burst a textile fabric. The experimental test is carried out in IFTH laboratory. The geometry of the tool and the deformed fabric is illustrated in Figure 20. The used balanced textile fabric (jersey) consisted of 77% polyamide fibres and 26% membrane elastane. The mechanical properties of the used fabric are given in [46]. The proposed remeshing approach is used to simulate the bursting of the fabric for 52mm of punch displacement. Numerical simulations obtained with adaptive remeshing are shown in Figure 21 for different remeshing steps corresponding to punch displacement u = 24, 36, 48 and 52mm. We can see that, the initial membrane resin is computed using an initial coarse mesh (400 membrane elements), the mesh is again refined uniformly and the adaptive mesh refinement procedure is activated where elements are created automatically in regions of large curvature to even more accurately represent the complex material flow (large stretching) around the punch radii. The final mesh of the resin has 12468 T3 elements and 12216 Q4 elements. The bursting load prediction versus the punch displacement is compared with the experimental results in Figure 22. The maximum force (80kN) and the displacement (65mm) corresponding to the initiation and the propagation of the damage inside the fabric. Good agreement between the predicted results and the experimental values. Noting that, the remeshing procedure improves the predicted results in comparison with the experimental values.

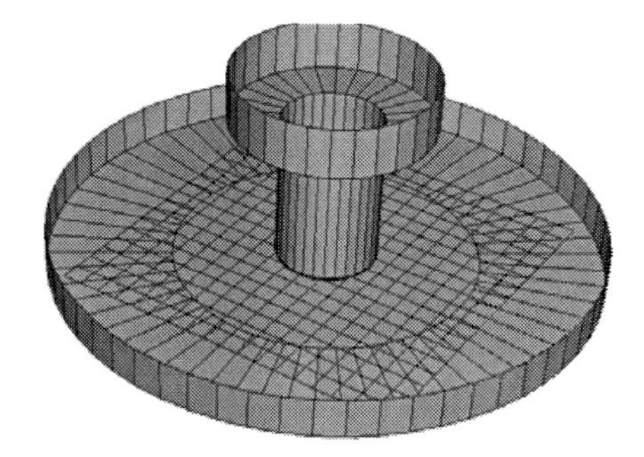

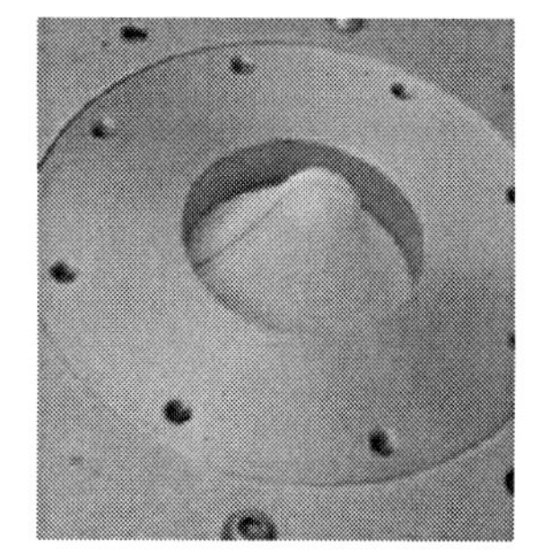

Figure 20. Geometry of textile burst and experimental test.

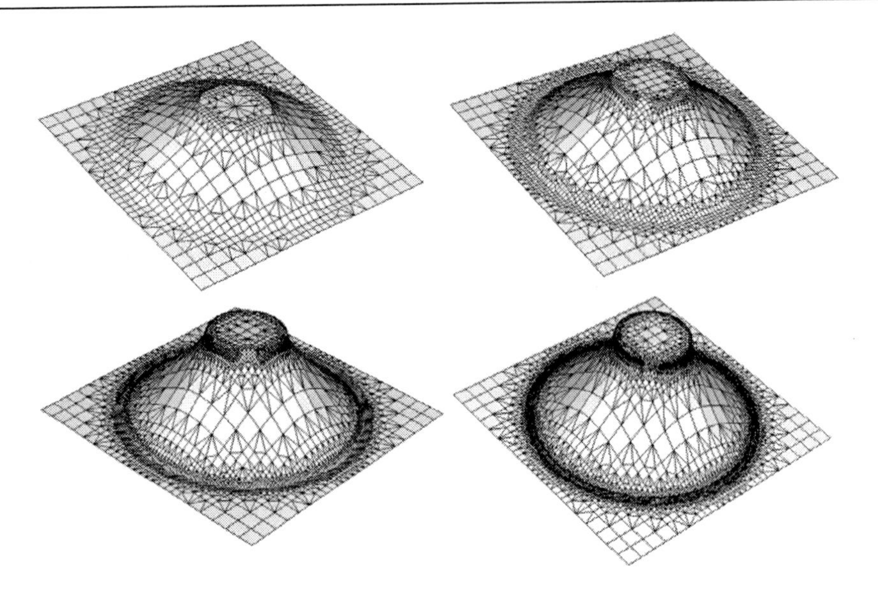

Figure 21. Fabric mesh at different punch displacement (24, 36, 48 and 52mm).

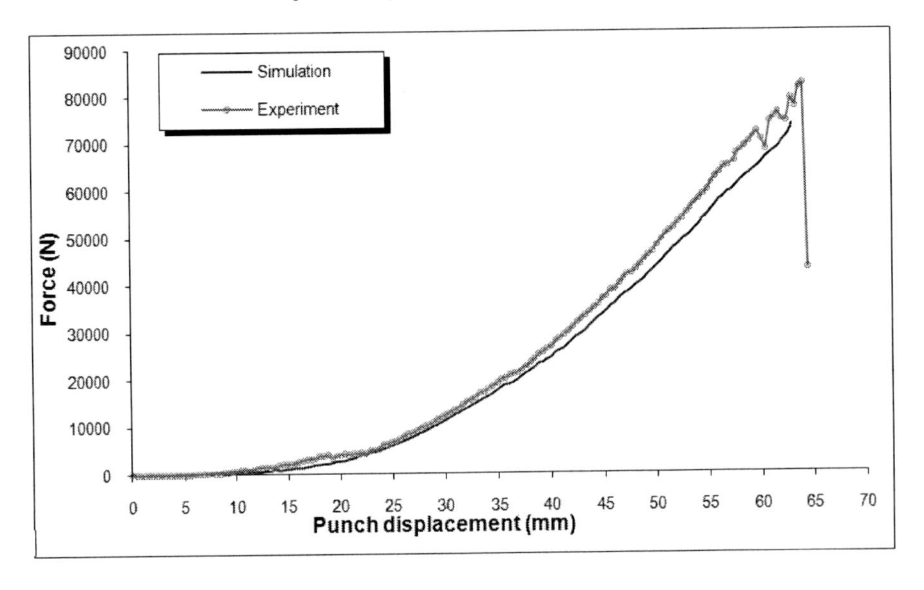

Figure 22. Bursting force versus punch displacement.

4.4.2. Stamping of Complex Shape

The second example concerns the simulation of composite glass fabric stamping with complex tools using the mechanical approach and the computational remeshing procedure. The initial shape of the taffetas woven fabric is (700 mm x 350 mm). The high tensile stiffness along the warp respectively weft yarn direction is introduced via truss elements that connect the nodes of the membrane element. The Young's modulus in warp and weft directions is 70.00GPa. The local fibre directions are tracked on the part via a 3D-digital image correlation technique that assesses the intersections of a raster pattern that is painted on the pre-consolidated part along the fibres directions. The geometry of the stamping simulation is shown in Figure 23. The first order triangular T3 and quadrilateral Q4 membrane element type was used for the stamping. The binding force was 30N and the punch stroke was 47mm.

The friction coefficient between the glass fabric and the tools is 0.3. The amounts of draw-in and fiber orientation were measured. The fibre direction is assumed (0°/90°). The final deformed shape of (0°/90°) fabric for a 47mm displacement of the punch is shown in figure 24. Numerical simulations obtained with adaptive remeshing are shown in Figures 24a, 24b, 24c and 24d for different remeshing steps corresponding to punch displacement u = 8, 20, 30 and 47mm. We can see that, the initial membrane resin is computed using an initial coarse mesh (5000 Q4 membrane elements), the mesh is again refined uniformly and the adaptive mesh refinement procedure is activated where elements are created automatically in regions of large curvature to even more accurately represent the complex material flow (large stretching) around the punch radii. The final mesh of the resin has 1530 T3 elements and 6131 Q4 elements. The shear angle distribution was plotted in Figure 24e. Figure 24e shows the adapted mesh of the warp and the weft to the curved shape of the punch. As shown in the Figure 25, maximum shear angle was observed at the highly curved area and such positions were almost same for all simulation cases. From the numerical simulation maximum shear angles was 35°.

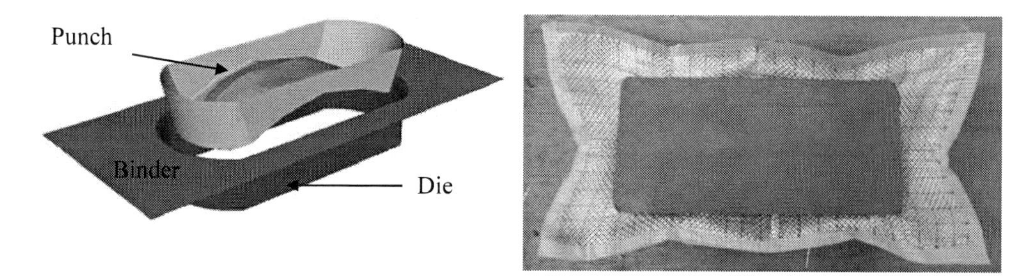

Figure 23. Experimental test of deep-drawing punch shape and final forming.

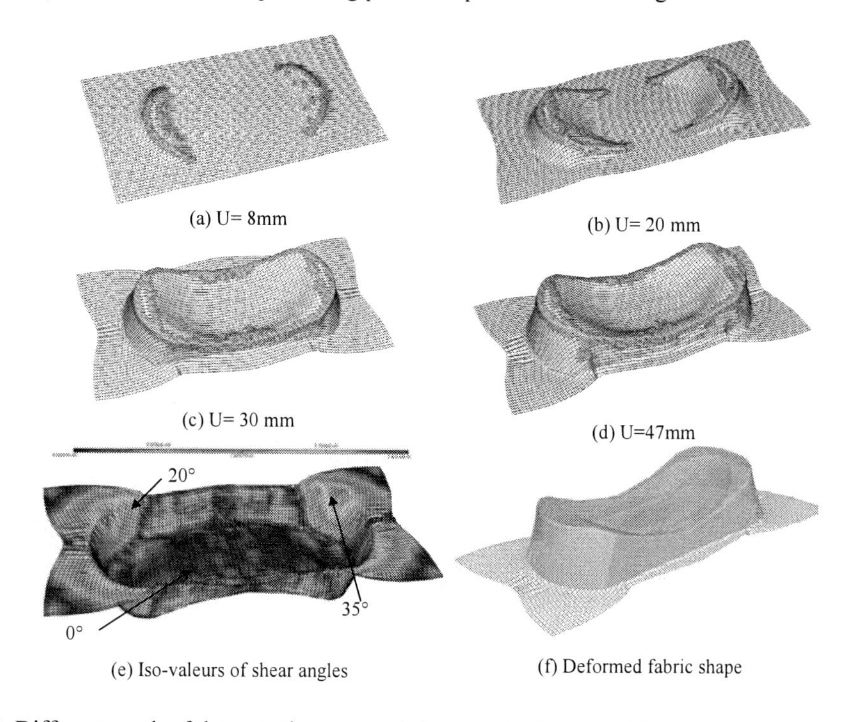

Figure 24. Different mesh of the complex part and shear angles between fibres.

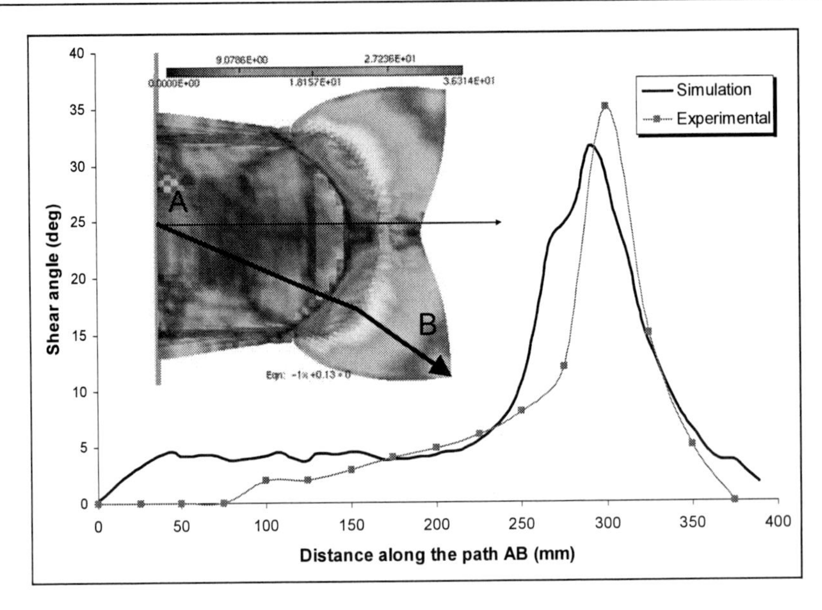

Figure 25. A comparison between experimental and simulated shear angles along the diagonal line.

Conclusion

An important step in the manufacturing processes of thin composite components is the deformation of woven fabric by deep-drawing. The prediction of the angular distorsion of the woven fabric during forming and the changes in fibre orientation are essential for the understanding of the manufacture process and the evaluation of the mechanical properties of the composite structures. An efficient mechanical approach has been presented to simulate accurately the deep-drawing of composite fabric. The mechanical approach is based on a mesostructural model. It allows us to take into account the mechanical properties of fibres and resin and the various dominating mode of deformation of woven fabrics during the forming process (specific deformations of fibre, structural effects of the fabric weaving and the viscosity effects behaviour of the resin). The geometrical approach is based on a modified "MOSAIC" algorithm, which is suitable to generate a regular quad mesh representing the lay-up of the curved surfaces (giving the exact fibre orientations).

The numerical analysis provides input data for the pre-processing of the computation of the final piece after polymerisation and gives the mechanical limits of the fabric during the forming process. Numerical examples concerning the deep-drawing of composite fabric demonstrated the efficiency of the proposed model. An adaptive remeshing technique for composite fabric forming process with refinement and coarsening procedures has been proposed. The implementation with continuum triangular and quadrilateral elements and truss element in the ABAQUS code allows us to validate the proposed approach for several types of problems. Numerical examples concerning the deep-drawing or draping of prepregs composite fabric demonstrated the efficiency of the proposed model.

REFERENCES

[1] Billoët JL., *Introduction aux matériaux composites à hautes performances*, Teknea, Paris, FR, 1993.

[2] Hou M., Ye L., Mai Y.W. (1997), Manufacturing process and mechanical properties of thermoplastic composite components. *J Mater Process Technol* 63:334–338

[3] Rudd CD and Long AC, *Liguid molding technologies*, Woodhead publishing Limited, 1997

[4] Hsiao SW, Kikuchi N (1999) Numerical analysis and optimal design of composite thermoforming process. *Comp Meth Appl Mech Eng* 177:1–34

[5] George Marsh, *Duelling with composites*, Reinforced Plastics, vol. 50(6), 2006, 18-23.

[6] Trochu, F.; Ruiz E., Achim V., Soukane S. (2006), Advanced numerical simulation of liquid composite molding for process analysis and optimization. *Composites Part A: Applied Science and Manufacturing*, 37, 890-902.

[7] Rajiv Asthana, Ashok Kumar, Narendra B. Dahotre (2006), Composites get in deep with new-generation engine, *Reinforced Plastics*, Volume 50, Issue 11, 26-29.

[8] Boisse P., *Mise en forme des renforts fibreux de composites,* Techniques de l'ingénieur, AM 3734, 2004, 1-10.

[9] Lomov SV, Ivanov DS, Verpoest I, Zako M, Kurashiki T, Nakai H, Hirosawa S (2007) Meso-FE modelling of textile composites: road map, data flow and algorithms. *Compos Sci Technol* 67:1870–1891

[10] Kawabata S., Niwa M., Kawai H., The Finite deformation theory of plain-weave fabrics, *J. Text. Inst*, Parts I, II and III, vol. 64, 1973, p. 21-83.

[11] Lim T.C., Ramakrishna S., Shang H.M., Optimization of the formability of knitted fabric composite sheet by means of combined deep drawing and stretch forming, *J. of Materials Processing Technology*, vol. 89–90, 1999, p. 99-103.

[12] Liu L., Chen J., Li X., Sherwood J., Two-dimensional macro-mechanics shear models of woven fabrics, *Composites Part A*, vol. 36, 2005, p. 105-114.

[13] Rozant O., Bourban P.E., Manson J.A.E., Drapability of dry textile fabrics for stampable thermoplastic performs, *Composites Part A*, vol. 31, 2000, p. 1167–1177.

[14] Hagege B, Boisse P, Billoët JL. Analysis and simulation of the constitutive behavior of fibrous reinforcements. In: *Proceedings of Esaform 7 conference, Trondheim, avril;* 2004. p. 317–20.

[15] Potluri P., Parlak I., Ramgulam R. and Sagar T.V., Analysis of tow deformations in textile preforms subjected to forming forces, *Composites Science and Technology*, vol. 66 (2), 297-305, 2006.

[16] Cherouat A., Gelin J.C., Boisse P., Sabhi H., Numerical modeling of glass woven fabric deep-drawing using finite element method, *Eur J Comput Mech*, vol. 4, 1995, p.159-182.

[17] Cherouat A., Billoët J.L., Mechanical and numerical modelling of composite manufacturing processes deep-drawing and laying-up of thin pre-impregnated woven fabrics, *J. Mat. Proc. Technology,* vol. 118, 460-471, 2001.

[18] Boisse P., Buet K., Gasser A., Launay J., Meso/macro-mechanical behaviour of textile reinforcements for thin composites, *Composites Science and Technology*, vol. 61(3), 2001, 395-401, 2001.

[19] Hamila. N., P. Boisse, Simulations of textile composite reinforcement draping using a new semi-discrete three node finite element, *Composites Part B: Engineering*, vol. 39(6), 999-1010, 2008.

[20] ElHami A., Radi B., Cherouat A., Treatment of the composite fabric's shaping using a Lagrangian formulation, *Mathematical and Computer Modelling*, vol 49(7-8), 2009, 1337-1349, 2009.

[21] Warby M.K., Whiteman J.R., Jiang W.-G., Warwick P., Wright T., *Finite element simulation of thermoforming processes for polymer sheets, Mathematics and computers in simulation* vol. 61, 209-218, 2003.

[22] Teik-Cheng Lim, S. Ramakrishna Modelling of composite sheet forming: a review, *Composites Part A: Applied science and manufacturing*, vool.33, 515-537, 2002.

[23] Pickett AK, Creech G, de Luca P (2005), Simplified and advanced simulation methods for prediction of fabric draping. *Eur J Comput Mech* 14(6–7):677–691

[24] Boisse P, Gasser A, Hagege B, Billoet JL (2005) Analysis of the mechanical behaviour of woven fibrous material using virtual tests at the unit cell level. *Int J Mater Sci* 40:5955–5962

[25] Cherouat A, Borouchaki H, Billoët JL (2005) Geometrical and mechanical draping of composite fabric. *Eur J Comput Mech*, vol. 14 (6–7), 693–708

[26] Mark C, Taylor HM, The fitting of woven cloth to surfaces. *J Text Inst vol.* 47; 477–488, 1956.

[27] Van Der Ween F (1991) Algorithms for draping fabrics on doubly curved surfaces. *Int J Numer Methods Eng*, vol. 31, 1414-1426, 1991

[28] Long AC, Rudd CD, A simulation of reinforcement deformation during the production of preform for liquid moulding processes. *Proc Inst Mech Eng Part B J Eng Manuf vol.* 208, 269-278, 1994

[29] Cherouat A, Borouchaki H, Billoët JL, Geometrical and mechanical draping of composite fabric. *Eur J Comput Mech* 14 (6–7): 693–708, 2005

[30] Borouchaki H., Cherouat A., Drapage géométrique des composites, *C.R. Acad. Sci. Paris, Série II B, mécanique des solides et des structures*, vol. 331, 2003, 437-442.

[31] Harrison P., Clifford M.J., Long A.C., Rudd C.D., A constituent-based predictive approach to modelling the rheology of viscous textile, *Composites Part A*, vol. 35, 2004, 915-931.

[32] S.G. Hancock and K.D. Potter, Inverse drape modelling—an investigation of the set of shapes that can be formed from continuous aligned woven fibre reinforcements , *Composites Part A: Applied Science and Manufacturing* vol. 36(7), 947-953, 2005.

[33] K. Vanclooster, S.V. Lomov, I. Verpoest, Experimental validation of forming simulations of fabric reinforced polymers using an unsymmetrical mould configuration *Composites Part A: Applied Science and Manufacturing,* vol. 40(4), 2009, 530-539

[34] P. Potluri, S. Sharma, R. Ramgulam, Comprehensive drape modelling for moulding 3D textile preforms, *Composites Part A: Applied Science and Manufacturing*, vol. 32(10), 2001, 1415-1424

[35] Spencer AJM (2000) Theory of fabric-reinforced viscous fluid. *Compos Part A* 31:1311–1321

[36] Yu WR, Pourboghrat F, Chung K, Zamploni M, Kang TJ (2002) Non-orthogonal constitutive equation for woven fabric reinforced thermoplastic composites. *Compos Part A* 33:1095–1105

[37] Xue P, Peng X, Cao J (2003) A non-orthogonal constitutive model for characterizing woven composites. *Compos Part A* 34:183–193

[38] Peng X, Cao J (2005) A continuum mechanics-based nonorthogonal constitutive model for woven composite fabrics. *Compos Part A* 36:859–874

[39] Hagège B, Boisse P, Billoët J-L (2005) Finite element analyses of knitted composite reinforcement at large strain. *Eur J Comput Mech* 14(6–7):767–776

[40] Ten Thije RHW, Akkerman R, Huetink J (2007) Large deformation simulation of anisotropic material using an updated Langrangian finite element method. *Comput Methods Appl Mech Eng* 196(33–34):3141–3150

[41] Durville D (2002) Modélisation par éléments finis des propriétés mécaniques de structures textiles: de la fibre au tissu. *Eur J Comput Mech* 11(2–3–4):463–477

[42] Sharma SB, Sutcliffe MPF (2004) A simplified finite element model for draping of woven material. *Compos Part A* 35:637–643

[43] Duhovic M, Bhattacharyya D (2006) Simulating the deformation mechanisms of knitted fabric composites. Compos Part A 37– 11:1897–1915 28 *Int J Mater Form* (2008) 1:21–29

[44] Ben Boubaker B, Haussy B, Ganghoffer JF (2007) Discrete models of woven structures. Macroscopic approach. *Compos Part B* 38:498–505

[45] Borouchaki H., Cherouat A., Billoët J.L., *GeomDrap New Computer Aided Design and Manufacturing for Advanced Textile Composites*, Version 1, 1999.

[46] BenNaceur I., Cherouat A., Borouchaki H., Bachmann J.M, Caractérisation et modélisation de l'aptitude à la déformation des structures souples, *Revue des Composites et des Matériaux Avancés,* vol. 13 (3), 231-240, 2003.

[47] Umer R., Bickerton S. and Fernyhough A., Modelling the application of wood fibre reinforcements within liquid composite moulding processes, Composites Part A: *Applied Science and Manufacturing*, vol. 39 (4), 624-639, 2008.

[48] Vilnis Frishfelds F., Staffan Lundström T. and Jakovics A., Bubble motion through non-crimp fabrics during composites manufacturing, *Composites Part A: Applied Science and Manufacturing,* vol. 39(2), 213 25, 2008.

[49] Long A. C., Process modelling for liquid moulding of braided performs, *Composites Part A: Applied Science and Manufacturing*, vol. 32(7), 941-953, 2001.

[50] Fan J.P., Tang C.Y., Tsui C.P., Chan L.C. and Lee T.C., 3D finite element simulation of deep drawing with damage development, *International Journal of Machine Tools and Manufacture*,vol. 46(9), 1035-1044, 2006.

[51] Jauffrès D., Morris C. D., Sherwood J.A. and Chen J., Simulation of the thermostamping of woven composites: mesoscopic modelling using explicit FEA codes, *Int J Mater Form*, vol. 2(1) 173-176, 2009.

[52] Borouchaki H., Cherouat A., Laug P., Saanouni K. Adaptative meshing for ductile fracture prediction in metal forming, *Comptes Rendus Mecanique*, vol. 330(10), 709-716, 2002.

[53] Giraud-Moreau L., Borouchaki H., Cherouat A., Remaillage adaptatif pour la mise en forme des tôles minces, C.R. Acad. Sci. Paris, Serie II B, *Mécanique des Solides et des Structures*, vol. 333(4), 371-378, 2005.

[54] Cho, J.-W., Yang, D.-Y., A mesh refinement scheme for sheet metal forming analysis, *Proc. of the 5th International Conference*, NUMISHEET'02, 2002, 307-312.

[55] Fourment, L., Chenot J.-L., "Adaptive remeshing and error control for forming processes", *Revue européenne des éléments finis* 3, 2, 1994, 247-279.

[56] Cherouat A. Borouchaki H, Giraud-Moreau L., Mechanical and geometrical approaches applied to composite fabric forming, *Int J Mater Form*, DOI 10.1007/s12289-010-0692-5.

[57] Bergsma O.K., Huisman J., Deep Drawing of fabric reinforced thermoplastic, 2nd *Inter. Conf. Comp. Aided Design in Composite Material Technology,* 323-333, 1988.

[58] Benoit Y., Berthé E., Chabin M., Dufort L., Mustapha Ziane M., De Luca P., *Predicting mechanical performance of composite parts though manufacturing simulations,* SAMPE EUROPE 2006.

[59] Abaqus theory, *User's Manual*, 2009.

[60] Gommers B., Verpoest I., Van Houtte P. "Modelling the elastic properties of knitted fabric-reinforced composites", *Composites science and technology*, vol. 56, 685-694, 1996.

[61] Peng X., Cao J., *A dual homogenization and finite element approach for material characterization of textile composites" composites Part: engineering*, vol 33, 45-56, 2002.

[62] Luo Y., Verpoest I. '' Biaxial tension and ultimate deformation of knitted fabric reinforcements" *Composites Part A : applied science and manufacturing* vol 33, 197-203, 2002.

[63] Padmanabhan, K. A. Metal forming at very low strain rates. *Encyclopedia of Materials: Science and Technology*, 2008, pp. 5384-5389.

[64] A. Cherouat and H. Borouchaki, *Present State of the Art of Composite Fabric Forming: Geometrical and Mechanical Approaches, Materials* 2009, vol. 2(4), 1835-1857.

[65] Cherouat, A; Radi, B.; El Hami, A., The Study of the Composites Fabric's Shaping using an Augmented Lagrangian Approach , *Multidiscipline Modeling in Materials and Structures*, vol. 5(2), 185-198, 2009.

[66] R.H.W. ten Thije R. Akkerman and J. Huétink, Large deformation simulation of anisotropic material using an updated Lagrangian finite element method, *Computer Methods in Applied Mechanics and Engineering*, vol. 196(33-34), 3141-3150, 2007.

[67] Gilormini P., Roudier P., *Abaqus and Finite Strain, Rapport interne n° 140, Cahchan,* France 1993

[68] Dafalias YF, Corotational rates for kinematic hardening at large plastic deformations. *J Appl Mech* 50:561–565, 1983.

[69] Crisfield MA, *Non-linear finite element analysis of solids and structures, II: Advanced topics*. Wiley, New York, 1991

[70] Green AE, Naghdi PM, A general theory of an elastic– plastic continuum. *Arch Ration Mech Anal* 18:251–281, 1965

[71] Dienes JK, On the analysis of rotation and stress rate in deforming bodies. *Acta Mech* 32:217-232, 1997.

[72] Zhu, Y. Y., Zacharia, T., and Cescotto, S., "Application of Fully Automatic Remeshing to Complex Metal Forming Analyses", *Comp. Struct.*, Vol. 62, No. 3, pp. 417-427, 1997.

[73] Gifford, L. N., "More on Distorted Isoparametric Elements", *Int. J. Numer. Meth. Engng.*, Vol. 14, pp. 290-291, 1979.

[74] Babuska, I., Zienkiewicz, O. C., Gago, J. and Olivera D. A., eds., *Accuracy Estimates and Adaptive Refinements in Finite Element Computations*, John Wiley, 1986.

[75] Baker, T. J., "Automatic Mesh Generation for Complex Three-Dimensional Regions Using a Constrained Delaunay Triangulation", *Engineering with Computers*, vol. 5, 1989, pp. 161-175.

[76] Zienkiewicz, O. C. and Zhu, J. Z., "A Simple Error Estimator and Adaptive Procedure for Practical Engineering Analysis", *Int. J. Numer. Meth. Engng.*, Vol. 24, pp. 337-357, 1987.

[77] Oden, J. T., Demkowicz, L., Rachowicz, W. and Westermann, T. A., "Towards a Universal h-p Adaptive Finite Element Strategy, Part 2. A posteriori Error Estimation", *Comp. Meth. App. Mech. Engng.*, vol. 77, pp.113-180, 1989

In: Resin Composites: Properties, Production and Applications ISBN: 978-1-61209-129-7
Editor: Deborah B. Song

Chapter 11

NATURAL RESINS: CHEMICAL CONSTITUENTS AND MEDICINAL USES

Lorenzo Camarda, Vita Di Stefano and Rosa Pitonzo*

Department of Chemistry and Pharmaceutical Technology,
University of Palermo, Palermo, Italy

ABSTRACT

Plants and their exudates are used worldwide for the treatment of several diseases and novel drugs continue to be developed through phytochemical research. There are more than 20,000 species of high vegetables, used in traditional medicines that are sources of potential new drugs. Following the modern medicine and drug research advancing, chemically synthesized drugs have replaced plants as the source of most medicinal agents in industrialized countries. However, in developing countries, the majority of the world's population cannot afford pharmaceutical drugs and use their own plant based indigenous medicines. Several exudates from plants are well-known in folk medicine since ancient time, and they are today employed also for practical uses.

Dragon's blood is a deep red resin, which has been used as a famous traditional medicine since ancient times by many cultures. Dragon's blood is a non-specific name for red resinous exudations from quite different plant species endemic to various regions around the globe that belong to the genera *Dracaena* (Africa) and *Daemonorops* (South-East Asia), more rarely also to the genera *Pterocarpus* and *Croton* (both South America).

Dracaena draco L. is known as the dragon's blood tree, and it's endemic to the Canary Islands and Morocco. Phytochemical studies of resins obtained from incisions of the trunk of *D. draco*, have led to the isolation of flavans, along with homoisoflavans, homoisoflavones, chalcones and dihydrochalcones. Dragon's blood has been used for diverse medical applications in folk medicine and artistic uses. It has astringent effect and has been used as a hemostatic and antidiarrhetic drug.

Frankincense, also known as *Olibanum*, is an old-known oleogum resin obtained from the bark of trees belonging to the genera *Boswellia*. There are 43 different reported species in India, Arabian Peninsula and North Africa. The importance of these plants is related to the use of extracts and essential oils of resin in traditional medicine like

* Department of Chemistry and Pharmaceutical Technology, University of Palermo, Via Archirafi, 32 90123 Palermo, Italy, Corresponding author: rosapitonzo@unipa.it

Ayurvedic and Chinese. Extracts from *B. serrata* resin are currently used in India for the treatment of rheumatic diseases and ulcerative colitis. Furthermore, the extracts and essential oils of frankincense have been used as antiseptic agents in mouthwash, in the treatment of cough and asthma and as a fixative in perfumes, soaps, creams, lotions and detergents. In ancient Egypt the resin was used in mummification balms and unguents. Today frankincense is one of the most commonly used resins in aromatherapy.

The biological activity of frankincense resins is due to the pentacyclic triterpenic acids, α- and β- boswellic acids and their derivatives, which showed a well documented anti-inflammatory and immunomodulatory activities.

Manna is an exudate from the bark of *Fraxinus* trees (Oleaceae). Originally it was only collected from trees with damaged bark, but later in southern Italy and northern Sicily plantations were established for manna production, in which the bark is intentionally damaged for exudation and collection of manna.

In July-August a vertical series of oblique incisions are made in the bark on alternate sides of the trunk. A glutinous liquid exudes from this cut, hardens as it oxidises in the air into a yellowish crystalline mass with a bittersweet taste, and is then harvested. Manna is still produced in Sicily, mainly in the Castelbuono and Pollina areas, from *Fraxinus ornus* and *Fraxinus angustifolia* trees. The main component of manna is mannitol; it also contains glucose, fructose, maltotriose, mannotetrose, minerals and some unknown constituents.

Manna is a mild laxative and an excellent purgative, it is suitable in cases of digestive problems, in atonic or spastic constipation. It's useful as expectorant, fluidifier, emollient and sedative in coughs; as a decongestant in chronic bronchitis, laryngitis and tonsillitis; in hypertonic solutions it acts as a dehydrating agent in the treatment of wounds and ulcers. It can be used as a sweetener in cases of diabetes as it does not affect glycemia levels or cause glycosuria; in addition it is also a cholagogue as it promotes the flow of the contents of the gall bladder and bile ducts and so stimulates bile production.

INTRODUCTION

Many higher plants produce economically important organic compounds such as oils, resins, tannins, natural rubbers, waxes, dyes, flavors and fragrances, pharmaceuticals. However, most species of higher plants have never been described, much less surveyed for chemical or biologically active constituents, and new sources of commercially valuable materials remain to be discovered.

Extractable organic substances accumulate in quantities sufficient in many plants, are economically useful as chemical feedstocks or raw materials for various scientific, technological, and commercial applications. For the sake of convenience, plant chemicals are often classified as either primary or secondary metabolites.

Primary metabolites are substances widely distributed in plant kingdom and are often concentrated in seeds and vegetative organs. These compounds are needed for physiological development because of their role in basic cell metabolism, and they are also used as industrial raw materials, foods, or food additives.

Secondary metabolites are compounds biosynthetically derived from primary metabolites but more limited in distribution in the plant kingdom, being restricted to a particular taxonomic group (species, genus, family, or closely related group of families). These compounds, accumulated by plants in smaller quantities than primary metabolites, have no apparent function in a plant's primary metabolism but often have an ecological role. The resins, *secretion*s of many plants, are used widespread as adhesives, ingredients of cosmetic preparations, as fragrances in daily rituals and religious ceremonies, as coating materials and as remedies in folk medicine [1].

In the first Pharmacopoeia written by the Greek botanist Dioscorides [2] there were around 600 advised remedies, which were prepared from mixtures of natural resins and balsams as well as some herbal preparations. The antiseptic activity of resins was assumed to be known long before as both the Egyptians used to burn a special mixture of them during plague and the Indians used gurjun balsam against leprosy.

It is known that Hindus, Babylonians, Assyrians, Persians, Romans, Chinese and Greeks as well as the people of old American civilisations like Incas, Mayas and Aztecs, used natural resins primarily for embalming. For the same purpose, Egyptians used styrax (*Liquidamber orientalis*), myrrh (*Commiphora* spp.), colophonium (*Pinus palustris*), cedar (*Cedrus* spp.) and labdanum (*Cistus ladaniferus*). During sacrification ceremonies resins were burned to prevent the influence of bad spirits on souls and Egyptians used olibanum (*Boswellia* spp.), myrrh, bdellium (*Commiphora wightii*), mastic (*Pistacia lentiscus*), styrax, santal (*Santalum album*), cinnamon (*Cinnamomum aromaticum*), aloe wood (*Aloe succotrina*), cedar and juniper (*Juniperus communis*).

Dragon's blood is a non-specific name refers to reddish resinous products, usually occur as granules, powder, lumps or sticks, which has been used as a famous traditional medicine since ancient times. The origin of the resin is believed to be from Indian Ocean island of Socotra, now part of Yemen [3] and several alternative sources are from Canary Islands, Madeira, and South East Asia and also from East and West Africa have been identified [4]. However, there is a great degree of confusion regarding the source and identity of this resin.

This particular name is applied to many red resins described in the medical literature, e.g. East Indian Dragon's blood (from the fruit of *Daemonorops draco* (Willd.) Blume), Socotran or Zanzibar Dragon's blood (exudates of *Dracaena cinnabari* Balf. f.), Canary Dragon's blood (exudates from the trunk of *Dracaena draco* (L.) L.), West Indian Dragon's blood (exudates of *Pterocarpus draco* L.), Mexican Dragon's blood (resin of *Croton lechleri* Müll. Arg.) and the Venezuelan Dragon's blood (resin of *Croton gossypifolium* Vahl) [5]. Thus, the name *Dragon's blood* in general is used for all kinds of resins and saps obtained from four distinct plant genera *Croton* (Euphorbiaceae), *Dracaena* (Liliaceae), *Daemonorops* (Palmaceae), and *Pterocarpus* (Fabaceae) [6].

Dracaena draco L. is known as the dragon's blood tree. *D. draco* subsp. *draco* was found in the Madeira, Canary and Cape Verde archipelagoes, and subsp. *ajgal* in Morocco. The resin exudes from the wounded trunk or branches of the tree (Figure 1).

According to a Greek myth, Landon, the hundred-headed dragon, guardian of the Garden of the Hesperides (the nymph daughters of Atlas, the titan who holds up earth and heaven) was killed by either Hercules (in his quest) or Atlas (as punishment) while bringing back three golden apples from the garden, depending upon the version of the myth. Landon's red blood flowed out upon the land and from it sprung up the trees known as *Dragon Trees* [7]. Dragon's blood was also called *Indian cinnabar* by Greek writers. The name Dragon's blood dates back to the 1st century B.C. when a Greek sailor wrote, in a shipping manual "Periplus of the Erythrean Sea", about an island called Dioscorida where the trees yielded drops of cinnabar. Plinius [8] also described that the resin got its name from an Indian legend based on Brahma and Shiva. Emboden [9] and Lyons [10] had also summarized the history and mythology of Dragon's blood. According to Lyons, the struggle between a dragon and an elephant that, at its climax, led to the mixing of the blood of the two creatures resulted in a magical substance, *Dragon's blood* imbued with medicinal properties.

Figure 1. Dragon's blood resin from Dracaena draco.

Dragon's blood has been used for several applications, including artistic uses and folk medicine.

The resin from *Dracaena* was used for ceremonies in India, as red varnish for wooden furniture in China, to color the surface of writing paper for banners and posters, especially for weddings and for Chinese New Year, as pigment in paint, enhancing the color of precious stones and staining glass, marble and the wood for Italian violins. Dragon's blood from *Dracaena draco* was detected in a broad variety of art objects dating from the fifteenth to the nineteenth centuries [11]. It was predominantly used in gold lacquers and Hinterglasmalerei (reverse-glass paintings), and relatively rarely in glazes on conventional paintings or lacquers on furniture. In all these cases, the artists took advantage of the special properties of dragon's blood because of a film-forming resin with a natural red colour that is nevertheless translucent. In lacquers, dragon's blood was mixed with other natural resins such as sandarac [12], larch turpentine [13] and mastic [14], and also with other red resins such as shellac or gum benzoin, which is a dark red balsamic resin obtained from several species of the genus Styrax [15].

Dragon's blood was used by early Greeks, Romans, and Arabs for its medicinal properties. Locals of Moomy city on Socotra island used the resin from *Dracaena draco* as a sort of cure-all, by using for general wound healing, coagulant, curing diarrhoea, lowering fevers, dysentery diseases, internal ulcers of mouth, throat, intestine and stomach, respiratory and stomach viruses and skin disorders such as eczema. Dioscorides and other early Greek writers described its medicinal uses. Dragon's blood resin has strong astringent properties and is used as a muscle relaxant [16]. It was also used to treat gonorrhea, stoppage of urine, watery eyes and minor burns [17].

Compounds in dragon's blood resins that could be associated with the red colour, were studied by Brockmann and Junge [18], who attributed the color to the group of compounds known as anthocyanins. Dracoflavylium (7,4'-dihydroxy-5-methoxyflavylium) was isolated and characterized from a commercial source of powdered dragon's blood resin that was obtained from a species *D. draco* [19]. The red colour of dragon's blood resin is associated with the red quinoid bases of the respective yellow flavylium cations [19].

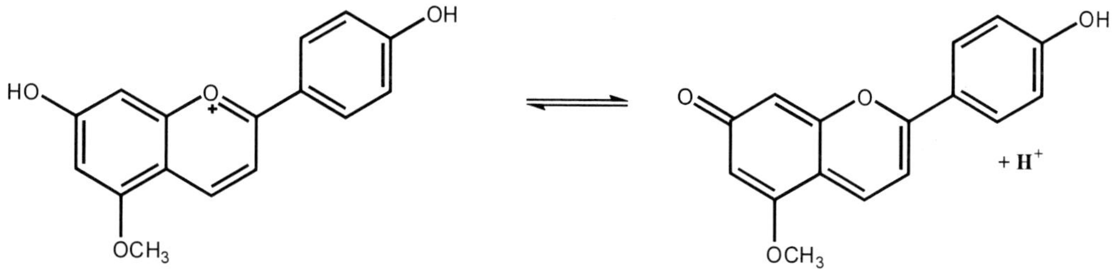

In solution, both forms are connected, and can undergo multiple structural transformations, in what can be described as a multistate system, reversibly interconverted by external stimuli, such as pH [19–24]. Since compounds with a flavylium chromophore have not been found in the red resins or exudates from the species of *Croton* and *Pterocarpus* called as dragon's blood [25], it suggests that this chromophore can be used as marker to identify resins origin obtained from *D. draco* [26]. Phytochemical studies carried out on *D. draco* resins, have led to the isolation of the most abundant resin constituents, belonging to the class of flavonoids: flavans **(1-5)**, along with homoisoflavans **(6-7)** and homoisoflavanones **(8-15)**. In addition were isolated chalcones and dihydrochalcones [27-30].

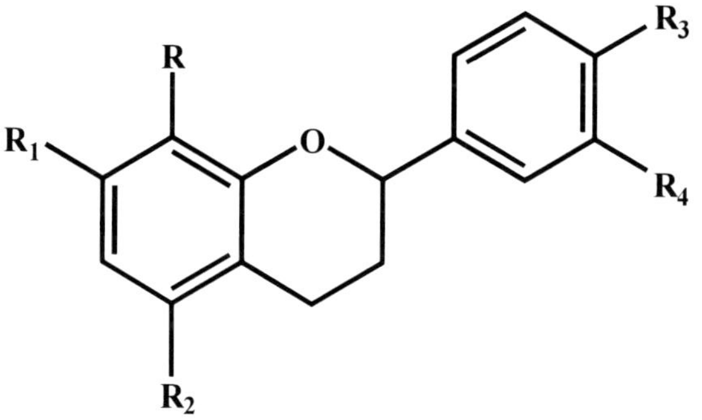

(1) R =CH_3, R_1 =OH, R_2 =H, R_3 =OCH_3, R_4 =OH;
(2) R =CH_3, R_1 =OH, R_2 =H, R_3 =OH, R_4 =OCH_3;
(3) R =H, R_1 =OH, R_2 =H, R_3 =OH, R_4 =OCH_3;
(4) R =CH_3, R_1 =OH, R_2 =H, R_3 =OH, R_4 =H;
(5) R =CH_3, R_1 =OCH_3, R_2 =OH, R_3 =OH, R_4 =H

Regarding the biological activity of the resin obtained from *D. draco*, no data is reported in the literature. Preliminary experimental data not yet published, concern the *in vitro* antibacterial and anti-biofilm activities of acetonic extracts obtained from *Dracaena draco* resin, collected from a specimen tree which grows in the Botanical Garden of the University of Palermo.

(6)	R =H,	R_1 =OH,	R_2 =H,	R_3 =H,	X =H_2
(7)	R =H,	R_1 =OH,	R_2 =H,	R_3 =OCH_3,	X =H_2
(8)	R =H,	R_1 =OH,	R_2 =H,	R_3 =H,	X =O
(9)	R =OH,	R_1 =H,	R_2 =OH,	R_3 =H,	X =O
(10)	R =H,	R_1 =OH,	R_2 =H,	R_3 =OH,	X =O
(11)	R =OH,	R_1 =OCH_3,	R_2 =OH,	R_3 =H,	X =O
(12)	R =OH,	R_1 =OH,	R_2 =OCH_3,	R_3 =H,	X =O
(13)	R =H,	R_1 =OH,	R_2 =CH_3,	R_3 =OH,	X =O
(14)	R =OH,	R_1 =CH_3,	R_2 =OH,	R_3 =H,	X =O
(15)	R =CH_3,	R_1 =OH,	R_2 =CH_3,	R_3 =OH,	X =O

The antimicrobial and anti-biofilm activities of acetone extracts were evaluated against different Gram-positive and Gram-negative bacteria strains. In particular was found a considerable antibacterial activity against *Staphylococcus aureus* and *Staphylococcus epidermidis,* with MIC values of 100 μg/mL, and an interesting anti-biofilm activity against preformed *S. aureus* biofilms, with MIC values ranging from 54 and 39% at concentrations between 200 and 50 μg/mL.

The efficacy of these acetonic extracts should be considered as an important factor in evaluating them as an interesting source of innovative plant derived antimicrobial agents that can be used in the development of new strategies to treat biofilms of medical relevance.

Flavonoids are one of the largest groups of secondary metabolites, and they play an important role in plants as defence and signalling compounds in reproduction, pathogenesis and symbiosis [6,31]. Plant flavonoids are involved in response mechanisms against stress, as caused by elevated UV-B radiations [32-35], infection by microorganisms [36] or herbivore attack [37]. They are pigment sources for flower colouring compounds [38] and play an important role in interactions with insects [39]. They also affect human and animal health because of their role in the diet, which is ascribed to their antioxidant properties [40] or their estrogenic action [41], and to a wide range of antimicrobial and pharmacological activities [42,43].

Some of the flavonoid compounds isolated and identified in *Dracaena draco* resin were also identified in its fruits samples collected from a specimen tree which grows in the Botanical Garden of the University of Palermo. In particular, were performed MS-MS experiments that afforded to identify some homoisoflavanone compounds **(8-15)** [44]. In addition, following minor constituents were isolated from roots, bark and aerial parts: steroidal saponins, along with known compounds including sterols, carotenes and aromatic compounds. [45-47].

Steroidal saponins are reported to show potent cytostatic activity against HL-60 cells with IC_{50} value being 1.3 and 2.6μg/mL, respectively compared with etoposide (IC_{50} 0.3μg/mL) used as a positive control [45-46,48-49]. The mechanism of these compounds' cytotoxicity was also evaluated and found to be via activation of apoptotic process.

Olibanum also called *Frankincense* is an *aromatic* oleogum *resin* obtained from the bark of trees belonging to the genera *Boswellia* of the Burseraceae family, that includes approximately 43 different species of small trees, that grow mainly in India, Arabian Peninsula and North Africa. Some of the well-known species are: *B. papyrifera* (Del.) Hochst (Ethiopia and Sudan), *B. neglecta* S. Moore (Kenya), *B. rivae* Engl. (Ethiopia), *B. frereana* Birdw. (Somalia), *B. carteri* Birdw. (Ethiopia and Eritrea), *B. orodata* Hutch. and *B. dalzielli* Hutch. (Nigeria), *B. sacra* Flueck. (Oman), *B. serrata* Roxb. (India) and *B. thurifera* Colebr. (Arabia e Somalia) [50].

Olibanum is produced in a restricted geographical area from uncultivated trees, and usually collected by small nomadic groups; the name is derived from the Arab word "al Luban", which means milk and is a reference to the milky sap that exudes from the tree upon incision (Figure 2).

Appreciated by ancient civilizations, *Boswellia* resins ranked along with gold and ivory, spices and textiles as valuables for trading and barter.

Figure 2. Frankincense oleogum resin sample.

Harvesting Frankincense is a time consuming process that begins in December, reaching a peak from March to May [51]. The trees start producing resin when they are about 8 to 10 years old. [52]

A deep incision is made into the bark (according to recent research more than 5 incisions causes considerable stress to the trees) and this wounding causes the tree to bleed a milky white substance that seals and heals the wound and prevents infection; after three months the resin has hardened enough to be scraped off the trunk.

The first mention about use of *Boswellia* resin as a drug is reported in the Ebers papyrus, (1500 B.C.) [51], that is the oldest list of prescriptions describing the value of resin in funerals, mummification and cremation procedures. Early Egyptian myth describes *frankincense* as representing the Horus's tears. Later texts of Greek and Roman origin describe the trade of these appreciated resins, which were exported to Rome, China and North Africa [52].

Between 35 and 25 B.C. Celsus recommended that frankincense be used to treat wounds, as a possible antidote to hemlock and to stop internal bleeding and superficial bruising if mixed with leek juice [52]. In the Babilonian Talmud (3rd–6th centuries B.C.) is reported that resin was administered in wine to prisoners condemned to death to benumb the senses [53]. Based on this text, some scholars assumed that the drink given to Jesus before crucifixion contained *Boswellia* resin [54]. In a syriac book of medicine (5th century) frankincense is indicated as useful in the catarrh, gout, colic treatments and gastrointestinal hemorrhage while in the work of Ibn Sina (Avicenna) of the 11th century is reported frankincense use in inflammation and infection of the urinary tract [52]. In the 17th century Culpepper, a herbalist practicing in the East End of London, used frankincense to treat stomach ulcers and as a topical unguent for bruising [55]. During this period distillates of the resin called "oils of olibanum", were widely employed by the barber surgeons, apothecaries and alchemists, for their appealing scent [56]. In Ethiopia resin was believed to have a tranquilizing effect [57] and in Kenya it was used for dressing wounds and, mixed with sesame oil, to reduce the loss of blood in the urine from schistosomiasis infestation. In India, the applications regarded dental and skin diseases, respiratory complaints and digestive troubles. In China resin was a constituent of several skin remedies, including those for bruises and infected sores [52].

In Arabian countries where frankincense traditional medical use is widespread, every part of the tree including root, bark, bud, flower and fruit, is still used. The powdered bark is made up into an astringent paste which is used as a soothing ointment and as a remedy for swelling; it is applied as remedy for wounds and burns, after mixing with water. The resin is chewed to treat dental infections . Buds and fruits provide a cleansing tonic for the digestive system. The resin is used as a decoction with *Cinnamon* and *Cardamom* to treat stomach aches.

Olibanum burnt it was also thought to act as an expectorant and in cases of colds, flu and diseases of the upper respiratory tract. The smoke is also a powerful insect deterrent and thus served as a prophylactic to prevent the bites of malaria carrying mosquitoes.

Today frankincense is widely employed in Catholic Christian ceremonies as well as other religious and secular traditions. It is also an important component in cosmetic industry [52] and it is widely marketed as a food supplement [58].

Phytochemical studies carried out on several *Boswellia spp.* resins led to the isolation of at least 200 constituents; the chemical composition of the resins varies depending on the species [59-60]. In general the rubbery fraction represents about one third of the natural product and it is composed of polysaccharides, the oil fraction ranges from 5% to 10%, and the non volatile terpenoic fraction represents about 40% including tetracyclic triterpenes with dammarane or tirucallane skeletons, and pentacyclic triterpenoids belonging to the oleanane, ursane and lupane classes [61]. Specific chemical markers of *Boswellia* resins are pentacyclic triterpenoic acids called as boswellic acids (BAs) and their derivates, and tirucallic acids (TAs) and their derivates.

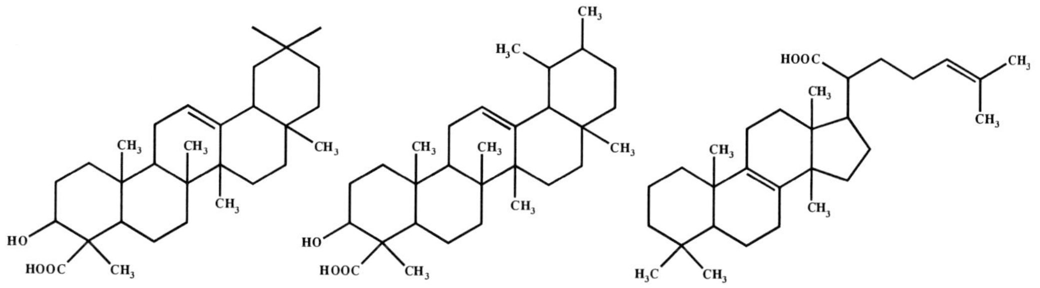

α-Boswellic acid β-Boswellic acid Tirucallic acid

Research results on the structure-activity relationship between anti-inflammatory activity and natural triterpenoids, indicated that an acid functional group increased the effect [62].

Boswellic acids (BAs) and their derivatives including acetyl-ß-boswellic acid (ABA), 11-keto-ß-boswellic acid (KBA) and acetyl-11-keto-ß-boswellic acid (AKBA), as tirucallic acids (TAs), show a well-documented anti-inflammatory activity [47-49]. The mechanism of action of the pentacyclic triterpene acids, indicate that the primary site of action of these compounds is inhibition of 5-lipoxygenase enzyme (5-LO) preventing the formation of leukotrienes (LTB4, LTC4, LTD4), both acute and chronic inflammation, so they did not affect the cyclooxygenase activities and therefore synthesis of prostaglandins [59, 66]. Studies about structure-activity relationships reported that the pentacyclic triterpene ring system is crucial to bind the inhibitor to the highly selective effector site.

It has been found that compounds with 11-keto function on the skeleton, provides the best inhibitory action on 5-LO.

Moreover, the acetoxy-derivatives potently inhibited the activities of human recombinant Akt1 and Akt2, thus, tirucallic acid derivatives represent a new compound class with antitumor properties [67]. Other molecular targets of BAs include human leukocyte elastase, CYP 2C8/2C9/3A4, topoisomerase I, topoisomerase IIa, and IKK α/β that explain anti-cancer activity. For example BAs, like activity of polyphenols, may have a role in inhibition of the human leukocyte elastase interfering with NF$_{\kappa}$B pathway [68]; although the anti-asthmatic reports on BAs revealed that leukotriene and elastase enzyme inhibition might be responsible for this effect [69-71].

Several clinical trials revealed a comparable efficacy of *Boswellia serrata* with synthetic drugs like sulfasalazine and mesalazine in the local treatment of active Crohn's disease and colitis ulcerosa, with a risk-benefit analysis in favour of BAs [59,72-73]. As for their effect as an anticomplement system, a mixture of BAs was found to inhibit the activity of C3 convertase in the classic complement pathway, which consequently suppressed the conversion of C3 into C3a and C3b [73]. Besides their renowned anti-inflammatory activity, boswellic acids have been extensively investigated with respect to their activity against tumor cells and chemopreventive effects. Literature data demonstrate that BAs induce cycle arrest [73-74] and downregulate several antiapoptotic genes [76-79].

In literature there are few studies about chemical composition and biological activity of Boswellia resin essential oils; their composition from different Boswellia spp. depends to the climate, harvest conditions and geographical sources of resins. Frankincense oils are prepared by steam distillation directly from resins and have been used as topical antiseptics, antimicrobial, healing and sedative properties and to alleviate the pain caused by rheumatism [80-82]; moreover they are commonly used in aromatherapy practices.

Boswellia spp. resin essential oils were characterized by a combination of GC and GC/MS analyses, that led to the identification of the chemotaxonomy marker components for each species. The main classes of compounds identified were monoterpene hydrocarbons (HM), oxygen monoterpenes (OM), sesquiterpene hydrocarbons (SH), oxygen sesquiterpenes (OS), diterpene hydrocarbons (DH), oxygen diterpenes (OD), ether derivates, alcohol and ester derivates.

The GC-MS profile of *B. serrata* resin oil is quite similar to that of *B. carteri*, except for the presence of α-thujene and methylchavicol, which should be considered as chemotaxonomy marker components. The main components identified in *B. carteri* oil were: α-pinene, myrcene, limonene and α-cedrene. In *B. papyrifera* oil, n-octyl acetate was the main constituent and n-octanol, incensole and incensole acetate were minor components. The oleogum resin oil of *B. rivae* contains

hydrocarbon and oxygenated monoterpenes mainly. In accordance with the literature limonene, 3-carene, α-pinene and trans-verbenol were the most abundant constituents, that should be considered as chemotaxonomy markers. Thus, it was observed in this oil the complete absence of the sesquiterpene and diterpene components [83,84]. On the other hand *B. frereana* oil is rich in the monoterepens as α-thujene, p-cymene and α-pinene, as well *B. neglecta* oil. In addiction *B. frereana* oil is poor in sesquiterpenes and totally devoid of the diterpenes of the incensole type [84].

Several studies reported in literature encourage the medical use of these oils as antimicrobial agents, their antibacterial activity was evaluated by determining MIC values against several fungi, Gram-positive and Gram-negative bacterial strains. The *B. carteri* resin oil demonstrated the highest degree of activity against the methicillin-resistant *S. aureus* and against *Pseudomonas aeruginosa*, while *B. rivae* resin oil showed the lowest MIC value against *E. coli*. Interestingly, although *B. serrata* resins are the most widely studied for their high content of boswellic acids, its essential oil was the least active against all of the tested bacterial strains [85-86].

B. rivae resin oil demonstrated a lower MIC value against *Candida albicans*, while *B. carteri* and *B. papyrifera* resin oils showed good MIC values against *C. albicans* and *Candida tropicalis*. A monoterpene hydrocarbon most notably presents in *B. carteri* and *B. rivae* oleogum resin oils was limonene. It is known that limonene inhibits the growth of several fungal and Gram positive bacterial strains [87].

The essential oils of *Boswellia* spp. have been also tested as anti-biofilm agents; in particular that of *B. papyrifera* showed considerable activity against both *S. epidermidis* and *S. aureus* biofilms. *Boswellia rivae* essential oil was very active against preformed *C. albicans* biofilms and inhibited the formation of *C. albicans* biofilms at a sub-MIC concentration. The efficacy of these oils is considered important against staphylococcal and *C. albicans* biofilms, that are pathogens in a protective physiological form intrinsically resistant to conventional antibiotics [88].

Manna is an exudate obtained from the bark of trees belonging to the genera *Fraxinus* L., distributed mostly in the temperate regions and the subtropics of the Northern hemisphere [89-90]. The classifications of Knoblauch, Taylor and Johnson place the *Fraxinus* species into the tribe Fraxineae of the subfamily Oleoideae of the Oleaceae family [89]. Knoblauch [91] describes 39 *Fraxinus* spp. divided in two sections: *Ornus* and *Fraxinaster*; Lingelsheim [92] recognizes 63 spp. grouped in the same two sections. In a recent classification of Oleaceae, the subfamily level is omitted, the family is splitted into five tribes and the genus *Fraxinus* consists of about 50 spp., including *Fraxinus ornus* and *Fraxinus angustifolia.*

Fraxinus genus are botanically described as trees or rarely shrubs, deciduous or rarely evergreen. Leaves are odd-pinnate, opposite or rarely whorled at branch apices; inflorescences are terminal or axillary toward end of branches; flowers are small, unisexual, bisexual, or polygamous. The corolla is white to yellowish, four-lobed, divided to base. The fruit is a samara with elongated wing [90].

The *Fraxinus* species have economical, commercial and medicinal importance [89,90,93]; the plants from this genus are widely used for timber [94].

Originally manna was only collected from trees with naturally damaged bark, but today in Sicily plantations the bark is expressly damaged for exudation and collection of manna.

The earliest reports on the production of manna in Sicily date from the second half of 1500, but cultivation on the island has been developed intensively only in the XVIII century. The areas of greatest production were Castelbuono, Cefalù, Geraci Siculo, Pollina and San Mauro Castelverde, located in Madonie National Park. Manna production from *Fraxinus ornus* and *Fraxinus angustifolia* trees, today is mainly restrict to the Castelbuono and Pollina areas.

The collection of manna is made during the summer; a vertical series of oblique incisions known locally as "*ntacche*" are made in the bark on alternate sides of the trunk, and in the main branches. In order not to damage the plant and to preserve the abundance of the harvest, incisions should be done by experienced hands can accurately use the appropriate tool known locally as "*mannaluoru o cutièddu â manna*", a kind of hatchet and pointed sharp.

From incisions in the trunk exude a glutinous liquid that hardens quickly to form a yellowish crystalline mass with a bittersweet taste named *manna* (Figure 3).

Three different types of manna are produced from *Fraxinus ornus* and *Fraxinus angustifolia* trees, the so called *manna in cannolo*, *manna in sorte* and *manna in rottame.*

Manna in cannolo is the most valuable type and can be consumed as it is; it is obtained when whitish liquid dripping form a stalactite of various lengths. Recently, to facilitate the formation of *manna in cannolo*, under the incision line is placed a small metal conduit whose end is fixed nylon held in tension by a small weight, in order to conveyed the exudate inside the metal conduit and drips down the wire on which hardens before to fall to the ground.

Manna in sorte occurs when the production is particularly abundant and climatic conditions are likely to slow down the solidification of the liquid along the trunk; so, the liquid drains to the ground where it is collected in the shovels of prickly pear where hardens slowly.

Manna in rottame is exudate that hardens on the tree trunk and is the less valuable type.

The manna harvest is done during hottest hours of the day to facilitate the detachment of the manna from bark and to prevent loss of liquid in the process of condensation.

Manna in cannolo is collected first and carefully placed in special wooden shelves. The residues that remained stuck to the trunk (*manna in rottame*) are scraped by a metal shovel. The harvested manna is dried in a well ventilated place, shaded the first day and on the sun in the days following. During drying, the product is stirred properly and protected from rain and damp night in order to avoid browning and molds proliferation.

Furthermore, the different types of manna, pending the sale, should be stored separately in sealed wooden boxes or other containers suitable for the purpose, in dry and dark.

Figure 3. Manna obtained from the bark of Fraxinus L. tree.

The Italian law December 24, 1928 No 3144, states that the name *manna* is assigned only to the product resulting from incision of the cortex of "Orniello or Amolleo" (*Fraxinus ornus*) and of "Ossifillo" (*Fraxinus angustifolia).* It is forbidden to prepare, sell, to be sold or put on the market manna still containing sucrose, starchy substances or foreign substances of any kind, except the natural impurities in the normal proportion for different types of manna. This law on supervision of agricultural products to protect the authenticity of the product, provides consumers and defines the duties of producer [95].

The main component of manna is D-mannitol, an hexavalent alcohol colorless, odorless and sweet taste also known as "manna sugar"; it also contains other sugars such as glucose, fructose, mannotriose, mannotetrose, minerals, water and minor constituents [96].

Manna and mannitol have pharmacological activity and are both listed in the Italian F.U. VIII Ed. In other pharmacopoeias (IX-XII) is given only the mannitol, whereas in the European Pharmacopoeia and the Pharmacopoeia of the United States is reported as such manna.

Manna is a mild laxative and an excellent purgative, it is suitable in cases of digestive problems, in atonic or spastic constipation. It's useful as expectorant, fluidifier, emollient and sedative in coughs; as a decongestant in chronic bronchitis, laryngitis and tonsillitis; in hypertonic solutions it acts as a dehydrating agent in the treatment of wounds and ulcers. It can be used as a sweetener in cases of diabete as it does not affect glycemia levels or cause glycosuria; in addition it is also a cholagogue as it promotes the flow of the contents of the gall bladder and bile ducts and so stimulates bile production [97].

In order to increase knowledge on the biodiversity of endemic *Fraxinus* spp. in Sicily, were carried out phytochemical studies on manna samples produced by trees belonging to different *Fraxinus angustifolia* subsp. *angustifolia* and *F. ornus* cultivars growing in the areas of main production, located in Madonie National Park.

In particular, were analyzed following cultivars locally named *Verdello*, *Nziriddu*, *Baciciu*, *Russu*, *Macigna* and *Sarvaggiu*, all belonging to *Fraxinus angustifolia* subsp. *angustifolia*, and the cultivar locally named *Serracasale* belonging to *Fraxinus ornus*.

Analytical investigations carried out have allowed to define the sugars composition (such as mannitol, monosaccharides and oligosaccharides) of dried samples of *manna in cannolo* produced by different cultivars, the chemical composition of the volatile fraction of several samples of fresh manna, and the identification of some minor constituents of coumarins type.

The analysis of the sugars composition in samples of manna were performed by GC-MS using methods described in the literature [98-100]. The abundance of sugars varies depending on the cultivar analyzed. Were analyzed samples of *manna in cannolo* collected in august 2005 in the area of Castelbuono, during the period of higher production.

Figure 4 shows a comparison between the cultivars of *Fraxinus angustifolia* and F. *ornus*, with regard to the contents in monosaccharides (sum of glucose and fructose), oligosaccharides (sum of mannotriose and mannotetrose) and mannitol.

It could be observed that the monosaccharides content is almost constant in all cultivars, with only a slight percentage increases for cultivars *Russu* (*F. angustifolia*) and *Serracasale* (*F. ornus*).

Regarding the content of oligosaccharides, the percentage increases significantly in the cultivars *Sarvaggiu* (*F. angustifolia*) and *Serracasale* (*F. ornus*). Mannitol, the main constituent of sugars, has a nearly constant percentage abundance in all cultivars analyzed [97].

The analysis of the chemical composition of the volatile fraction were performed by GC-MS. Manna fresh samples produced by trees belonging to *Nziriddu*, *Verdello* and *Baciciu* cultivars, were collected during the summer of 2004 and immediately analyzed using the SPME (Solid Phase Microextracion). Identification of the individual components was based on matching with NIST 2005 mass spectra library database and comparison with spectra of authentic samples and literature data. The most abundant compounds identified are esters including methyl palmitoleate, ethyl palmitate and ethyl oleate, and alcohols such as farnesol [97]. Because of the volatility of these compounds, their abundance percentage decreases over the residence time of the manna on the trunk of the trees. It is therefore essential for the SPME analysis to use freshly harvested samples of manna.

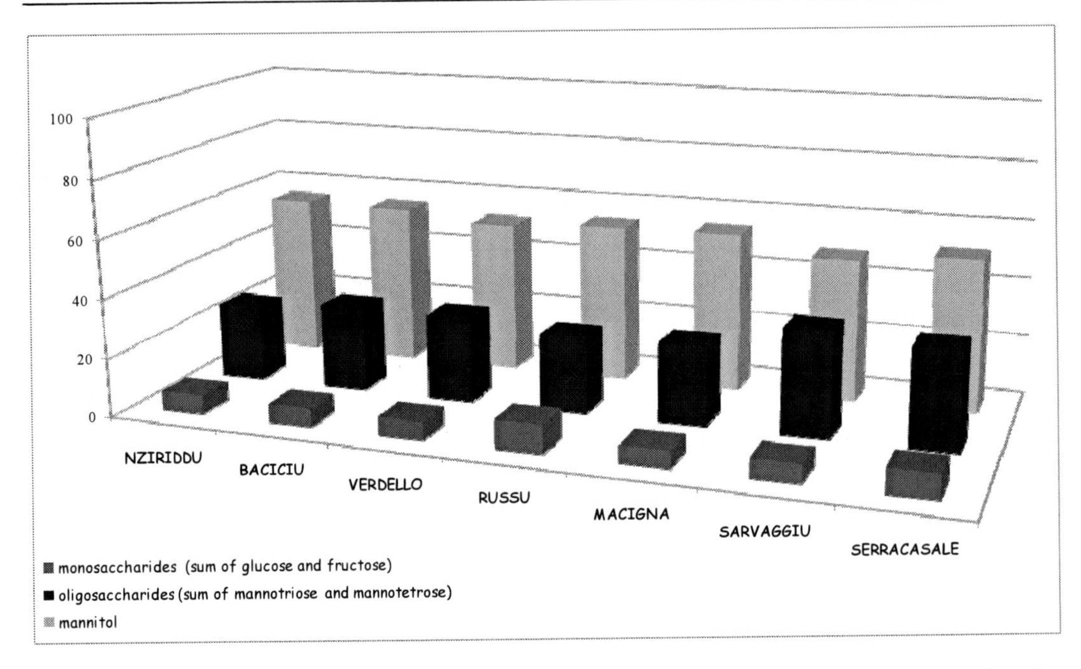

Figure 4. comparison between the cultivars of Fraxinus angustifolia and F. ornus, with regard to the contents in monosaccharides , oligosaccharides and mannitol.

It is reported in the literature that the presence of coumarins, secoiridoids, and phenylethanoids is a characteristic feature of *Fraxinus* species. These chemical constituents were isolated from bark, leaves and flowers of *Fraxinus angustifolia* and *ornus* trees. The secoiridoids occur mainly in the form of glucosides and esters of hydroxyphenylethyl alcohols. Lignans, flavonoids and simple phenolic compounds are also common, but they appear to have more limited distribution. The occurrence of coumarins distinguishes the genus *Fraxinus* from the other genera in Oleaceae. Traditionally, the genus has always been associated with investigations on coumarins [94].

Even the analytical investigations carried out on samples of manna have shown the presence of coumarin constituents. In particular, from a sample of *manna in rottame* were isolated compounds **1** and **2**, whose structure are shown below [101].

Preliminary experimental data not yet published carried out by HPLC-DAD-ESI-MS, have led to the identification of three coumarins [30]. Figure 5 shows the reconstructed chromatogram on the basis of the total ion current (TIC) of compounds 1-3 recorded at a wavelength of 320 nm.

OH
OH
O
H_3CO
OCH_3
O
O

1

$OCOCH_3$
H_3CO
O
O
O
$OCOCH_3$

2

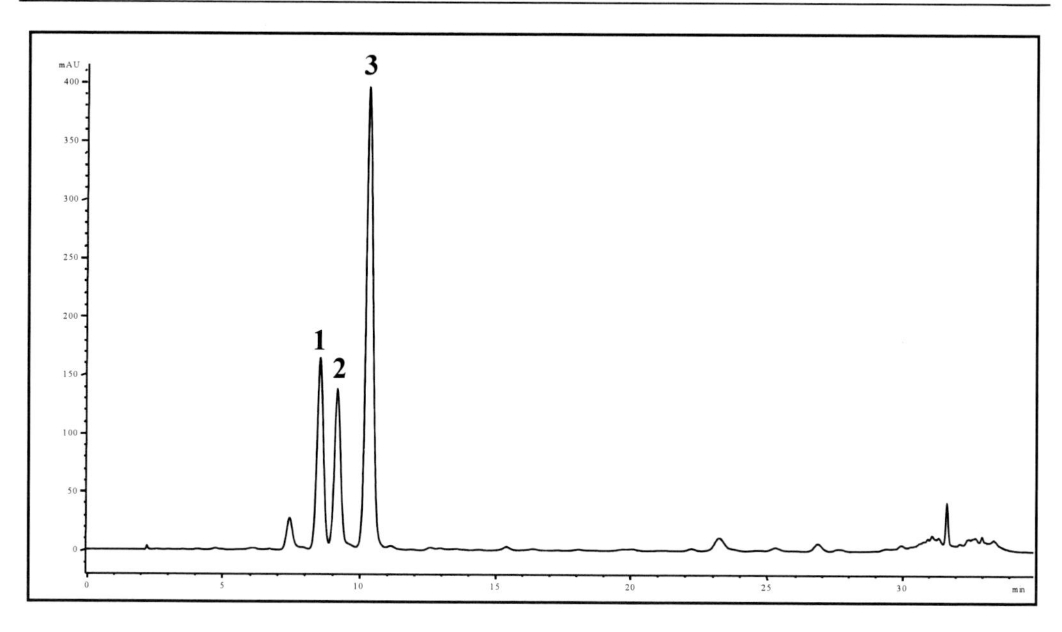

Figure 5. reconstructed chromatogram based on the total ion current (TIC) of compounds 1-3 recorded at a wavelength of 320 nm.

In order to indicate a structure for the three compounds, were performed MS-MS experiments. Through these analysis it was possible to identify the compounds as fraxidin, isofraxidin and fraxinol, whose structures are shown below.

H_3CO, H_3CO, OH, O, O

H_3CO, HO, OCH_3, O, O

OCH_3, HO, H_3CO, O, O

Fraxidin Isofraxidin Fraxinol

CONCLUSION

Although resins discussed in this chapter have proved to be popular alternative or complementary medicine used in the treatment of many diseases, clinical trial evaluation of these claims using currently accepted protocols is needed. The reported resins offer huge potential as a possible pharmacological application but it's necessary a further investigation to verify whether purified compounds isolated may have better therapeutic potential as compared to crude extracts.

Since there is considerable variation in the chemical composition among resins referred, quality control/assurance needs to be established for the traditional medical trade.

Advances in chromatographic and spectroscopic techniques only now allow us isolation and structural analysis of potent biologically active plant constituents that are present in too low quantity to have been previously characterized. The aim of this work is an effort to highlight the potential and problems related to the sources and possibilities of isolating new pharmaceutically active molecules, using traditional knowledge in our search, for new and effective drugs molecules.

These new chemicals will serve to enhance the continued usefulness of higher plants and their products as renewable resources of chemicals.

REFERENCES

[1] Langenheim, J. *Plant Resins: Chemistry, evolution, ecology, and ethnobotany*. Timber Press: Portland; 2003.

[2] PhD thesis of Basar, S. *Phytochemical Investigations On Boswellia Species*. Istanbul, Turkey; 2005

[3] Angiosperm Phylogeny Group. *Annals of the Missouri Botanical Garden*, 1974, 85, 531.

[4] Alexander, D; Miller, A. Socotra's misty future. *New Scientist*, 1995, 147, 32.

[5] Sollman, T. A sketch of the medical history of Dragon's Blood. *Journal of the American Pharmaceutical Association,* 1920, 9, 141–144.

[6] Gupta, D; Bleakley, B.; Gupta, RK. Dragon's blood: Botany, chemistry and therapeutic uses. *Journal of Ethnopharmacology,* 2008, 115, 361–380.

[7] ELHAH. The Eleventh Labor of Hercules: The Apples of The Hesperides, 2007.

[8] Plinius Secundus. *The Historie of the Worlde*. London, 1601.

[9] Emboden, WA. *Bizarre Plants: Magical, Monstrous and Mythical.* Studio Vista: London; 1974.

[10] Lyons, G. In search of dragons or: the plant that roared. *Cactus and Succulent Journal,* 1974, 44, 267–282.

[11] Baumer, U; Dietemann, P. Identification and differentiation of dragon's blood in works of art using gas chromatography/mass spectrometry. *Anal Bioanal Chem*, 2010, 397, 1363–1376.

[12] Koller, J; Baumer, U; Schmid, E; Grosser, D. Sandarac. In: Walch K, Koller J (eds) Baroque and Rococo Lacquers, *Arbeitshefte des Bayerischen Landesamtes für Denkmalpflege*. Lipp, München; 1997.

[13] Koller, J; Baumer, U; Grosser, D; Walch, K. Larch turpentine and Venetian Turpentine. In: Walch K, Koller J (eds) Baroque and Rococo Lacquers, *Arbeitshefte des Bayerischen Landesamtes fürDenkmalpflege.* Lipp, München; 1997.

[14] Koller, J; Baumer, U; Grosser, D; Schmid, E. Mastic. In: Walch K, Koller J (eds) Baroque and Rococo Lacquers, *Arbeitshefte des Bayerischen Landesamtes für Denkmalpflege.* Lipp, München; 1997.

[15] Pastorova, I; de Koster, CG; Boon, JJ. Analytical Study of Free and Ester Bound Benzoic and Cinnamic Acids of Gum Benzoin Resins by GC–MS and HPLC–frit FAB–MS. *Phytochem Anal* 1997, 8, 63–73.
[16] Milner, JE. *The Tree Book*. Colllns & Brown Ltd., London; 1992.
[17] Parkinson, J. *Theatricum Botanlcum*. London. 1640
[18] Brockmann, H; Junge, H. Die Konstitution des Dracorhodins, eines neuen Farbstoffes aus dem "Drachenblut". *Ber. Dtsch. Chem. Ges. B* 1943, 76, 751-763.
[19] Melo, MJ; Sousa, MM; Parola, AJ; Seixas de Melo, JS; Catarino, F; Marcalo ¸ J; Pina, F. Identification of 7,4'-Dihydroxy-5-methoxyflavylium in "Dragon's Blood": To Be or Not To Be an Anthocyanin. *Chem. Eur. J.* 2007, 13, 1417-1422.
[20] Pina, F. *J. Chem. Soc. Faraday Trans.* 1998, 94, 2109.
[21] Pina, F; Maestri, M; Balzani, V. in: H.S. Nalwa (Ed.), Handbook of Photochemistry and Photobiology, vol. 3, Supramolecular Photochemistry, American Scientific Publishers; 2003.
[22] Pina, F; Melo, MJ; Parola, AJ; Maestri, M; Balzani, V. pH-Controlled photochromism of hydroxyflavinium ions. *Chem Eur J.* 1998, 4, 2001.
[23] Pina, F; Maestri, M; Balzani, V. *Chem Commun* 1999, 107.
[24] McClelland, RA; Gedge, S. Hydration of the Flavylium Ion. *J Am Chem Soc* 1980, 102, 5838.
[25] Maxwell, CA; Philips, DA. Concurrent synthesis and release of nod-gene-inducing flavonoids from alfalfa roots. *Plant Physiol.* 1990, 93, 1552-1558.
[26] Sousa, MM; Melo, MJ; Parola, A; b, J de Melo, JSS; Catarino, F; Pina, F; Cooke, FEM; Simmondse, MSJ; Lopes, JA. Flavylium chromophores as species markers for dragon's blood resins from *Dracaena* and *Daemonorops* trees. *Journal of Chrom A.* 2008, 1209, 153–161.
[27] Camarda, L; Merlini, L; Nasini, G. Dragon's blood from *Dracaena draco*, structure of novel homoisoflavonoids. *Heterocycles* 1983, 20, 39–43.
[28] Gonzáles, AG; León, F; Sánchez-Pinto, L; Padrón, JI; Bermejo, J. Phenolic Compounds of Dragon's Blood from *Dracaena draco*. *J Nat Prod.* 2000, 63, 1297–1299.
[29] Gonzáles, AG; León, F; Hernández, JC; Padrón, JI; Sánchez-Pinto, L; Bermejo Barrera, J. Flavans of dragon's blood from *Dracaena draco* and *Dracaena tamaranae*. *Biochem Syst Ecol.* 2004, 32, 179–184.
[30] PhD thesis of Pitonzo, R. Identification and determination of analytes in environmental matrices and in plants: the search for new analytical methods sensitive and selective. Palermo, Italy; 2008
[31] Barz, W; Welle, R; in: H.A. Stafford (Ed.), *Flavonoid Metabolism*, CRC Press, Boca Raton, FL, USA; 1990.
[32] [Strack, D. in: H.-W. Heldt (Ed.), *Plant Biochemistry*, Academic Press, New York, NY, USA; 1997.
[33] [Rosenberg Zand, RS;. Jenkins, DJA; Diamandis, EP. Flavonoids and steroid hormone-dependent cancers. *J. Chromatogr. B.* 2002, 777, 219.
[34] Olsson, LC; Veit, M; Weissenböck, G; Bornman, JF. *Phytochemistry.* Differential flavonoid response to enhanced UV-B radiation in *Brassica napus*.1998, 49, 1021.
[35] Reuber, S; Bornman, JF; Weissenböck, G. *Physiol Plant.* 1996, 97, 160.
[36] Middleton, EM; Teramura, AH. The Role of Flavonol Glycosides and Carotenoids in Protecting Soybean from Ultraviolet-B Damage. *Plant Physiol.* 1993, 103, 741.

[37] Wang, CY; Huang, HY; Kuo, KL; Hsieh, YZ. *J Chromatogr A* 1998, 802, 225.
[38] Goto, T; Kondo, T; *Angew Chem* 1991, 30, 17.
[39] Biggs, DR; Lane, GA. *Phytochemistry*. 1978, 17, 1683.
[40] Rice-Evans, CA; Miller, NJ; Paganga, G. *Trends Plant Sci.* 1997, 2, 152.
[41] Miksicek, RJ. *Mol Pharmacol.* 1993, 44, 37.
[42] Wollenweber, E. in: V. Cody, E. Middleton Jr., J.B. Harborne, A. Beretz (Eds.), Plant Flavonoids in Biology and Medicine. II. *Biochemical, Cellular and Medicinal Properties,* Liss, New York, NY, USA; 1988.
[43] Weidenborner, M; Jha, HC.Antifungal activity of flavonoids and their mixtures against different fungi occurring on grain. *Mycol Res.* 1994, 98, 1376.
[44] Camarda, L; Di Stefano, V; Pitonzo, R. Homoisoflavonoids analysis in *Dracena draco* fruits by HPLC-DAD-ESI-MS. 1st French-Italian Conference on Mass Spectrometry. 2008, 123-124.
[45] Gonzalez, AG; Hernandez, JC; Leon, F; Padron, JI; Estevez, F; Quintana, J; Bermejo, JJ. Steroidal saponins from the bark of *Dracaena draco* and their cytotoxic activities. *Nat Prod.* 2003, 66, 793–798.
[46] Hernández, JC; León, F; Quintana, J; Estévez, F; Bermejo, J. Icogenin, a new cytotoxic steroidal saponin isolated from *Dracaena draco*. *Bioorganic and Medicinal Chemistry*. 2004, 12, 4423–4429.
[47] Hernández, JC; León, F; Estévez, F; Quintana, J; Bermejo, J. A Homo-Isoflavonoid and a Cytotoxic Saponin from *Dracaena draco. Chemistry & Biodiversity,* 2006, 3, 62-68.
[48] Mimaki, Y; Kuroda, M; Ide Atsushi Kameyama, A; Yokosuka, A; Sashida, Y.. Steroidal saponins from the aerial parts of *Dracaena draco* and their cytostatic activity on HL-60 cells. *Phytochemistry.* 1999, 50, 805–813.
[49] Darias, V; Bravo, L; Rabanal, R; Sanchez Mateo, C; Gonzalez Luis, RM; Hernandez Perez, AM. New contribution to the ethnopharmacological study of the Canary Islands. *Journal of Ethnopharmacology* 1989, 25, 77–92.
[50] Vollesen, K. *Burseraceae.* In I. Hedberg and S. Edwards (eds.), Flora of Ethiopia, Volume 3. National Herbarium, Addis Ababa University, Addis Abeba and Uppsala University, Uppsala, 1989.
[51] Marshall, S. *Frankincense*: festive pharmacognosy. *Pharm. J.,* 2003, 271, 862-864.
[52] Michie, CA; Cooper, E. Frankincense and Myrrh as remedies in Children. *J. R. Soc. Med.*, 1991, 84 (10), 602-605.
[53] Shachter, J; Epstein, I. The Babylonian Talmud, Seder Nezikin, *Sanhedrin tractate,* The Suncino Press, New-York, 1935.
[54] Koskenniemi, E; Nisulab, K; Topparic, J. *J. Study New Testament*, 2005, 27, 379-391.
[55] Culpeper, N. *A Physical Directory, or, A translation of the London Dispensatory Made by the College of Physicians in London*. London,1649.
[56] Griggs, B. *Green pharmacy*. London: Robert Hale Press, 1981.
[57] Getahon, A. (1976) Some Common Medicinal and Poisonous Plants Used in Ethiopian Folkmedicine. *Mimeographed*, Addis Abeba University, Addis Abeba.
[58] Khan, IA; Abourashed, EA. *Leung's Encyclopedia of Common Natural Ingredients: Used in Food, Drugs and Cosmetics*, 3rd Edition, Wiley, 2009.
[59] Ammon, HTP. Boswellic acids in chronic inflammatory diseases. *Planta Med.* 2006, 72(12),1100-1116.

[60] Hamm, S; Bleton, J; Connan, J; Tchapla, A. A chemical investigation by headspace SPME and GC-MS of volatile and semi-volatile terpenes in various olibanum samples. *Phytochemistry*, 2005, 66, 1499-1514.

[61] Mathe C; Culioli G; Archier P; Vieillescazes C. Characterization of archaeological frankincense by gas chromatography–mass spectrometry. *Journal of Chromatography A*, 2004, 1023, 277–285.

[62] Recio MC; Giner RM; Manez S; Rios JL. Structural Requirements for the Anti-inflammatory Activity of Natural Triterpenoids, *Planta Med.*, 1995, 61, 182-185.

[63] Bannoa N; Akihisa T; Yasukawa K; Tokuda H; Tabata K; Nakamura Y; Nishimura R; Kimura Y; Suzuki T. Anti-inflammatory activities of the triterpene acids from the resin of Boswellia carteri, *Journal of Ethnopharmacology*. 2006, 107, 249–253.

[64] Singh, GB; Atal, CK. Pharmacology of an extract of salai guggal ex-Boswellia serrata, a new nonsteroidal anti-inflammatory agent. *Agents Actions,* 1986, 18, 407–412.

[65] Safayhi, H; Sailer, ER. Anti-inflammatory actions of pentacyclic triterpenes. *Planta Medica* 1997,63, 487-493.

[66] Ammon HPT; Mack T; Singh GB; Safayhi H. Inhibition of leukotriene B_4 formation in rat peritoneal neutrophils by an ethanolic extract of the gum resin exudate of Boswellia serrata. *Planta Med* 1991, 57(3), 203-207.

[67] Estrada AC; Syrovets T; Pitterle K; Lunov O; Buchele B; Schimana-Pfeifer J; Schmidt T; Morad SAF; Simmet T. Tirucallic Acids Are Novel Pleckstrin Homology Domain-Dependent Akt Inhibitors Inducing Apoptosis in Prostate Cancer Cells, *Mol. Pharmacol.*, 2010, 77(3), 378-387.

[68] Poeckel, D; Werz, O. Boswellic acids: biological actions and molecular targets. *Curr. Med. Chem.* , 2006, 13, 3359-3369.

[69] Banno, N; Akihisa, T; Yasukawa, K; Tokuda, H; Tabata, K; Nakamurab, Y; Nishimura, R; Kimura, Y; Suzuki, T. Anti-inflammatory activities of the triterpene acids from the resin of Boswellia carteri, *Journal of Ethnopharmacology* 2006, 107, 249–253.

[70] Takada, Y; Ichikawa, H; Badmaev, V; Aggarwal, BB. Acetyl-11- keto-ß-boswellic acid potentiates apoptosis, inhibits invasion, and abolishes osteoclastogenesis by suppressing NF-kappa B and NF-kappa B-regulated gene expression. *J Immunol.* 2006, 176, 3127–3140.

[71] Badria, FA; Mohammed, EA; El-Badrawy, MK; El- Desouky, M. Natural leucotriene inhibitor from Boswellia: a potential new alternative for treating bronchial asthma. *Altern. Complement. Ther.* 2004, 10 (5), 257–265.

[72] Gerhardt, H; Seifert, F; Buvari, P; Vogelsang, H; Repges, R; Therapy of active Crohn disease with *Boswellia serrata* extract H 15. *Z Gastroenterol.* 2001, 39, 11-17.

[73] Gupta, I, Parihar, A, Malhotra, P, Gupta, S, Lüdtke, R, Safayhi, H, Ammon, HP. Effects of gum resin of Boswellia serrata in patients with chronic colitis. *Planta Med.* 2001,67, 391-395.

[74] Liu, JJ; Nilsson, A; Oredsson, S; Badmaev, V; Duan, RD; Keto- and acetyl-keto-boswellic acids inhibit proliferation and induce apoptosis in Hep G2 cells via a caspase-8 dependent pathway. *Int J Mol Med.* 2002,10, 501-505.

[75] Syrovets, T; Gschwend, JE; Büchele, B; Laumonnier, Y; Zugmaier, W; Genze, F; Simmet, T. Inhibition of IkappaB kinase activity by acetyl-boswellic acids promotes apoptosis in androgen-independent PC-3 prostate cancer cells in vitro and in vivo. *J. Biol. Chem.* 2005, 280, 6170-6180.

[76] Zhao, W; Entschladen, F; Liu, H; Niggemann, B; Fang, Q; Zaenker, KS; Han, R. Boswellic acid acetate induces differentiation and apoptosisin highly metastatic melanoma and fibrosarcoma cells, *Cancer Detection and Prevention*, 2003, 27(1), 67-75.
[77] Huang ,MT; Badmaev, V; Ding, Y; Liu, Y; Xie, JG; Ho, CT. Anti-tumor and anti-carcinogenic activities of triterpenoid, β-boswellic acid. *BioFactors,* 2000, 13, 225–230.
[78] Burlando, B; Parodi, A; Volante, A; Bassi, AM. Comparison of the irritation potentials of *Boswellia serrata* gum resin and of acetyl-11-keto-b-boswellic acid by in vitro cytotoxicity tests on human skin-derived cell lines. *Tox. Lett.* 2008, 177, 144-149.
[79] Xia, L; Chen, D; Han, R; Fang, Q; Waxman, S; Jing, Y. Boswellic acid acetate induces apoptosis through caspase-mediated pathways in myeloid leukemia cells. *Mol. Cancer Ther.* 2005, 4, 381-388.
[80] Shealy, CN. *The Illustrated Encyclopaedia of Healing Remedies.* Element books, Australia 1998.
[81] Stevensen, CJ. Aromatherapy in dermatology. *Clinics in Dermatology* 1998, 16, 689–694.
[82] Wootton, S. *Aromatherapy and Natural Health.* GE Fabbri Ltd, London, UK 2005.
[83] Norihiro, B; Toshihiro, A; Ken, Y; Harukuni, T; Keiichi, T; Yuji, N et al. *J Ethnopharmacol*, 2006, 107, 249.
[84] Camarda, L; Dayton, T; Di Stefano,, V; Pitonzo, R; Schillaci, D. Oleogum resin essential oils from *Boswellia* spp. (Burseraceae): chemical composition analysis and antimicrobial activity. *Annali di Chimica* 2007, 97, 837–844.
[85] Gupta, I; Parihar, A; Malhotra, P; Singh, GB; Luettke, R; Safayhi, H. et al. *Eur J Med Res*, 1997, 2, 1
[86] Hammer, KA; Carson, CF; Riley, TV. *J Appl Microbiol*, 1999, 86, 985.
[87] Shao, Y; Ho, CT; Chin, CK; Badmaev, V; Ma, W; Huang ,MT. *Planta Med*, 1998, 64, 328.
[88] Schillaci, D; Arizza, V; Dayton, T; Camarda, L; Di Stefano, V. In vitro anti-biofilm activity of *Boswellia* spp. Oleogum resin essential oils. *Letters in Applied Microbiology* 2008, 47, 433–438.
[89] Wallander, E; Albert, VA. Phylogeny and classification of Oleaceae based on rps16 and trnL-F sequence data. *Am J Bot*. 2000;87:1827.
[90] Flora of China. http://flora.huh.harvard.edu.
[91] [Knoblauch, E. Oleacea. In: Engler A, Prantle K, editors. Die Natürlichen Pflanzenfamilien IV, vol. 2. Leipzig: Verlag vonWilhelm Engelmann; 1897.
[92] Classification of *Fraxinus* L. sensu Lingelsheim. *http://www2.botany.gy.se1920.*
[93] Stoyanov, N. *Our medicinal plants*, vol. 1. Sofia: Nauka i izkustwo; 1973.
[94] Kostova, I; Iossifova, T. Chemical components of *Fraxinus* species. *Fitoterapia.* 2007, 78, 85–106
[95] *www.ilfrassino.it*
[96] Lazzarini, E; Lonardoni, AR. *La manna salute dalla natura. Edizioni Mediterranee*, Roma; 1984.
[97] Schicchi, R; Camarda, L; Spadaro, V; Pitonzo, R. Caratterizzazione chimica della manna estratta nelle Madonie (Sicilia) da cultivar di *Fraxinus angustifolia* e di *Fraxinus ornus* (Oleaceae). *Quad Bot Amb Appl* 17/2, 139-162 (2006).

[98] Oddo, E; Saiano, F; Bellini, E; Alonzo, G; Analisi del contenuto in mannitolo di manna da due specie di frassino coltivate in Sicilia. *Quad Bot Ambientale Appl.* 1997, 8, 61-63.

[99] Redwell, R.J. Fractionation of plant extracts using ion-exchange Sephadex. *Anal Biochem*. 1980, 107, 44-50.

[100] Medeiros, PM; Simoneit, BRT. Analysis of sugars in environmental samples by gas chromatography-mass spectrometry. *Journal of Chromatography A*. 2007, 1141, 271-278.

[101] Camarda, L; Giammona, G; Listro, O; Palazzo, S. Ricerche sui costituenti minori della manna:isolamento e caratterizzazione di una nuova cumarina. *Boll Chim Farmaceutico*, 1989, 128, 225-228.

In: Resin Composites: Properties, Production and Applications ISBN: 978-1-61209-129-7
Editor: Deborah B. Song

Chapter 12

Resin Composite Laminates: Conservative and Esthetic Restorations

Jefferson Ricardo Pereira, Graciela Talhetti Brum, Simone Xavier Silva Costa, Maria Stela do Nascimento Brasil and Janaina Salomon Ghizoni
Dental School – University of Southern Santa Catarina - UNISUL, Tubarão – SC, Brazil

Introduction

Many traditional concepts in dentistry have changed since the 1950s. In 1955, Buonocore was the forerunner of a new phase in dentistry. With the advent of the technique that involved etching enamel, the implementation of bonding procedures began to be incorporated into daily practice; this enabled the execution of more conservative restorative techniques that required minimal removal of healthy tooth structure. Later, in 1963, Bowen developed composite resin restorations with esthetic characteristics including conservation of tooth structure. The improvement of the physical and mechanical properties, and esthetics of composite resins and made it possible to directly manufacture restorations on teeth by freehand; although the cosmetic results were very satisfactory, the process requires a great deal of manual skill on the part of the professional who performs it.

Generally speaking, laminates are an esthetic coating material that cover the entire enamel of anterior teeth and have color and/or an altered form. They can be divided into direct and indirect laminates. The laminates applied directly by the professional (those that do not require the services of a dental lab) normally require composite resin restorative material; indirect laminates are fabricated by the dental laboratory on a working model and can be made of resin composite or ceramic (Baratieri, 1998).

When comparing the advantages and disadvantages of direct and indirect techniques, the indirect technique usually requires more chair-time, greater wear on tooth structure, a temporary restoration, molding, and an adhesive cementation laboratory phase; all of these make this procedure more complex and costly for the dentist and the patient. Nevertheless, the fabrication of the indirect laminate is all performed by the lab technician outside of the oral cavity, which allows precise determination of the correct anatomical shape, points of contact, and desired color. Moreover, laminates allow for direct implementation in a single session; usually, more conservative cavity preparations are subjected to medium- and long-term repairs with ease. The production of a laminate of direct composite resin is perhaps one of the most difficult tasks to accomplish; it is difficult to achieve satisfactory results for both the patient and the professional. These difficulties result from choosing the correct color, using opaque agents, determining the correct anatomy of the tooth, and applying different layers and colors of resins to obtain adequate shade, opacity and translucency (Baratieri, 2000).

In clinical situations of low complexity, the choice of technique and restorative material is quite varied and can be left to the patient and the professional. In cases where the enamel structure does not compromise its entire vestibular length, the best choice of restorative material and technique is direct composite resin to preserve as much healthy tooth enamel as possible (Silva e Souza Jr. *et al.*, 2000).

The clinician often finds himself with the question of whether to use a direct laminate or an indirect laminate. One can understand that the final results are very similar, and the final decision will be based on the following aspects:

1. Degree of dental browning;
2. Cost;
3. Professional skill;
4. Degree of demand from the patient.

Of course, each clinician must choose the most appropriate course to a final resolution for each specific clinical case. Therefore, this chapter will directly address the topic of direct laminates, discussing all of the aspects involved in this restorative technique in order to advise professionals regarding in which clinic conditions they should be applied to achieve esthetic and functional excellence.

Particulars of Direct Laminates

Generally, laminates can be directly applied to correct alterations in color, shape, or position on the buccal surface of any tooth. However, because it is a highly esthetic restorative method, it is generally performed on the anterior teeth.

Indications:

- Specific cases of fractured anterior teeth;
- Previous anatomical transformations;
- Teeth with hypoplastic or congenital malformations;
- Teeth with one or more large and poor restorations involving the labial surface;
- Teeth requiring realignment of the dental arch;
- Teeth that change color;
- Diastema closure;

- Lengthening of the clinical crown.

Prior to starting treatment, it is necessary that the professional is aware of the esthetics and particularities of each case so that treatment can be performed safely and predictably.

As a general statement, we can say that a direct composite laminate should be planned in those cases where two-thirds or more of the vestibular structure is compromised by changing color, shape, or texture. These changes may be remedied through more conservative techniques (Silva e Souza Jr. 2000; Carvalho, 1998; Mondelli , Mondelli, 2001).

Contraindications of Indiret Laminates

Laminates cannot be contraindicated in an accurate and definitive manner because dental materials are constantly changing. However, some clinical situations, such as the following, limit this type of restoration.

- The presence of enamel on the cavosurface angle: the margins of the cavity preparation should preferably be bounded by enamel to promote an adequate peripheral seal around the cavosurface angle because adhesion to dentin is less stable than adhesion to enamel; (Perdigão, Lopes, 1999; Lopes et al., 2002).
- Harmful habits: bruxism and biting habits such as biting on pencils, pens, and nails have the potential to exceed the limits of the properties of the restorative material;
- Poor occlusion: the presence of an occlusion butt should be observed in addition to the existing occlusal contacts during maximal and habitual intercuspation excursive movements, both in lateral movement and in protrusion. It is important to note whether there is enough space for the excursive movements of the restorative material, as excessively thin edges of composite may fracture more easily;
- Remaining dental structure: in teeth with vital pulp that have large restorations or large amounts of unsupported enamel and dentin, the strength of the remaining dental laminates should not be undermined;
- Periodontal problems and/or caries: Patients with periodontal disease should not be subjected to treatment with laminates before such problems are controlled. In addition, the gingival morphology should be checked, as treatment with direct resin composite laminates is closely connected with esthetic factors. The professional should not disregard the influence that the gingiva have on the dimensions of the tooth crown and the harmony of the smile, especially in those patients with high smile lines (Conceição, 2007).

It should be noted that generally the adhesives procedures are highly delicate; in unfavorable situations, as in cases of patients at high risk of caries, these procedures are contraindicated (Mondelli *et al.,* 2003).

Due to the increasing quality of restorative materials and the growing concern among patients about appearance, some situations occur over indications. In some cases, with the aim of encouraging more esthetic, achieving a uniform color, texture and form, are made various laminates in various teeth despite a lack of indication. Of course it is much easier to achieve a harmonious and natural result when treating teeth in several laminates, but this technique is not considered correct (Silva e Souza Jr. *et al.*, 2003).

Advantages and Disadvantages of the Direct Technique

As mentioned earlier in this chapter, the advantages of making composite resin laminates are the following: the possibility for and timeliness of repairs, proper safety and effectiveness, certain cases can be executed without any preparation; there is no need for the laboratory, it requires less clinical time, no temporary is needed, and the preparation is generally more conservative than that required for indirect veneers. Therefore, it is a procedure that is available to all students, general practitioners and specialists.

Its disadvantages include the following: the direct laminates have higher physical and mechanical resistance than the indirect laminates made of ceramics as well as lower color stability and an increased time spent by the clinical professional. In addition, the final esthetic result, an accurate reproduction of the optical characteristics and surface texture of the natural teeth are superior to the indirect ceramic laminates.

Composite Resins Used in the Fabrication of Direct Laminates

The final esthetic results of the direct laminates of resin composite have improved due to the increased availability of materials with a greater diversity of hue, chroma, value and translucency, making it the most natural type of direct restoration (Lopes et al., 2001).. Composite resins are commonly classified according to the size, distribution and percentage by weight/volume of the particles of charge. The use of each type of composite resin varies according to the clinical situation in which it is applied.

Resins of Microparticles

The emergence of composite resin microparticles caused a very significant change in the restorations of anterior teeth because this material has particles with an average size load of approximately 0.04 μm. Employing this type of material, it is possible to obtain restorations with surface smoothness and brightness comparable to that of tooth enamel. Moreover, this type of composite resin has a reduced amount of cargo, about 40 to 50% by weight (Silva e Souza Jr. *et al.*, 2001).

From a clinical perspective, the reduction in the amount of cargo is relevant for two reasons: the first refers to the reduction of some mechanical properties, which makes the laminate more susceptible to wear and surface degradation; the second is related to higher water absorption because the amount of organic matrix is higher than in the hybrid materials and high charge density. As a result of this, the water-soluble pigments are more easily absorbed, resulting in a rapid color change (Mondelli, 2001). Moreover, the composite microspheres are susceptible to chipping, especially when they are deployed in areas of high stress.

Despite the preference for composite microparticles by many professionals, there are now composite resins with satisfactory high cargo density and surface texture. The composites that exhibit these characteristics are actually called hybrids, microhybrids and, more recently, nanoparticles (Baratieri *et al.,* 2001), which will be discussed later in this chapter.

HYBRID COMPOSITE RESINS AND MICROHYBRIDS

he hybrid composite resins and inorganic particle microhybrids have adjusted to an average size of 0.6 μm - 2.0 μm. The hybrid resins have the highest concentration of charged particles with an average diameter of approximately 1.0 to 1.2 μm, while the resin microhybrid particles have diameters of less than 1.0 μm. This type of resin still has the peculiarity of having associated microparticles (0.04 μm), and this association helps to increase the concentration of cargo in the same volume of organic resin because the microparticles are unable to occupy positions filled by larger particles. This increases the wear resistance of the restoration and improves polishing (Chain *et al.,* 2002).

The microhybrids resins can be used safely for making direct laminates without the necessity of using a final layer of microfill to improve the surface smoothness and brightness of a restoration.

RESINS NANOPARTICLES

More recently, composites containing nanoparticles have emerged that employ nano-sized particles (resin nanoparticle) or aggregates of these particles (nanohybrids).

The purpose of this new generation of composites is to incorporate nano-sized particles (average particle size from 5 to 75 nanometers) into the organic matrix, resulting in greater strength and better surface smoothness. These resins maintain polishing and superior brightness when compared to the materials of microparticle composites and also exhibit similar characteristics of resistance when compared to current hybrid composites or microhybrids (Reis et al., 2007; Costa et al, 2007).

This category of resin can also be used for making direct laminates without requiring the final layer of microfill to improve surface smoothness and brightness.

In essence, every type of composite resin described above can be used for direct laminates; however, the hybrid composites and, nanoparticle microhybrids are more resistant to wear, staining and degradation of the surface and edges of the microparticles. These have the advantage of being closer to our initial esthetic goal but can provide more rapid modifications to color and texture (Silva e Souza Jr., *et al.,* 2003).

Longevity of Direct Laminates

Direct resin laminates may have unlimited shelf life; however, on average, they behave well for about 10 years (Baratieri *et al.*, 2001). According to Hirata (2011), the estimated lifetime of direct restorations with composite resin cannot be estimated due to the multiple factors involved in these cases.

The composite resin is placed directly onto the tooth by the dentist himself, whereas a laboratory produces a more refined porcelain laminates that ultimately provides more polishing than direct laminates. As a result, smokers or those who frequently consume drinks containing dyes such as cola drinks, tea, coffee and red wine should be informed of the potential for reduced longevity of the restoration because these pigments are absorbed by the resin matrix of composites. If the patient cannot suspend or limit the use of these chromogenic substances, indirect porcelain veneers may be the best option (Blank, 2002).

When the dentist and the patient are informed of all of the limitations of direct composite resin laminates, their longevity can be increased. The fact that the laminate may be directly liable for repairs prolongs the longevity of this procedure and makes it conservative because the periodic replacement of worn restorations leads to healthy dental structure.

Following Surgery

This section will explain the clinical protocol to be followed by topic, and shortly thereafter, we will address the details regarding the specific steps of the procedure:

1. Prophylaxis in the region to be restored and pre-selection of color with reference ranges;
2. Test or selection of the color with a small amount of the pre-selected resin applied directly to the tooth and cured without the prior stages of accession; 3. Cavity preparation;
3. Isolation of the operative field, preferably with the aid of retractors and mouth retractor wire;
4. Acid etching for 15 seconds, followed by copious washing for 30 seconds, drying with absorbent paper and the selection of an adhesive system following the manufacturer's instructions;
5. Incremental insertion of layers of composite resin following the optical characteristics of each dental substrate;
6. Finish initial after de restoration;
7. Final polishing after a minimum of 24 hours.

Select Color

The organizational system of color most used in dentistry was proposed in 1915 by physicist Albert Munsell. In this system, color can be represented by three dimensions: hue, chroma and value.

Hue is the attribute that allows the observer to distinguish a particular color from different color families. It is represented in scales as the reference standard VITA CLASSICAL letters (A, B, C and D). Chroma determines the degree of saturation of a color. It is determined by the standard reference range of numbers (1, 2, 3, 4, 5). The value determines the brightness of a color or the grey that it contains (Miyashita et al., 2004).

Optical Properties of Teeth to be Considered for Selecting the Color of the Resins

When screening for color, different optical properties between the enamel and dentin have to be considered. At present, resins have a wide range of colors that have different levels of translucency to obtain optical characteristics as close as possible to the natural color. The enamel resins have high translucency however dentin resins have low specific translucency ensuring opacity characteristics to the restorations. In most cases, it is necessary to use a layered technique, which involves using different thicknesses of resin composite to mimic the optical characteristics and color of dentin and enamel.

In cases involving dark teeth, it often becomes necessary to use opacifiers (opaque resins opacifiers or pigments) before applying the specific resin to dentin. Importantly, when using these resources, one must be careful to not to make the restoration become artificial and the tooth lacks its vitality. In addition to the features of varying translucency, one should also consider properties such as fluorescence and opalescence. The opalescence and fluorescence properties are inherent to most current composites and provide greater brightness, depth and life to the restorations (Marques et al., 2005; Kano, Gondo, 2008).

- Ambient lighting conditions for color selection

The environment should be well lit with special fluorescent lamps called "daylight" lamps, which exhibit a temperature between 4,500 and 5,000 Kelvin (K) and represent the color of daylight (Goldstein et al., 2000). Another option would be to perform this procedure near a natural source of light such as a window. Ideally, this light should come through a blind with a neutral color that is half open to prevent over-brightness. Conventional incandescent lamps have a low color temperature, around 2,800 K, emitting excessive radiation with a wavelength in the range of yellow that tends to increase the perception of yellow (Silva e Souza Jr. et al., 2003).

- Pre-selection of color with reference scales

The professional should always seek scales that are compatible with the restorative system selected. As a general rule, the operator must use the color chart provided by the manufacturer of the resin that will be employed. An interesting option includes the production of a reference scale by the professional with the resins used in his office. The universal color scale or VITA CLASSICAL VITA 3D-MASTER can also be employed at this restorative stage.

Before making use of the color scale, one must make a preliminary assessment and seek to identify the hue, saturation and approximate value, not forgetting to consider the different saturation levels between the neck and body portions of the tooth. By performing this pre-selection, one can choose a correspondent color, in the scale, to verify the correct tooth color (Mondelli et al., 2003).

It is also important that the scale and the teeth are wet while selecting the color so that there is an appropriate interaction of the incident and reflected light with the ambient brightness.

- Select the color using the composite resin

After pre-selection of the color using scales, one must apply the restorative material directly onto the tooth or adjacent teeth. The tooth surface should be dry and without prior application of phosphoric acid or the adhesive system. The restorative material should be placed and light cured for the amount of time recommended by the manufacturer. Then, one should dampen the composite resin and the teeth to observe the color.

Another alternative that can be used is the mock up. In this procedure, after the pre-selection of color with reference scales, runs up the steps of the restoration without adhesive system stratifying layers of resin with the preselected color. The patient is released, and in a future session, there will be an esthetic result. This procedure has several benefits such as estimating the translucency of each layer and having a preview of the final outcome of treatment. Thus, one can estimate the color, dimensions, shape, proportion, translucency and other aspects related to the planning of the event (Silva e Souza Jr., 2003).

Cavity Preparation

In certain cases, such as with lingually inclined teeth with abnormal shapes and teeth without color change, the laminates can be placed without any preparation (Baratieri, 2002).

The cavity preparation may or may not be realized; preferably, wear is limited to the enamel, ranging from 0.2 to 2.0 mm of the gingival chamfer terminating at or slightly subgingival. However, it is not always possible to avoid exposure to dentin, which depends on factors such as the alignment of teeth in the dental arch and the degree of browning (devitalized teeth) (Galan Júnior, Namem, 1999).

The preparation can be made, preferably in an isolated operative field. The production of grooves for guidance is appropriate, especially when the dentist has less clinical experience. Initially, running up the orientation grooves on cervical and proximal faces, using a spherical diamond tip at high speed, numbers 1012 to 1014, depending on the prepare depth.. One should use half of the tip at high speed, resulting in a chamfer (Vieira et al. 1994; Gomes et al. 1996; Iorio, 1999; Baratieri 2001).

With regard to the proximal extension of the preparation, when there is no change in tooth color, the proximal extension need not advance to the proximal contact. In the region of subcontact, the preparation need not be extended towards the palatal face; however, due to the dynamic area of visibility, even in teeth without color change, the preparation occasionally has to be extended to approximately half of the palatal area contact. When there is a change in tooth color, the preparation has to be extended into the contact region and subcontacts so as to leave no visible tooth structure in this region after the tooth has been restored (Baratieri et al., 2001).

Subsequently, we produced vestibular grooves that were oriented to define the depth of the preparation; they were horizontal and vertical. The horizontal grooves were created with a diamond-specific type wheel, such as the numbers 4141 or 4142, whose selection depended on the desired depth. Tip 4141 generates grooves of 0.8 mm, and tip 4142 offers depth of 1.0 mm. The next step will be taken with cylindrical diamond burs with rounded ends, such as, the 2135 bur. The union of the grooves was formed. Alternatively, grooves of vertical orientation may be created after the use of spherical tip. In this case, a good alternative is the use of a silhouette technique with to produce grooves on the mid-buccal surface of the tooth and the depth of a conical diamond-tip rounded end, such as the number 2135 bur, and subsequently wear half of the union of the ridges and then wear extension of the other portion. It should be noted that the tip should be positioned on two inclinations following the contour of the buccal surface (Silva e Souza Jr. et al., 2003).

Insertion of Composite

The operative field can be isolated with the help of oral retractors, cotton rolls, wire retractors, a powerful saliva-sucker, or rubber dam. Bonding procedures are performed according to the adhesive system that was selected. It is important for this step that the adhesive tape, a matrix of cellulose acetate (polyester strip) or polytetrafluoroethylene (thread sealing tape), is positioned in the interdental spaces of the teeth to be faceted to protect them from contact with the materials used.

The categories of composites that can be used have been discussed earlier in this chapter. One option is to use a microhybrid composite microparticle internally and another externally. However, this option is being used more infrequently because of the good results obtained from the individual resins and nanoparticle microhybrids.

The resin increments can be placed on the surface of the tooth with the aid of a special syringe and capsules or with the aid of a spatula to composites. Thus, after the placement of each increment on the tooth surface, the resin should be spread with the help of brushes and a palette knife. The use of synthetic hairbrushes in various formats is very important to obtain smoothness, texturing and sculpting. Each increment should be cured for the time recommended by the manufacturer of the resin composite with a unit of photopolimerizer in the patterns of light intensity required to promote adequate polymerization (> 300 mW/cm2).

In this step, we should be attentive to the detail and the optical characteristics of the adjacent teeth to be copied through the use of specific resins to enamel and dentin, opacifier agents (when necessary) and the effect of resins with different degrees of value.

Finishing and Polishing

There is consensus among the authors that the stage of finishing and polishing is very important because it improves and maximizes the final esthetic result and longevity of restorations. A rough surface favors plaque buildup (which has repercussions on the periodontium), a greater likelihood of surface staining, infiltration and recurrent caries.

This step should, wherever possible, be postponed to the following appointment to allow the composite to undergo water absorption, hygroscopic expansion and complete polymerization. This will allow for a better marginal seal. If this step is performed in the same appointment, the material will be subject to permanent changes because of the heat generation during the procedure (Kano, Gondo, 2008).

Immediately after the restoration, you must complete the initial removal of excess immediately. Scalpel blade number 12 is the instrument used in this step. If one cannot remove the gross excess with the scalpel blade, a fine-grained diamond burr and extra-thin or multilaminated tips should be used. In this initial stage, the rotary instruments should be used minimally due to heat generation (Araújo et al., 2005).

While finishing and polishing, fine-grained diamond burrs and extra-thin or multilaminated tips are employed to determine the texture of the buccal surface, always considering the adjacent teeth. Sanding discs in sequential granulation are also employed during this step. Here, one must take into account the flat areas and the convexity of the buccal surface. A surface texture with a natural optical effect is obtained when the ambient light is focused on the restoration. Then, it should be used for specific rubber abrasive finishing and polishing of composites, following the manufacturer's instructions and with the grain. Finally, it is used the felt disc associated with specific folders for polishing composite resins. Tips, specifically silicon carbide tips, can also be used for this purpose.

Esthetic Factors in Direct Fabrication of Laminates

The current societal esthetics require the restoration of anterior teeth to blend into the natural tooth structure and be unnoticeable to the patient and other people in their surroundings. Various materials and techniques have been suggested in the literature to achieve a natural look, but each patient must be considered individually because the concept of beauty is subjective and dynamic, depending on cultural and individual factors. There are no rules that can be applied in all cases; however, certain principles help the clinician in developing esthetic rehabilitations and should be considered when working in the anterior region (Baratieri, 1998).

The following describes the general characteristics that can be observed during the course of anterior tooth.

1. Face

Dental treatment must be integrated and in balance with the facial characteristics of each individual. The face, in general, is classified as square, oval or triangle. A relationship with the shape of teeth can be observed in some patients, but other details that make an esthetic analysis must be considered to define the final shape of the tooth to be restored (Goldstein, 2000).

Lines that divide the face into thirds are used by professionals to assess the harmony of an individual's facial structure compõem (Conceição, 2007). The midline of the face can be seen in a frontal view. The midline follows an imaginary line that runs vertically through the glabella, the tip of the nose, philtrum and chin point. Any deviation from the midline will be noted. Other lines such as the interpupillary line, incisal plane and the outline of the gingival margin are analyzed, and they usually need to be parallel to result in a pleasant smile. The type of smile influences on esthetics and it can be considered low, medium and high. A smile is considered low when less than 75% of the clinical crown is exposed. Medium is when 75 to 100% of the upper anterior teeth are exposed (Jan Hatjó, 2008). In a high smile, the crowns of all of the teeth and the gingival are exposed. More than 4 mm of exposed gingiva is considered to be an unattractive smile by both professionals and patients alike (Kokich, 1999).

2. Periodontium

The concern of periodontal situation is no longer exclusively the prevention and treatment of diseases but now also includes the search for esthetic alternatives for the contour of the gingival tissue. Therefore when analyzed, the cervical gingiva must be continuous, without abrupt irregularities and must maintain a harmonious relationship with the lip line. In the region of the lateral incisors, the configuration of the cervical gingiva presents a coronal shift compared to embarking upon central and canines. Changes that may affect the esthetic harmony can be corrected with a plastic gingiva.

The gingival zenith should also be observed and taken into account during the fabrication of restorations on the buccal surface of anterior teeth. Gingival zenith is located distally and is the highest point of the gum of the tooth. Moreover, the interdental papilla should promote closure of the interdental space behind the contact point, contributing to a pleasant smile (Conceição, 2007; Kyrillos, 2005).

3. Dental Characteristics

3.1. Teeth Together

The harmony of the dental arch can be analyzed by the midline and through it we should see the symmetry of the face and teeth. Teeth properly positioned and aligned resulting in an significantly esthetic pleasing. The incisal line of the upper front teeth, for example, must accompany the convexity of the lower lip. However, individual characteristics such as crowded teeth, or bad positioned much inclined to the buccal or lingual alone can cause dental disharmony (Conceição, 2007).

Restorative dentistry can correct any irregularities although it may take several procedures, and in some situations, it is necessary to collaborate with other disciplines such as orthodontics and periodontics (Rufenacht, 2003).

3.2. Shape and Size of Incisal Embrasures

Cervical embrasures in young people are filled by gingiva but in elderly or those suffering from periodontal disease, they have become more evident. The cervical embrasures have the form of an inverted "V" and they are enlarged. Typically, the incisal embrasures are narrow and straight in the region mesio-incisal maxillary central incisors, more rounded and asymmetrical between the distal central incisor and mesial lateral incisor and increasingly wide between the distal lateral incisor and upper canine (Baratieri, 2001; Conceição, 2007).

The shape and size of incisal embrasures change over time and influence the appearance of teeth. The change of cervical/incisal embrasures, according to the morphological changes resulting from aging or deleterious habits (attrition, for example), cause the disappearance of incisal embrasures (Baratieri, 2001; Conceição, 2007).

3.3. Proportion

The correct proportion of teeth in a dental arch can be achieved by implementing a restoration or through small multi-plasties in the enamel. When evaluating a dental arch, we start with the measurements of the height and width of the central incisors because they are largely determine the composition of a smile. A proportion where the width is 75 to 80% of the height of the tooth can be considered smooth for the central incisors; in lateral incisors the proportion is generally 60 to 65% (Conceição, 2007; Jan Hatjó, 2008).

3.4. Balance

Balance seems more important than symmetry. Ideally, the teeth on the left side of the arch must have the same weight in the smile as the teeth on the right side of the arch. This means that the other components of esthetics such as tooth size, visibility, shape, color, position, surface texture, alignment, slope, incisal embrasures, contact points, furrows and ridges and angles of development should be arranged in a harmonious way. Just as the teeth in the same arch must be in balance with each other, the upper teeth should be in balance with the lower teeth (Baratieri, 1998).

3.5. Tooth size

While the teeth should be in proportion to each other, they must also be in proportion to the face. A variation in tooth size in relation to the dimensions of the face may affect the achievement of good esthetic results. The anterior teeth length and average anterior teeth width of the male, are usually larger than females (Guillen, 1994; Starret, 1999). However, the average exposure of the upper central incisor during smiling has been calculated to be 1.91 mm to 3.40 mm for men and women. The average visible length of maxillary central incisors decreases with increasing age, whereas the length of the lower incisors increases (Baratieri, 2001).

An important benchmark in determining the size of the maxillary incisors is that they usually have the same item-cervical length of canines ((Rufenacht, 2003) because the lateral incisors are, on average, 1.5 mm smaller than the central incisors (Jan Hatjó, 2008). This detail becomes very important when the six upper anterior teeth are being submitted for the manufacturer of composite resin laminates.

3.6. Shape

The ideal shape for a dental restoration is that of the corresponding natural tooth. If one does not have such a reference, photos and personal characteristics of the patient (such as age and sex) can be used (Baratieri, 1998).

When the tooth shape is altered, the direction of the reflection of ambient light that covers the tooth is also changed. Flat and smooth surfaces reflect light more directly to the observer and thus seem wider, broader and closer. On the other hand, rounded and irregular surfaces reflect light to the sides, reducing the amount of light reflected directly to the viewer and appearing closer, smaller and more distant (Silva e Souza Jr. et al., 2000).

3.7. Surface texture

The surface texture of teeth undergoes changes with aging. Typically, young teeth have more surface roughness. Given the physiological wear over the years, the surface texture of the anterior teeth appears to be more smooth and regular. As mentioned earlier in this chapter, the restoration of the surface elements can be accomplished by the use of diamond tips for small- and medium-grain on low speed after the final polishing of the restoration.

One must remember that smooth surfaces reflect more light, and that the greater the amount of reflected light, the took shows wider, brighter and closer than they appear before are apparent, making them appear artificial (Fernandes *et al.*2002).

Employment of Optical Illusion in Esthetic Restorations

An object that has many acute angles can be easily defined, while an object with curved surfaces is difficult to determine. The variation of the shape and angle of the tooth surface involves the reflection of light and It can simulate the natural appearance of teeth (Goldstein, 2000; Mondelli et al., 2003).

Remodeling cosmetics can be employed to modify the shape or contour of teeth. By changing the reflection of light, it creates an optical illusion that can change the length and the wide of those teeth. This feature can be used in various clinical situations by creating cosmetic illusions, thus solving many esthetic problems (Goldstein, 2000; Mondelli et al., 2003; Araújo et al., 2005).

The flat area found on the buccal surface of anterior teeth can influence both the observation of the width and the length of the anterior teeth, especially the upper ones. When the dentist will be done a laminate teeth in a large teeth, reduce or close diastema teeth or in very narrow teeth, for example, the conoid lateral incisors, this check is very interesting. In the first case (large teeth), it is important to reduce the flat area on the mesio-distal aspect.

Furthermore, using a slightly darker color on the body and restoring with a more grayish color in the mesial and distal thirds may assist in reducing the apparent width of a tooth, as darker colors are less visible.

When the intention is to increase the width of a tooth, the flat area must be increased mesio-distally. Although teeth may have the same actual width, they may have different apparent widths depending on the width of the flat area (Baratieri et al., 2001; Araújo et al., 2005)

The apparent length of teeth can also be changed by manipulating the flat area. A tooth can be apparently reduced by lowering the flat area. The buccal surface of anterior teeth is divided into thirds: cervical, central, and incisal. By changing these parts, especially enlarging or reducing the central region, the area can be set flat. Therefore, a lower flat area in the cervico-incisal direction results in a lower apparent length of the tooth. The opposite strategy can also be applied to give the impression of increased length (Araújo et al., 2005).

Case Report

Patient 1

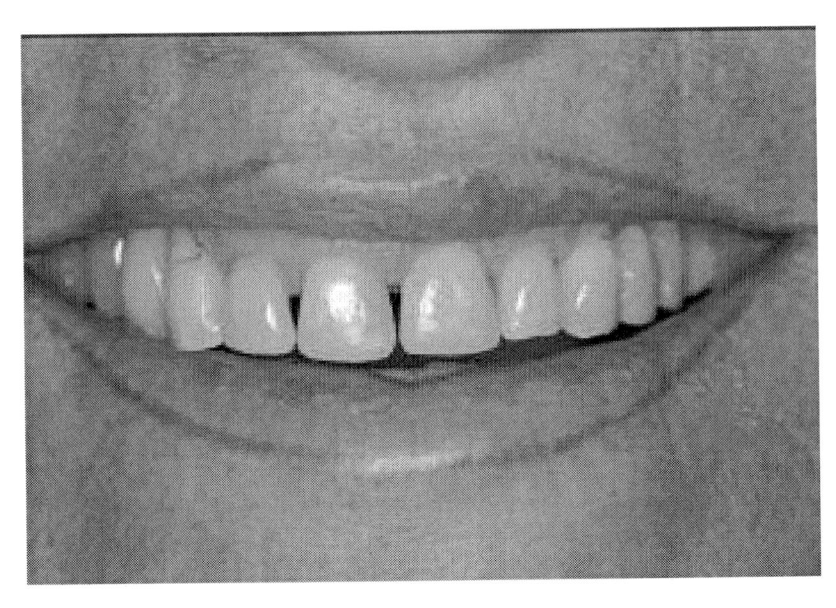

Figure 1. The patient came to office complaining about aesthetics of your anterior teeth.

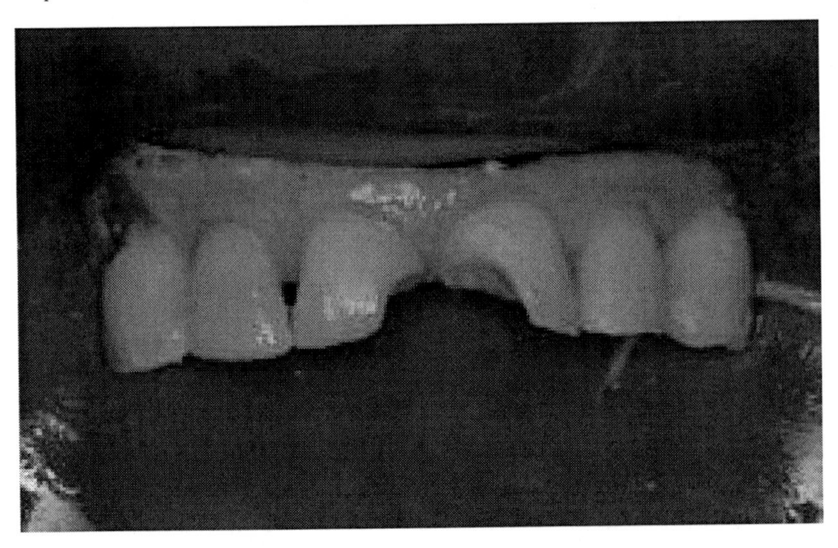

Figure 2. It was planned to perform reconstructions with direct laminates. The old restorations were removed.

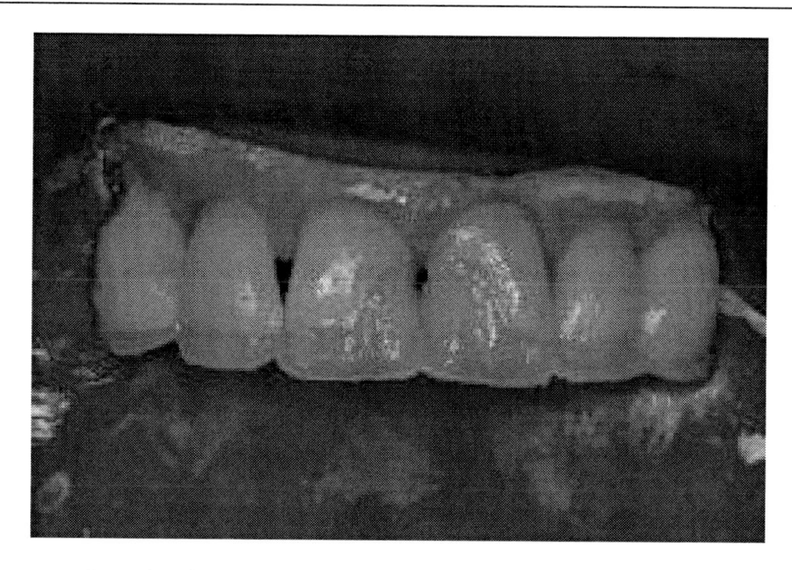

Figure 3. It was performed palatal shell of composite resin to assist in the teeth reconstructions. After that was done the application of dentin composite resin.

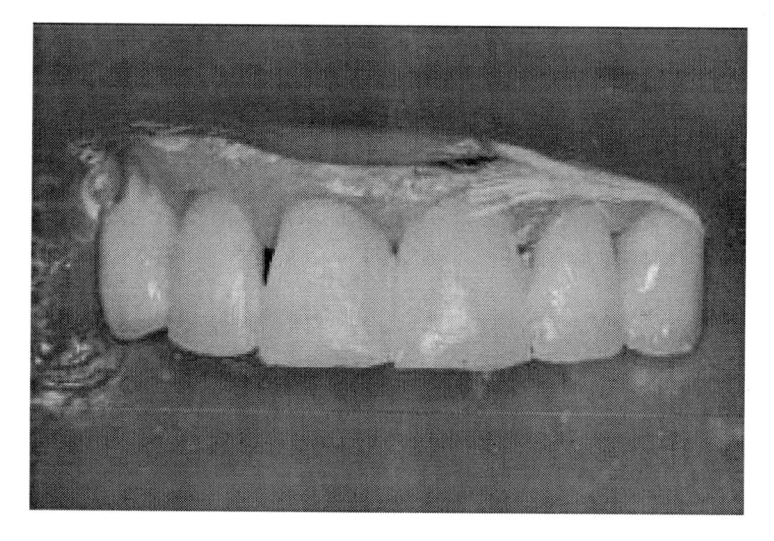

Figure 4. It was performed the application of the enamel composite resin.

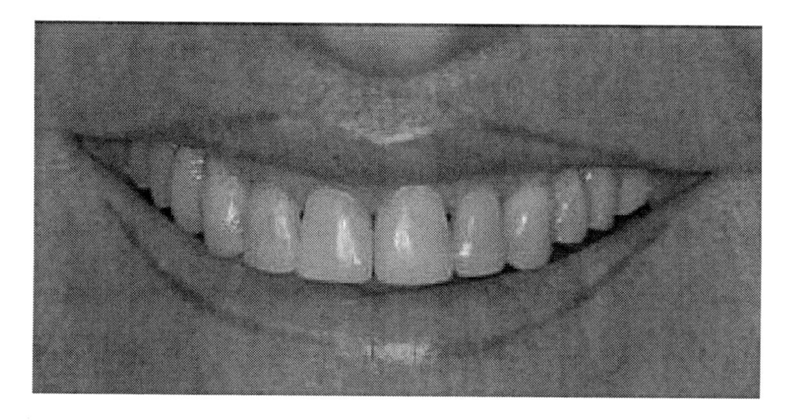

Figure 5. Aesthetic ends reaching the expected result.

Patient 2

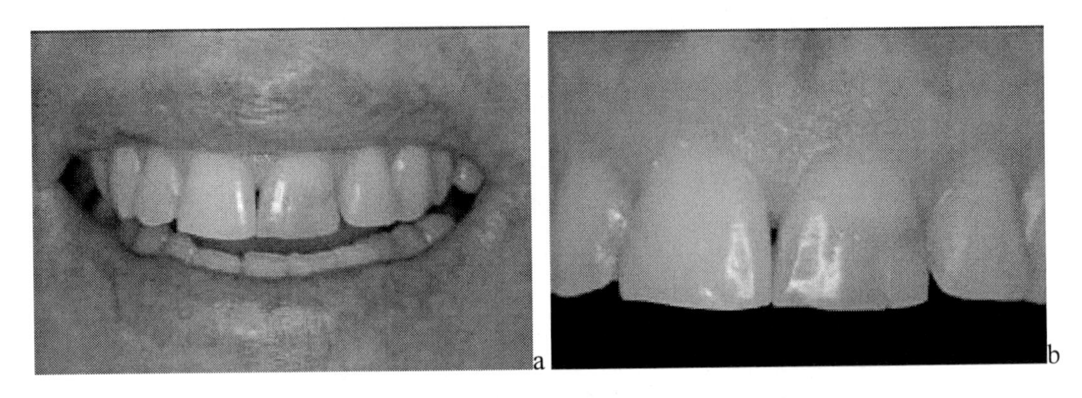

Figure 6a and 6b. Patient came to the clinic complaining of anterior tooth darkened.

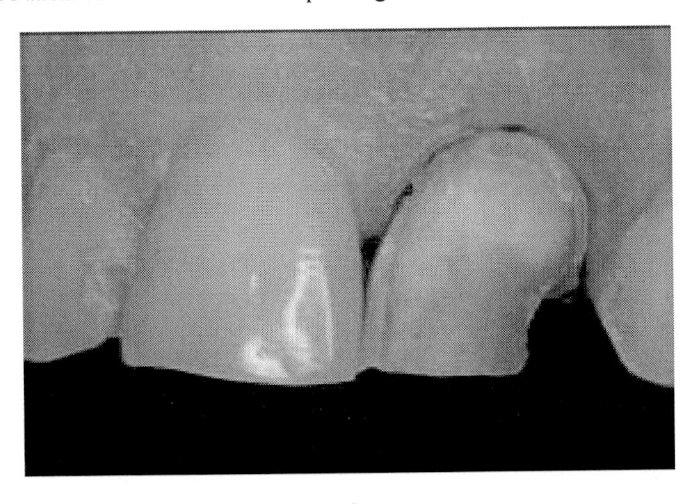

Figure 7. Teeth prepared to receive a direct laminate.

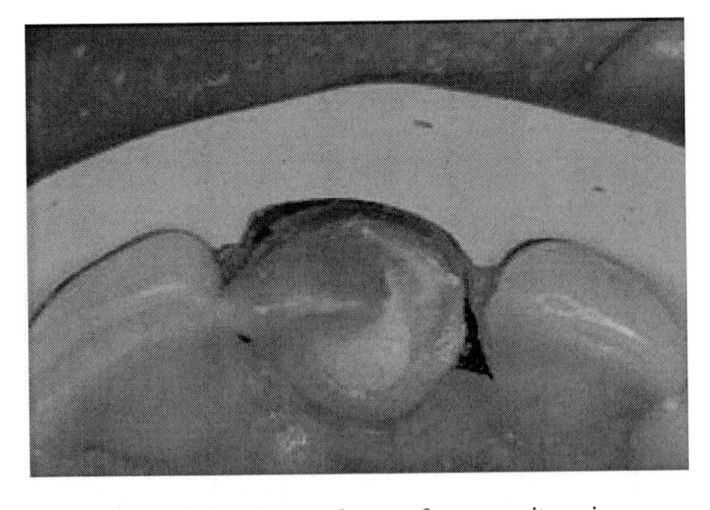

Figure 8. Matrix proof to observe if there is enough space for composite resin.

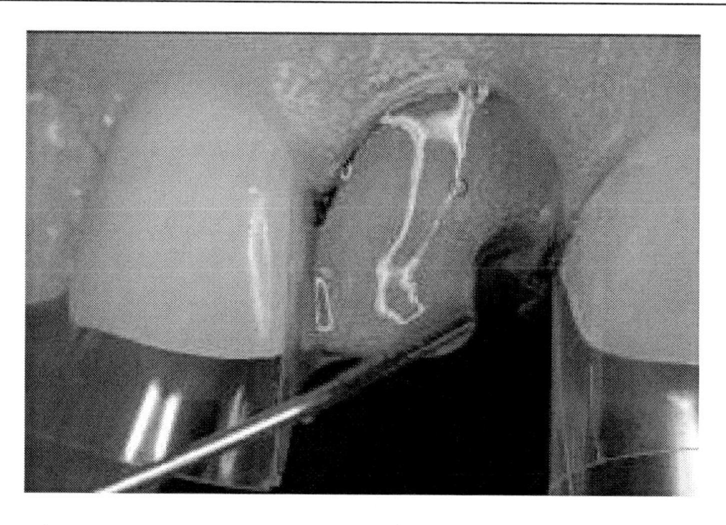

Figure 9. Acid attack was carried out for 30 seconds.

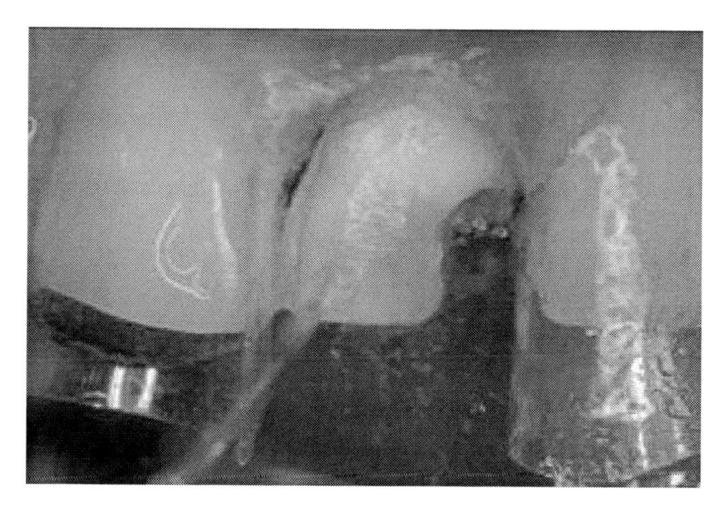

Figure 10. The tooth was washed for 20 seconds with abundant water.

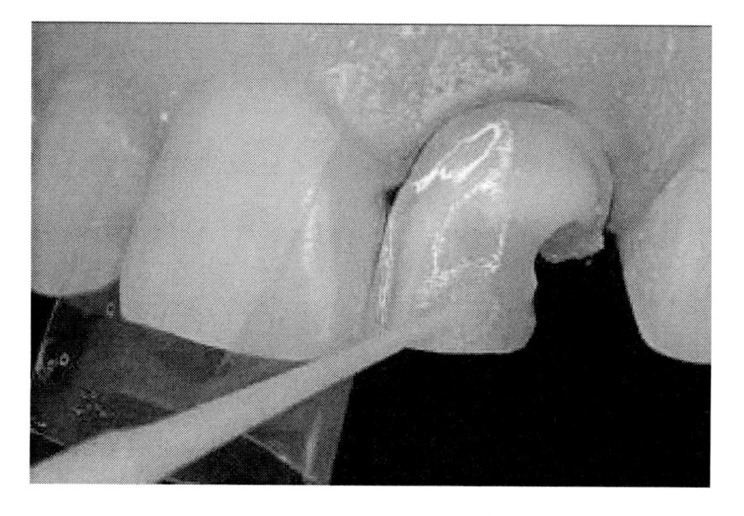

Figure 11. The primer was applied and dried with mild air jets.

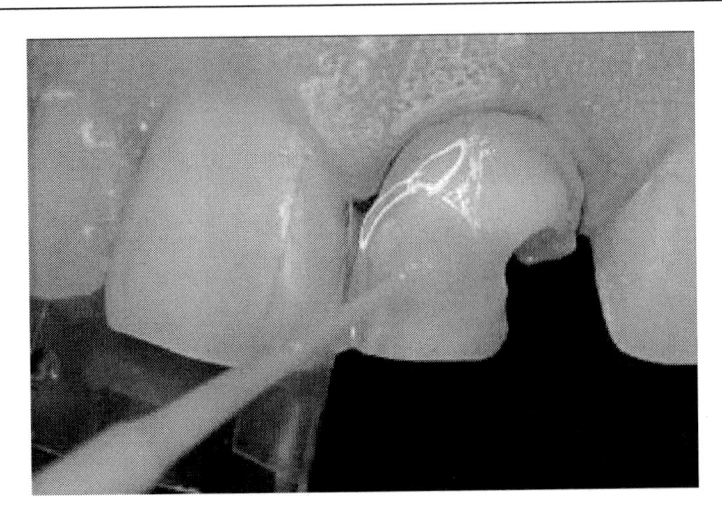

Figure 12. The adhesive was applied and dried with mild air jets.

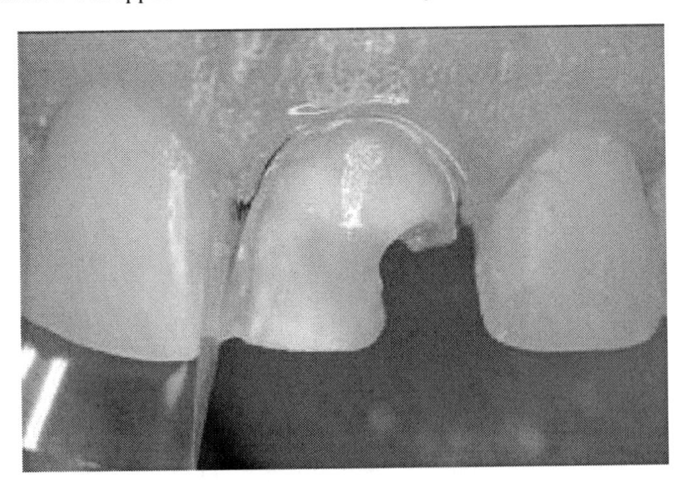

Figure 13. Adhesive Photopolimerization.

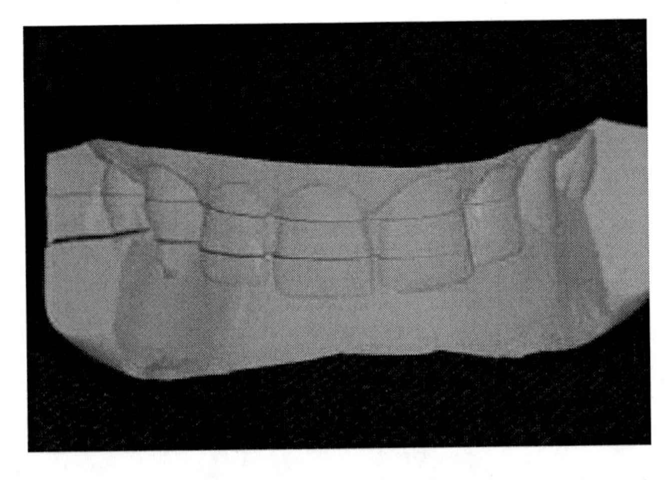

Figure 14. Silicone guide to control the thickness of the layers of composite resin. It was made after diagnosis waxing.

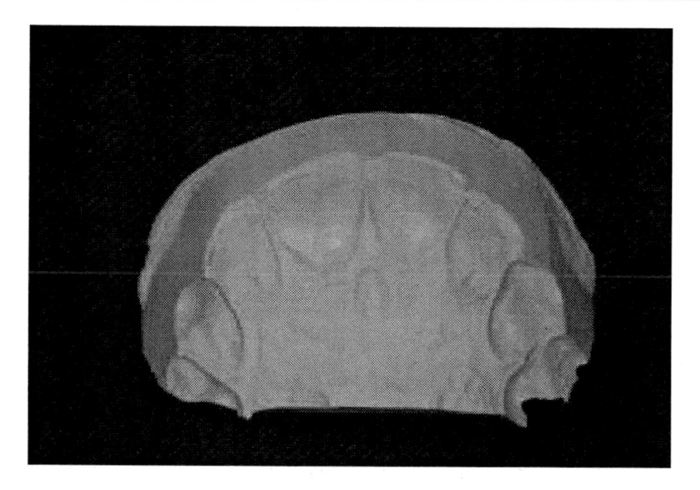

Figure 15. Silicone palatal shell to assist in the teeth reconstructions. It was made after diagnosis waxing.

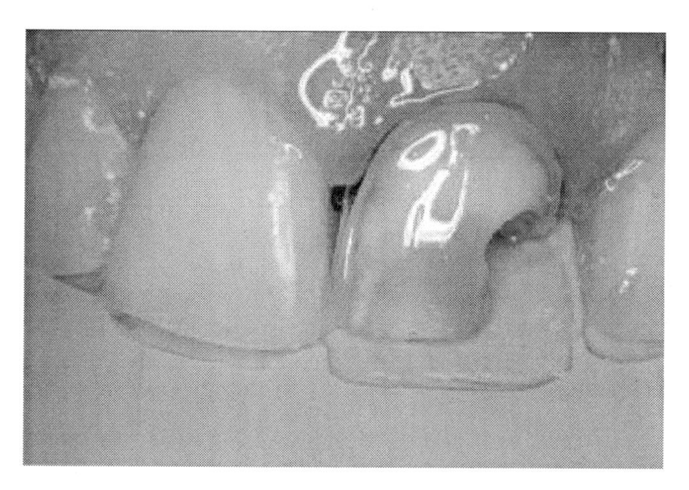

Figure 16. Proof of silicone palatal shell.

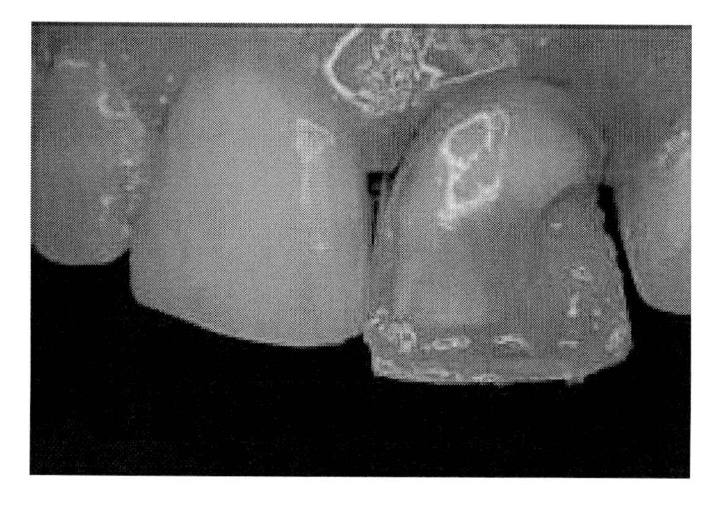

Figure 17. Palatal shell of composite resin.

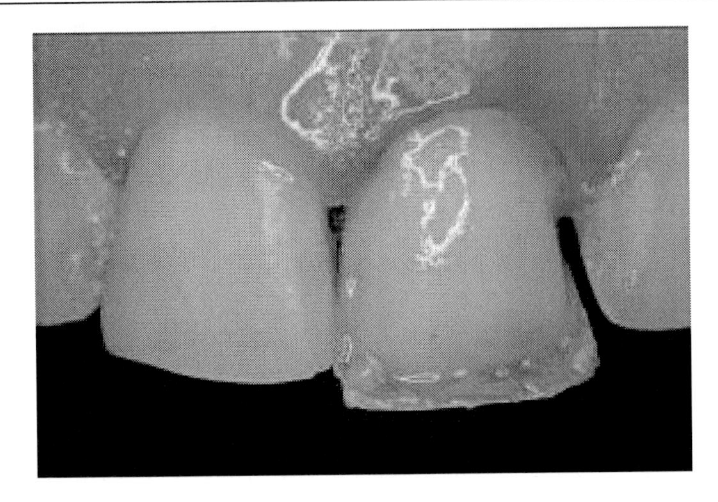

Figure 18. A thin layer of opacifiers was used.

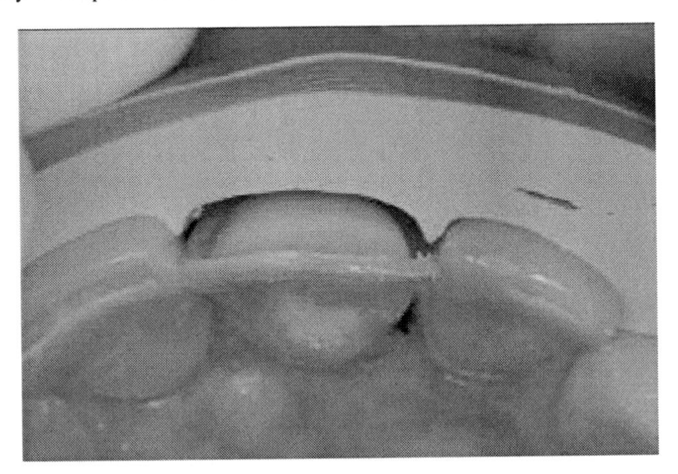

Figure 19. Space available to use composite resin.

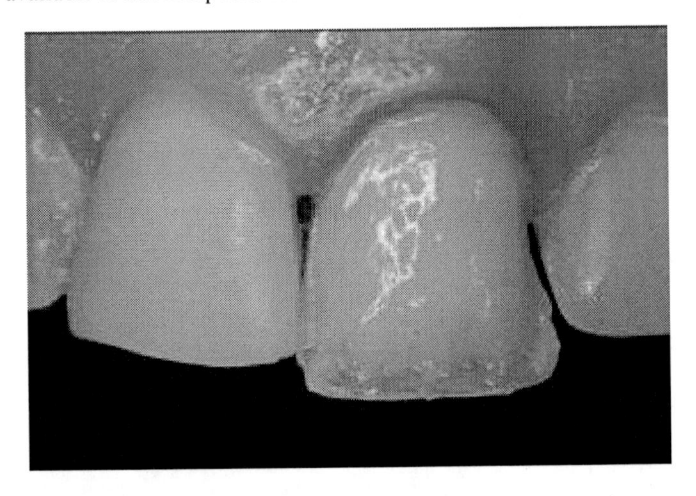

Figure 20. It was used composite resin simulating dentin.

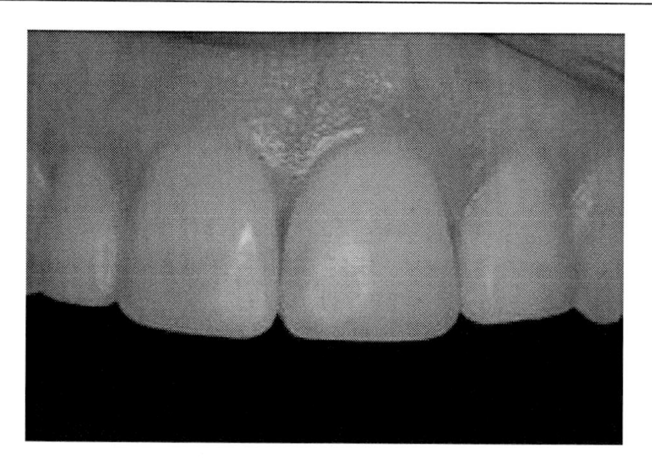

Figure 21. Restored teeth before polished.

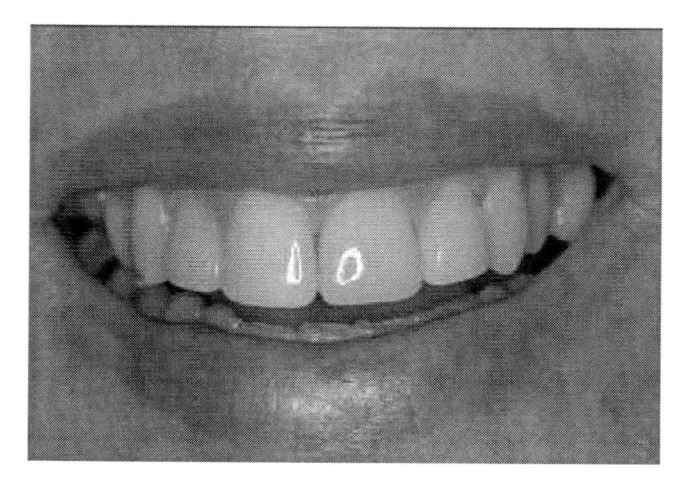

Figure 22. The end of the treatment reached the expected aesthetic outcome.

Bibliographical References

Araújo *et al. Estética para o clínico geral*. São Paulo: Artes Médicas, 2005.

Baratieri, L.N. *Estética: restaurações adesivas diretas em dentes anteriores fraturados*. São Paulo: Santos, 1998.

Baratieri, L.N. *Dentística: procedimentos preventivos e restauradores*. São Paulo: Santos, 2000.

Baratieri, L.N. *Odontologia Restauradora: fundamentos e Possibilidades*. 1. ed. São Paulo: Santos, 2001.

Blank Jt. Case seection criteria and a simplified technique for placing and finishing direct composite veneers. *Compend Contin Educ Dent*. 2002 Sep;23(9 Suppl 1): 10-7.

Carvalho, R.M. Adesivos Dentinários: Fundamentos para aplicação clínica. *Revista Dent Rest,* p. 62-96, 1998.

Chain *et al.* Materiais restauradores estéticos poliméricos e cerâmicos do novo século. In: *Odontologia integrada*. Rio de Janeiro: Pedro I, v. 1, p.102-137, 2002.

Conceição. Laminado de porcelana. In: *Dentística-Saúde e estética*. Porto Alegre: Artmed, p. 283-296, 2007.

Costa, S. X. S.; Becker, A. B. B.; Rastelli, A. N. S.; Andrade, M. F. Contorno Cosmético com Resina Composta Nanoparticulada: *Relato de Caso Clínico. R Dental* Press Estét, v. 4, n. 3, p. 24-33, jul./ago./set., 2007.

Fernandes, *et al. Odontologia Estética*. São Paulo: Santos, 2002.

Galan Júnior; J. Namen, F.M. Facetas diretas e indiretas na restauração de dentes anteriores: vantagens e desvantagens. In: Vanzillotta P.S. *et al. Odontologia integrada: atualização multidisciplinar para o clínico e o especialista*. Rio de Janeiro: Ed. Pedro I, 1999.

Gillen RJ, Schwartz RS, Hilton TJ et al. An analysis of selected normative tooth proportion. *Int J. Prosthodont*. 1994;7: 410-7.

Goldstein RE. *A estética em odontologia*. 2 ed. São Paulo:Ed. Santos; 2000.

Hirata, R. *Restaurações estéticas e transformações anteriores*. In: HIRATA, R. Tips: dicas em Odontologia estética. São Paulo: Artes médicas. 2011, p. 207-385.

Iorio, P.A.C. Preparos e Restaurações com facetas estéticas. In: *Dentística clínica-Adesiva estética*. São Paulo: Santos,p.362, 1999.

Jan Hatjó. *Anteriores - A Beleza Natural dos Dentes Anteriores* – São Paulo: Santos, 2008.

Kano, P., Gondo, R. O uso de compósitos em dentes anteriores. In: Baratieri, L.N. et al. *Soluções clínicas: fundamentos e técnicas*. Florianópilis: Ponto, p. 215-250, 2008.

Kyrillos, M. Moreira, M. *Sorriso Modelo - O Rosto em Harmonia*. São Paulo: Santos, 2005.

Kokich, V. O., Asuman Kiyak, H. and Shapiro, P. A. (1999), Comparing the Perception of Dentists and Lay People to Altered Dental Esthetics. *Journal of Esthetic and Restorative Dentistry,* 11: 311–324.

Lopes GC, Baratieri LN, Andrada MAC, Vieira LCC. Dental adhesion: present state of the art and future perspective. *Quintessence* Int. 2002 Mar;33(3):213-24.

Lopes GC, Pezzini R, Baratieri LN, Monteiro Jr S, Vieira LCC. Uma intervenção conservadora na restauração estética de dentes escurecidos após traumatismo. *JBC – J Bras Clin*. 2001 Jul-Ago; 28(5): 350-4.

Marques, S.M.L, Ribeiro, F.S.V, Machado, F., Ramos Jr, L., Carvalho, M.C.F.S, Costa, S.C et al. *Estética com resinas compostas: percepção, arte e naturalidade*. São Paulo: Santos, 2005.

Miyashita, E; Fonseca ,AS. *Odontologia Estética-O Estado da Arte*. São Paulo: Artes Médicas, 2004. 768p.

Mondelli, R.F., Mondelli, J.,et al. Reabilitação estética do sorriso com facetas. *Bio Odonto Revista odontológica*,set/out..2001.

Mondelli, J. *Estética e cosmética em clínica integrada restauradora*. São Paulo: Ed. Santos, p.546-562, 2003.

Perdigão J, Lopes M. Dentin bonding questions for the new millenium. *J Adhes Dent*. 1999 Autum;1(3): 191-209

Reis, A. et al. Resinas compostas. In: Reis, A.; Loguercio, A. D. *Materiais dentários restauradores diretos: dos fundamentos à aplicação clínica*. São Paulo: Santos, 2007. p. 137-180.

Rufenach, CR. *Princípios da integração estética*. 1 ed. São Paulo, quintessence, 2003.

Silva e Souza Jr. MH, Carvalho RM, Mondelli RFL. *Odontologia estética: fundamentos e aplicações clínicas: restaurações de resina composta*. 1 ed. São Paulo: Ed. Santos; p.171-190, 2000.

Silva e Souza Jr MH da, Carvalho RM de, Mondelli, RFL, et al. *Odontologia estética: fundamentos e aplicações clínicas: restaurações indiretas sem metal: resinas compostas e cerâmica.* 1ed. São Paulo: Ed. Santos, 2001.

Silva e Souza Jr. MH, Silva CM, Araújo JLN. Facetas vestibulares de resina composta. *Bio Odonto Revista odontológica*, p.9-96, 2003 jul/ago.

Sterrett JD, Oliver T, Robinson F, Fortson W, Knaak B, Russell CM. Width/length ratios of normal clinical crowns of the maxillary anterior dentition in man. *J Clin Periodontol.* 1999 Mar; 26 (3):153-7.

Vieira, G.F. *et al. Facetas Laminadas*. São Paulo: Santos, 1994.

INDEX

A

B

C

E

F

G

H

I

J

K

L

M

N

O

P

Q

R

T

U

V

W

X

Y

Z